Nutraceutical and Functional Food Regulations in the United States and Around the World

Second Edition

Food Science and Technology International Series

A complete list of books in this series appears at the end of this volume.

Nutraceutical and Functional Food Regulations in the United States and Around the World

Second Edition

Edited by

Debasis Bagchi, PhD MACN CNS MAIChE
Department of Pharmacological and Pharmaceutical Sciences
University of Houston College of Pharmacy
Houston, TX, USA

AMSTERDAM • BOSTON • HEIDELBERG • LONDON
NEW YORK • OXFORD • PARIS • SAN DIEGO
SAN FRANCISCO • SINGAPORE • SYDNEY • TOKYO
Academic Press is an imprint of Elsevier

Academic Press is an imprint of Elsevier
32 Jamestown Road, London NW1 7BY, UK
225 Wyman Street, Waltham, MA 02451, USA
525 B Street, Suite 1800, San Diego, CA 92101-4495, USA

Notice
No responsibility is assumed by the publisher for any injury and/or damage to
persons or property as a matter of products liability, negligence or otherwise, or
from any use or operation of any methods, products, instructions or ideas
contained in the material herein. Because of rapid advances in the medical
sciences, in particular, independent verification of diagnoses and drug dosages
should be made

British Library Cataloguing-in-Publication Data
A catalogue record for this book is available from the British Library

Library of Congress Cataloging-in-Publication Data
A catalog record for this book is available from the Library of Congress

ISBN: 978-0-12-405870-5

For information on all Academic Press publications
visit our website at http://elsevierdirect.com

Typeset by MPS Limited, Chennai, India
www.adi-mps.com

Printed and bound by CPI Group (UK) Ltd, Croydon, CR0 4YY
14 15 16 17   10 9 8 7 6 5 4 3 2 1

This Book is Dedicated to My Beloved Nephew, Neel Nanda

In Memoriam

Dr. Hirobumi Ohama, Ph.D., contributed an esteemed chapter in this book, and unfortunately passed away before the publication of the second edition. During the last two decades, Dr. Ohama San contributed immensely to the field of Nutraceutical and Functional Food Regulations in Japan and around the world. He also contributed significantly in the fields of nutraceutical and functional food research. Dr. Ohama San was an excellent person, a great friend, a wonderful human being, and an outstanding scientist. On behalf of all of the contributors of this book, I convey our sincere regards and respect to our beloved, late Dr. Ohama San. May his soul rest in peace.

Contents

PART I INTRODUCTION

PART IV REGULATIONS AROUND THE WORLD

PART V REGULATIONS ON PET FOOD

PART VI VALIDATION APPROACH

Preface

Innovation and marketing of nutraceuticals and food supplements are now the fastest growing segments. Currently, the rising costs and toxicity of some pharmaceuticals are driving the population around the world to move forward with less expensive nutraceuticals and functional foods.

However, the critical point is that nutraceuticals and herbal ingredients, including the entry of new functional foods, are important because of their acceptance as novel and unique forms that demonstrate benefits for human health. Because of the rapid expansion in the nutraceutical market, an imbalance exists between the increasing number of claims and the amount of new products. The development of policies to regulate their application and safety is warranted. Also, it is necessary to check their efficacy through extensive research and rigorous quality control to ensure human safety.

Several regulatory constraints are enforced in the United States and around the world, and a number of interesting chapters are included in this book to cover these salient features. The European Food Safety Authority (EFSA) has enforced the need to provide scientific requirements for health claims and to find new regulatory issues for health food products. Accordingly, the EFSA asked its Dietetic Products, Nutrition, and Allergies panel to draft additional guidelines on scientific assessment of these claims. A new approach for a strict substantiation of health issues in selected fields, Regulation (EC) No. 1924/2006 established that health claim applications should only be authorized after a high level of scientific validation conducted by EFSA using appropriate measures during the evaluation. The legislation should protect consumers and define specific research areas with appropriate outcome measures to assess the quality, relevance, and adequacy of studies conducted for scientific validation of health claims.

In Japan, Food for Specified Health Use (FOSHU) regulation is being enforced. FOSHU is classified as a food "of which a functional component was identified, and of which the biological regulation function has been confirmed in clinical trials, and of which a health claim was permitted by the Ministry of Health, Labor, and Welfare." For example, three categories of FOSHU exist concerning bone health: foods helping mineral absorption, foods helping mineral absorption and conditioning the intestines, and foods for people concerned about bone health. Thus, in Japan, incorporating FOSHU appropriately into diet and nutrition could help bone health. Similar enforcement has been introduced by the other countries.

According to market statistics, the global functional food and nutraceutical market is growing at a rate that is outpacing the traditional processed food market. In 2012, the Council for Responsible Nutrition (CRN) reported that 68% of Americans take nutritional or dietary supplements based on the data released from its annual consumer survey. The CRN further reported that these data are consistent with previous years' statistics of 69% in 2011, 66% in 2010, and 65% in 2009. According to the results from the 2012 CRN Consumer Survey on Dietary Supplements, approximately 76% of users classify themselves as "regular" users, while 18% are occasional users, and 6% are "seasonal" users. In 2011, 77% of users classified themselves as "regular" users, while 74 and 73% of users classified themselves as "regular" in 2010 and 2009, respectively [1]. According to a new report by Global Industry Analysis, the global nutraceutical market will be worth $243 billion by 2015 [2].

A highly impressive group of professionals has contributed immensely to the successful accomplishment of this second edition. It formalizes an expert panel and provides descriptions of health food regulatory aspects from North America, the UK, Europe, Australia, Asia, Africa, and other selected countries in the Pacific Rim. Other topics include marketing insight, current good manufacturing compliance (cGMP), generally recognized as safe (GRAS) status, analytical validation, adverse event reporting, validation, intellectual property, branding, trademark, and regulatory approvals as well as the impact of World Trade Organization (WTO) regulations on the global food supply chain, all of which are reviewed in 30 different chapters divided into eight different parts.

In **Part I: Introduction**, A.L. Almada and A. Das and C.K. Sen discuss the scope and significance of global regulations on nutraceuticals and functional foods. Almada further elaborates on market opportunities and future directions of nutraceuticals and functional foods. A. Shao highlights the salient features on global entry regulations. Part II: Manufacturing Compliance and Analytical Validation was written by leading analytical researchers D. Chahar and D.K. Nath, and C.S. Yeevani. In this section they detail the impact of GMP and cGMP compliance on consumer confidence in dietary supplements. Part III: Importance of Safety Assessment was written by G.A. Burdock and I.G. Carabin and G. Houlahan. Burdock and Carabin highlight the importance of GRAS to the functional food and nutraceutical industries. Houlahan reviews the role of NSF International in the dietary supplement and nutraceutical industries.

Part IV: Regulations Around the World was provided by a selected panel of world renowned professionals who specialize in their respective country's legislation on dietary supplements and functional food regulations. J.E. Hoadley and J.C. Rowlands highlight the Food and Drug Administration (FDA) perspectives on health claims for food labels in the United States. S. Agarwal, a leading nutrition scientist, and S. Hordvik and S. Morar, renowned regulatory legal counsels, focus their discussions on the nutritional claims for functional foods and supplements in the United States. P.P. Fu and Q. Xia demonstrate the importance of critical assessment of safety and quality assurance of herbal dietary supplements. C.A. Lewis and M.C. Jackson provide an overview on understanding medical foods under FDA regulations. J.R. Harrison and E.R. Nestmann review the basic principles of Canada's regulations and examples of major issues faced by the Natural Health Products Directorate.

European regulations are discussed by two eminent and highly professional teams. First, in Chapter 13 by P. Coppens and S. Pettman, and again in Chapter 14 by O.P. Gulati, P.B. Ottaway, and P. Coppens, who emphasize the health claims made on botanical-sourced, functional and fortified foods and food supplements in the European Union. Japanese regulations are also detailed by two eminent groups. In Chapter 15 M. Shimizu outlines the history and current status of Japanese regulations and in Chapter 16 H. Ohama, H. Ikeda, and H. Moriyama explain Japanese regulations for health foods, as well as the intricate aspects of FOSHU regulations.

Australian regulations are discussed by D. Ghosh, while the Russian regulations are discussed by V.A. Tutelyan, B.P. Sukhanov, A.A. Kochetkova, S.A. Sheveleva, and E.A. Smirnova. The nutraceutical and functional food regulations in India are outlined by two well-established professional teams comprised of R.K. Keservani, A.K. Sharma, Md.F. Ahmad, and M.E. Baig (Chapter 19) and A.V. Krishnaraju, K. Bhupathiraju, K. Sengupta, and T. Golakoti (Chapter 20). C. Hu extensively discusses the Chinese regulations. Korean regulations are highlighted by J.Y. Kim, S.J. Kim, and H.J. Lee. T. Bahorun, V.S. Neergheen-Bhujun, M. Dhunnoo, and O.I. Aruoma discuss the regulatory status of botanical drugs, nutraceuticals, and functional food in the African continent. J. Zawistowski provides an extensive review on the regulatory system for functional foods in the eight countries of the Pacific Rim: Taiwan, Hong Kong, South Korea, Malaysia, Indonesia, Singapore, the Philippines and Thailand as well as their diverse acts, regulations, and guidelines, while T.C. Lau focuses exclusively on Malaysian regulations. O.I. Aruoma summarizes the impact of the WTO and food regulation on the food supply chain.

Part V: Regulations on Pet Food is written by N. McGee, J. Radosevich, and N.E. Rawson, who are renowned experts on pet food regulations. They provide a thorough overview on the stringent pet food regulations. Part VI: Validation Approach is a new section that was introduced because of reader demand. This is a very important and critical issue. D.K. Nath and C.S. Yeevani emphasize the critical insight and a stepwise solution in their chapter.

In Part VII: Adverse Event Reporting, A. Shao highlights a detailed chapter on global adverse event reporting regulations on the functional foods as well as dietary, food, and health supplements, keeping in mind that global adverse event reporting regulations are becoming very important for the nutraceutical and functional food industries. Part VIII: Intellectual Property, Branding, Trademark and Regulatory Approvals in Nutraceuticals and Functional Foods provides an update on these subjects by L.K. Chong, L.J. Udell, and B.W. Downs.

Although nutraceuticals and functional foods have significant promise in the promotion of human health and disease prevention, health professionals, nutritionists, and regulatory toxicologists should strategically work together to derive appropriate regulations to provide the optimal health and therapeutic benefits to mankind.

References

[1] <http://www.crnusa.org/prpdfs/CRNPR12-ConsumerSurvey100412.pdf>.

[2] <http://www.prweb.com/releases/nutraceuticals/dietary_supplements/prweb4563164. htm>.

Preface to the First Edition

Historically, plants have been used as a valuable source of prophylactic agents for the prevention and treatment of diseases in humans and animals. Hippocrates conceptualized the relationship between the use of appropriate foods for health and their therapeutic benefits in his renowned quote:

Let food be thy medicine and medicine be thy food.

Hippocrates (460−377 BC)

This intimate relationship between food and drugs is even recognized in the American legal definition of a drug:

§201(g)(1) The term "drug" means…articles intended for use in the diagnosis, cure, mitigation, treatment, or prevention of disease in man or other animals; and articles (other than food*) intended to affect the structure or any function of the body of man or other animals… (Federal Food Drug and Cosmetic Act as Amended) [1].
* Emphasis added

Within the last decade, consumers have made increasing reference to "nutraceuticals" and "functional foods," recognizing the relationship between nutrition and health, to the point of avoiding an over reliance on pharmaceuticals and regarding prescription drugs as often unnecessary, too expensive, and of dubious benefit once all the risks are considered. This, combined with a more widespread understanding of how diet affects disease, health care costs and an aging population, has created a market for functional foods and natural health products.

According to market statistics, the global functional food and nutraceutical market is growing at a rate that is outpacing the traditional processed food market. A poll conducted by the Council for Responsible Nutrition reported that 52% of Americans identify themselves as regular users of dietary supplements in 2007, up from 46% in 2006 [2]. In the United States alone, consumer expenditures on dietary supplements and functional foods reached a reported $22.4 billion and $31.4 billion sales, respectively, more than double the amount spent in 1994 [3].

Natural products industries face diverse challenges [4]. In an attempt to define efficacy and support marketing claims for the products, extensive safety studies including acute, subacute, subchronic, chronic, and long-term toxicity studies; genotoxicity; reproductive toxicology; teratogenicity; and molecular mechanisms of action both *in vitro* and *in vivo* should be complemented with supplementation studies in animal models and clinical trials for human efficacy for specified indication.

This book, *Nutraceuticals and Functional Foods Regulations in the United States and Around the World*, formalizes an expert panel and provides descriptions of health food regulatory aspects from North America, the UK, Australia, New Zealand, Asia, Brazil, Africa, and other select countries in the Pacific Rim. Other topics include marketing insight, current good manufacturing compliance (cGMP), analytical validation, intellectual property, branding, trademark, and regulatory approvals. Furthermore, the impact of World Trade Organization (WTO) regulations on the global food supply chain is reviewed. A special section on the safety assessment and assurance of nutraceuticals through obtaining generally recognized as safe (GRAS) status and the use of traceability technologies and nanotechnology is also discussed.

A highly impressive group of professionals has contributed immensely to the successful accomplishment of this special issue.

In Part I: Introduction, A.L. Almada and Professors O. Hänninen and C.K. Sen discussed the scope and significance of global regulations on nutraceuticals and functional foods. Almada elaborated on market opportunities and future directions of nutraceuticals and functional foods. Part II: Regulatory Hurdles for Marketing was written by C. Noonan and W.P. Noonan and A.V. Maher. Noonan and Noonan reviewed the common reasons for rejection of New Dietary Ingredient submissions and discussed some tactics to help improve the odds for successful marketing in the United States. Maher explained how the Federal Trade Commission (FTC) evaluates the adequacy of scientific substantiation and provides numerous examples of how the FTC has applied the standard in law enforcement actions.

Part III: Manufacturing Compliance and Analytical Validation was provided by leading analytical researchers Drs. R. Crowley, L.H. Fitzgerald, and D. Sullivan. Crowley and Fitzgerald emphasized the impact of cGMP compliance on consumer confidence in dietary supplements. Sullivan and Crowley gave an overview of the development and validation of analytical methods for health foods and dietary supplements. In Part IV: Importance of Safety Assessment Drs. G.A. Burdock and I.G. Carabin and J.C. Griffiths and Dr. P.A. Lachance discussed the importance of GRAS to the functional food and nutraceutical industries. Lachance reviewed the concept of traceability of nutraceuticals and pharmaceuticals to prevent the covert introduction of chemical and microbiological hazards as agents of terrorism or counterfeiting.

Part V: Regulations Around the World was provided by a selected panel of world renowned professionals specializing in their respective country's legislation on dietary supplements and functional food regulations. Drs. J.E. Hoadley and J.C. Rowlands highlighted the Food and Drug Administration's (FDA) perspectives on health claims for food labels in the United States. Dr. S. Agarwal, a leading nutrition scientist, and S. Hordvik and S. Morar, renowned regulatory legal counsels, focused their discussion on the nutritional claims for functional foods and supplements in the United States. Dr. O.I. Aruoma summarized the impact of WTO and food regulation on the food supply chain.

Drs. S. Martyres, M. Harwood, and E.R. Nestmann reviewed the basic principles of Canada's regulations and examples of major issues faced by the Natural Health Products Directorate. Dr. P. Coppens et al. and Drs. O.P. Gulati and P.B. Ottaway, two renowned professional teams, discussed in detail the legislation governing enrichment of foods and health claims made on botanical-sourced, functional and fortified foods and food supplements in the European Union. Dr. S.A. Ruckman provided an understanding of European and UK food law and how they apply specifically to nutraceuticals and functional foods.

Drs. D. Ghosh, L.R. Ferguson, and M. Skinner presented a succinct discussion on the roles of the Therapeutic Goods Administration and the Medicine and Medical Devices Safety Authority in evaluating complementary and alternative medicines in Australia and New Zealand. Drs. H. Ohama, H. Ikeda, and H. Moriyama provided a detailed overview on Japanese regulations for health foods and foods, as well as intricate aspects of Foods for Specified Health Uses regulations. Korean FDA representatives Professor J.Y. Kim of Seoul National University of Science and Technology and Dr. D.B. Kim of Korean Ministry of Food and Drug Safety, in collaboration with Professor H.J. Lee of Seoul National University, gave a thorough update on Korean Regulatory aspects. A. Roberts and R. Rogerson discussed the regulations and requirements for regulatory approval within the Chinese market.

V.A. Tutelyan and B.P. Sukhanov reviewed biologically active food supplements as well food supplement quality, safety, efficacy, and regulating the registration process in the Russian Federation. Drs. K. Shelke and C. Hewes gave an in-depth narration of the historical background and the current status of Indian nutraceuticals and regulation foods regulations. The authors concluded the chapter with an outlook of the regulatory challenges that lie ahead in the nutraceuticals and functional foods industry in India.

Drs. T. Bahorun and O.I. Aruoma focused on the regulatory status on botanical drugs, nutraceuticals, and functional food in the African continent. Drs. M.C. de Figueiredo Toledo and F.M. Lajolo described the legislation that is relevant in the marketing of supplements and functional foods in Brazil. Dr. J. Zawistowski provided an extensive review on the regulatory system for functional foods in the eight countries of the Pacific Rim: Taiwan, Hong Kong, South Korea, Malaysia, Indonesia, Singapore, Philippines and Thailand as well as their diverse acts, regulations, and guidelines.

Part VI. Intellectual Property, Branding, Trademark and Regulatory Approvals in Nutraceuticals and Functional Foods provided an update on these subjects by L.K. Chong, L.J. Udell, and B.W. Downs. Further discussion on the challenges of intellectual property and branding are reviewed by the legal group at Marshall, Gerstein, & Borun LLP.

Although nutraceuticals and functional foods have significant promise in the promotion of human health and disease prevention, health professionals, nutritionists, and regulatory toxicologists should strategically work together to derive appropriate regulations to provide the optimal health and therapeutic benefits to mankind.

References

[1] 21 U.S.C. 343(r)(4)).
[2] <http://www.nutraingredients-usa.com/news/ng.asp?n=80363-crn-ipsos-survey>.
[3] Nutrition Business Journal, 2007 Industry Overview, July/August 2007.
[4] Nutrition Business Journal Feb/March, XII (2/3):1−31, 2007.

List of Contributors

Sanjiv Agarwal
NutriScience LLC, East Norriton, Pennsylvania, USA

F. Ahmad
F-34, Okhla, New Delhi, India

Anthony L. Almada
Vitargo Global Sciences, Inc., Dana Point, California, USA

Okezie I. Aruoma
American University of Health Sciences, Signal Hill, California, USA

Theeshan Bahorun
ANDI Centre of Excellence for Biomedical and Biomaterials Research, University of Mauritius, Réduit, Republic of Mauritius

Mirza E. Baig
Pfizer Ltd. Haryana, India

Kiran Bhupathiraju
Laila Nutraceuticals, India

George A. Burdock
Burdock Group, Orlando, Florida, USA

Ioana G. Carabin
Island ENT, Key West, Florida, USA

Digambar Chahar
Consulting Regulatory Affairs Scientist, Mississauga, Ontario, Canada

Leighton K. Chong
Udell Associates, Castro Valley, California, USA

Patrick Coppens
EAS Strategic Advice, Brussels, Belgium

Amitava Das
Davis Heart and Lung Research Institute, Department of Surgery, The Ohio State University Medical Center, Columbus, Ohio, USA

Mayuri Dhunnoo
ANDI Centre of Excellence for Biomedical and Biomaterials Research, University of Mauritius, Réduit, Republic of Mauritius

Bernard W. Downs
Udell Associates, Castro Valley, California, USA

Peter P. Fu
National Center for Toxicological Research, Jefferson,
Arkansas, USA

Dilip Ghosh
Nutriconnect, Sydney, Australia

Trimurtulu Golakoti
Laila Nutraceuticals, India

Om P. Gulati
Scientific & Regulatory Affairs, Horphag Research Management S.A., Geneva,
Switzerland

John R. Harrison
JRH Toxicology, Ontario, Canada

James E. Hoadley
EAS Consulting Group, Alexandria, Virginia, USA

Stein Hordvik
Hordvik's Consulting, Elkhorn, Nebraska, USA

Greta Houlahan
NSF International, Ann Arbor, Michigan, USA

Chun Hu
Nutrilite Health Institute, Buena Park, California, USA

Hideko Ikeda
Biohealth Research Ltd., Tokyo, Japan

Michelle C. Jackson
Venable LLP, Washington, DC, USA

Raj K. Keservani
School of Pharmaceutical Sciences, Rajiv Gandhi Proudyogiki Vishwavidyalaya,
Bhopal, India

Ji Yeon Kim
Department of Food Science and Technology, Seoul National University of
Science and Technology, Seoul, South Korea

Seong Ju Kim
Korea Food and Drug Administration, South Korea

Alla A. Kochetkova
Institute of Nutrition of the Russian Academy of Medical Sciences, Department of Foods for Special Dietary Uses, Moscow, Russia

Alluri V Krishnaraju
Laila Nutraceuticals, India

Deb Kumar Nath
Apotex Inc., Toronto, Canada; Global QA, Apotex Inc, Toronto

Teck-Chai Lau
Department of International Business, Universiti Tunku Abdul Rahman (UTAR), Malaysia

Hyong Joo Lee
WCU Biomodulation, Department of Agricultural Biotechnology, College of Agriculture and Life Sciences, Seoul National University, Seoul, South Korea

Claudia A. Lewis
Venable LLP, Washington, DC, USA

Nikita McGee
AFB International, St. Charles, Missouri, USA

Sandra Morar
McGrath North Mullin & Kratz PC LLO, Omaha, Nebraska, USA

Hiroyoshi Moriyama
The Japanese Institute for Health Food Standards, Bunkyo-ku, Tokyo, Japan

Vidushi S. Neergheen-Bhujun
ANDI Centre of Excellence for Biomedical and Biomaterials Research, University of Mauritius, Réduit, Republic of Mauritius

Earle R. Nestmann
Health Science Consultants Inc., Ontario, Canada

Hirobumi Ohama
Biohealth Research Ltd., Tokyo, Japan

Peter Berry Ottaway
Berry Ottaway & Associates Limited, Hereford, United Kingdom

Simon Pettman
EAS Strategic Advice, Brussels, Belgium

Jennifer Radosevich
AFB International, St. Charles, Missouri, USA

Nancy E. Rawson
AFB International, St. Charles, Missouri, USA

J. Craig Rowlands
The Dow Chemical Company, Toxicology and Environmental Research and Consulting, Midland, Michigan, USA

Chandan K. Sen
Davis Heart and Lung Research Institute, Department of Surgery, The Ohio State University Medical Center, Columbus, Ohio, USA

Krishanu Sengupta
Laila Nutraceuticals, India

Andrew Shao
Herbalife International of America, Inc., California, USA

Anil K. Sharma
School of Pharmaceutical Sciences, Rajiv Gandhi Proudyogiki Vishwavidyalaya, Bhopal, India

Svetlana A. Sheveleva
Institute of Nutrition of the Russian Academy of Medical Sciences, Department of Food Microbiology and Microecology, Moscow, Russia

Makoto Shimizu
The University of Tokyo, Bunkyo-ku, Tokyo, Japan

Elena A. Smirnova
Institute of Nutrition of the Russian Academy of Medical Sciences, Department of Foods for Special Dietary Uses, Moscow, Russia

Boris P. Sukhanov
Institute of Nutrition of the Russian Academy of Medical Sciences, Department of Nutrition Enzymology, Moscow, Russia

Victor A. Tutelyan
Institute of Nutrition of the Russian Academy of Medical Sciences, Director, Moscow, Russia

Lawrence J. Udell
Udell Associates, Castro Valley, California, USA

Edward Wyszumiala
NSF International, Ann Arbor, Michigan, USA

Qingsu Xia
National Center for Toxicological Research, Jefferson, Arkansas, USA

Chandra S. Yeevani
Apotex Inc., Toronto, Canada; Global QA, Apotex Inc, Toronto, Canada

Jerzy Zawistowski
University of British Columbia, Faculty of Land and Food Systems, Food, Nutrition and Health, Vancouver, British Columbia, Canada

Introduction

Nutraceuticals and Functional Foods: Aligning with the Norm or Pioneering Through a Storm

1

Anthony L. Almada

Vitargo Global Sciences, LLC, Dana Point, California

The nutrition industry thrives with a global footprint, enjoying substantial economic momentum. In this chapter I define "nutraceuticals"—still nutrition industry/scientific community parlance and lacking a regulatory definition—as supplements or complements to the diet that are comprised of bioactives that occur in food or are produced *de novo* in human metabolism, botanicals, or their constituents, intended to impart a physiological or medicinal effect after ingestion; and vitamins and minerals, delivered in forms that differ from conventional foods or beverages, e.g., solid dosage forms like capsules, tablets, or ethanolic liquid extracts. "Functional foods and beverages"—another nutrition industry/scientific community descriptor lacking a regulatory parent—I define as the above, yet delivered in conventional food and beverage formats, e.g., a single-serving beverage, a fermented dairy product, or a food bar. Collectively I will refer to these two classes as "NFx." Biologically, NFx are intended to impart a physiological or medicinal effect, *without or in addition to a nutritional effect*.

1.1 Myths devolve into faith and dogma

The consumer promise of an NFx holds the allure of improved or preserved physiologic or metabolic function. The spectrum of consumer expectation ranges from disease/condition prophylaxis to disease/condition treatment, to even cure. The quest for "natural," the drive to engage in "self-care," and the almost free and boundless access to the virtual, omniscient libraries called the Internet and social media messaging are forging a new breed of companies and consumers. Armed and dangerous, with a modicum of evidence or simply a sugar-coated science tale, bioactives are birthed, brands are born, products are launched, and tribes are formed.

Nutraceutical and Functional Food Regulations in the United States and Around the World.
DOI: http://dx.doi.org/10.1016/B978-0-12-405870-5.00001-3

In the world of NFx, from a seedling of science often emerges a fad or a trend. In 1981 a Canadian research team developed the glycemic index (GI), a tool conceived to provide a physiological metric related to a carbohydrate-containing food, supplement, or beverage and its influence on blood glucose dynamics [1]. An NFx exhibiting a "high" GI value is widely believed to deliver its glucose to the circulation at a "faster" rate, while an NFx displaying a "low" GI value is believed to deliver its glucose at a slower rate. However, the GI does *not* reflect the rate of digestion and absorption of glucose entering the blood [2–5]. Rather, it is simply the net result of how much glucose entered the blood (from the gut; hepatic gluconeogenesis) less how much exited the blood (primarily disposal into tissues, especially skeletal muscle and liver) over a certain time or time interval. Despite this, numerous influencers and marketers of NFx state that GI is correlated to the rate of entry of glucose into the blood or that a low GI NFx is equivalent to a "slow burning" carbohydrate [6–8], despite zero or contravening, *direct* evidence in place [2–5].

A more recent entrant into the faith-fueled NFx category is the gluten-free moniker. Despite the identifier, a gluten-free NFx is *not* free or devoid of gluten but exhibits ≤20 parts per million of gluten in the final consumable form. In the United States, from 2012 forward the Food and Drug Administration issued 17 product withdrawals due to undeclared wheat allergen in foods [9]. Over the same interval Health Canada issued 13 undeclared gluten recalls [10]. This, coupled with the observation that most "gluten-free" certifications do *not* typically require quantitative, post-manufacturing testing for gluten on *each* batch of any certified product, imparts a questionable degree of confidence to the gluten-sensitive or celiac individual (who typically pays a premium for "gluten-free"). What if a manufacturing or labeling error or omission was committed, or an intentional act of domestic bioterrorism was perpetrated, within a manufacturing facility, rendering an NFx labeled as gluten-free to be effectively gluten-*rich*?

In the previous edition of this book I wrote:

> *From this author's perspective the NFx products that enjoy (or have enjoyed) robust, sustainable revenues are those defined by one of two phenomena: 1) consumer experience, and 2) high circulation, incremental media editorial messaging.*

Reflecting upon this passage I see the lack of emphasis upon the operative phrase "one of two" and word "sustainable" as problematic. Numerous NFx action words/phrases (low glycemic; gluten-free; açai; probiotic, antioxidant; super fruit; anti-inflammatory; coconut water; omega-3) enjoy moderate to broad consumer awareness, linked to salutary or desirable effects. However, few enjoy robust and sustainable (or incremental) revenues, with less being buttressed by an even semi-compelling evidence base.

1.2 Value addition or illusion?

Our own market and scientific research still indicates that *less than one out of a thousand* finished NFx products—not individual bioactive ingredients but *the final form purchased and ingested by a consumer*—enjoys at least a single randomized controlled trial (RCT) performed against a placebo or a positive control (e.g., an appropriate drug), and demonstrating a statistically significant outcome superior to that of the placebo, or equivalent or even superior to a positive drug comparator. Add in the additional criterion of such an RCT published in a reviewed journal and the percentage of qualifying NFx falls precipitously. What is more unsettling is our finding that of the ~250−300 RCTs conducted on NFx products in North America annually only a fraction are made public (via presentation or publication). This underscores the prevalence of publication bias and reveals the far more haunting truth that data *suppression* is not uncommon. Sadly, the "gagging" of investigators related to dissemination of clinical trial data on NFx products reaches into the ranks of publicly traded and multinational NFx marketing entities. The "burial" of this data is achieved by increasing numbers of studies performed with contract research and clinical service organizations, where data ownership (and intellectual property rights) is exclusive to the study sponsor.

The emergence and adoption of clinical trials registries (http://clinicaltrials. gov/; http://www.controlled-trials.com) has augmented the integrity of reporting while reducing publication bias, albeit in biomedical journals that require timely clinical trial registration (before enrolling the first participant within a 60 day grace period). A large proportion of the journals that entertain a large number of NFx publications do not require registration of a clinical trial as an inclusion criterion for publication (and editors of the "elite" journals may assert that the rigor and sample size of many NFx manuscripts do not warrant inclusion in their journals. . .). Given that a notable number of registered clinical trials on NFx do not publish their results after completion suggests that clinical trial registration alone is insufficient to convey transparency of research outcomes. Nevertheless, the adoption of clinical trial registration requirements [11] by the editors of journals that *de facto* cater to NFx manuscripts would raise the bar.

The importance and utility of NFx is defined not by revenues, share price, the annualized number of product introductions, nor calculated annual growth rate of the sector. NFx should serve the consumer by making a favorable, measurable impact upon her/his biology, congruent with the claimed benefit of the product and perhaps beyond. A probiotic beverage or solid dose that improved laxation *and* reduced LDL cholesterol could become a product that enjoys legions of brand zealots. The provision of NFx products, in its purest form, can be viewed as a public health initiative, centered upon offering consumer goods that meet four criteria: (1) excellent tolerability and hedonics, at least equal to the conventional food or beverage category leader (for solid dose forms, excellent tolerability, acceptable dose frequency, and ease of use/swallowing); (2) demonstration of consumer-relevant and scientific community-relevant efficacy through an incremental series of clinical trials; (3) aligning with existing, demographically dense unmet or undermet needs; and (4) remaining within a retail price comfort zone.

Aversion to "good for you" foods and beverages can be a significant deterrent for many consumers. Creating an equal or superior sensory experience with a functional food or beverage can have a dramatic impact upon consumer trial and repeat purchasing. The post-ingestive experience also has significant import, especially in relation to gastrointestinal symptoms. NFx products that are intended to have a gastrointestinal effect (e.g., prebiotics or synbiotics; fiber; anti-*Helicobacter pylori* agents) have the highest likelihood of causing gastrointestinal distress. Solid dose forms need to be delivered in a manner that does not require multiple dosing throughout the day nor large dose formats (e.g., large tablets or capsules). Liquids need to be easily measured and palatable, with masking/debitterizing agents considered for use. Powders need to be easily mixable and non-hygroscopic under normal storage conditions. The unrelenting growth in the interest in natural and sugar-free also demands NFx compositions to seek natural, non- or low-caloric bulk and/or high-intensity sweetener systems that exhibit high gut tolerability and palatability.

One cannot emphasize the critical need for a series of clinical trials on an NFx. One of the elements lacking among consumers of NFx products is confidence: in the claims, in the taste, and in the brand. A specific consumer product buttressed by even one well-designed RCT—where the number of servings or dose used in the study matches the labeled use directions and is not cost prohibitive—can exalt a brand into a new level of equity with consumers. It is a widely held misperception that sponsoring an RCT requires a budget in excess of $100,000. Harkening back to our finding of less than 0.1% prevalence of an NFx with at least one RCT demonstrating efficacy, even an RCT pilot study with a sample size of 12−20 (or half this, if employing a cross-over design) would confer distinction. A follow-up study—fueled by incremental revenues from the sales of the NFx—with a larger sample size and longer duration would allow for the creation of a foundation of evidence. Considering the allocation of resources by NFx companies to marketing, advertising, promotions, and sales it is typically highly feasible for even early stage companies to capitalize a clinical research program. It is not a matter of economics; it requires an executive commitment to invest in evidence-based NFx goods. Unlike pharma and life sciences industries, the vast majority of individuals leading NFx business entities is science illiterate and averse and has escalated the ranks through sales, marketing, or business development capacities. This is coupled with an ignorance of how to monetize product-specific clinical science, which further magnifies the resistance to invest.

A pivotal element in the monetization of product-specific clinical science is the public communication of positive results (oral or poster presentation at a national or international scientific/medical conference) and/or publication of the RCT (in a reviewed journal, ideally at or above the median impact factor for journals in the relevant category). This can become the axis of a strategic communications campaign that yields editorial mention in print (newspapers, magazines), online (health/medical news focused websites), and on air (television, radio). Editorial mention of science linked to a branded NFx can generate dialogue among consumers, elevate awareness, and potentially increase demand and trial purchase velocity. The advent of consumer-generated content, e.g., blogging and social media (Facebook, Twitter, Instagram) has fostered a novel form of media that has significant impact upon purchase behavior. A recent Nielsen online survey among over 28,000 consumers in 56 countries found a global average of 70% to trust online consumer reviews (second only to friends and family recommendations) as a reliable source of information, while sponsored ads on social networks are believed by over one-third of the global respondents [12].

The path to strategic innovation mandates both a keen awareness of consumer needs and linked demographics, those that are unmet and undermet. Latent needs (not readily apparent or acknowledged by the consumer), albeit a potentially viable pursuit, often require more intensive education and awareness generation. An unmet need would be an NFx that *reproducibly* and *preferentially* reduces body fat mass without the use of any stimulants (there does not appear to be any NFx that enjoys any solid evidence of such). An undermet need would be an NFx that prevents or rapidly treats seasonal allergic episodes (a plethora of products exist but their efficacy is highly variable or questionable). Demographics and market size are important elements to consider for innovation direction. NFx products targeted to orphan applications, e.g., scleroderma or McArdle's disease, are altruistic and admirable but may produce unsatisfactory yields if they are positioned as a primary revenue generator. These pursuits may also require higher retail pricing to obtain a satisfactory return on investment.

Cost to the consumer—the price they must pay to begin enjoying the benefits of an NFx—must be determined before allocation of any resources. This will vary from country to country. The price will have a segmenting effect upon which socioeconomic demographic the product needs to be positioned. One interesting phenomenon that could happen with higher/premium priced NFx products is the bandwagon effect [13]. This is described as consumer demand increasing as the number of consumers buying the good increases. This is often achieved by the conspicuous consumption or overt recommendation of a good by a celebrity or media luminary. Part of the monetization strategy of NFx innovation and commercialization warrants the prudent selection of influential users or endorsers of a specific product, especially an endorser who aligns with the brand by personal/medical need (a celebrity living with diabetes or with a history of acne) or by vocation (a current or former professional footballer or cricketeer, or Olympian gold medalist, plagued with injuries, advocating an NFx that has increased their mobility, pain-free performance, etc.)

1.3 Landmines and weapons of brand destruction

An NFx is birthed, after a 16 month gestational period wherein a small, single RCT was conducted (capturing acute satiety and increased caloric expenditure effects, along with a blunting of glycemic response to a mixed meal). Results are written up and presented at a national biomedical conference, aligned with strategic media events and a follow along national advertising and promotions campaign, in print, on air, on demand, online, and in person. The manuscript for the RCT is nearing the completion of its editorial review prior to imminent publication. Eight months later cumulative sales have eclipsed the $10 million mark, with annualized projections exceeding $15 million. After the winter holiday an early morning priority envelope is delivered directly to the CEO's desk, from a prestigious law firm. It remains unopened in her absence, for 3 hours. Just before noon, as the CEO walks into the office for the first time in the New Year, she courses by the office of the General Counsel. The GC, visibly ashen and distressed, is asked by the CEO if he is okay. Silently he hands her a lawsuit that the company had been served moments before her entry. It is a class action lawsuit, alleging fraudulent claims (weight loss and fat loss; improved glycemic control). She takes the suit to her desk and notices the urgent parcel. Hoping for better news she opens the envelope and the enclosed two page letter. Atop its first page she is transfixed by the words "Cease and Desist." Further down she notes a word very common to her company, the same word that represents one of the three (presumed) bioactives in her company's NFx that is enjoying meteoric success. Patent infringement.

Either of the two vignettes in this scenario, although compressed into one day, has stricken a staggering number of mid-market to large NFx companies. From allegedly unsubstantiated claims for a breakfast cereal asserting to enhance the cognitive function of children [14] to alleged mischaracterization of fat content and "healthiness" [15], to alleged infringement of otherwise dormant patents with new, protective parents [16], looming threats appear to await the less than hyperdiligent NFx marketer. Assembling an evidence base that closely aligns with an NFx message/claim profile, and performing thorough freedom to operate searches (and maintaining vigilance on patents that are queuing up to be granted) are essential prophylactic measures for any NFx.

1.4 Opportunity awaiting: pioneering upstream

The nutrition industry and academic communities continue to have a fascination with various "-omics" platforms, from genomics to lipidomics to proteomics to metabolomics. This focus appears to maintain a myopic vista of consumer relevance. For the consumer, a personalized omics profile offers a fascination to the innovator consumer only, which is a very small fraction of the population at large. The illusion of personalized nutrition via personalized biometric profiling has no relevance to what is a valued return on investment by the consumer: a novel, validated path to *superior* health and biofunctioning. What merits integration into -omics-centric business models in the nutrition industry is comparator trials that pose -omics-guided NFx intervention against conventional NFx intervention, with consumer-relevant outcome measures. For example, -omics biometric profiling of a cohort of adults with chronic facial acne would lead to a select suite of NFx products that would be expected to be effective. Via a cross-over design (with an appropriate washout period) the subjects would receive the -omics-guided NFx and maintain their normal diet or pursue a nascent, evidence-based dietary intervention, at different times. If the NFx intervention proved superior in the outcome measures, which are easily quantifiable in this example, the economic and biological ramifications of this could have tectonic-plate shifting industrial and consumer impact. As of the writing of this second edition no such achievement has been made public. Hopefully its revelation is not too distant.

Far less cost- and time-intensive, perhaps less risky, but requiring executive fortitude and "change meister" prowess, is the path of disruptive pioneering. One example is illustrated by probiotics, widely believed to require live/viable organisms to exert enteric or systemic effects. The probiotics "cartel," across the value chain, has invested an enormous amount of intellectual and financial capital to reinforce this. In contradistinction, one finds that one of the first "biotic" agents used was derived from heat-killed *Lactobacillus acidophilus* organisms extracted from human stools (Lacteol), which enjoys a variety of published RCTs [17,18] as do other heat-killed organisms [19,20]. Coupling the emerging evidence on "dead" probiotics with the finding that a micro fraction of an oral dose of an evidence-based (live) probiotic strain is delivered viable to the lower gut, relative to its heat-killed counterpart [21] may enable a pioneer-led brand to survive the Herculean march uphill against the strong current of the live probiotics cartel.

Another direction related to NFx and the harnessing of caffeine is to exploit analytical methodology in an attempt to demonstrate that natural caffeine is *not* present in a cadre of products, but rather synthetic caffeine, despite labeling otherwise [22]. Such a pioneering and competitive resistance-engendering path, if successful, could have far greater impact than in NFx products.

1.5 **Conclusion**

The promise of nutraceuticals and functional foods is prodigious. The challenge of creating finished goods that create sustainable and robust revenues, and provide sufficient investment returns, coupled with large legions of brand zealots, is equally substantial. The paucity of NFx products that enjoy a reputable evidence base affords a compelling opportunity for industrial and academic alliances that integrate consumer relevance, from concept to container. Of equal importance is the need to insert an intellectual property strategy from the inception point, both offensive (adequate working capital and vigilance) and defensive (preemptive). Delivering on the promise of nutritional goods that exert preventive, therapeutic, or quality of life enhancement can inject consumer confidence into a sector viewed with diffidence by both consumers and the medical community, and buoy industrial growth with science-driven revenues and intellectual assets. Elegantly articulated in a recent paper by a Kellogg's employee and a long-standing NFx clinical trial principal investigator and industry consultant [23], for NFx to be a sustainable offering and pursuit through the value chain, standardization and harmonization on evidence criteria, regulatory elements, and consumer engagement fronts are essential.

References

[1] Jenkins DJ, Wolever TM, Taylor RH, Barker H, Fielden H, Baldwin JM, et al. Glycemic index of foods: a physiological basis for carbohydrate exchange. Am J Clin Nutr 1981;34:362−6.

[2] Schenk S, Davidson CJ, Zderic TW, Byerley LO, Coyle EF. Different glycemic indexes of breakfast cereals are not due to glucose entry into blood but to glucose removal by tissue. Am J Clin Nutr 2003;78:742−8.

[3] Eelderink C, Moerdijk-Poortvliet TCW, Wang H, Schepers M, Preston T, Boer T, et al. The glycemic response does not reflect the in vivo starch digestibility of fiber-rich wheat products in healthy men. J Nutr 2012;142:258−63.

[4] Eelderink C, Schepers M, Preston T, Vonk RJ, Oudhuis L, Priebe MG. Slowly and rapidly digestible starchy foods can elicit a similar glycemic response because of differential tissue glucose uptake in healthy men. Am J Clin Nutr 2012;96:1017−24.

[5] Almada AL. Carbohydrate and muscle glycogen metabolism: exercise demands and nutritional influences. In: Bagchi D, Nair S, Sen CK, editors. Nutrition and enhanced sports performance. muscle building, endurance, and strength. London: Elsevier; 2013. p. 333−40.

[6] <www.nestle.co.nz/asset-library/documents/.../glycemic_index.pdf> [accessed 14.08.13].

[7] <http://www.extendbar.com/blog/low-glycemic-snack-foods> [accessed 12.08.13].

[8] Glycemic index diet. What's behind the claims <http://www.mayoclinic.com/health/glycemic-index-diet/MY00770> [accessed 18.08.13].

[9] <http://www.fda.gov/Safety/Recalls/default.htm> [accessed 14.08.13].

[10] <http://www.hc-sc.gc.ca/ahc-asc/media/advisories-avis/index-eng.php?cat = 4> [accessed 14.08.13].

[11] Huser V, Cimino JJ. Evaluating adherence to the International Committee of Medical Journal Editors' policy of mandatory, timely clinical trial registration. J Am Med Inform Assoc 2013;20:e169−74.

[12] Nielsen Newswire. Consumer trust in online, social and mobile advertising grows. <http://www.nielsen.com/us/en/newswire/2012/consumer-trust-in-online-social-and-mobile-advertising-grows.html>; 2012 [accessed 20.08.13].

[13] Leibenstein H. Bandwagon, snob, and Veblen Effects in the theory of consumers' demand. Quart J Econ 1950;64:183−207.

[14] Ellison JP. Recent Kellogg class action settlement is a reminder that litigation over advertising claims often comes in several waves. <http://www.fdalawblog.net/fda_law_blog_hyman_phelps/2013/05/recent-kellogg-class-action-settlement-is-a-reminder-that-litigation-over-advertising-claims-often-c.html>; 2013 [accessed 20.08.13].

[15] Manatt Phelps and Phillips, LLP. (2013). Got settlement? Muscle Milk suit settles for $5.3 million. <http://www.lexology.com/library/detail.aspx?g = 2345287b-24f0-4453-8d45-10b44d04315e> [accessed 20.08.13].

[16] Stump E, Almada A. Patent trolls strike supplements. Nutr Bus J 2013;18:36−7.

[17] Halpern GM, Prindiville T, Blankenburg M, Hsia T, Gershwin ME. Treatment of irritable bowel syndrome with Lacteol Fort: a randomized, double-blind, cross-over trial. Am J Gastroenterol 1996;98:1579−85.

[18] Salazar-Lindo E, Figueroa-Quintanilla D, Caciano MI, Reto-Valiente V, Chauviere G, Colin P, for the Lacteol Study Group. Effectiveness and safety of Lactobacillus LB in the treatment of mild acute diarrhea in children. J Pediatr Gastroenterol Nutr 2007;44:571−6.

[19] Shinkai S, Toba M, Saito T, Sato I, Tsubouchi M, Taira K, et al. Immunoprotective effects of oral intake of heat-killed *Lactobacillus pentosus* strain b240 in elderly adults: a randomised, double-blind, placebo-controlled trial. Br J Nutr 2013;109:1856−65.

[20] Hirose Y, Murosaki S, Yamamoto Y, Yoshikai Y, Tsuru. T. Daily intake of heat-killed *Lactobacillus plantarum* L-137 augments acquired immunity in healthy adults. J Nutr 2006;136:3069−73.

[21] Wutzke KD, Berg D, Haffner D. The metabolic fate of doubly stable isotope labeled heat-killed *Lactobacillus johnsonii* in humans. Eur J Clin Nutr 2008;62:197−202.

[22] Zhang L, Kujawinski DM, Federherr E, Schmidt TC, Jochmann MA. Caffeine in your drink: natural or synthetic? Anal Chem 2012;84:2805−10.

[23] Marinangelia CPF, Jones PJH. Gazing into the crystal ball: future considerations for ensuring sustained growth of the functional food and nutraceutical marketplace. Br J Nutr 2013;26:12−21.

Nutritional Supplements and Functional Foods: Functional Significance and Global Regulations

2

Amitava Das and Chandan K. Sen

Davis Heart and Lung Research Institute, Department of Surgery,
The Ohio State University Medical Center, Columbus, Ohio

2.1 Introduction

Two hundred years ago, processed sugar was only available in drug stores [1]. Today, it is one of the most successfully marketed nutritional chemicals, addicting consumers from a time when they are very young. On one hand, we are eating food material of compromised nutritive value because of milling and processing [2]. On the other hand, there is not enough time for a healthy balanced meal. To compensate for these losses, dietary supplements and functional foods are a must in today's lifestyle [3–6].

Dietary supplements refer to the preparations intended to compensate for the nutrients that otherwise might not be sufficiently present in the diet. Dietary supplements are categorized as food in some countries, while in other countries they are considered as drugs or natural health products. On the other hand, the term functional foods refers to supplements that have new or more ingredients added, resulting in enhanced function and improvement of health [7]. These types of foods are comprised of processed foods as well as foods enriched with health-promoting additives. The search for functional foods, or functional food ingredients, is one of the leading trends in today's food industry.

Bread baked from the old varieties of wheat and rye, oats porridge, fish and fermented milk products, and berries and fruits (some of which are also fermented like wines), as well as olive oil, were the main items of the daily diet in European countries. However, the present diets contain little or none of these. Instead they include refined carbohydrates (especially sucrose, which is almost of analytical grade) and in excess of the n-6 series of fatty acids in vegetable oils. Similar trends dominate in the Americas and Asia-Oceania. The previously necessary heavy dynamic work has also been replaced by sedentary occupations with significant persistent static muscle activation and lifestyle with elevated sympathetic tonus

Nutraceutical and Functional Food Regulations in the United States and Around the World.
DOI: http://dx.doi.org/10.1016/B978-0-12-405870-5.00002-5

and an imbalanced endocrine status. The result is the incidence of metabolic syndrome on an epic scale. The metabolic syndrome, being overweight and obese, represents a major burden on the health care system in many countries [8]. Glucotoxicity and lipotoxicity are more common than any other toxicities and harmful to the secretion of insulin, e.g., the key regulator of our metabolism [9].

Understanding the role of foods in the maintenance of health and cure of diseases has its roots in ancient culture. Ancient documents of the Indian Ayurveda, Chinese scriptures, the Judean and Christian Old Testament (the books of Moses and Daniel), and Greek (Hippocrates), Islamic (Koran), and also Finnish (Kalevala) cultures provide information on this understanding. Unfortunately, the rather recent developments of milling and other industrial refining methods of foods have obscured the old knowledge with the commercial trends in marketing. "White is beautiful" is misleading in this context. The refined and processed foods contain less and often only minuscule amounts of the nutrients that plants provide. In nature, phytochemicals present as a mixture that actually renders it to be functionally active, which is in contrast to the "one active principle" hypothesis. In other words, the functional activity of the phytochemicals is not because of the component alone but is triggered by the presence of other selected components that constitute the mixture. Dietary phytochemicals play a major role in regulating human health and disease [10]. As people can now easily compose their diets by avoiding the necessary and protective chemicals that nature provides, global epidemics of overweight and obesity in unprecedented scale are spreading not only in advanced and well-off nations [11] but also some economically privileged social layers in developing countries while other groups may be starving [12].

The term nutraceutical, a combination of the words "nutrition" and "pharmaceutical," was coined by Dr. Stephen L. DeFelice, founder and chairman of the Foundation of Innovation Medicine, Crawford, New Jersey [13], and it refers to the food or food ingredients that have defined physiological effects. They do not easily fall into the legal categories of food or drug and often reside in a gray area between the two [184]. Generally these products cover health promotion, "optimal nutrition," the concept of improved performance—both physically and mentally—and reduction of disease risk factors [185]. The terminology "physiologically functional foods" or simply "functional foods" was conceived over three decades ago in the context of nutrition during space travel [14]. The modern concept of functional food for the general population was proposed by Japanese academic society in the early 1980 s, and the legislation for functional foods was first implemented as Foods for Specified Health Use (FOSHU) [15,16]. The purpose of the Functional Food Science in Europe (FUFOSE) Concerted Action was to reach consensus on scientific concepts of functional foods in Europe by using the science base which provides evidence that certain nutrients affect physiological functions in a positive manner [186]. The conclusion proposed "a working definition" of functional foods: foods can be regarded as functional if they can be satisfactorily demonstrated to affect beneficially one or more target function in the body, beyond adequate nutritional effects, in a way relevant to an improved state of health and well-being and/or reduction of risk of disease. Functional foods must remain foods and they must be able to attain their effects in amounts normally consumed in a diet [186]. Evidence from human studies, based on markers related to biological response or on intermediate endpoint markers of disease, could provide a sound scientific basis for messages and claims about the functional food products. Two types of claims are proposed that relate directly to these two categories of markers: enhanced function claims (type A) and reduced risk of disease claims (type B). Next, a new European Union (EU) Concerted Action undertook to build upon the principles defined within FUFOSE [186]. The process for the assessment of scientific support for claims (PASSCLAIM) initiative was thus established [17–29]. The European branch of the International Life Sciences Institute (ILSI; established in 1978, ILSI is a worldwide foundation that aims to help public health by advancing the understanding of scientific issues related to nutrition, food safety, toxicology, and the environment. The institute brings together scientists from academia, government, industry, and the public sector to seek solutions for public health issues) [15,30] and the PASSCLAIM initiative are directed at assessing scientific support for claims on foods. The health-promoting properties can be mentioned in marketing in some countries, but not the lessening of the disease risk. In Japan, FOSHU products are for sale under proper legislature. FOSHU ranges over pharmacology, medicine, and food and nutrition. Critics have argued that even FOSHU contains highly purified or concentrated functional ingredients present in ordinary foods and thus, it is very important to take safety issues into consideration [31]. FOSHU is the only type of food product (not ingredients) that can carry health claims and is composed of functional ingredients that affect the structure/function (physiological functions) of the body [31]. These food products are intended to be consumed for the maintenance/promotion of health or special health uses by people who wish to control specified health conditions, such as gastrointestinal conditions and blood pressure [31]. Therefore, FOSHU products target healthy people and people in a preliminary stage of a disease or a borderline condition. When the products are manufactured or distributed, permission or approval from the government is required after rigorous evaluation of the safety and effectiveness of proposed specified health uses [16,31]. Functional foods are drawing more attention in public discussions and presumably soon there will be a standard global policy governing the sale and consumption of functional foods. Clearly, more research is needed to support that process. Research should also be carried out in different cultural settings, as the dietary habits are quite heterogeneous globally. Study outcomes are expected to be determined by the entire dietary matrix of which the functional food tested would represent only a minor component.

The aim of this chapter is to present an overview of the significance of traditional food items and their components as targets of research for functional foods and dietary supplements. A few select examples are discussed. Regulations in different countries that are directed at the sale and consumption of functional foods are concisely discussed.

2.2 Health behaviors and food markets

Health behavior refers to any action taken by a person to sustain, achieve, or reinstate good health and to prevent illness and reflects a person's health beliefs [32]. According to the Centers for Disease Control and Prevention a major fraction of the morbidity and mortality arising out of chronic diseases is attributed to four health risk behaviors: dearth of physical activity, poor nutrition, usage of tobacco, and excessive alcohol consumption [33]. Studies suggest that life expectancy increases if one or more of the healthy lifestyle behaviors are pursued [34]. Health behavior can be changed by appropriate interventions [35−37]. A good example is eastern Finland where cardiovascular morbidity and mortality were the highest in the world according to statistics reported three decades ago. The North Karelia Project [38,39] has been successful in steadily minimizing cardiovascular morbidity and mortality in eastern Finland during the course of the last three decades [40]. There are different theories about how the health behavior of people can be modified. As people are different and their willingness to adopt changes depends on their age and culture, different approaches must be run in parallel. Table 2.1 is a listing of those different approaches that may be used to change the health behaviors.

2.2.1 Possibilities to promote food health-promoting properties

The better functionality of foods can be promoted in different ways [15]. A better understanding of the body's molecular machinery helps in establishing the relationships between nutrition, lifestyle, and genetic predisposition including its

Table 2.1 Health Behaviors Can be Changed by Emphasizing Different Aspects

Health belief	Change takes place as the person believes to be at risk and (s)he believes that the change will help
Consumer	Market, i.e., price, availability, and presentation determine peoples' choices
Behavioral intention	Personal examples motivate others to change
Innovation diffusion	Novel ideas and practices gradually penetrate step by step first into a minority population and then eventually become adopted by the majority
Communication persuasion	Change will take place if the information provided is understood and interesting
Self-control	People study themselves and take respective measures
Self-efficacy	How determined the person is to change for personal reason
Problem behavior	Social immunization against poor food selections
Social marketing	Social acceptability is the starting point in marketing
Social learning	Individual is both the subject and object of the change

Modified from [41].

effects on health and quality of life [41]. Novel technologies, such as functional genomics and metabolomics, provide cues to study human variability in nutrient requirements and biological responses to foods [41]. As an example, the extensive research carried out in the field of functional food has led to the identification of anti-inflammatory and anti-carcinogenic activity, cardiovascular disease prevention, obesity control, and the diabetes alleviation properties of anthocyanins, which are believed to be important to plants as their color attracts animals [42]. There are many natural varieties of plants grown as part of agriculture. Better understanding of their nutritionally valuable components may revitalize their cultivation, as has taken place in the case of ancient wheat varieties. One can also introduce sets of genes to plants as has been done by adding a carotene-producing chain in rice. The opposite manipulation can also be done to lower or to remove an unhealthy component. Modification of foods and drinks can take place also by fortification, which is done by adding vitamins and minerals. One can also remove components that are proven to be harmful such as by removing cholesterol from eggs. In the production of the final food or drink, as in ancient times, fermentation has been used to increase digestibility and also to help storage. These bacteria have also been helpful by increasing resistance to infection for the consumers. The biology of probiotics and prebiotics is well understood. Drinks can also be fortified with valuable chemicals as is done for products made for athletes and the fitness-oriented general population.

2.2.1.1 Fiber

Though the health-promoting properties of whole-grain foods was recognized since the fourth century BC, it was in the 1970s when fiber and other bioactives were identified as the constituents to which these properties could be attributed [43]. The "fiber hypothesis" was developed based on observational studies conducted in the 1970s on African populations who ate high-fiber foods, and were interestingly found to be free of many Western diseases such as cardiovascular disease and colon cancer [44,45].

The main components of natural plant foods are often different kinds of fiber [47−49]. Fibers are valuable as distension of the stomach is one of the key regulatory signals contributing to satiety and termination of food intake. Dietary fiber also increases the volume of the gut contents. This is necessary for gut motility and functions including absorption and defecation. Dietary fiber components have many roles and this means that they are also perhaps the most important group of starting material in the development of functional foods and supplements [50]. Dietary fibers and whole grains have been used in treatment of childhood constipation as well [51]. Dietary fiber consists of the remnants of edible plant cells, polysaccharides, lignin, and associated substances resistant to (hydrolysis) digestion by the alimentary enzymes of humans [187]. Addition of fiber-rich flours or pure dietary fiber to bread has been reported in the literature with an intention of lowering the dietary glycemic index [52]. In Japan, the food tables list the dietary fiber content of animal as well as plant tissues, while many countries accept saccharides of less than DP-10 as dietary fiber (inulin, oligofructose, Fibersol-2, polydextrose, fructo-oligosaccharides, galacto-oligosaccharides, etc.)[187]. These shorter chain oligosaccharides do not precipitate as dietary fiber in the standard Association of Official Analytical Chemists method, which is accepted by the U.S. Food and Drug Administration (FDA), the U.S. Department of Agriculture, and the Food and Agriculture Organization (FAO) of the World Health Organization (WHO) for nutrition labeling purposes [187]. In the UK, the term dietary fiber has been replaced in nutrition labeling by non-starch polysaccharides [187]. Therefore, the American Association of Cereal Chemists (AACC) commissioned an *ad hoc* committee of scientists to evaluate continuing validity of the currently used definition and, if appropriate, to modify and update that definition [187]. Dietary fiber can be considered a functional food when it imparts a special function to that food aside from the normal expected function and similarly when the dietary fiber is used as an additive to foods [187]. As an example, dietary fiber contributes to colonic health, bifidobacterial or lactobacillus stimulation in the gut, coronary artery health, cholesterol reduction, glucose metabolism, insulin response, decrease in blood lipids, prevention of cancer, etc [187].

Fibers also serve as substrates of numerous metabolic reactions in the gut microflora. Some of the components are short chain fatty acids. These fatty acids represent necessary energy sources for the colonic mucosal cells. They also contribute to metabolic regulation in the liver, e.g., in the endogenous synthesis of cholesterol [53]. Fibers are starting materials of lignan production in gut microflora [54]. Lignans are important regulators of the microflora itself and of mucosal cell renewal. Absorbed lignans contribute to the responses of metabolism of, say, steroid hormones and also check the propagation of hormone-related cancer cells [54]. Lignans and other polyphenols are considered to have major beneficial effects on cardiovascular health [55,56]. A vegetarian diet is shown to alter the profiles of the gut microflora with significantly lower numbers of pathogenic bacteria [57]. In addition, consumption of polydextrose or soluble corn fiber was found to have a beneficial shift in the gut microbiome of adults [58]. In particular, fiber seems to be highly valuable [55,59]. In several central and north European countries, dark rye bread has been commonly used in all meals. Rye bread lowers the postprandial glucose response [59]. In addition, rye bread decreases serum total and low-density lipoprotein (LDL) cholesterol in men with moderately elevated serum cholesterol [60]. Rye is a good source of dietary fiber [46,61,62]. We seem to have so much data on the benefits of rye that it could be confidently designated as a functional food item as in bread.

Oatmeal has traditionally been eaten regularly at breakfast in several European countries and also in North America [63−65]. The soluble fiber in the form of β-glucan is abundant in oats [66−68]. It can reduce or prevent postprandial hyperglycemia and lipidemia, i.e., diminish glucotoxicity and lipotoxicity after a standard meal like breakfast. There are several studies on the effects of dietary β-glucan isolated from oats on blood glucose and insulin levels [69−71]. Oat-derived β-glucan also reduces total blood and LDL cholesterol concentration in hypercholesterolemic subjects [69−72]. We have so much data on oat products that they could perhaps meet the requirements necessary for nomination as functional foods [73].

Probiotics, i.e., living microbial food supplements, and prebiotics, i.e., non-digestible food ingredients, are a subject of current attention. Both popular concepts target the gastrointestinal microbiota. While in the Western world the intake of probiotics has been recommended for a long time, prebiotics in general, and non-digestible oligosaccharides in particular, have only recently received attention [74−77,188].

2.2.1.2 Xylitol

Trees use polymerized carbohydrates as their backbones. Actually, that reservoir has the clear potential of managing world famine significantly if taken into wider use. Unfortunately, the digestibility of cellulose is low in the human gut, but the cellulase technology is well known. Ruminants can use cellulose with the aid of their rumen bacteria. Some of the sugars of trees are widely used as food components. Birch trees contain a lot of pentose sugars. In Finland, the alcohol xylitol has been extensively studied for four decades. Xylitol, like other polyol sweeteners, is a naturally occurring sugar alcohol. Xylitol is sweet tasting, although the taste is a bit different from that of sucrose. When xylitol is added to candies and chewing gums instead of sucrose, bacterial growth in the mouth remains low compared to sucrose-containing products. In this way, xylitol may effectively manage oral health [78−82]. In addition, xylitol is also reported to prevent acute otitis media in children up to 12 years of age [83]. Although there are substantial data on the health benefits of xylitol, this natural sugar cannot be categorized as functional food. Xylitol traces in candies or chewing gums fall short of meeting the basic requirement of functional foods.

2.2.1.3 Polyphenols

Seeds, berries, and fruits as well as roots have protective polyphenols in addition to other valuable components [84,85]. The Bible tells us about the balsam. Albeit anecdotally, the Finnish folk healing practices have benefited from the use of resins of pine and spruce in treating dermal wounds. Research on the effects of dietary polyphenols on human health has developed considerably in the past decade. In totality, the evidence supports a role for polyphenols in the prevention of degenerative diseases, particularly cardiovascular diseases and cancers. In addition to its biological functions, kiwifruit-based polyphenols were used in the production of gluten-free bread [86]. The antioxidant properties of polyphenols have been extensively studied, but it has become clear that the mechanisms of action of polyphenols go beyond the modulation of oxidative stress [87]. Whether pro-oxidant, antioxidant, or any of the many other biological effects potentially exerted by polyphenols account for or contribute to the health benefits of diets rich in plant-derived foods and beverages remains to be definitively proven [88]. Thus, the case for polyphenols as functional food is weak at present.

2.2.1.4 Antioxidants

In the 1970s and 1980s, as the field of free radical biology was unfolding, there were hundreds of papers published demonstrating how oxygen-derived free radicals could be damaging to human health [89−92]. The field expanded so rapidly that it led to the establishment of dedicated journals such as Free Radical Biology & Medicine in the late 1980s. The academic community quickly turned toward solutions. The result has been a sharp rise in the study of agents that can protect against the detrimental effects of free radicals, i.e., antioxidants [93−96]. An antioxidant is a substance that, when present at low concentrations compared to those of an oxidizable substrate, significantly delays or prevents oxidation of that substrate [116]. Based on *in vitro* or biochemical studies, many substances have been suggested to act as antioxidants *in vivo*, but few have been proved to do so. Antioxidant therapy became the new buzzword of the mid-late 1980s and a quick glance at the literature would provide the impression that there was hardly any aspect of human disease that did not respond to antioxidant therapy in an experimental setting [97−106]. While the basic scientists were trying to grasp the delicacies involved in free radical antioxidant biology *in vivo*, quite a few clinical trials were launched to test the efficacy of antioxidants in a clinical setting [107]. Meaningful clinical trials rely on a comprehensive understanding of the fundamental principles that underlie the practical hypothesis being tested. Clearly, this represented a major weakness in many of the trials testing the efficacy of antioxidants in a clinical setting. The first trial outcomes proved to be disappointing [108]. Prospective epidemiological investigations suggested a reduction in cardiovascular risk associated with increased intake of antioxidant vitamins, particularly vitamin E. However, completed randomized trials did not support this finding [109]. In spite of the wide spectrum of research in the field of antioxidants, there is a dearth of direct experimental evidence from randomized trials correlating antioxidants with health benefits [110]. In general, antioxidants were regarded as safe [111]. That notion quickly changed with the reporting of large, randomized clinical trials testing β-carotene in primary prevention—no beneficial effect and potential for harm associated with the use of β-carotene. At the time, inconclusive and insufficient epidemiological and clinical trial data with regard to the role of vitamin C in cardiovascular protection was noted. Caution was issued against the widespread use of antioxidant vitamins in cardiovascular protection [112]. The antioxidant paradox thus emerged [113]. Two important lines of development in the mid-1990s supported the contention that free radical biology was more complex than most imagined: first, that an antioxidant can actually behave like a pro-oxidant depending on the local and environmental conditions [114−123] and second, that free radicals and their derivatives could actually serve beneficial physiological functions by acting as a cellular messenger [124,125]. A new interdisciplinary field of oxidation−reduction (redox) signaling in mammals thus emerged [126−128]. Redox-based regulation of signal transduction and gene expression emerged as a fundamental regulatory mechanism in cell biology. Electron flow through side chain functional CH_2-SH groups of conserved cysteinyl residues in proteins was noted to account for their redox-sensing properties [126]. A new journal Antioxidants & Redox Signaling (www.liebertpub.com/ars) was established to support this novel interdisciplinary specialty. During the last decade over 3000 papers have been published in redox signaling. Although the search for the right antioxidant nutrients for the preservation of human health is still on, it is clear that regulation of tissue redox status may be effectively utilized for therapeutic purposes [129−133]. The case for redox-active compounds, including antioxidants, as functional food is moderately strong. Antioxidants such as resveratrol [134,135], curcumin [136−139], and tocotrienol [140−142] are sharply on the rise. In addition, edible berries show encouraging results [143−145]. Since fruits and vegetables are rich sources of antioxidants, food packages meeting the FDA criteria may now carry the claim "Diets low in fat and high in fruits and vegetables may reduce the risk of some cancers" [146]. When taken in combination fruits and vegetables have been shown to exert synergism in their antioxidant potentials, which is beneficial for chronic diseases [147].

2.2.1.5 Unsaturated fatty acids

Fish has traditionally been available in regular meals in several Asian (such as Japan and coastal India), European (such as Finland and Norway), and Mediterranean countries and also in Greenland and North American native populations. The low incidence of coronary heart diseases in Greenland Eskimos opened our eyes to the benefits of fish diets and fish oils [148,149]. Fish diet diminishes the cardiovascular risk as shown in numerous studies [150,151]. n-3 Fatty acids are beneficial for cardiovascular health, reducing platelet aggregation, serum triglyceride levels, and the risk of sudden death from myocardial infarction [150−152]. Also fresh water fish available in inland lakes and rivers contain n-3 fatty acids which, according to the current understanding, explain the beneficial effect of fish-containing diets [153−155]. n-3 Fatty acids from fish oils have been shown to attenuate postprandial hyperlipidemia [156]. A combination of physical exercise and n-3 polyunsaturated fatty acids (PUFAs) intake is potently helpful [157]. n-3 PUFAs of marine origin limit induced obesity in mice by reducing cellularity of adipose tissue [158]. Dietary long chain n-3 PUFAs prevent sucrose-induced insulin resistance in rats [159]. The current Western diet contains high amounts of n-6 polyunsaturated fatty acids, which leads to a poor n-3 to n-6 PUFA ratio. Arachidonic acid effectively competes with eicosapentaenoic acid [160]. n-3 Series fatty acids are found also in some plants, e.g., flax, lipids [161]. Flax and lipids represent an important source of α-linolenic acid, which the body can use as starting material of eicosanoids, at least in adults. Fish oil capsules are marketed as dietary supplements to prevent cardiovascular diseases. n-3 Fatty acid consumption helped in lowering blood levels of triglycerides and cholesterol in patients with diabetes and metabolic syndrome [162,163]. Apart from cardiovascular diseases, unsaturated fatty acids are gaining importance in inflammation biology as well.

2.2.1.6 Micronutrients

Micronutrients are nutrients needed for life in small quantities [164–173]. They include chemical elements and chemical compounds such as minerals and vitamins. Microminerals or trace elements include at least iron, cobalt, chromium, copper, iodine, manganese, selenium, zinc, and molybdenum. They are dietary minerals needed by the human body in very small quantities (generally <100 mg/day) as opposed to macrominerals, which are required in larger quantities. Note that the use of the term "mineral" here is distinct from the usage in the geological sciences. Vitamins are organic chemicals that a given living organism requires in trace quantities for good health, but which the organism cannot synthesize, and therefore must obtain from its diet. Vitamins exhibit diverse biochemical functions ranging from hormone-like functions as regulators of mineral metabolism to regulators of cell and tissue growth and differentiation while some vitamins function as antioxidants.

More than 2 billion people (i.e., one in three persons worldwide) suffer from micronutrient deficiency, a form of malnutrition (http://www.micronutrient.org). The most common deficiencies can have devastating consequences:

1. Vitamin A deficiency: Nearly 3 million preschool children in developing countries are blind because of a vitamin A deficit.
2. Iron-deficiency anemia: Results in one out of four maternal deaths in the developing world. The cost of fortifying flour with iron is 20 cents per person per year (World Bank estimate).
3. Iodine deficiency: Is the world's leading cause of mental retardation—more than 2 billion children suffer from lowered IQ and retardation due to iodine deficiency. The costs of providing iodized salt are estimated at 10 cents per person per year.
4. Vitamin E, zinc, and manganese deficiencies are also serious causes for concern.

Micronutrient malnutrition is practically unknown in developed countries owing to inexpensive interventions such as food fortification, supplementation, and dietary diversification. South Asia and Sub-Saharan Africa are among the areas with the most affected people.

2.2.1.7 Plant sterols

High blood cholesterol levels indicate a risk of cardiovascular diseases. Blood cholesterol is either endogenous or exogenous by origin. Blood cholesterol levels are affected by diet and, in particular, by the type and amount of fat intake. Vegan foods are cholesterol free and may reduce the risk of cardiovascular diseases [174]. The beneficial aspects of consuming vegan foods have been cited elsewhere in this chapter. In recent years, vegetable oil spreads containing plant sterols/stanols (as their fatty acid esters) have been developed. Various clinical trials on spreads with added plant sterols/stanols have revealed that they have much higher cholesterol-lowering properties than conventional vegetable oil spreads [189]. Plant sterols decrease both dietary and biliary cholesterol absorption in the small intestine, with a consequential increase in excretion of cholesterol [175,189]. Stanol-enriched plant margarines were first marketed in Finland, but they are now staple foods in several countries. In the United States, plant sterols have received Generally Recognized as Safe (GRAS) status indicating that they are generally recognized as safe.

2.2.1.8 Probiotics

Fermented milk products have been used through the ages as staple foods by many nations. In India, cows have a special status in society for their contributions to human health. Previously, most of the milk consumed was fermented for practical reasons as storage possibilities were poor. Fermentation prevents the spoilage by harmful bacteria and prolongs the shelf-life of milk products. Lactobacilli play a substantial role in food biotechnology and influence our quality of life by their fermentative and probiotic properties. The normal human microflora is a complex and usually stable ecosystem. The intestinal microflora is important to the host with regard to several metabolic functions and in resistance to bacterial infections. Administration of antimicrobial agents may upset the normal microflora, leading to a reduction in colonization resistance and alterations in metabolic activities of the intestinal bacteria. Antimicrobial therapy may also lead to gastrointestinal disturbances and a reduction or elimination of lactobacilli in the intestinal microflora. Lactobacilli are a component of the normal gram-positive anaerobic microflora which through the production of lactic and acetic acids, hydrogen peroxide, and antimicrobial substances possibly contribute to the maintenance of colonization resistance. Numerous studies have indicated a protective effect of lactobacilli against potential pathogens in the gastrointestinal tract [190]. It is due to the protective effect of lactobacilli that certain antibiotics are given in combination with lactobacilli.

Also several other foods are used after fermentation. The traditional rye bread is a fermentation product. The effect of fermentation on oats and fish is less known, although again previously fermentation was very common, but very little is known about what really happens [176]. Fermented fish is still a national tradition in Swedish cuisine.

There is much research data on fermented foods, especially milk products. These data can be also developed further and can be included in daily diets in amounts that probably make the title functional food justified. Tourists traveling to countries with poor hygienic standards are recommended to take lactobacillus preparations with them.

2.3 Research needs: safety and efficacy

First and foremost, long- and short-term safety of dietary supplements and functional foods must be established in preclinical and clinical settings [177−181]. In the USA, the FDA regulates dietary supplements under a different set of regulations than those covering "conventional" foods and drug products (prescription and over the counter) [191]. As per the Dietary Supplement Health and Education Act of 1994 (DSHEA), the dietary supplement manufacturer is responsible for ensuring that a dietary supplement is safe before it is marketed. The FDA is empowered to take action against any unsafe dietary supplement product after it reaches the market [191]. Generally, manufacturers do not need to register their products with the FDA nor get FDA approval before producing or selling dietary supplements. Manufacturers must make sure that product label information is truthful and not misleading [191]. The FDA's post-marketing responsibilities include monitoring safety, e.g., voluntary dietary supplement adverse event reporting and product information, such as labeling, claims, package inserts, and accompanying literature. The Federal Trade Commission regulates dietary supplement advertising [192].

If award of functional food status to specific dietary components is to follow the same level of rigor that new medical drugs are subjected to, it is clear that many more appropriately controlled human studies are required. However, it is important to appreciate that double-blind intervention trials are challenging to execute given their obvious differences in visual presentation. Epidemiological studies have shown that the Adventists have significantly lower health risks than the other Americans as a number of items of their diet are different [182]. The duration of the studies must be long enough and also the populations large enough to power the study appropriately. The food must also contain the study component in such quantities that are relevant to real life, but also to given measurable effects in the markers selected. These studies should be conducted in different cultural settings as the chemical environments (matrixes) are important for dietary chemicals to reveal their properties. The studies should account for the physical activity state of the subjects because dietary outcomes are clearly linked to such states. The background pathophysiology of the metabolic syndrome is probably due to glucotoxicity and lipotoxicity, i.e., the fluctuation of blood levels that strains the endocrine system, especially during the secretion of insulin from the pancreas. Glucose and fatty acid metabolism are interlinked at several regulatory levels [9]. Health claims of foods should be supported by appropriate and acceptable research results. There are opinions that this evidence should be as rigorous as in the case of drugs. This level may be difficult to achieve as foods are mixtures of known and unknown chemicals and all foods are eaten as a part of the wholeness, which is difficult to describe completely. The combination of traditional experiences and research results will lead to the development of healthy foods and new preventive and perhaps therapeutic approaches.

2.3.1 DSHEA

The FDA receives many questions about the labeling of dietary supplements which questions are a consequence of the activity in this area over the past several years [193]. Some of the important events relating to the labeling of dietary supplements include:

1. The Nutrition Labeling and Education Act of 1990 amended the Federal Food, Drug, and Cosmetic Act (the act) in a number of important ways. Notably, by requiring that most foods, including dietary supplements, bear nutrition labeling.
2. The DSHEA amended the act, in part, by defining "dietary supplements," adding specific labeling requirements for dietary supplements, and providing for optional labeling statements.
3. On September 23, 1997 (62 FR 49826), the FDA implemented the DSHEA by publishing several key regulations on the statement of identity, nutrition labeling, ingredient labeling, and nutrient content and health claims for dietary supplements. On June 5, 1998 (63 FR 30615), the FDA amended the regulations pertaining to the nutrition labeling of extracts used in dietary supplements.
4. On January 15, 1997 (62 FR 2218), the FDA published regulations that require a label warning statement on dietary supplements with added iron. These regulations also required the unit-dose packaging of supplements containing 30 mg or more, but this requirement has been eliminated as a result of a court challenge in January 2003.
5. On July 11, 2003 (68 FR 41434), the FDA published a final regulation that amended the labeling requirements for dietary supplements, as well as for conventional foods, which would make the declaration of trans fat mandatory in nutrition labeling. This regulation requires that, when present at 0.5 g or more, trans fat is listed in the Supplement Facts panel of dietary supplements on a separate line under the listing of saturated fat by January 1, 2006.

The FDA's guidance documents do not establish legally enforceable responsibilities but describe the agency's current thinking on a topic and should be viewed only as recommendations, unless specific regulatory or statutory requirements are cited [194]. The use of the word "should" in agency guidance documents means that something is suggested or recommended, but not required [194].

Americans spent $6.5 billion on dietary supplements in 1996, nearly doubling the 1990 total of $3.2 billion. This booming industry, which consists largely of natural health products, is attracting significant attention of the FDA and the medical community. The key questions that arise in this context are whether this is a legitimate effort by the FDA and the medical community to protect the health and safety of the public. Skeptics argue that this is a concerted effort by the FDA and medical community to protect the financial interests of the pharmaceutical drug industry and the medical community. Both sides have a point making this an interesting debate. Billions of dollars spent on supplements would otherwise be spent on drugs and doctors. The FDA is now considering modifications to dietary supplement labeling for the purpose of "clarifying for manufacturers what types of claims they may and may not use on labels of dietary supplements under the DSHEA act (Public Law 103–417, 103rd Congress; http://www.fda.gov/opacom/laws/dshea.html) of 1994" according to the FDA (April 27, 1998, fact sheet) [195].

The DHSEA established the framework for FDA regulation of supplements and it not only provides dietary supplement manufacturers more freedom to market products but also provides information to consumers than they had previously [195]. This freedom, in large part, is the reason that supplement sales have increased dramatically in the past few years [195].

2.3.2 Codex: harmonizing food and supplement rules between all nations of the world

The biggest movement to develop harmonized or uniform health food standards is the Rome-based Codex Alimentarius ("food code") Commission (Codex), which was created in 1962 by two United Nations (UN) organizations, the FAO and the WHO. The goal was to design the Codex as an "international mechanism for promotion of health and economic interests of consumers while encouraging fair international trade in food," according to the U.S. Codex Office [196]. Codex comprises of more than 150 member countries and international organizations that meet and exchange information and ideas related to food safety and trade issues [196]. Its members are also members of the FAO and WHO and represent 98% of the world's population, according to the FAO [196]. Simply stated, the Codex Alimentarius is a collection of standards, codes of practice, guidelines, and other recommendations [197]. Some of these texts are very general and some are very specific. Some deal with detailed requirements related to a food or group of foods while others deal with the operation and management of production processes or the operation of government regulatory systems for food safety and consumer protection [197].

The Codex Alimentarius, or the food code, has become the global reference point for consumers, food producers and processors, national food control agencies, and the international food trade [198]. The code has had a huge impact on the thinking of not only the food producers and processors but also on the awareness of the end users—the consumers [198]. Its influence extends to every continent and its contribution to the protection of public health and fair practices in the food trade is immense [198]. The Codex Alimentarius system presents a unique opportunity for all countries to join the international community in formulating and harmonizing food standards and ensuring how they are implemented globally [198]. It also allows them to have a role in the development of codes governing hygienic processing practices and recommendations relating to compliance with those standards [198]. The significance of the food code for consumer health protection was underscored in 1985 by the UN Resolution 39/248, whereby guidelines were adopted for use in the elaboration and reinforcement of consumer protection policies [198]. The guidelines advise that:

> When formulating national policies and plans with regard to food, Governments should take into account the need of all consumers for food security and should support and, as far as possible, adopt standards from the...Codex Alimentarius or, in their absence, other generally accepted international food standards.

Currently, Codex Alimentarius contains more than 200 standards, encompassing issues like labeling, additives, methods of analysis and sampling, food import and export inspection and certification, pesticides in foods, and contaminants. The code also deals with nutrition and foods for special dietary uses, which includes dietary supplements [198].

The Commission for Nutrition and Foods for Special Dietary Uses (CNFSDU) of Codex, hosted by Germany, is charged with the following responsibilities [199]:

1. To study specific nutritional problems assigned to it by the Commission and advise the Commission on general nutrition issues
2. To draft general provisions, as appropriate, concerning the nutritional aspects of all foods
3. To develop standards, guidelines, or related texts for foods for special dietary uses, in cooperation with other committees where necessary
4. To consider, amend if necessary, and endorse provisions on nutritional aspects proposed for inclusion Codex standards, guidelines, and related texts

It does not address the broad category of dietary supplements, which includes herbals, amino acids, concentrates, metabolites, and many other non-essential nutrients.

In the early 1990s, the Codex Committee on Nutrition and Foods for Special Dietary Uses (CCNFSDU) began discussions on guidelines for vitamin and mineral supplements. At the 26th CCNFSDU session (Bonn, Germany, November 1−5, 2004), the Committee completed work on the Draft Guidelines for Vitamin and Mineral Food Supplements and submitted them for adoption by the Codex Alimentarius Commission (CAC). The guidelines were adopted at the 28th CAC Session that was held in Rome on July 4−9, 2005. The guidelines apply only to supplements that contain vitamins and/or minerals, where these products are regulated as foods. The guidelines address the composition of vitamin and mineral supplements, including the safety, purity, and bioavailability of the sources of vitamins and minerals. The guidelines do not specify upper limits for vitamins and minerals in supplements but provide criteria for establishing maximum amounts of vitamins and minerals per daily portion of supplement consumed, as recommended by the manufacturer [200]. The criteria specify that maximum amounts should be established by scientific risk assessment based on generally accepted scientific data and taking into consideration, as appropriate, the varying degrees of sensitivity of different consumer groups [200]. The guidelines also address the packaging and labeling of vitamin and mineral supplements [200]. The US government has published its position (http://www.cfsan.fda.gov/~dms/dscodex.html). The USA supports consumer choice and access to dietary supplements that are safe and labeled in a truthful and non-misleading manner [200]. The DSHEA ensures that a broad array of dietary supplements are available to US consumers [200]. The Codex Guidelines for Vitamin and Mineral Food Supplements do not, in any way, affect the availability of supplement products to US consumers but on the contrary, the absence of science-based Codex guidelines could adversely affect the ability of US manufacturers to compete in the international marketplace [200]. In its present form, it is clear that Codex Alimentarius falls much short of the global governance needed to regulate the marketing and consumption of dietary supplements. Individual countries such as Japan seem to be much ahead of the rest of the world.

In 1991, the Japanese Ministry of Health and Welfare (MHW) established the FOSHU labeling regulation [201]. It was implemented by adding a new category of FOSHU to "Foods for Special Dietary Uses" in the Nutrition Improvement Law. The word functional, however, was not used since it appears in the definition of "pharmaceuticals" where things intended to affect the structure or functions of the body are described as pharmaceuticals [201]. FOSHU approval by the MHW is a permission system for labeling to claim that a food helps maintain or is suitable for a health condition, when the claim is substantiated based on medical and nutritional science [201]. Therefore, scientific documents are required to obtain FOSHU approval, namely clinical and nutritional documentation demonstrating the health efficacy of the food or the ingredient, clinical and nutritional evidence on dietary intake, validation of safety and stability, and analytical methods and the results for identification [201]. Initially not many industries were interested in FOSHU because of these burdens, but, in recent years, MHW took deregulation measures to encourage industries and so the FOSHU market has been expanding [201]. From December 1999, the total number of approved FOSHU was 171 and according to a survey, total FOSHU sales were around 130 billion yen in 1997, but it increased to 220 billion yen in 1999 [201].

Originally, foods that have the function to modulate the body function that contributes to the prevention of a disease were termed "functional foods" [201]. However, it is now widely understood to implicate foods that claim such biological effects beyond ordinary nutritional effects based on scientific validation [201]. In Japan, there is a variety of so-called health foods but whether they are functional foods or not should be judged one by one from the above viewpoint [201]. Among them, vitamins and minerals have been paid special attention as so-called dietary supplements [201]. Although many scientific papers report on their functionality, there may be some debate over whether they can be called functional foods when the form of the product is an important element of the definition of functional foods because dietary supplements are usually recognized in the form of tablets or capsules [201]. In 1999, the market size of health foods was approximately 750 billion yen [201]. The momentum in Japan seems to be in the right direction providing the rest of the world with an opportune platform for further development.

2.3.3 International alliance of dietary food supplement associations

Since its creation in 1998, the International Alliance of Dietary Food Supplement Associations (IADSA) has developed into an alliance of more than 50 dietary supplement associations spread over six continents [202]. There are at present more than 9500 companies who are part of the IADSA member associations. The efficacy of IADSA as an organization is based on its ability to communicate information and ideas around the globe to regulators, consumer organizations, scientific bodies, and national associations and companies. The 2007 report on nutrition, healthy aging, and public policy has been recently released [183]. An IADSA workshop on International Perspectives on Dietary Supplements Regulation was held in Yokohama, Japan, on April 17, 2007. Nearly 300 delegates, including regulators and government officials from the Association of South East Asian Nations (ASEAN), China, the EU, Japan, Mexico, the United States, scientific bodies, trade associations, and industry attended the workshop. In the final session Dr. Yozo Hayashi, Director General of The Japan Health Food & Nutrition Food Association and Dr. Tohru Inoue from the Biological Safety Research Center of the Japanese National Institute of Health Sciences shared with the delegates their conclusions on the development of global regulation for food products and their vision of the future application and benefits of food supplements and functional foods. In his concluding remarks, IADSA Chairman Randy Dennin said:

> *The Workshop demonstrated that while there continue to be many different approaches to regulating dietary supplements, the principles that form the basis of these are increasingly consistent throughout the world. IADSA will build on this global workshop with increased dialogue to encourage further detailed discussions at national levels.*

The major activities of IADSA include [202]:

1. Providing a fast flow of regulatory and policy information on dietary supplements, ensuring that there is an awareness and understanding of new developments
2. Coordinating strategy and action on global regulatory issues, particularly in relation to Codex Alimentarius initiatives on vitamin and mineral supplements, additives, and health claims
3. Widening and deepening the network of associations around the world by helping the establishment of new dietary supplement associations and supporting existing national associations
4. Organizing global and regional events to promote dialog on the scientific and regulatory issues underpinning the dietary supplement market.

Decisions and policies from global regulatory bodies have become increasingly influential for dietary supplement companies. Such bodies include Codex Alimentarius, the WHO, and the FAO. IADSA works closely with these international bodies to ensure that the views of the dietary supplement industry are taken into account in the development of policy [203].

Acknowledgment

Supported in part by National Institutes of Health Grant NS42617 to C.K.S.

References

[1] Brillat Savarin A. Physiologie du gout (J-F Revel) Paris; 1965.

[2] Tovey FI, Hobsley M. Milling of wheat, maize and rice: effects on fibre and lipid content and health. World J Gastroenterol 2004;10:1695−6.

[3] Burdock GA, Carabin IG, Griffiths JC. The importance of GRAS to the functional food and nutraceutical industries. Toxicology 2006;221:17−27.

[4] Choi YM, Bae SH, Kang DH, Suh HJ. Hypolipidemic effect of lactobacillus ferment as a functional food supplement. Phytother Res 2006;20:1056−60.

[5] Olmedilla-Alonso B, Granado-Lorencio F, Herrero-Barbudo C, Blanco-Navarro I. Nutritional approach for designing meat-based functional food products with nuts. Crit Rev Food Sci Nutr 2006;46:537−42.

[6] Sieber CC. Functional food in elderly persons. Ther Umsch 2007;64:141−6.

[7] <http://www4.agr.gc.ca/AAFC-AAC/display-afficher.do? id = 1171305207040&lang = eng>.

[8] Fitch K, Pyenson B, Iwasaki K. Metabolic syndrome and employer sponsored medical benefits: an actuarial analysis. Value Health 2007;10(Suppl. 1):S21−8.

[9] Poitout V, Hagman D, Stein R, Artner I, Robertson RP, Harmon JS. Regulation of the insulin gene by glucose and fatty acids. J Nutr 2006;136:873−6.

[10] Jatoi SA, Kikuchi A, Gilani SA, Watanabe KN. Phytochemical, pharmacological and ethnobotanical studies in mango ginger (Curcuma amada Roxb.; Zingiberaceae). Phytother Res 2007.

[11] Hill JO, Peters JC, Wyatt HR. The role of public policy in treating the epidemic of global obesity. Clin Pharmacol Ther 2007.

[12] Yoon KH, Lee JH, Kim JW, et al. Epidemic obesity and type 2 diabetes in Asia. Lancet 2006;368:1681−8.

[13] Kalra EK. Nutraceutical-definition and introduction. AAPS pharmSci 2003;5(3):E25.

[14] Dymsza HA. Nutritional application and implication of 1,3-butanediol. Fed Proc 1975;34:2167−70.

[15] Ashwell M. Concepts of Functional Foods. ILSI Europe Concise Monograph Series 2002.

[16] Ohama H, Ikeda H, Moriyama H. Health foods and foods with health claims in Japan. Toxicology 2006;221:95−111.

[17] Aggett PJ, Antoine JM, Asp NG, et al. PASSCLAIM: consensus on criteria. Eur J Nutr 2005;44(Suppl. 1):I5−30.

[18] Asp NG, Contor L. Process for Assessment of Scientific Support for Claims on Food (PASSCLAIM): overall introduction. Eur J Nutr 2003;42(Suppl. 1):I3−i15.

[19] Contor L, Asp NG. Process for the assessment of scientific support for claims on foods (PASSCLAIM) phase two: moving forward. Eur J Nutr 2004;43(Suppl. 2):II3−6.

[20] Cummings JH, Antoine JM, Azpiroz F, et al. PASSCLAIM − gut health and immunity. Eur J Nutr 2004;43(Suppl. 2):II118−73.

[21] Cummings JH, Pannemans D, Persin C. PASSCLAIM − Report of First Plenary Meeting including a set of interim criteria to scientifically substantiate claims on foods. Eur J Nutr 2003;42(Suppl. 1):I112−9.

[22] Howlett J, Shortt C. PASSCLAIM − report of the second plenary meeting: review of a wider set of interim criteria for the scientific substantiation of health claims. Eur J Nutr 2004;43(Suppl. 2):II174−83.

[23] Mensink RP, Aro A, Den Hond E, et al. PASSCLAIM − Diet-related cardiovascular disease. Eur J Nutr 2003;42(Suppl. 1):I6−27.

[24] Prentice A, Bonjour JP, Branca F, et al. PASSCLAIM − Bone health and osteoporosis. Eur J Nutr 2003;42(Suppl. 1):I28−49.

[25] Rafter J, Govers M, Martel P, et al. PASSCLAIM − diet-related cancer. Eur J Nutr 2004;43(Suppl. 2):II47−84.

[26] Riccardi G, Aggett P, Brighenti F, et al. PASSCLAIM − body weight regulation, insulin sensitivity and diabetes risk. Eur J Nutr 2004;43(Suppl. 2):II7−46.

[27] Richardson DP, Affertsholt T, Asp NG, et al. PASSCLAIM − Synthesis and review of existing processes. Eur J Nutr 2003;42(Suppl. 1):I96−111.

[28] Saris WH, Antoine JM, Brouns F, et al. PASSCLAIM − Physical performance and fitness. Eur J Nutr 2003;42(Suppl. 1):I50−95.

[29] Westenhoefer J, Bellisle F, Blundell JE, et al. PASSCLAIM − mental state and performance. Eur J Nutr 2004;43(Suppl. 2):II85−117.

[30] Erdman Jr JW, Balentine D, Arab L. et al. Flavonoids and Heart Health: proceedings of the ILSI North America Flavonoids Workshop, May 31−June 1, 2005, Washington, DC. J Nutr 2007;137:718S−737S.

[31] Saito M. Role of FOSHU (food for specified health uses) for healthier life. Yakugaku Zasshi 2007;127:407−416c.

[32] <http://medical-dictionary.thefreedictionary.com/health + behavior>.

[33] <http://www.cdc.gov/features/livelonger/>.

[34] Ford ES, Zhao G, Tsai J, Li C. Low-risk lifestyle behaviors and all-cause mortality: findings from the National Health and Nutrition Examination Survey III Mortality Study. American Journal of Public Health, published online ahead of print August 18, 2011.

[35] Kinzie MB. Instructional design strategies for health behavior change. Patient Educ Couns 2005;56:3−15.

[36] Colella C, Laver J. Setting the stage for changing health behavior. Nurse Pract 2005;30:68−70.

[37] Nieuwenhuijsen ER, Zemper E, Miner KR, Epstein M. Health behavior change models and theories: contributions to rehabilitation. Disabil Rehabil 2006;28:245−56.

[38] Pietinen P, Nissinen A, Vartiainen E, et al. Dietary changes in the North Karelia Project (1972−1982). Prev Med 1988;17:183−93.

[39] Puska P, Koskela K, Pakarinen H, Puumalainen P, Soininen V, Tuomilehto J. The North Karelia Project: a programme for community control of cardiovascular diseases. Scand J Soc Med 1976;4:57−60.

[40] Pietinen P, Lahti-Koski M, Vartiainen E, Puska P. Nutrition and cardiovascular disease in Finland since the early 1970s: a success story. J Nutr Health Aging 2001;5:150−4.

[41] <http://www.nestle.com/media/newsandfeatures/insight-life-sciences>.

[42] He J, Giusti MM. Anthocyanins: natural colorants with health-promoting properties. Annu Rev Food Sci Technol. 2010;1:163−87.

[43] Slavin J. Whole-grains and human health. Nautr. Res. Rev. 2004;17:99−110.

[44] Trowell H. Ischemic heart disease and dietary fiber. Am. Clin. Nutriition 1972;25:926.

[45] Trowell H. The development of the concept of dietary fiber in human nutrition. A. J. Clin. Nutr. 1978;31:S3.

[46] Katina K, Laitila A, Juvonen R, et al. Bran fermentation as a means to enhance technological properties and bioactivity of rye. Food Microbiol 2007;24:175−86.

[47] Dikeman CL, Fahey GC. Viscosity as related to dietary fiber: a review. Crit Rev Food Sci Nutr 2006;46:649−63.

[48] Rock CL. Primary dietary prevention: is the fiber story over? Recent Results Cancer Res 2007;174:171−7.

[49] Rose DJ, DeMeo MT, Keshavarzian A, Hamaker BR. Influence of dietary fiber on inflammatory bowel disease and colon cancer: importance of fermentation pattern. Nutr Rev 2007;65:51−62.

[50] Keenan MJ, Zhou J, McCutcheon KL, et al. Effects of resistant starch, a non-digestible fermentable fiber, on reducing body fat. Obesity (Silver Spring) 2006;14:1523−34.

[51] Stewart ML, Schroeder NM. Dietary treatments for childhood constipation: efficacy of dietary fiber and whole grains. Nutr Rev 2013;71(2):98−109.

[52] Scazzina F, Siebenhandl-Ehn S, Pellegrini N. The effect of dietary fibre on reducing the glycaemic index of bread. Br J Nutr 2013;109(7):1163−74.

[53] Wong JM, de Souza R, Kendall CW, Emam A, Jenkins DJ. Colonic health: fermentation and short chain fatty acids. J Clin Gastroenterol 2006;40:235−43.

[54] Adlercreutz H, Heinonen SM, Penalvo-Garcia J. Phytoestrogens, cancer and coronary heart disease. Biofactors 2004;22:229−36.

[55] Hallmans G, Zhang JX, Lundin E, et al. Rye, lignans and human health. Proc Nutr Soc 2003;62:193−9.

[56] Linko AM, Juntunen KS, Mykkanen HM, Adlercreutz H. Whole-grain rye bread consumption by women correlates with plasma alkylresorcinols and increases their concentration compared with low-fiber wheat bread. J Nutr 2005;135:580−3.

[57] Zimmer J, et al. A vegan or vegetarian diet substantially alters the human colonic faecal microbiota. Eur J Clin Nutr 2012;66(1):53−60.

[58] Hooda S, et al. 454 pyrosequencing reveals a shift in fecal microbiota of healthy adult men consuming polydextrose or soluble corn fiber. J Nutr 2012;142 (7):1259−65.

[59] Hagander B, Bjorck I, Asp NG, et al. Rye products in the diabetic diet. Postprandial glucose and hormonal responses in non-insulin-dependent diabetic patients as compared to starch availability in vitro and experiments in rats. Diabetes Res Clin Pract 1987;3:85−96.

[60] Leinonen KS, Poutanen KS, Mykkanen HM. Rye bread decreases serum total and LDL cholesterol in men with moderately elevated serum cholesterol. J Nutr 2000;130:164−70.

[61] Bertram HC, Duarte IF, Gil AM, Knudsen KE, Laerke HN. Metabolic profiling of liver from hypercholesterolemic pigs fed rye or wheat fiber and from normal pigs. High-resolution magic angle spinning 1H NMR spectroscopic study. Anal Chem 2007;79:168−75.

[62] Landberg R, Linko AM, Kamal-Eldin A, Vessby B, Adlercreutz H, Aman P. Human plasma kinetics and relative bioavailability of alkylresorcinols after intake of rye bran. J Nutr 2006;136:2760−5.

[63] Andersson M, Ellegard L, Andersson H. Oat bran stimulates bile acid synthesis within 8 h as measured by 7alpha-hydroxy-4-cholesten-3-one. Am J Clin Nutr 2002;76:1111−6.

[64] Brighenti F, Casiraghi MC, Ciappellano S, Crovetti R, Testolin G. Digestibility of carbohydrates from rice-, oat- and wheat-based ready-to-eat breakfast cereals in children. Eur J Clin Nutr 1994;48:617−24.

[65] Pick ME, Hawrysh ZJ, Gee MI, Toth E, Garg ML, Hardin RT. Oat bran concentrate bread products improve long-term control of diabetes: a pilot study. J Am Diet Assoc 1996;96:1254−61.

[66] Murphy EA, Davis JM, Brown AS, Carmichael MD, Ghaffar A, Mayer EP. Oat beta-glucan effects on neutrophil respiratory burst activity following exercise. Med Sci Sports Exerc 2007;39:639−44.

[67] Queenan KM, Stewart ML, Smith KN, Thomas W, Fulcher RG, Slavin JL. Concentrated oat beta-glucan, a fermentable fiber, lowers serum cholesterol in hypercholesterolemic adults in a randomized controlled trial. Nutr J 2007;6:6.

[68] Reyna-Villasmil N, Bermudez-Pirela V, Mengual-Moreno E, et al. Oat-derived beta-glucan significantly improves HDLC and diminishes LDLC and non-HDL cholesterol in overweight individuals with mild hypercholesterolemia. Am J Ther 2007;14:203−12.

[69] Biorklund M, van Rees A, Mensink RP, Onning G. Changes in serum lipids and postprandial glucose and insulin concentrations after consumption of beverages with beta-glucans from oats or barley: a randomised dose-controlled trial. Eur J Clin Nutr 2005;59:1272−81.

[70] Braaten JT, Wood PJ, Scott FW, Riedel KD, Poste LM, Collins MW. Oat gum lowers glucose and insulin after an oral glucose load. Am J Clin Nutr 1991;53:1425−30.

[71] Tapola N, Karvonen H, Niskanen L, Mikola M, Sarkkinen E. Glycemic responses of oat bran products in type 2 diabetic patients. Nutr Metab Cardiovasc Dis 2005;15:255−61.

[72] Braaten JT, Wood PJ, Scott FW, et al. Oat beta-glucan reduces blood cholesterol concentration in hypercholesterolemic subjects. Eur J Clin Nutr 1994;48:465−74.

[73] Theuwissen E, Mensink RP. Simultaneous intake of beta-glucan and plant stanol esters affects lipid metabolism in slightly hypercholesterolemic subjects. J Nutr 2007;137:583−8.

[74] Boehm G, Stahl B, Jelinek J, Knol J, Miniello V, Moro GE. Prebiotic carbohydrates in human milk and formulas. Acta Paediatr Suppl 2005;94:18−21.

[75] Cashman KD. A prebiotic substance persistently enhances intestinal calcium absorption and increases bone mineralization in young adolescents. Nutr Rev 2006;64:189−96.

[76] Corcoran BM, Ross RP, Fitzgerald GF, Stanton C. Comparative survival of probiotic lactobacilli spray-dried in the presence of prebiotic substances. J Appl Microbiol 2004;96:1024−39.

[77] Schoonen M, Smirnov A, Cohn C. A perspective on the role of minerals in prebiotic synthesis. Ambio 2004;33:539−51.

[78] Burt BA. The use of sorbitol- and xylitol-sweetened chewing gum in caries control. J Am Dent Assoc 2006;137:190−6.

[79] Granstrom TB, Izumori K, Leisola M. A rare sugar xylitol. Part I: the biochemistry and biosynthesis of xylitol. Appl Microbiol Biotechnol 2007;74:277−81.

[80] Granstrom TB, Izumori K, Leisola M. A rare sugar xylitol. Part II: biotechnological production and future applications of xylitol. Appl Microbiol Biotechnol 2007;74:273−6.

[81] Kitchens DH. Xylitol in the prevention of oral diseases. Spec Care Dentist 2005;25:140−4.

[82] Ly KA, Milgrom P, Rothen M. Xylitol, sweeteners. and dental caries. Pediatr Dent 2006;28:154−63 discussion, 192−158.

[83] Azarpazhooh A, et al. Xylitol for preventing acute otitis media in children up to 12 years of age. Cochrane Database Syst Rev 2011;(11)):CD007095.

[84] Duthie GG, Gardner PT, Kyle JA. Plant polyphenols: are they the new magic bullet? Proc Nutr Soc 2003;62:599−603.

[85] Manach C, Scalbert A, Morand C, Remesy C, Jimenez L. Polyphenols: food sources and bioavailability. Am J Clin Nutr 2004;79:727−47.

[86] Sun-Waterhouse D, et al. Kiwifruit-based polyphenols and related antioxidants for functional foods: kiwifruit extract-enhanced gluten-free bread. Int J Food Sci Nutr 2009;60(Suppl 7):251−64.

[87] Scalbert A, Johnson IT, Saltmarsh M. Polyphenols: antioxidants and beyond. Am J Clin Nutr 2005;81:215S−7S.

[88] Halliwell B. Dietary polyphenols: good, bad, or indifferent for your health? Cardiovasc Res 2007;73:341−7.

[89] Dormandy TL. Free-radical reaction in biological systems. Ann R Coll Surg Engl 1980;62:188−94.

[90] Emanuel NM. Kinetics and free-radical mechanisms of ageing and carcinogenesis. IARC Sci Publ 1985;:127−50.

[91] Harman D. Prolongation of life: role of free radical reactions in aging. J Am Geriatr Soc 1969;17:721−35.

[92] Slater TF. Free-radical mechanisms in tissue injury. Biochem J 1984;222:1−15.

[93] Cordier JF. Oxidant-antioxidant balance. Bull Eur Physiopathol Respir 1987;23:273−4.

[94] Frankel EN. The antioxidant and nutritional effects of tocopherols, ascorbic acid and beta-carotene in relation to processing of edible oils. Bibl Nutr Dieta 1989:297−312.

[95] Godin DV, Wohaieb SA. Nutritional deficiency, starvation, and tissue antioxidant status. Free Radic Biol Med 1988;5:165−76.

[96] Halliwell B. How to characterize a biological antioxidant. Free Radic Res Commun 1990;9:1−32.

[97] Clemens MR. Antioxidant therapy in hematological disorders. Adv Exp Med Biol 1990;264:423−33.

[98] Grimes JD, Hassan MN, Thakar J. Antioxidant therapy in Parkinson's disease. Can J Neurol Sci 1987;14:483−7.

[99] Hearse DJ. Prospects for antioxidant therapy in cardiovascular medicine. Am J Med 1991;91:118S−21S.

[100] Jackson MJ, Edwards RH. Free radicals and trials of antioxidant therapy in muscle diseases. Adv Exp Med Biol 1990;264:485−91.

[101] Muller DP. Antioxidant therapy in neurological disorders. Adv Exp Med Biol 1990;264:475−84.

[102] Rice-Evans CA, Diplock AT. Current status of antioxidant therapy. Free Radic Biol Med 1993;15:77−96.

[103] Schiller HJ, Reilly PM, Bulkley GB. Tissue perfusion in critical illnesses. Antioxidant therapy. Crit Care Med 1993;21:S92−102.

[104] Uden S, Bilton D, Guyan PM, Kay PM, Braganza JM. Rationale for antioxidant therapy in pancreatitis and cystic fibrosis. Adv Exp Med Biol 1990;264:555−72.

[105] Yoshikawa T, Naito Y, Kondo M. Antioxidant therapy in digestive diseases. J Nutr Sci Vitaminol (Tokyo) 1993;39:S35−41 Suppl.

[106] Youn YK, LaLonde C, Demling R. Use of antioxidant therapy in shock and trauma. Circ Shock 1991;35:245−9.

[107] Hennekens CH. Antioxidant vitamins and cancer. Am J Med 1994;97:2S−4S discussion 22S−28S.

[108] Clifton PM. Antioxidant vitamins and coronary heart disease risk. Curr Opin Lipidol 1995;6:20−4.

[109] Jha P, Flather M, Lonn E, Farkouh M, Yusuf S. The antioxidant vitamins and cardiovascular disease. A critical review of epidemiologic and clinical trial data. Ann Intern Med 1995;123:860−72.

[110] <http://www.foodinsight.org/Resources/Detail.aspx? topic = Functional_Foods_Fact_Sheet_Antioxidants>.

[111] Meyers DG, Maloley PA, Weeks D. Safety of antioxidant vitamins. Arch Intern Med 1996;156:925−35.

[112] Lonn EM, Yusuf S. Is there a role for antioxidant vitamins in the prevention of cardiovascular diseases? an update on epidemiological and clinical trials data. Can J Cardiol 1997;13:957−65.

[113] Bland JS. The pro-oxidant and antioxidant effects of vitamin C. Altern Med Rev 1998;3:170.

[114] Halliwell B. The antioxidant paradox. Lancet 2000;355:1179−80.

[115] Bowry VW, Ingold KU, Stocker R. Vitamin E in human low-density lipoprotein. When and how this antioxidant becomes a pro-oxidant. Biochem J 1992;288(Pt 2):341−4.

[116] Halliwell B. Vitamin C: antioxidant or pro-oxidant in vivo? Free Radic Res 1996;25:439−54.

[117] Laughton MJ, Halliwell B, Evans PJ, Hoult JR. Antioxidant and pro-oxidant actions of the plant phenolics quercetin, gossypol and myricetin. Effects on lipid peroxidation, hydroxyl radical generation and bleomycin-dependent damage to DNA. Biochem Pharmacol 1989;38:2859−65.

[118] Podmore ID, Griffiths HR, Herbert KE, Mistry N, Mistry P, Lunec J. Vitamin C exhibits pro-oxidant properties. Nature 1998;392:559.

[119] Spencer JP, Jenner A, Butler J, et al. Evaluation of the pro-oxidant and antioxidant actions of L-DOPA and dopamine in vitro: implications for Parkinson's disease. Free Radic Res 1996;24:95−105.

[120] Truscott TG. Beta-carotene and disease: a suggested pro-oxidant and anti-oxidant mechanism and speculations concerning its role in cigarette smoking. J Photochem Photobiol B 1996;35:233−5.

[121] Wolff SP, Spector A. Pro-oxidant activation of ocular reductants. 2. Lens epithelial cell cytotoxicity of a dietary quinone is associated with a stable free radical formed with glutathione in vitro. Exp Eye Res 1987;45:791−803.

[122] Wolff SP, Wang GM, Spector A. Pro-oxidant activation of ocular reductants. 1. Copper and riboflavin stimulate ascorbate oxidation causing lens epithelial cytotoxicity in vitro. Exp Eye Res 1987;45:777−89.

[123] Yeh S, Hu M. Antioxidant and pro-oxidant effects of lycopene in comparison with beta-carotene on oxidant-induced damage in Hs68 cells. J Nutr Biochem 2000;11:548−54.

[124] Carlson JC, Sawada M. Generation of free radicals and messenger function. Can J Appl Physiol 1995;20:280−8.

[125] Llamas R. [Hydrogen peroxide, the 2d messenger for insulin action on the metabolism of glucose and fat bodies in adipose tissue. Its effects on adrenaline lipolysis]. Gac Med Mex 1984;120:109−12.

[126] Sen CK. Redox signaling and the emerging therapeutic potential of thiol antioxidants. Biochem Pharmacol 1998;55:1747−58.

[127] Sen CK, Packer L. Antioxidant and redox regulation of gene transcription. Faseb J 1996;10:709−20.

[128] Sun Y, Oberley LW. Redox regulation of transcriptional activators. Free Radic Biol Med 1996;21:335−48.

[129] Biaglow JE, Miller RA. The thioredoxin reductase/thioredoxin system: novel redox targets for cancer therapy. Cancer Biol Ther 2005;4:6−13.

[130] Friedlich AL, Beal MF. Prospects for redox-based therapy in neurodegenerative diseases. Neurotox Res 2000;2:229−37.

[131] Kinnula VL, Fattman CL, Tan RJ, Oury TD. Oxidative stress in pulmonary fibrosis: a possible role for redox modulatory therapy. Am J Respir Crit Care Med 2005;172:417−22.

[132] Pennington JD, Wang TJ, Nguyen P, et al. Redox-sensitive signaling factors as a novel molecular targets for cancer therapy. Drug Resist Updat 2005;8:322−30.

[133] Roy S, Khanna S, Nallu K, Hunt TK, Sen CK. Dermal wound healing is subject to redox control. Mol Ther 2006;13:211−20.

[134] Athar M, Back JH, Tang X, et al. Resveratrol: A review of preclinical studies for human cancer prevention. Toxicol Appl Pharmacol 2007.

[135] Holme AL, Pervaiz S. Resveratrol in cell fate decisions. J Bioenerg Biomembr 2007.

[136] Campbell FC, Collett GP. Chemopreventive properties of curcumin. Future Oncol 2005;1:405−14.

[137] Maheshwari RK, Singh AK, Gaddipati J, Srimal RC. Multiple biological activities of curcumin: a short review. Life Sci 2006;78:2081−7.

[138] Singh S, Khar A. Biological effects of curcumin and its role in cancer chemoprevention and therapy. Anticancer Agents Med Chem 2006;6:259−70.

[139] Thangapazham RL, Sharma A, Maheshwari RK. Multiple molecular targets in cancer chemoprevention by curcumin. Aaps J 2006;8:E443−e449.

[140] Packer L, Weber SU, Rimbach G. Molecular aspects of alpha-tocotrienol antioxidant action and cell signaling. J Nutr 2001;131:369S−73S.

[141] Sen CK, Khanna S, Roy S. Tocotrienol: the natural vitamin E to defend the nervous system? Ann NY Acad Sci 2004;1031:127−42.

[142] Theriault A, Chao JT, Wang Q, Gapor A, Adeli K. Tocotrienol: a review of its therapeutic potential. Clin Biochem 1999;32:309−19.

[143] Erlund I, Freese R, Marniemi J, Hakala P, Alfthan G. Bioavailability of quercetin from berries and the diet. Nutr Cancer 2006;54:13−7.

[144] Freese R. Markers of oxidative DNA damage in human interventions with fruit and berries. Nutr Cancer 2006;54:143−7.

[145] Juranic Z, Zizak Z. Biological activities of berries: from antioxidant capacity to anti-cancer effects. Biofactors 2005;23:207−11.

[146] Food and Drug Administration—Center for Food Safety and Applied Nutrition Code of Federal Regulations: Title 21, V 2. Available at: <http://www.cfsan.fda.gov/~lrd/cf101-78.html>.

[147] Liu RH. Potential synergy of phytochemicals in cancer prevention: mechanism of action. J. Nutr 2004;134:3479S−85S.

[148] Dyerberg J. Coronary heart disease in Greenland inuit: a paradox. Implications for western diet patterns. Arctic Med Res 1989;48:47−54.

[149] Jorgensen KA, Hoj Nielsen A, Dyerberg J. Hemostatic factors and renin in Greenland Eskimos on a high eicosapentaenoic acid intake. Results of the Fifth UmanaK Expedition. Acta Med Scand 1986;219:473−9.

[150] Kris-Etherton PM, Harris WS, Appel LJ. Fish consumption, fish oil, omega-3 fatty acids, and cardiovascular disease. Arterioscler Thromb Vasc Biol 2003;23:e20−30.

[151] Wang C, Harris WS, Chung M, et al. n-3 Fatty acids from fish or fish-oil supplements, but not alpha-linolenic acid, benefit cardiovascular disease outcomes in primary- and secondary-prevention studies: a systematic review. Am J Clin Nutr 2006;84:5−17.

[152] Sands SA, Reid KJ, Windsor SL, Harris WS. The impact of age, body mass index, and fish intake on the EPA and DHA content of human erythrocytes. Lipids 2005;40:343−7.

[153] Agren JJ, Hanninen OO. Effect of moderate freshwater fish diet on erythrocyte ghost phospholipid fatty acids. Ann Med 1991;23:261−3.

[154] Sen CK, Atalay M, Agren J, Laaksonen DE, Roy S, Hanninen O. Fish oil and vitamin E supplementation in oxidative stress at rest and after physical exercise. J Appl Physiol 1997;83:189−95.

[155] Vidgren HM, Agren JJ, Schwab U, Rissanen T, Hanninen O, Uusitupa MI. Incorporation of n-3 fatty acids into plasma lipid fractions, and erythrocyte membranes and platelets during dietary supplementation with fish, fish oil, and docosahexaenoic acid-rich oil among healthy young men. Lipids 1997;32:697−705.

[156] Agren JJ, Hanninen O, Julkunen A, et al. Fish diet, fish oil and docosahexaenoic acid rich oil lower fasting and postprandial plasma lipid levels. Eur J Clin Nutr 1996;50:765−71.

[157] Smith BK, Sun GY, Donahue OM, Thomas TR. Exercise plus n-3 fatty acids: additive effect on postprandial lipemia. Metabolism 2004;53:1365−71.

[158] Ruzickova J, Rossmeisl M, Prazak T, et al. Omega-3 PUFA of marine origin limit diet-induced obesity in mice by reducing cellularity of adipose tissue. Lipids 2004;39:1177−85.

[159] Ghafoorunissa Ibrahim A, Rajkumar L, Acharya V. Dietary (n-3) long chain polyunsaturated fatty acids prevent sucrose-induced insulin resistance in rats. J Nutr 2005;135:2634−8.

[160] Whelan J, Li B, Birdwell C. Dietary arachidonic acid increases eicosanoid production in the presence of equal amounts of dietary eicosapentaenoic acid. Adv Exp Med Biol 1997;400B:897−904.

[161] Schuman BE, Squires EJ, Leeson S. Effect of dietary flaxseed, flax oil and n-3 fatty acid supplement on hepatic and plasma characteristics relevant to fatty liver haemorrhagic syndrome in laying hens. Br Poult Sci 2000;41:465−72.

[162] Caterina RD, Madonna R, Bertolotto A, Schmidt EB. Omega-3 fatty acids in the treatment of diabetic patients: biological rationale and clinical data. Diabetes Care 2007;30(4):1012−26.

[163] Satoh N, Shumatsu A, Kotani K, Sakane N, Yamada K, Suganami T, et al. Purified eicosapentaenoic acid reduces small dense LDL, remnant lipoprotein particles, and C-reactive protein in metabolic syndrome. Diabetes Care 2007;30(1):144−6.

[164] Drain PK, Kupka R, Mugusi F, Fawzi WW. Micronutrients in HIV-positive persons receiving highly active antiretroviral therapy. Am J Clin Nutr 2007;85:333−45.

[165] Karp SM, Koch TR. Mechanisms of micronutrient deficiency. Dis Mon 2006;52:208−10.

[166] Neumann CG. Symposium: food-based approaches to combating micronutrient deficiencies in children of developing countries. Background. J Nutr 2007;137:1091−2.

[167] Rennie KL, Livingstone MB. Associations between dietary added sugar intake and micronutrient intake: a systematic review. Br J Nutr 2007;97:832−41.

[168] Rennie KL, Livingstone MB. Systematic review: associations between dietary added sugar intake and micronutrient intake. Br J Nutr 2007;:1−10.

[169] Shenkin A. The key role of micronutrients. Clin Nutr 2006;25:1−13.

[170] Shenkin A. Micronutrients in health and disease. Postgrad Med J 2006;82:559−67.

[171] Visioli F, Hagen TM. Nutritional strategies for healthy cardiovascular aging: focus on micronutrients. Pharmacol Res 2007;55:199−206.

[172] Volpe SL. Micronutrient requirements for athletes. Clin Sports Med 2007;26:119−30.

[173] Webb P, Nishida C, Darnton-Hill I. Age and gender as factors in the distribution of global micronutrient deficiencies. Nutr Rev 2007;65:233−45.

[174] Key TJ, Appleby PN, Rosell MS. Health effects of vegetarian and vegan diets. Proc Nutr Soc 2006;65:35−41.

[175] John S, Sorokin AV, Thompson PD. Phytosterols and vascular disease. Curr Opin Lipidol 2007;18:35−40.

[176] Darbre A, Norris FW. Vitamins in germination; determination of free and combined inositol in germinating oats. Biochem J 1956;64:441−6.

[177] Gardiner P. Dietary supplement use in children: concerns of efficacy and safety. Am Fam Physician 2005;71:1068−71.

[178] Knight J. Safety concerns prompt US ban on dietary supplement. Nature 2004;427:90.

[179] Morrow JD, Edeki TI, El Mouelhi M, et al. American Society for Clinical Pharmacology and Therapeutics position statement on dietary supplement safety and regulation. Clin Pharmacol Ther 2005;77:113−22.

[180] Yen PK. Food and supplement safety. Geriatr Nurs 2005;26:279−80.

[181] Ziker D. What lies beneath: an examination of the underpinnings of dietary supplement safety regulation. Am J Law Med 2005;31:269−84.

[182] Melby CL, Toohey ML, Cebrick J. Blood pressure and blood lipids among vegetarian, semivegetarian, and nonvegetarian African Americans. Am J Clin Nutr 1994;59:103−9.

[183] Richardon DP, Group IS. Nutrition, health ageing and public policy. Brussels: International Alliance of Dietary Food Supplement Associations (IADSA); 2007. pp. 1−71.

[184] Kauhanen J, Myllykangas M, Salonen JT, Nissinen A. Kansanterveystiede. 2nd edn. Porvoo: WSOY; 1998.

[185] Gulati OP, Berry Ottaway P. Legislation relating to nutraceuticals in the European Union with a particular focus on botanical-sourced products. Toxicology 2006;221 (1):75–87.

[186] Contor L. Functional food science in europe. Nutr Metab Cardiovasc Dis 2001;11 (4 Suppl):20–3 2001 Aug.

[187] Prosky L. When is dietary fiber considered a functional food? Biofactors 2000;12 (1–4):289–97.

[188] Swennen K, Courtin CM, Delcour JA. Non-digestible oligosaccharides with prebiotic properties. Crit Rev Food Sci Nutr 2006;46(6):459–71.

[189] Ntanios FY, Duchateau GS. A healthy diet rich in carotenoids is effective in maintaining normal blood carotenoid levels during the daily use of plant sterol-enriched spreads. Int J Vitam Nutr Res 2002;72(1):32–9.

[190] Lidbeck A, Nord CE. Lactobacilli and the normal human anaerobic microflora. Clin Infect Dis 1993;16(Suppl 4).S181–7.

[191] <http://www.fda.gov/Drugs/GuidanceComplianceRegulatoryInformation/ImportsandExportsCompliance/ucm297872.htm>.

[192] <http://ods.od.nih.gov/Health_Information/ODS_Frequently_Asked_Questions.aspx>.

[193] <http://www.fda.gov/Food/GuidanceRegulation/GuidanceDocumentsRegulatoryInformation/DietarySupplements/ucm2006823.htm>.

[194] <http://www.fda.gov/downloads/Regulatoryinformation/Guidances/UCM291085.pdf>.

[195] <http://voicesweb.org/archive/sn/fdaherbs1198.html>.

[196] <http://www.naturalproductsinsider.com/articles/2000/04/solidarity.aspx>.

[197] <http://www.fao.org/docrep/008/y7867e/y7867e04.htm>.

[198] <http://www.fao.org/docrep/008/y7867e/y7867e01.htm>.

[199] <https://www.ccnfsdu.de/>.

[200] <http://www.fda.gov/Food/GuidanceRegulation/GuidanceDocumentsRegulatoryInformation/DietarySupplements/ucm113860.htm>.

[201] <http://www.nutraingredients.com/Research/Functional-foods-in-Japan>.

[202] <http://www.nmif.no/filestore/Pdf/TheGlobalalliance.pdf>.

[203] <http://www.iadsa.org/page.php?key = general,56dac28b5c1e06e0307e219ccf86fa31b285cc01,0,1>.

Global Market Entry Regulations for Nutraceuticals, Functional Foods, Dietary/Food/Health Supplements

3

Andrew Shao

Herbalife International of America, Inc.

3.1 Introduction

The policy that underlies the regulation for any consumer product should be one that appropriately balances benefit and consumer access and choice with safety (risk). This principle should apply regardless of the regulatory category, whether a drug, device, food, supplement, or nutraceutical. Requiring that a new promising treatment for cancer be free of any side effects whatsoever may minimize risk to the public, but such a treatment would likely fail to ever reach the market, thus doing consumers little, if any good. In contrast, allowing an untested medical device on the market that provides little benefit but poses high safety concerns, allows for open consumer access, but does not appropriately balance risk.

Starting with market entry, global regulations should aim to strike an appropriate balance between consumer access and risk. For nutraceuticals, functional foods, and supplements, the specific market entry requirements depend on the overall regulatory framework employed within a given country. For most countries, these frameworks are based on a food or drug, including an over-the-counter-based approach. Products that are regulated under a food-like category tend to be subjected to notification or registration-based systems, whereas products regulated under a drug-like category tend to be subjected to a pre-market approval system. This chapter compares and contrasts the different approaches used globally and whether these approaches help to achieve the appropriate access/risk balance.

Nutraceutical and Functional Food Regulations in the United States and Around the World.
DOI: http://dx.doi.org/10.1016/B978-0-12-405870-5.00003-7

3.2 Market entry requirements

For most countries worldwide, foods (including conventional foods and functional foods) and supplements (including nutraceuticals) are regulated as a category of food. In some regions or countries, there are a specific set of regulations governing supplements (e.g., United States, European Union, EU, Association of South East Asian Nations, ASEAN) and nutraceuticals (India), which stem from a food-based regulatory paradigm. For most of these countries, some form of a notification- or registration-based system is required to bring new products to market (Table 3.1). This approach is in contrast to the pre-market approval approach required for drugs in most countries, and appropriately fits the category of foods, as these pose inherently low safety risks relative to drugs.

3.2.1 Positive and negative ingredient lists

The safety and acceptability of new products into the marketplace starts first with the ingredients. In most regions of the world, the safety and acceptability of excipients used in a given formula is governed via established food additive requirements. In the United States [1] (including Generally Recognized as Safe, GRAS [2]) and EU [3] there are well-established food additive regulations, and for those countries that lack their own established approaches, many rely on standards set by the Codex Alimentarius [4]. Therefore, when evaluating the entry of new products to the marketplace, most regulators focus their attention on the safety and acceptability of the active or driver ingredients.

In many countries, regulators rely on established formal "positive" and/or "negative" ingredient lists, or derivations thereof to help guide their evaluations. These

Table 3.1 Global Notification or Registration for Foods/Supplements

Country/region	Notification/registration
Argentina	Registration
ASEAN	Notification or registration
Australia	Notification/listing
Brazil	Registration
Canada	Registration
Chile	Notification
China	Registration
Colombia	Registration
United States	Notification or none
Japan	None
Mexico	Notification
Russia	Registration
EU	Notification or none

lists, as their name implies, indicate which ingredients are considered acceptable for use in foods and/or supplements and which are not allowed. Some countries will rely on only a positive list, others only a negative list; some countries rely on both and some do not rely on a list but employ a broader approach. The basis for how ingredients are added to these lists varies widely around the world, and the formal role the lists play varies as well. An ingredient may be added to a negative (not allowed in foods) list based on well-established safety issues, e.g., a botanical with well-known toxic effects, or an ingredient may be added to the list based on the perception (often ill conceived) or speculation of adverse effects. Ingredients known to be inherently safe, such as vitamins and minerals, will appear on most positive lists, or due to their inherent safety may not appear on a list at all. Most regulatory agencies keep separate lists for botanicals allowable for use in foods and supplements. New botanicals or other novel ingredients may also be allowed and added to positive lists on a case-by-case basis following their formal evaluation. Whether an ingredient is considered new or novel depends on the specific policies and regulations of the given country or region.

In Australia the Therapeutic Goods Administration (TGA) regulates supplements under a Complementary Medicines category and maintains a positive list of allowable ingredients, including excipients and actives [5]. This list includes vitamins, minerals, and botanicals. China [6] is an example of a country that utilizes both a positive and negative list. China's positive list includes ingredients that can be added to foods, drugs, and functional foods, as well as a list of banned ingredients. Canada employs a unique monograph system—ingredients that comply with already established monographs have an easier path to market as Natural Health Products (NHPs), while those that have not yet been subject of a monograph must go through a more formal approval process [7].

In the EU, the European Commission has published a list of vitamin and mineral ingredients allowable in foods and food supplements [8]. Ingredients that have not been used for human consumption to a significant degree before May 1997 are considered novel foods, requiring the marketer to apply for authorization [9]. There is an online catalog that uses a traffic light system to highlight whether the ingredient is novel, not novel, can only be used in food supplements, or it is not certain yet [10].

For other physiologic substances, including botanicals, the regulations in the EU for acceptable ingredients have not yet been harmonized. As a result, different positive lists have been developed by some individual Member States. For example, Italy [11], Denmark [12], and the Czech Republic [13] have established their own versions of positive lists of substances that can be added to foods and/or supplements. Russia is an example of a non-Member State with a positive list [14].

In the United States, the U.S. Federal Food and Drug Administration (FDA) maintains a list of ingredients that are GRAS for use in foods [2]. In many cases, these ingredients are acceptable for use in dietary supplements as well, although the agency does not maintain a specific positive list of ingredients for supplements. Regarding ingredients not allowed for use in supplements, the FDA has

Table 3.2 Countries/regions Employing Positive and/or Negative Ingredient Lists

Market/region	Positive List	Negative List	Other
Argentina			
Australia	✓		
Brazil	✓		
Canada			Monograph system
Chile			
China	✓	✓	
Colombia			
Mexico	✓		
Paraguay			
Russia	✓		
United States			GRAS, new/old dietary ingredients
Uruguay			
Venezuela	✓	✓	
EU	✓		Novel food
Italy	✓		
Denmark	✓		
Czech Republic	✓		
ASEAN			Restricted list

published several rulings [15,16] and/or warning letters [17−19] indicating certain ingredients are not acceptable for inclusion in dietary supplements or are unsafe for the public [20]. In the United States, the FDA must be notified of New Dietary Ingredients (NDI) while "old" dietary ingredients, or grandfathered ingredients (ingredients marketed prior to October 1994), need not be notified. However, no formal list of grandfathered ingredients exists in the United States.

In South America, the situation varies from country to country, with some not relying on a list (e.g., Argentina, Chile, Colombia), others a positive list (e.g., Brazil), and others both (e.g., Venezuela).

In the ASEAN, the region is aiming to establish a restricted list of substances allowed in traditional medicines and health supplements [21]. A brief summary of the different types of lists utilized in different countries or regions is summarized in Table 3.2.

3.2.2 Botanicals

Botanicals can present unique regulatory, quality, and safety challenges and thus tend to be more restricted in many markets. Lack of understanding on the part of regulators, lack of transparency in the supply chain, and a less robust evidence base relative to vitamins and minerals serve to undermine the confidence of regulators in this category of ingredients. Thus, the appearance of botanical ingredients on positive lists is quite limited in many countries. In Brazil, the presiding regulatory agency over foods and drugs, Agência Nacional de Vigilância Sanitária (National Sanitary Surveillance Agency, ANVISA), only allows water or ethanol-extracted botanicals in foods, and each product is evaluated on a case-by-case basis [22]. In Mexico, the Federal Commission for the Protection against Sanitary Risk (COFEPRIS) maintains a positive and negative list for botanicals allowed in foods [23]. The agency has recently proposed to add certain botanical ingredients with a long history of safe use to a negative list, including inulin and aloe [24], based on the belief that these have documented therapeutic uses (and thus are medicines, not acceptable for use in foods). In the EU, the European Food Safety Authority (EFSA) has published a botanicals compendium [25], which includes a list of botanicals reported to contain toxic, addictive, psychotropic, or other substances of concern. While this document does not have official regulatory standing, it could be interpreted as a form of a negative list for the EU. However, other EU Member States have published and continue to update their own positive and negative botanical lists. Italy maintains a list of permitted botanicals with related health benefits [26] and one of non-permitted botanicals [26]. Belgium has one list including permitted and non-permitted botanicals and mushrooms with related maximum allowable levels for some botanicals [27]. Germany maintains a list, which has been used by the regulators as a guidance document for assessing botanicals in Germany [28].

In these countries, whether a botanical appears on a positive or negative list for foods depends largely on if it is perceived as medicinal or not. A plant may be judged to be medicinal if there are scientific data showing an influence on physiological functions. Where applicable an assessment of the minimum dosage for such activity is determined. A number of acceptable scientific sources to support this assessment may be consulted, including compendial monographs. Another check is whether the ingredient is novel, in which case the EU novel food catalog is consulted as a reference. Finally, the botanical may be checked against conformity with other approvals, e.g., additives and if the product would be considered as safe.

If formal lists are to be in place, the preference should be for a negative list (i.e., limiting the ingredients that are *not* allowed), which tends to be much shorter and easier to maintain, and also allows consumers access to a greater variety of products. If allowable ingredients are restricted to a positive list, this limits somewhat the industry's ability to bring novel products to market. Given the inherent safety of most food and supplement ingredients, a negative list most appropriately balances consumer access with risk.

3.2.3 Notification versus registration

Although there are some differences between countries, a notification-based system usually represents a means of informing the regulatory authorities of a product's introduction to the marketplace or intent to introduce a new product. Notifications tend to be comprised of information on the manufacturer, product form, formulation, and label and these tend to be subjected to minimal review by regulators. Product ingredients must be derived from a positive list (if one exists), and the entry of the product to the marketplace does not signify or constitute an "approval" or "authorization" from regulators, but they do have the authority to question or object to a notification. The notification results in the addition of the new product on a list that regulators maintain for enforcement purposes. The notification approach tends to be the most efficient for getting new products to market, for both industry and regulators, as minimal pre-market resources are necessary. This approach also allows regulators to focus their resources on enforcement of critical post-market activities, such as good manufacturing practices, and adverse event reporting. Notification is thus a proportionate approach to the low level of risk posed by these categories of food products.

Registration-based approaches differ from notification-based approaches in that they require more detailed information in the form of a registration dossier, intense review by regulators, and a more lengthy review process. A typical registration dossier requires much more detailed information than is found in a notification, including details on the specifications of the finished product, evidence supporting the safety and efficacy of the ingredients, a Certificate of Analysis, stability testing (to support shelf-life dating), the product label, a certificate of free sale, and good manufacturing practice (GMP) certification or compliance statement. While this approach allows regulators to closely scrutinize new products, in most countries that require registration, it is a slow and extremely resource-intensive process, ranging from months up to years. A lengthy registration time tends to be due to a lack of resources and the obligation regulators have, since a positive registration equates to regulator approval or authorization. A drawback of this approach is the length of time to market for products, regardless of the risk they may pose, and even fewer resources for enforcement in other areas. A comparison of notification versus registration appears in Table 3.3.

The EU is an example of a region that utilizes the notification approach [29]. While the specific requirements differ between countries, of the 27 Member States, 23 have opted to incorporate a notification-based approach for new product entry. Austria, Holland, Sweden, and the UK are among those for which notification for new products is optional.

In the ASEAN, the 10 member country region is in the process of finalizing harmonized regulations for traditional medicines and health supplements [30]. Currently, both registration and notification-based procedures are utilized, depending on the country (Table 3.4). As with other regions, the specific requirements in each country vary [31].

The intent is to harmonize regulations by 2013 and begin implementation with completion by 2015. For market entry for health supplements, the region will require a comprehensive product dossier as part of a registration-based process, with the emphasis on quality, safety, and efficacy.

The United States represents a somewhat unique situation where no product notification is required unless the product contains an NDI [32]. If a new product

Table 3.3 Notification versus Registration Submissions for Foods, Supplements, and Nutraceuticals

	Notification	Registration
Ingredients in accordance with applicable positive and negative lists	Required	Required
Applicable GMPs	Manufacturer expected to adhere to	Official certification or statement required with dossier
Label according to regulation	Required	Required
Detailed product specifications	N/A	Required
Certificate of analysis	N/A	Required
Claim substantiation[a]	Manufacturer expected to have on file	Required
Shelf-life substantiation	N/A or manufacturer expected to have on file	Required
Certificate of free sale	N/A	Required
Implications	Surveillance and control regime	Reduces post-surveillance and control regime
	Helps the authority to process a high volume of products	Requires high government resources
	Immediate access to market	Slow access to market

[a]In those countries where claims are allowed.

does not contain an NDI, then no notification is required. If a new product contains an NDI, then a notification to the U.S. FDA is required 75 days before the product enters the market. The basis of the notification is substantiation that the manufacturer or marketer believes the NDI is safe in the product under its intended conditions of use [33].

In Latin America, the market entry requirements for foods and supplements vary, with notification-based approaches employed in Mexico and Chile, while registration-based approaches are followed in Colombia, Brazil, and Argentina. In countries such as Brazil (ANVISA [34]), China (China State Food and Drug Administration, SFDA [6]), and Taiwan (Taiwan Food and Drug Administration, TFDA [35]) the regulators require animal and/or human clinical studies as a requirement of the product registration requirement. This requirement tends to be limited to those products for which efficacy claims are intended.

In Australia and Canada, supplements and nutraceuticals are regulated more closely as a drug than food category. The TGA in Australia regulates supplements

Table 3.4 Current Health Supplement Market Entry Requirements for Various ASEAN Member Countries

	Brunei	Cambodia	Indonesia	Malaysia	Philippines	Singapore	Thailand	Vietnam
Pre-marketing registration	No	Yes	Yes	Yes	Yes	No	Yes	Yes
Time line for evaluation								
<2 months								
2–3 months			✓		✓			✓
6 months							✓	
6–12 months		✓		✓				

under a Complementary Medicines category [36]. The pre-market approval requirements for this category are comparable to registration requirements for foods and supplements in other countries [37], although other requirements, such as GMPs tend to be more drug-like than food-like. In Canada, Health Canada (HC) regulates supplements under an NHP category [38]. Due to the allowance of medicinal or therapeutic claims for NHPs, HC takes a more rigorous approach to new product entry. The licensing process in Canada for NHPs is risk-based, with products posing the highest risk requiring the greatest pre-market review by the agency.

While a variety of individual systems exists in various countries and regions worldwide, with a few exceptions the approaches for bringing new food, supplement, or nutraceutical products to market tend to follow three basic approaches. The notification-based approach most optimally balances pre-market resources, consumer access, and consumer safety. However, a key aspect to assuring product safety and quality in the marketplace, regardless of the pre-market requirements, is robust post-market surveillance.

References

[1] FDA. (2012). Food additives. Retrieved January 15, 2013, from <http://www.fda.gov/Food/FoodIngredientsPackaging/FoodAdditives/default.htm>.

[2] FDA. (2012). Generally Recognized as Safe (GRAS). Retrieved January 15, 2013, from <http://www.fda.gov/Food/FoodIngredientsPackaging/GenerallyRecognizedasSafeGRAS/default.htm>.

[3] European Commission. Food additives and flavourings. Retrieved January 15, 2013, from <http://ec.europa.eu/food/fs/sfp/flav_index_en.html>.

[4] JECFA. About Codex. Codex Alimentarius Retrieved January 15, 2013, from <http://www.codexalimentarius.org/scientific-basis-for-codex/jecfa/en/>; 2013.

[5] Therapeutic Goods Administration. (2007). Substances that may be used in Listed medicines in Australia. Retrieved January 15, 2013, from <http://www.tga.gov.au/pdf/cm-listed-substances.pdf>.

[6] SFDA. Regulations on Supervision of Functional Foods (Draft for Approval). Retrieved January 15, 2013; 2009.

[7] Natural Health Products Directorate. Compendium of Monographs. Drugs and Health Products Retrieved January 15, 2013, from <http://www.hc-sc.gc.ca/dhp-mps/prodnatur/applications/licen-prod/monograph/index-eng.php>; 2009.

[8] European Commission. Commission Regulation (EC) No 1170/2009 of 30 November 2009 amending Directive 2002/46/EC of the European Parliament and of Council and Regulation (EC) No 1925/2006 of the European Parliament and of the Council as regards the lists of vitamin and minerals and their forms that can be added to foods, including food supplements. Official Journal of the European Union L314/36, 2009.

[9] European Commission. Novel foods and novel food ingredients. Food and Feed Safety Retrieved January 15, 2013, from <http://ec.europa.eu/food/food/biotechnology/novel-food/index_en.htm>; 2012.

[10] European Commission. Novel Food catalogue—Search. Food and Feed Safety Retrieved January 15, 2013, from <http://ec.europa.eu/food/food/biotechnology/novelfood/nfnetweb/mod_search/index.cfm>

[11] Ministero della Salute. Altri Nutrienti E Altre Sostanze Ad Effetto Nutritivo O Fisiologico. Retrieved January 15, 2013, from <http://www.salute.gov.it/imgs/C_17_pagineAree_1268_listaFile_itemName_4_file.pdf>.

[12] Ministerialtidende. Bekendtgørelse om tilsætning af visse andre stoffer end vitaminer og mineraler til fødevarer. Retrieved January 15, 2013, from <https://www.retsinformation.dk/Forms/R0710.aspx?id = 137299>; 2012.

[13] SBÍRKA ZÁKONŮ. Kterou se stanoví požadavky na doplňky stravy a na obohacování potravin, 2008.

[14] Customs Union Commission. Uniform sanitary and epidemiological and hygienic requirements for products subject to sanitary and epidemiological supervision (control), 2010.

[15] FDA. Final rule declaring dietary supplements containing ephedrine alkaloids adulterated because they present an unreasonable risk. 2004;69:6788—6854.

[16] CFSAN, C. f. F. S. a. N., Office of Food Additive Safety. FDA Response to OVOS Natural Health Homotaurine Petition. Department of Health and Human Services, Food and Drug Administration, Washington, DC; 2011.

[17] FDA. FDA Advises Dietary Supplement Manufacturers to Remove Comfrey Products From the Market 2001.

[18] FDA. February 28, 2003 Warning Letter to Powerhouse Supplements 2003.

[19] FDA. Letter to Health Professionals regarding safety concerns related to the use of botanical products containing aristolochic acid, 2001.

[20] FDA. Consumer Advisory: Kava-Containing Dietary Supplements May be Associated With Severe Liver Injury, 2002.

[21] ASEAN. Guiding Principles for Inclusion of Active Substances into the Restricted List for Traditional Medicines and Health Supplements (TMHS), 2010.

[22] ANVISA. Novos Ingredientes Aprovados. Retrieved January 15, 2013, from <http://portal.anvisa.gov.br/wps/content/Anvisa + Portal/Anvisa/Inicio/Alimentos/Assuntos + de + Interesse/Novos + Alimentos + e + Novos + Ingredientes/29bd7700401adec6b403b654e035b7cb>; 2009.

[23] COFEPRIS. What is the COFEPRIS? Retrieved January 15, 2013, from <http://www.cofepris.gob.mx/Paginas/Idiomas/Ingles.aspx>; 2011.

[24] Asociacion Nacional de la Industria de los Suplementos Alimenticios. ANAISA Noticias. Retrieved January 15, 2013, from <http://www.anaisa.mx/>; 2013.

[25] European Food Safety Authority, E. Compendium of botanicals reported to contain naturally occuring substances of possible concern for human health when used in food and food supplements. EFSA J 2012;2012:**10**.

[26] Ministero della Salute. Disciplina dell'impiego negli integratori alimentari di sostanze e preparati vegetali (G.U. 21-7-2012 serie generale n. 169), 2012.

[27] Belgian Health, F. C. S. a. E. Food supplements − Enriched foodstuffs. Retrieved January 15, 2013, from <http://www.health.belgium.be/eportal/foodsafety/foodstuffs/foodsupplements/index.htm?fodnlang = en#Plants>; 2012.

[28] The Federal Office of Consumer Protection and Food Safety (BVL). Entwurf einer Liste für die Kategorie "Pflanzen und Pflanzenteile". Retrieved January 15, 2013, from <http://www.bvl.bund.de/SharedDocs/Downloads/01_Lebensmittel/stoffliste/stoffliste_pflanzen_pflanzenteile.html?nn = 1406620>; 2010.

[29] European Commission. Directive 2002/46/EC Of The European Parliament And Of The Council, 2002.

[30] ASEAN. Harmonization of Standards and Technical Requirements in ASEAN. Retrieved January 15, 2013, from <http://www.asean.org/news/item/harmonization-of-standards-and-technical-requirements-in-asean>; 2012.

[31] ASEAN. Profile of Definition, Terminology, and Technical Requirement of Traditional Medicines and Health Supplements among ASEAN Member Countries 2006.

[32] FDA. Draft Guidance for Industry: Dietary Supplements: New Dietary Ingredient Notifications and Related Issues. Retrieved January 15, 2013, from <http://www.fda.gov/food/guidancecomplianceregulatoryinformation/guidancedocuments/dietarysupplements/ucm257563.htm>; 2011.

[33] FDA. New Dietary Ingredients in Dietary Supplements—Background for Industry. Retrieved January 15, 2013, from <http://www.fda.gov/Food/DietarySupplements/ucm109764.htm>; 2012.

[34] ANVISA. Registration of Products: Manual covering procedures for registration and exemption from registration of imported products. Retrieved January 15, 2013, from <http://www.anvisa.gov.br/eng/food/registration.htm>; 2000.

[35] TFDA. Food and Drug Administration, Department of Health. Retrieved January 15, 2013, from <http://www.taiwan.gov.tw/ct.asp?xItem = 25613&ctNode = 1957& mp = 999>; 2012.

[36] TGA. The regulation of complementary medicines in Australia—an overview: Pre-market assessment. Retrieved January 15, 2013, from <http://www.tga.gov.au/industry/cm-basics-regulation-overview.htm#pre>; 2012.

[37] TGA. Australian Guidelines for Complementary Medicines (ARGCM) Part II: Listed Complementary Medicines, 2011.

[38] Health Canada. Natural Health Products. Retrieved January 15, 2013, from <http://www.hc-sc.gc.ca/dhp-mps/prodnatur/index-eng.php>; 2012.

Manufacturing Compliance and Analytical Validation

Natural Health Products and Good Manufacturing Practices

Digambar Chahar

Consulting Regulatory Affairs Scientist, Mississauga, Ontario, Canada

4.1 Overview

Natural Health Products (NHPs) are a valuable tool in the health and well-being of consumers so they have a choice of safe and effective remedies for their personal health. Manufacturers of natural products have adopted or are in the process of adopting Good Manufacturing Practices (GMPs) to ensure that all products are manufactured under a GMP environment, and ensured that the safety and quality of their products meet the required regulations.

NHPs (dietary supplements) are the products that contain "dietary ingredients" intended to supplement the diet, e.g., vitamins, minerals, herbs, or other botanicals and amino acids and substances such as enzymes, extracts, or concentrates in the form of tablets, capsules, soft gels gel caps, liquids, or powder and bars.

4.1.1 Who does it apply to?

- Manufacturers, packagers, labelers, and distributors
- Warehouse/storage facilities

NHP/dietary supplements must be manufactured, stored, and distributed following GMPs.

4.1.2 GMPs: what are they?

GMPs are measures/rules designed to ensure an effective overall approach to product quality and safety (SISPQC) and standards and practices for product testing, manufacturing, storage, handling, and distribution.

What does SISPQC stand for?

- Safety
- Integrity

Nutraceutical and Functional Food Regulations in the United States and Around the World.
DOI: http://dx.doi.org/10.1016/B978-0-12-405870-5.00004-9

- Strength
- Purity
- Quality
- Composition

The implementation of GMP/Hazard Analysis and Critical Control Points (HACCP) programs is the assurance to consumers that consistencies for testing and checks have been maintained throughout the process of manufacturing, packaging, and distribution of products.

4.1.3 Key elements of GMP

- Place
- People
- Process
- Product

4.1.4 Places

- Premises must be clean and orderly; surfaces should allow for effective cleaning and are designed to prevent contamination of NHPs.
- Equipment surfaces must allow for effective cleaning, must function in accordance with intended use, and must be designed to prevent contamination of NHP.

4.1.5 People

- Personnel must be qualified by education, training, or experience to perform their respective functions.
- Quality Assurance is responsible to release each product for sale, release raw material and packaging components for use, approve all operational procedures, master formulas and specifications, review all returns prior to resale, and investigate each complaint.

4.1.6 Processes

- Sanitation programs include documented procedures for effective cleaning of the premises, equipment, handling of substances, and the health and hygienic behavior of personnel.
- Operation includes a system for the following:
  - Review and release of all raw materials and packaging materials.
  - Water that is potable and meets the Guidelines for Canadian Drinking Water Quality or applicable regulations.
  - Complete batch records for each batch allowing for traceability.
  - Each finished package is identified with a lot number and expiry date.

- Labels are secured and controlled.
- Quality agreements are in place with contracts.

4.1.7 Products

- Specifications are available for every product and include purity, quantity, and identity of medicinal ingredients, potency, and test methods.
- Every lot must be tested for conformance with its finished product specifications.
- Importers may follow a reduced testing program in which
  - The first lot of each product is fully tested.
  - For each subsequent lot:
    - A Certificate of Analysis with actual test results is reviewed.
    - The lot is positively identified upon receipt.
    - Transportation and storage conditions do not adversely impact the product.
- Full confirmatory testing is conducted on at least one lot per year dosage form, per supplier.
- Stability results demonstrate that the product will meet specifications under the recommended storage conditions.
- Samples are retained under the recommended storage conditions for 1 year past the expiry date of product. There are sufficient samples available to perform complete testing of the product.
- Records must be retained for 1 year past the expiration date of the product to which they refer.
- Recall procedure is documented and effective by performing a Mock Recall.
- Sterile NHPs must be manufactured and packaged in a separate enclosed area under the supervision of a trained person using scientifically proven methods to ensure sterility.

4.1.8 Benefits of GMP/HACCP

There are many benefits of implementing GMP/HACCP programs, which will substantially increase the quality of the product, and increase revenues and customer satisfaction. Some of the benefits of being in compliance to GMP/HACCP include:

- Operating cost drops as rework and penalties due to non-compliance reduce and efficiencies increase.
- Increase the customer satisfaction; employees, stockholders, regulators, and competitors develop sustainable respect for an organization that demonstrates commitment to NHP safety.
- GMP/HACCP covers all safety and written procedures (standard operating procedure, SOP), which makes employees more efficient and reduces errors.

4.2 GMP procedures: considerations for manufacturers

Since the final GMP rule for dietary supplements was implemented in 2007, the Food and Drug Administration has made it very clear that GMP inspections should be a major priority for dietary supplement manufacturers. As evidenced by the number of violations uncovered to date, there has been a severe lack of procedures in place to ensure the quality of the manufactured product.

The new protocols will provide manufacturers with the comprehensive elements needed to have a fully compliant GMP system in place. Every aspect of the manufacturing process should have an accompanying SOP in place, including:

- Personnel
- Requirements for the cleaning and sterility of equipment and utensils
- Process control systems for the varying stages of manufacture
- Proper packaging, labeling, and storage conditions
- Master Manufacturing Records and batch records
- Process for dealing with product complaints
- Complete recordkeeping

Under the GMP rules, it is the manufacturer's responsibility to ensure that third-party contractors hired for additional work conduct their operations in accordance with GMPs. Third parties can include contract manufacturers and third-party testing laboratories, and should be reviewed in advance by manufacturers to confirm that they have processes in place to ensure the quality of their work, including the appropriate test methods for the product.

The GMPs for dietary supplements specify that test methods should be scientifically valid, referring to their ability to repeatedly deliver reliable results. Although there are many validated methods available for contract or in-house testing, many times these methods may have been developed for, and validated by, a specific food product, and are not necessarily adaptable to dietary supplements. Even with accredited methods, such as those found in Association of Official Analytical Chemists, pharmacopeias, or compendiums, the testing facility must verify that these methods are appropriate for their products.

When using a contract laboratory, a manufacturer can ensure the competency of the lab by reviewing their processes and SOPs that would affect the consistency of the test results, and by scheduling an on-site visit to the testing facility. The manufacturer needs to show that it did its due diligence with their chosen lab, and to be certain that the laboratory was capable of properly testing raw materials and finished products. It is important to keep in mind that it may be difficult to find one lab that is able to perform the full spectrum of tests required. When choosing testing labs, be sure to determine the specialty of each. This screening will ensure that the finished product is efficacious, safe, and consistently of high quality.

While there is a definite action toward compliance within the manufacturing and testing of dietary supplements, there are still those that rely on falsifying data in lieu of proper testing. To determine if this practice is going on, it is advisable for manufacturers to provide the lab with known samples to see if the lab's data matches up with the known results. Some samples that would be helpful in this screening would include:

- Purchasing a reference standard (a highly characterized material) and send only the sample to the lab. Ask the lab to identify the purity of the reference standard material to see if it can accurately confirm the material's concentration.
- Requesting the lab to test for the presence of a different compound, one that the manufacturer knows should not be present. If the laboratory reports not finding that compound, then that is a good indication of their honesty.
- Having the lab test a finished product for which the manufacturer knows the amount of the ingredients present, and compare the data with the lab's analysis.

Since GMP standards are now permeating throughout the entire product life cycle, the compliance of dietary supplements is increasing, providing greater security to consumers. Manufacturers will benefit from the confidence that their products are produced and tested to the highest standards when it comes to health.

A good quality system management is the key to success in manufacturing the quality product including the GMP/HACCP programs.

4.2.1 Overview of quality management system

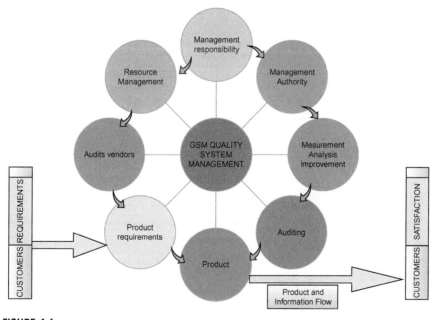

FIGURE 4.1

Overview of the quality mangement system.

4.2.2 **Overview of GMP/HACCP process flow**

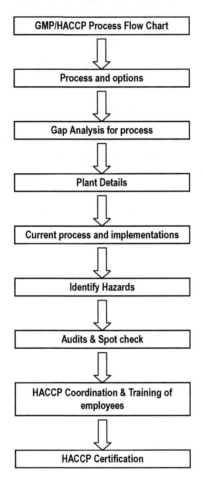

FIGURE 4.2

Overview of GMP/HACCP process flow.

4.3 **Conclusion**

The manufacture of pharmaceuticals and nutraceuticals is a regulated business with increasing emphasis on quality and standard setting; it is very important to understand the requirement and importance of documentation practices to supply safe drug/food products. It is anticipated that the approach/regulations described here will be a useful reference for maintaining or developing the quality in manufacturing. Again, documentation is a very important part of GMP and will enhance visibility of quality assurance and quality control function.

Further reading

<http://www.nutritionaloutlook.com/article/dietary-supplement-testing-how-detect-%E2%80%9Cdry-labbing%E2%80%9D-3-10235>.

<http://www.nutritionaloutlook.com/article/testing-dietary-supplements-face-unpredictable-microorganisms-2-10229>.

<http://www.qualityandcompliance.com/pdfs/QuickNote%20-%20NHP%20GMP_February%202009.pdf>.

<http://www.gulfsouthmachine.com/Quality%20Control/Quality%20System%20Management.jpg>.

<http://www.fda.gov/food/dietarysupplements/default.htm>.

<http://hc-sc-gc.ca/dhp-ps/prodnatur/imdex-eng.php>.

Current Good Manufacturing Practices for Nutraceuticals

5

Deb Kumar Nath and Chandra S. Yeevani

Apotex Inc., Toronto, Canada

5.1 Introduction

There are some basic manufacturing regulations or guidelines that are available to the nutraceutical industry for producing quality products. However, in most cases it could be noted that these regulations are not adequate in providing guidance for the manufacturing facilities, testing, etc. The nutraceutical industry is looking for more resources from the industry experts, especially from pharmaceutical industry segments, for valuable guidance to ensure safety of the products delivered to consumers.

In addition, the utilization of nutraceutical products is increasing as a number of patients are moving toward alternate medicinal paths from the conventional medication and prescription drugs, such as multivitamins, anti-diabetics, psoriasis treatment, etc. Therefore, some of the pharmaceutical industry guidelines can be adapted by the nutraceutical industry as well. Let us see the potential implementation of minimal guidelines by taking note of ICH 9, ICH 10, and Quality by Design (QbD) guidelines. Eventually, regulators around the world sooner or later will adapt these regulations to ensure the safety of nutraceutical products.

Nutraceutical industries are equally responsible for ensuring the safety of their consumable products. For example, the combination of nutraceutical and prescription drugs can be crucial or often may lead to adverse reactions if not controlled properly.

The following sections will provide minimum requirements in terms of building, facilities, quality control, material systems, and documentation requirements. Please note that this is not a complete guideline and we recommend the industry to refer to local and international requirements specific to certain specialty products.

This list can be used to develop and establish the internal standards to increase the compliance level of nutraceuticals, wherever applicable.

1. International Conference on Harmonization (www.ich.org)
    a. Stability Q1A−Q1F
    b. Analytical Validation Q2

Nutraceutical and Functional Food Regulations in the United States and Around the World.
DOI: http://dx.doi.org/10.1016/B978-0-12-405870-5.00005-0

    c. Impurities Q3A-Q3D

    d. Specification Q6A−Q6B

    e. Pharmaceutical development—Q8

    f. Quality Risk Management—Q9

    g. Pharmaceutical Quality Systems—Q10

2. Good Manufacturing Practices (GMPs)

    a. EudraLex—Volume 4 GMP Guidelines

    b. WHO Guidelines on GMPs for Herbal Medicine

    c. ICH Q7A

    d. Pharmaceutical Inspection Cooperation Scheme (PIC/S).

    e. Code of Federal Regulations 21CFR 210—current GMP (cGMP) in Manufacturing, Processing, Packing, or holding of Drugs; General—U.S. Food & Drug Administration (FDA)

    f. Code of Federal Regulations 21CFR 211 cGMP for Finished Pharmaceuticals—FDA

3. General references

    a. Parental Drug Association (PDA; www.pda.org)

    b. International Society of Pharmaceutical Engineering (ISPE; www.ispe.org)

4. Pharmacopeias

    a. United States Pharmacopeia (http://www.uspnf.com)

    b. British Pharmacopoeia (http://www.pharmacopoeia.co.uk)

    c. European Pharmacopoeia (http://www.edqm.eu)

    d. Food Chemicals Codex (http://online.foodchemicalscodex.org)

    e. Pharmaceutical Excipients (Medicines Complete; http://www.medicinescomplete.com/mc/)

    f. Pharmaceutical Additives Handbook

5.2 GMPs [1]

5.2.1 Premises [2,5]

The building and other facilities for the manufacturing of nutraceuticals in the form of powders, tablets, capsules, and liquids should have the following minimum requirements to produce quality products meeting customer requirements and other regulations [2]. Maintenance activities of plant and buildings should be planned, conducted, and documented in a manner that prevents contamination risks. Facilities designed for handling laboratory animals must be isolated from the manufacturing areas and comply with the relevant regulations in force.

5.2.1.1 Personnel and material flow

Manufacturing and testing facilities should be designed to allow unidirectional flow of the materials and products throughout the manufacturing, packaging, storage, and testing process. This is required to ensure against the potential mix-ups and cross-contamination of the materials. To comply with these requirements, the firm needs to design the proper physical segregation of material and products in every stage of manufacturing and storage. For example, adequate design of air-locks for material and personnel entry would separate GMP areas and non-GMP areas and would prevent direct access from manufacturing and packaging areas to the outside environment.

5.2.1.2 Walls and ceiling
The doors, walls, ceilings, and floors should not contain any holes or gaps. Wooden doors and walls should not be used and material that sheds particles should be avoided. Ideally, epoxy flooring should be used to permit effective cleaning. Production areas where dust is generated must have collection systems and procedures for disposal of collected dust. Any extraction systems used should avoid potential cross-contamination.

5.2.1.3 Heating, ventilation, and air conditioning system
Adequate ventilation should be provided in manufacturing and packing areas where there is potential exposure of product to environment and people. As such there are no regulatory requirements to use terminal HEPA filters in air-handling units. However, it is recommended to use either HEPA filters or an industry standard adequate filtration system that is equivalent in multiproduct facilities to avoid cross-contamination during recirculation of air within the manufacturing rooms. In absence of proper air filtration systems within the air-handling units, risk assessment documentation is required to support the design of the HVAC.

It is recommended that lighting used in warehouse and production areas is made of explosion-proof quality and should be shielded.

5.2.1.4 Utilities
Compressed air and any type of inert gases used in production should be of adequate quality and a procedure should be in place to evaluate the quality of such air/gas at a specified frequency. Any liquids and gases used in production must be clearly identified as to their content.

5.2.1.5 Water system
A water system is required for liquid preparations. At a minimum, it is recommended to use purified or distilled water complying with USP standards. The firm is advised to qualify the water system for its intended purpose. There must be adequate sanitization practices followed for the water in use. Additionally, a system should be in place for periodic microbiological and chemical analysis of the purified water produced in-house or if purchased from external source.

Premises used in the manufacturing of nutraceuticals should be totally segregated and prevented from utilizing the same facilities used for manufacturing of synthetic hormonal, cytotoxic, and mutagenic drugs; antibiotics; and special types of products that might lead to potential cross-contamination.

Security access in production and warehouse areas needs to be designated to authorized personnel only.

5.2.1.6 Pest control

The nutraceutical industry is proven to indulge in manufacturing of certain chemicals that could potentially attract insects from the surrounding environment. The firm is advised to have adequate measures in place to prevent infestation of pests, rodents, and insects into the production and operation affiliated areas. Pest control activities could be contracted to external service providers.

Documentation required:

- Minimum GMP documentation required includes floor plans indicating material, personnel flow, and room pressure differential
- Cleaning and sanitation records
- Pest control records
- Temperature and humidity records

5.2.2 Equipment [2,5]

There should be a list of processing equipment, including accessories that are in contact with the product [2]. The equipment used for production, packaging, and storage of a product should be designed to meet the required quality characteristics and should be located in areas that permit the installation, operation, cleaning, maintenance, and qualification activities.

The material considered for the design and construction of equipment and accessories that are in direct contact with components and in-process or finished products should not be reactive and additive or have any untoward impact on the product quality. Any substance required for the operation of equipment, such as lubricants, coolants, oils, etc., should not be in contact with the components of the formula, primary packaging of the product, or the product itself. These substances should be acquired on the basis of an approved specification. In the event of such substances coming into contact with the product, they must be at least of food grade quality and officially approved by the local health authorities.

Equipment that is unsuitable for its intended use should either be removed or clearly labeled for quarantine status.

It is recommended that the computer systems installed on computers to control the manufacturing process are validated.

5.2.2.1 Preventative maintenance

There should be preventive maintenance procedures for critical equipment used in operations. A written preventive maintenance program at an established frequency should be in place for critical equipment.

5.2.2.2 Equipment ID

All equipment used in production should have a unique identification numbering system.

5.2.2.3 Log books

Equipment usage logs should be maintained for critical process operations and should include product identification, date of operation, cleaning, and personnel involved with such activity.

5.2.2.4 Qualification of major equipment

The equipment should be qualified for the process and product to be manufactured. There should be a standard written procedure describing the qualification process of major equipment.

5.2.2.5 Calibration program of instruments

Any automatic, mechanical, and/or instrument used in the monitoring and control of critical process parameters should be calibrated and inspected according to a written program. The calibration and inspection operations should be documented and controlled. There should be an established calibration frequency, calibration method, approved limits for accuracy and precision, and identification of the equipment or instrument. In the event that the calibration of equipment and instruments are performed by an external service provider, it should comply with the provisions laid down by internal company policies. Standards used for calibration should be traceable to national and international standards.

5.2.2.6 Cleaning

Equipment and utensils should be cleaned and maintained in accordance with an established procedure and program. This should contain a description of cleaning methods, level of cleanliness required, and frequency of such cleaning activities. Records should be maintained.

5.2.3 Personnel [2,5]

The duties and responsibilities of personnel should be in writing and signed by each employee [2].

5.2.3.1 Qualification

The personnel responsible for the manufacture and control (testing) of products, including temporary staff, should be qualified for the function assigned to them. The qualification criteria should be based on experience, education, and training for the role.

5.2.3.2 Job responsibilities

There should be a written procedure in place describing the job accountability of the aligned job functions. A periodic review and evaluation of job performance should in place for all levels in production-related jobs.

5.2.3.3 Training

There should be a written program for training and continuous training of personnel in the functions assigned to them.

5.2.3.4 Health and hygiene

There should be a written program in place describing the health and hygiene requirements to be followed by personnel responsible for manufacturing and testing activities.

5.2.3.5 Organization chart with reporting structure

There should be a detailed and updated organizational chart that clearly identifies the units responsible for production and quality of the highest order and they should report to each other.

5.2.4 Quality assurance [2–5]

There should be a written document describing the system of quality management according to established policies and quality objectives. A quality unit should be responsible for approval of product specifications and standard operating procedures.

There should be a system and written procedures in place to ensure that all deviations or non-conformances to specifications, procedures, and methods of analysis are investigated, evaluated, and documented. Analytical results confirmed to be out of specification could be considered as deviations or non-conformances.

There should be a written procedure for handling product complaints.

There should be a written procedure outlining the process of batch record review and release/rejection of raw materials and finished products.

It is recommended to have a written procedure that includes identification, documentation, review, and approval of changes related to manufacturing processes and methods of analysis.

The quality unit is advised to have a system in place for selection, evaluation, and qualification of suppliers and vendors. There should be a system to ensure that all suppliers are assessed before being approved and included in the list of suppliers of raw materials and packaging materials utilized for manufacturing of finished products.

Documentation required:

1. Deviation logs
2. Change control logs
3. List of SOP (SOP index)
4. Master signature list

5. Trend reports
6. Risk assessments
7. Documentation control (batch records, investigations, analytical reports, methods, specifications SOPs, etc.)

5.2.5 Sanitation program [2,5]

The manufacturing establishment should have written procedures designed to ensure its safe and sanitary operation and maintenance. The cleaning procedures should include the cleaning and sanitizing requirements for the establishment and for all equipment and utensils in it. A list of all cleaning and sanitizing agents used in the cleaning process should exist, including their concentrations and uses. The cleaning intervals for manufacturing areas as well as for equipment should be established.

5.2.6 Operations [2,5]

There should be established procedures outlining the clear identification and separation by physical or control systems during manufacturing operations of multiple products in the same facility.

There should be adequate measures in the manufacturing area to prevent cross-contamination.

Access to the manufacturing areas should be restricted to authorized personnel only.

Operations should be carried out according to the production process and should be registered in controlled documents for each batch or lot of the product. The production process should have set parameters and established process controls that are required to ensure that the product remains within specification as previously established. The critical steps in manufacturing should be witnessed and records verified by a second individual.

There should be a defined time period for each critical stage of the production process. The finished product storage conditions and maximum storage period should be established for each product.

It is recommended to assign a product code to individual products manufactured in the facility. The production record should be numbered with a unique batch or identification number. Such records should be dated and signed during issuance. In the event of continuous production the product code together with the date and time could serve as the unique identifier.

Packaging operations should be performed with the materials specified in the corresponding packaging order and as per the instructions laid down in a specific packaging procedure.

5.2.7 Specifications [3,4]

Develop and implement written specifications for all raw and finished products [2]. Ensure specifications are maintained and every change is approved by the quality assurance person prior to use. Set up and follow written procedures that describe the tests to be conducted to ensure the identity and purity of finished products. If required, potency must be tested as per predefined, approved specifications.

Procedure should be in place to ensure each lot of raw and finished products are in compliance with specifications prior to release [3].

5.2.8 Stability [3,4]

Stability studies should be carried out in at least three pilot batches of the same product that have been manufactured by applying the same manufacturing method that simulates the process that will be used in the manufacture of commercial production lots.

Stability studies should be carried out in the same container-closure system or a representative system to the one proposed for its final storage and distribution.

The conditions of the stability study and its duration should be adequate to cover the storage, distribution, and usage of the product. The frequency of analysis of stability samples should be established for the product under study.

5.2.9 Samples [3,4]

There should be a written program in place describing the sampling practices followed for incoming materials and intermediate, in-process, and finished products. The sampling plan should comply with an international standard, e.g., military standard and ANSI standard. Based on a specified acceptable level the firm is advised to approve or reject a production lot.

5.2.10 Records [2,5]

There should be a written procedure established describing the good documentation practices that need to be followed during documentation of data records [2].

Data should be recorded by the person who performed the activity and once the activity is completed. Data recorded should be clear, legible, and ineradicable. Any correction should be made in such a manner that the original data is not blacked out. The corrected data should be signed and dated by the person making the correction. For electronic records, a system should be in place for tracking and for audit traceability to identify at least the changes made, date of change, and individual who made the changes.

There should be a standard written procedure for the retention period of documents at the firm. This includes production records, testing records, distribution records, and all other records that are deemed to be critical for the control of quality of the product.

5.2.11 Recall [2,5]

There should be written procedures that define controls to ensure the effective recall of a product, including notification to regulatory authorities, if applicable. The procedure should include the following items:

- Personnel responsible for initiating and coordinating recall activities
- Outlining the steps for implementing a recall
- Distribution records should be maintained for lot traceability
- Notification to regulatory agencies, if applicable

5.3 Conclusion

Although the nutraceutical industry is not fully regulated, as a health care provider, it is everyone's responsibility to protect the health interest of consumers by implementing cGMPs. This will certainly help not only to improve the level of compliance but also to avoid short falls and rejection rates within manufacturing processes, which subsequently will result in higher profitability. It has been observed that recent recalls in the food industry most likely happen because of not following good personal sanitization or good cleaning practices. The nutraceutical industry should use the vast knowledge base of the pharmaceutical industry to enhance consumer protection as well as industry growth. In this chapter a brief description of each key area within the cGMPs was given to keep the balance between cost and manufacturing processes associated to implement cGMP as well.

References

[1] Good manufacturing practices, Guidance document, Natural Health Products Directorate, Health Canada; 2006.
[2] Good Manufacturing Practice (GMP) Guidelines/Inspection Checklist, February 12, 1997; Updated April 24, 2008 (http://www.fda.gov/Cosmetics/GuidanceComplianceRegulatory Information/GoodManufacturingPracticeGMPGuidelinesInspectionChecklist/default.htm).
[3] Compendium of Food Additive Specifications, 74th meeting, WHO.
[4] Food Chemicals Codex (http://online.foodchemicalscodex.org/online/login).
[5] General references: Good Manufacturing Practices (GMP) Guidelines - 2009 Edition, Version (GUI-0001), Health Canada and 21 CFR part 210 and 211.

Importance of Safety Assessment

Breaking Down the Barriers to Functional Foods, Supplements and Claims[1]

6

George A. Burdock* and **Ioana G. Carabin**[†]

Burdock Group, Orlando, Florida [†]Island ENT, Key West, Florida

6.1 Introduction

Historically, food was something we ate as the result of its having nutritional, hedonic, or satiating value (Table 6.1). Today, however, most consumers recognize additional categories of foods, including the added-value categories of "functional foods" and "dietary supplements,"[2] which are located somewhere along a continuum of substances that have more benefit than simple foods and probably thought of as generally equivalent to "officially recognized" vitamins and minerals (e.g., those with a Recommended Daily Intake). The consumer's perceived value of a functional food or supplement also likely bears directly on the consumer's self-image. That is, a hypothetical functional food/supplement consumer might see himself (or herself) as normally performing within a range of 85–95% of his total ability (that is, he is at 85–95% efficiency); this is a range in which he feels in equilibrium with his surroundings. Below about 80% efficiency, there is no longer the "good feeling" about oneself (i.e., no longer in equilibrium), and even lower, the status of feeling "sick," prompting a need for some sort of therapeutic intervention (i.e., a drug). The middle of this range, ∼90%, is the best the consumer could expect to feel, given healthy lifestyle habits and a wide variety of acceptable foods. However, with the stress and pressure of work, family, and social commitments, a limited food selection of less healthy choices, as well as restrictions on physical activity, maintaining a goal of 90% efficiency becomes nearly impossible. However, the ready availability of functional foods and supplements to the consumer puts the goal of 90% efficiency within reach.

There are likely a variety of other reasons for consuming these added-value substances and may include a concern about the high cost of prescription drugs, a quest

---

[1]Portions of this article were previously published by the same authors as "The Importance of GRAS to the Functional Food and Nutriceutical Industries" in *Toxicology* 221: 12–27, 2006.
[2]Hereinafter as often referred to simply as "supplements" and possibly by others as "nutraceuticals."

Nutraceutical and Functional Food Regulations in the United States and Around the World.
DOI: http://dx.doi.org/10.1016/B978-0-12-405870-5.00006-2

Table 6.1 Evolution of Health Care	
2000 BC	Here, eat this root.
850 AD	That root is heathen; here, say this prayer.
1000 AD	That prayer is superstition; here, drink this potion.
1940 AD	That potion is snake oil; here, swallow this pill.
1965 AD	That pill is ineffective; here, take this antibiotic.
2014 AD	That antibiotic does not work anymore; here, eat this root.
Adapted from [12].	

for more natural remedies, or simply engaging in preventative measures (e.g., fiber, vitamins, and minerals), but there is no doubt that consumption of functional food and supplements are on the rise. According to a press release of the respected Leatherhead Food Research organization, the value of the worldwide functional foods market (defined as food and drink products making a specific health-type claim) was $24.2 billion in 2010.[3] According to Partnership Capital Growth, U.S. consumer sales of functional foods and beverages reached $30.4 billion at the wholesale level in 2009 and overall sales of supplements in the United States reached $26.9 billion at the retail level in 2009.[4] Therefore, by any measure, the functional food/supplement market is nothing short of enormous.

6.2 **Terminology**

To food formulators and marketers, the designations of functional foods or supplements have a slightly different meaning—as those foods that contain some health-promoting component(s) beyond traditional nutrients. To the Food and Drug Administration (FDA), terms like functional foods or nutraceuticals have no meaning in regulation or the law and, in fact, are regarded as "fanciful" (Table 6.2). This seeming inflexible posture by the FDA (i.e., not being seen as current with popular terminology) is the result of the mandate to the Agency from Congress to protect the public from the "snake oil" salesmen and "health crusaders" of the early 20th century, whose claims for various foods and quack remedies seem at best, silly to us today, and at worst, dangerous to the consumer [1].[5] That is, for foods, the FDA must ensure that the substance does not touch upon the definition of a drug (so

---

[3]http://www.leatherheadfood.com/long-may-the-growth-in-functional-foods-continue (site accessed 25May13).
[4]http://www.pcg-advisors.com/marketstatistics (site accessed 29May13).
[5]See also, "The American Chamber of Horrors." http://www.fda.gov/AboutFDA/WhatWeDo/History/ProductRegulation/ucm132791.htm (site accessed 25May13).

Table 6.2 Examples of "Regulatory" versus "Fanciful"[a] Terms

Regulatory Terms	Fanciful Terms
Drug	Nutraceutical
Food ingredient (FA[b] or GRAS[c])	Aquaceutical
Food for special dietary use	Herbal supplement
Medical food	Natural food
Nutrient supplement[d]	Cosmeceutical
Dietary supplement	Functional food
Cosmetic	
Claim	
Nutrient claim	
Dietary guidance claim	
Health claim (SSA)	
Qualified Health Claims (QHC)	
Structure Function Claims (SFC)	

[a]"...an arbitrary or fanciful name which is not false or misleading in any particular" (21 CFR 133.148(e)(2)).
[b]FA, food Additive.
[c]GRAS (FFDCA §201(s)).
[d]21 CFR 170.3(o)(20) [substances] "...that are necessary for the body's nutritional and metabolic processes."

there is no potential to defraud the consumer), according to the definition of a drug in the Federal Food Drug and Cosmetic Act (FFDCA):

> ...*articles intended for use in the diagnosis, cure, mitigation, treatment, or prevention of disease in man or other animals [or] articles* (**other than food**) *intended to affect the structure or any function of the body of man or other animals...*(*§201(g)(1)(B) FFDCA*) (***emphasis added***).

Sometimes, however, when an article of food is a nutrient, it can act like a drug when, for instance, it prevents a disease (such as scurvy), cures or prevents malnutrition or starvation, or treats a disease such as hypovitaminosis; therefore, the "carve out" exempting food from regulation as a drug was provided by Congress (i.e., "other than food"). Regulations govern the designation of nutrients as all must be approved as such by the FDA.[6] Likewise, in the definition of a drug (FFDCA §201(g)), there is a provision that a substance is not necessarily a drug simply because the label contains a health claim—this provision provides flexibility for certain foods, which may contain a health claim. Claims permitted for food are shown in Table 6.3.

---

[6]See 21 U.S.C. 343(r)(4): that is, according to the Nutrition Labeling and Education Act (NLEA, 1990), a nutrient must have an established amount that exerts a beneficial effect, such as an RDI. However, in as much as the food components being studied today for their beneficial effects such as fiber or broccoli with high amounts of sulforanes, are not (yet) recognized "officially" as beneficial and therefore may not meet the NLEA standard for a nutrient.

Table 6.3 Five Types of Health-Related Statements or Claims are Allowed on Food and Dietary Supplement Labels

1. Nutrient content claims indicate the presence of a specific nutrient at a certain level.
2. Structure and function claims describing the effect of dietary components on the normal structure or function of the body.
3. Dietary guidance claims describing the health benefits of broad categories of foods.
4. QHCs conveying a developing relationship between components in the diet and risk of disease, as approved by the FDA and supported by the weight of credible scientific evidence available.
5. Health claims confirming a relationship between components in the diet and risk of disease or health condition, as approved by the FDA and supported by significant scientific agreement.

There are as well, "foods for special dietary use" (21 CFR Part 105) and "medical foods"[7]; however, there are no efficacy standards for either of these categories, nor is there a requirement for pre-market approval by the FDA, at least for the latter if determined as being generally recognized as safe (GRAS).

Recently, the FDA has given voice to a possible new group of "bioactive" food ingredients [2]. Examples of this new category from the GRAS notification list include: vegetable oil sterol esters, phytostanol esters, lactoferrin, fructo-oligosaccharides, fish oils (including tuna oil), diacylglycerol, and inulin [2]. Another example of a bioactive food ingredient that received food additive approval several years ago is Olestra, a cooking oil that is not absorbed during the digestive process and therefore does not contribute to caloric intake. Interestingly, Olestra is not a "Food for Special Dietary Use" (21 CFR 105.3), for which it would seem to be eminently qualified (at least as an anti-obesity agent), nor does it have a health claim for which it would also seem to be qualified. By the same reasoning, margarine could be considered to contribute a beneficial effect, as a non-cholesterol-containing butter substitute, as would margarine substitutes that do not contain trans fats. The FDA certainly has had ample opportunities to become creative in its interpretation and application of claims for functional foods, but the industry has failed to take up the issue in a forceful manner.

6.3 The struggle: players and issues

At this point, each of the stakeholders with an interest in functional foods and supplements (i.e., consumers, industry, and the FDA) have taken positions

[7]Medical foods are regarded as foods for the maintenance of a patient as opposed to a treatment. Further, medical foods are not those foods included within a healthy diet intended to decrease the risk of disease, such as reduced-fat foods or low-sodium foods, nor are they weight loss products (http://www.cfsan.lfda.gov/~dms/ds-medfd.html), (site accessed 25 May 13).

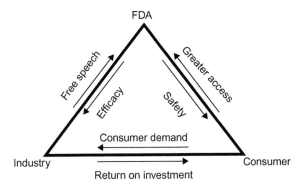

FIGURE 6.1

Functional food players and the impasse.

creating a tension with the other two and a stalemate has resulted (Figure 6.1). Consumers are demanding greater access to products for which they see a benefit and industry is demanding that its right of free speech not be abridged by the FDA. The push-back by the FDA is fulfillment of its mandate to protect the consumer, by ensuring the product is safe and by protecting the consumer from fraud as the result of false or misleading claims on the label.[8] How did we get to this stalemate and how can it be resolved?

6.3.1 Ontogeny and current status of the regulations

A generation or two ago, an aging population accepted getting older and retiring to less active pursuits, thereby ceding center stage to the next generation, but in the present youth-oriented culture, the current aging generation (the "Baby Boomers") are less willing to give in to the aging process and are actively seeking some palliative treatment. In addition, the sheer number of aging Americans (21% of the population will be "senior citizens" by 2050),[9] presents a forbidding burden to the health care system, which has historically been targeted to treatment of disease, rather than prevention. The answer to both forces is to engage in prevention through healthier lifestyles and dietary choices before the inevitable onset of age-associated disease "…let food be thy medicine…" (Hippocrates, 460–377 BC).

---

[8]The Federal Trade Commission (FTC) and FDA have complementary jurisdiction over the marketing of products, including functional foods and dietary supplements. Under the terms of a 1971 liaison agreement (MOU 225-71-8003), the FTC has primary responsibility for regulating advertising and the FDA has primary responsibility for labeling. (http://www.fda.gov/AboutFDA/PartnershipsCollaborations/MemorandaofUnderstandingMOUs/DomesticMOUs/ucm115791.htm) (site accessed 26 May 13).

[9]http://knowledgetoday.wharton.upenn.edu/2013/03/can-the-u-s-meet-its-aging-populations-health-care-needs/ (site accessed 26 May 13).

In the 1980s, the Japanese Ministry of Health, Labor, and Welfare (MHLW) recognized the gravity of the problem of an aging population and the lack of focus on preventative measures though dietary control and responded with Foods for Specified Health Uses (FOSHU).

FOSHU refers to foods containing [an] ingredient with functions for health and officially approved to claim its physiological effects on the human body. FOSHU is intended to be consumed for the maintenance/promotion of health or special health uses by people who wish to control health conditions, including blood pressure or blood cholesterol. In order to sell a food as FOSHU, the assessment for the safety of the food and effectiveness of the functions for health is required, and the claim must be approved by the MHLW.[10]

Similarly intended legislation was passed by Congress with the NLEA in 1990 (allowing health claims for foods), which was followed, in 1997, by the FDA Modernization Act (FDAMA; an attempt by Congress to liberalize the FDA's narrow interpretation of the NLEA), and finally, the Dietary Supplement Health and Education Act (DSHEA) in 1994. Each amendment to the FFDCA permitted certain types of claims. As public interest increased and distinctions began to be made between products, the terms *functional foods* and *nutraceuticals* were coined. As indicated earlier, while neither term is supported by regulation or the law, each has a certain cachet that supports a persuasive concept.

6.3.2 Health claims and NLEA and FDAMA

The NLEA (1990), which permitted *health claims* for food, had long been resisted by the FDA. The agency protested that any health claim would render the product an unapproved drug and would up-end the hard-fought victories since the turn of the century against false and misleading claims by charlatans. Further, the agency complained that the wording of the 1938 Amendment to the FFDCA tied the hands of the agency by prohibiting health claims. However, the agency faced considerable pressure to change its attitude, including in 1984, with the Kellogg's All-Bran claim to reduce cancer risk. In great measure, the proposed NLEA had the potential to take pressure off the FDA and allow a legitimate pathway for claims for food, but the agency was intransigent. Despite agency protest, the NLEA was passed and the agency was faced with a law it never wanted and was slow to implement it, not finalizing the regulation until 3 years after passage of the law.

Since implementation of the law, debate has centered on the proof of efficacy required for a health claim to meet the standard of the *Significant Scientific Agreement* (SSA), the precise definition of which is not included in the law, but a concept inherent with the FDA as part of a continuum of scientific discovery that passes, in terms of agreement, through the following process [3] (Figure 6.2). Importantly, and like drugs, only a government agency can determine whether the

---

[10]http://www.mhlw.go.jp/english/topics/foodsafety/fhc/02.html (site accessed 26 May 13).

FIGURE 6.2

The continuum of scientific discovery.

standard was met, but the FDA retained final say in the matter of what met the SSA standard. Further, a health claim requires promulgation of a regulation by the FDA (including, at least theoretically, a proposed rule, comment period, and a final rule). In January of 1993, the FDA adopted final regulations implementing NLEA health claims, but never established specific criteria for SSA, stating only that it would make case-by-case determinations. The FDA also indicated that it would not permit disease (i.e., health) claims for which a difference of scientific opinion exists.[11] Because this standard was simply set too high, Congress passed FDAMA in 1997 as an attempt to liberalize the interpretation the FDA had mandated for health claims, by allowing the claims to be based on "authoritative statements" of qualified scientific bodies much like judgments of safety could be made by expert panels for GRAS substances. In response, the FDA met Congress half-way by allowing the National Academy of Science or National Institutes of Health-type statements such as those describing the benefits of fluoride in drinking water, low trans fats in heart disease, or the presence of choline in a particular food. However, these are not "health claims" per se, but "nutrient content claims," which the FDA calls "FDAMA claims." In as much as the FDA has ceded territory for these FDAMA claims, the agency still maintained control of the process, indicating that the FDA must be notified of the claim and, if the FDA does not act to prohibit or modify the claim, it may be used 120 days after receipt of the notification by the agency.[12]

In December of 1999, the FDA made the pivotal decision that there must be significant scientific agreement about the substance/disease relationship *rather than* the actual claim being made. Although this guidance was subsequently withdrawn,[13] it was supplanted by guidance issued in January of 2009,[14] which refers to 21 CFR 101.14 (Health claims: general requirements) and specifically:

§*(b) Eligibility (1) The substance must be associated with a disease or health-related condition for which the general U.S. population, or an identified U.S.*

---

[11]Federal Register 58:2501-02, 1993.

[12]Guidance for Industry: Notification of a Health Claim or Nutrient Content Claim Based on an Authoritative Statement of a Scientific Body (June 11, 1998). (http://www.fda.gov/Food/GuidanceRegulation/GuidanceDocumentsRegulatoryInformation/LabelingNutrition/ucm056975.htm) (site accessed 26 May 13).

[13]Federal Register 74:3060, 2009, January, 16, 2009.

[14]Guidance for Industry: Evidence-Based Review System for the Scientific Evaluation of Health Claims—Final (January 2009). http://www.fda.gov/Food/GuidanceRegulation/GuidanceDocumentsRegulatoryInformation/LabelingNutrition/ucm073332.htm# (site accessed 26 May 13).

population subgroup (e.g., the elderly) is at risk, or, alternatively, the petition submitted by the proponent of the claim otherwise explains the prevalence of the disease or health-related condition in the U.S. population and the relevance of the claim in the context of the total daily diet and satisfies the other requirements of this section.

This decision pushed beyond what was required in the law (FFDCA 403(r)(3)(B)(i)), which required only significant scientific agreement about the claim, not a linkage between the substance and the disease:

The Secretary shall promulgate regulations authorizing claims of the type described in subparagraph (1)(B) only if the Secretary determines, based on the totality of publicly available scientific evidence (including evidence from well-designed studies conducted in a manner which is consistent with generally recognized scientific procedures and principles), that there is significant scientific agreement, among experts qualified by scientific training and experience to evaluate such claims, that the claim is supported by such evidence.

The agency's interpretation has had a chilling effect by forcing petitioners perilously close to the edge of a drug claim, leaving only a narrow gap for claims for reduced incidence of disease in otherwise healthy populations. For example, because a claim based on reducing the symptoms of arthritis in arthritis sufferers became a drug claim, manufacturers were forced to test the ability of the substance to decrease the incidence of arthritis in the general population—significantly, the population could not already have arthritis. This requirement to test in a general population resulted in a quantum leap in the cost of proving the claim, because it required a very large sample size, a much longer study duration, and complex subject exclusion criteria [4,5]. Further, the FDA continues to cling to the *weight of evidence* concept, contrary to the urging from the Courts that *credible evidence* be permitted as the basis of a claim. As illustrated in Figure 6.3, the chances are slim for achieving SSA based on anything other than the "gold standard" of an interventional, randomized, controlled clinical trial.[15]

Although there are now 12 health claims, claims languished at the Agency until 1993 when the first claim (calcium and osteoporosis) was approved (Table 6.4) and notably, no new health claims have been approved in over a decade (although some of the original health claims have been amended). It would be fair to say that none of these health claims were "tough calls" and, for many, the claims were obviously valid even before new supporting data were generated in response to the requirement for SSA.

In comparison, the Japanese FOSHU list of foods or ingredients is quite lengthy (Table 6.5).

---

[15]Note: Inasmuch as the guideline in which this illustration appears has been withdrawn, it remains a succinct description of the thought process by which the agency ranks studies.

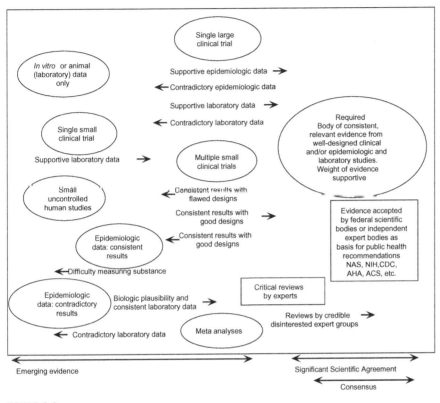

FIGURE 6.3

Schematic for assessing strength and consistency of scientific evidence leading to significant scientific agreement.

6.3.3 Qualified health claims

As noted previously, the award of a health claim regulation is rarely, or at best, slowly, obtained by petitioners. This is because the very agency that had resisted passage of the NLEA health claim process was made the gatekeeper for approvals of the very claims it had resisted making before NELA was passed. The bar for SSA status was simply set too high and, in response, a series of lawsuits was launched against the FDA (i.e., *Pearson v. Shalala* and others) asserting the right of commercial free speech on food labels. As the result of decisions by the court, the FDA was forced to provide a mechanism accommodating free speech. In response, the FDA unveiled the Task Force Final Report: Consumer Health Information for Better Nutrition Initiative in July 2003, allowing for Qualified Health Claims (QHCs), which were theoretically something less than a significant agreement or general agreement (see Figure 6.2) under the rubric of "emerging

Table 6.4 Significant Scientific Agreement	
21 CFR	**Substance and Disease or Health-related Condition**
101.72	Calcium and osteoporosis
101.73	Dietary lipids and cancer
101.74	Sodium and hypertension
101.75	Dietary saturated fat and cholesterol and risk of coronary heart disease
101.76	Fiber-containing grain products, fruits and vegetables, and cancer
101.77	Fruits, vegetables, and grain products that contain fiber, particularly soluble fiber, and risk of coronary heart disease
101.78	Fruits and vegetables and cancer
101.79	Folate and neural tube defects
101.80	Dietary non-cariogenic carbohydrate sweeteners and dental caries
101.81	Soluble fiber from certain foods and risk of coronary heart disease
101.82	Soy protein and risk of coronary heart disease
101.83	Plant sterol/stanol esters and risk of coronary heart disease

evidence." For QHC, the FDA has chosen not to officially approve any claims, but instead to "exercise enforcement discretion" relative to claim language the FDA has mandated using its qualifications.

At present, there are six categories for which the agency has allowed claims, but the claims permitted have been worded by the FDA to accommodate its estimate of the strength of the scientific evidence supporting the claims (Table 6.6). Needless to say, in an agency as conservative as the FDA, the claims (and accompanying FDA disclaimer) are sometimes only those from which a competitor could benefit, as exemplified by the claim for green tea and cancer:

> *Drinking green tea may reduce the risk of breast or prostate cancer. FDA does not agree that green tea may reduce that risk because there is very little scientific evidence for the claim.*[16]

The FDA's grant of permission to use claims, such as for green tea (above), does not provide much encouragement to manufacturers to generate the quality and quantity of data required to convince the agency of the validity of a claim. At this point, the standard will continue to be set by Health Claims.

In as much as the FDA strains to protect consumers from the fraud of "false and misleading" claims, the agency is also guilty of misleading consumers. That

---

[16]Letter Responding to Health Claim Petition dated January 27, 2004: Green Tea and Reduced Risk of Cancer Health Claim (Docket number FDA-2004-Q-0427) dated February 24, 2011. (http://www.fda.gov/Food/IngredientsPackagingLabeling/LabelingNutrition/ucm072774.htm) (site accessed 27 May 13).

Table 6.5 FOSHU

Health Claim and Functional Ingredients	Number of Products Approved	Type of Products in the Market
Food that improves gastrointestinal conditions		
Prebiotics: oligosaccharides, raffinose, lactulose, arabinose Probiotics: lactobacillus, bifidobacterium dietary fiber	336	Soft drinks, yogurt, cookies, table sugar, soy bean curd, vinegar, chocolate, powdered soup, fermented milk, miso soup, cereal
Foods for those with high serum cholesterol		
Soy protein and peptide, alginate, chitosan, sitosterol ester	20	Soft drinks, meat balls, sausage, soy milk, soup, cookies, margarine
Foods for those with high blood pressure		
Peptides	42	Soft drinks, soup, drinkable products containing probiotic bacteria, soybean products
Foods for those with high serum triacylglycerol		
Diacylglycerol and sitosterol	9	Cooking oil
Foods related to mineral absorption and transport		
Casein, calcium citrate, isoflavone	17	Soft drinks, fermented soybean (natto), jelly
Non-cariogenic foods		
Mannitol, xylitol, palatinose	6	Chocolate, chewing gum
Foods for hyperglycemics		
Wheat albumin, globulin digest, polyphenols	4	Candy, soup, soft drinks

is, consumer survey information [6] has demonstrated that the use of a letter grade for claims, embedded in disclaimers or point—counterpoint statements (as seen in the green tea claim) are often interpreted wrongly by the public. Three categories of problems have emerged:

1. Some consumers felt that the claim reflected the *quality* of the product, rather than the strength of the evidence supporting the claim.
2. Anything less than a "B" grade was unacceptable.
3. The statements had the paradoxical effect such that structure/function claims (SFCs; which do not require pre-market approval) were looked upon more favorably than QHCs, presumably because of fewer "weasel-words" and the outright distancing by the agency of an approved claim (see the second sentence in the green tea claim above).

Table 6.6 QHCs Subject to Enforcement Discretion[a]

Atopic dermatitis
Whey-protein partially hydrolyzed infant formula and reduced risk of atopic dermatitis
Cancer risk
Tomatoes and/or tomato sauce and prostate, ovarian, gastric, and pancreatic cancers Calcium and colon/rectal cancer and calcium and recurrent colon/rectal polyps Green tea and cancer Selenium and cancer Antioxidant vitamins and cancer
Cardiovascular disease risk
Nuts and heart disease Walnuts and heart disease Omega-3 fatty acids and coronary heart disease B vitamins and vascular disease Monounsaturated fatty acids from olive oil and coronary heart disease
Cognitive function
Phosphatidylserine and cognitive dysfunction and dementia
Diabetes
Chromium picolinate and diabetes
Hypertension
Calcium and hypertension, pregnancy-induced hypertension, and preeclampsia
Neural tube birth defects
Folic acid (0.8 mg) and neutral tube birth defects
[a]*http://www.fda.gov/Food/IngredientsPackagingLabeling/LabelingNutrition/ucm072756.htm*(site accessed 28May13).

6.4 DSHEA: SFCs and a new safety standard

6.4.1 A closer look at the definition

On a similar front, in the early 1980s, the use of dietary supplements began to proliferate, quickly followed by FDA attempts to curtail them through several measures, including a challenge that dietary supplements were actually unapproved food additives [7]. Congress responded again to the public outcry with the passage of DSHEA, which declared dietary supplements not to be food additives, but foods, for which SFCs could be made. The rationale by which dietary supplements were actually foods is evident from the list of examples provided by Congress (Table 6.7).

Table 6.7 Definition of a Dietary Supplement (FFDCA)

Chapter II: definitions

SEC. 201 For the purposes of this chapter

(ff) The term "dietary supplement"

(1) means a product (other than tobacco) intended to supplement the diet that bears or contains one or more of the following dietary ingredients:

(A) a vitamin; (B) a mineral; (C) an herb or other botanical; (D) an amino acid; (E) a dietary substance for use by man to supplement the diet by increasing the total dietary intake; or (F) a concentrate, metabolite, constituent, extract, or combination of any ingredient described in clause (A), (B), (C), (D), or (E)

Clearly, this list is one of substances that either should be or could be, a legitimate addition to the diet. Particularly insightful is the phrase "a dietary substance for use by man to supplement the diet by increasing the total dietary intake"; this confirms that Congress viewed supplements as substances that should or could have been part of the diet in the first place but, importantly, did not require the substances to have established nutritive value. Implied by its omission is the requirement that the supplement must be extracted from a "food"; that is, the supplement could be produced by synthetic means (e.g., the concept of "natural identical" at one time used in past European regulations to designate synthetic congeners of natural substances). However, the agency clearly interprets the law differently, as stated in its July 2011 Draft Guidance for Industry, in response to the query, "Is a synthetic copy of a constituent or extract of an herb or other botanical a dietary ingredient?":

> *No. A synthetic copy of a constituent of a botanical was never part of the botanical and thus cannot be a "constituent" of the botanical that qualifies as a dietary ingredient under section 201(ff)(1)(F) of the FD&C Act (21 U.S.C. 321(ff)(1)(F)). Similarly, a synthetic version of a botanical extract is not an "extract" of a botanical under section 201(ff)(1)(F) because it was not actually extracted from the botanical.[17]*

The basis for this decision that synthetic copies cannot be considered as dietary supplements was first expressed in the final rule on ephedrine alkaloids[18]:

> *...Furthermore, ephedrine hydrochloride and other synthetic sources of ephedrine cannot be dietary ingredients because they are not constituents or extracts*

---

[17]Draft Guidance for Industry: Dietary Supplements: New Dietary Ingredient Notifications and Related Issues, July 2011. (http://www.fda.gov/Food/GuidanceRegulation/GuidanceDocumentsRegulatoryInformation/DietarySupplements/ucm257563.htm) (site accessed 27 May 13).

[18]Federal Register 69:6788- 6854, February 11, 2004. Final rule declaring dietary supplements containing ephedrine alkaloids adulterated because they present an unreasonable risk.

Table 6.8 Supplement Claims

SEC. 403. [343] (r)(6) For purposes of paragraph (r)(1)(B), a statement for a dietary supplement may be made if:

- "the statement claims a benefit related to a classical nutrient deficiency disease and discloses the prevalence of such disease in the United States, describes the role of a nutrient or dietary ingredient intended to affect the structure or function in humans,
- [the statement] characterizes the documented mechanism by which a nutrient or dietary ingredient acts to maintain such structure or function [of the body], or
- [the statement] describes general well-being from consumption of a nutrient or dietary ingredient."
- [However, these claims are allowed only if]:
- "(B) the manufacturer of the dietary supplement has substantiation that such statement is truthful and not misleading, and
- (C) the statement contains, prominently displayed and in boldface type, the following: "'This statement has not been evaluated by the Food and Drug Administration. This product is not intended to diagnose, treat, cure, or prevent any disease.'"
- "... the manufacturer shall notify the Secretary no later than 30 days after the first marketing of the dietary supplement with such statement that such a statement is being made."

---

of a botanical, nor do they qualify as any other type of dietary ingredient. For these reasons, products containing synthetic ephedrine cannot be legally marketed as dietary supplements (See section 201(ff)(1) and 201(ff)(3)(B) of the act (21 U.S.C. 321(ff)(1) and (ff)(3)(B)))....

Undoubtedly, the issue of synthesis of supplements has yet to be resolved. Last, the supposition that the dietary supplement should or could have been part of the diet is reflected in the claims allowed. That is, DSHEA allows three types of "statements of nutritional support" on labels without obtaining the FDA's approval (Table 6.8), with the caveats, however, that (1) the statement is truthful, not misleading, and can be documented; (2) a statement follows relating to the fact that the FDA does not endorse the claim; and (3) that the substance is not a drug (according to the definition of a drug).[19] That is, only SFCs may be made. These claims relate only to the *maintenance* of a physiologic function or structure (e.g., healthy bones or healthy urinary tract function), which falls short of a *nutrient* claim. DSHEA also requires that "the statement be truthful and not misleading."

Historically, foods per se were permitted to bear SFCs, but such claims were made on the basis of the nutrients they contained and for which a daily requirement had been established, such as a recommended daily allowance [8]; for

---

[19]This statement avoids the conundrum of "when is a food a drug" (e.g., cure disease (such as starvation), mitigate or prevent a disease (such as a vitamin deficiency), and further, that, "[a] food, dietary ingredient, or dietary supplement for which a truthful and not misleading statement is made in accordance with section 403(r)(6) of this title is not a drug ... solely because the label or the labeling contains such a statement."

example, calcium in milk or potassium in orange juice. However, in an FDA Guidance Letter to industry dated January 2007, the FDA indicated a willingness to extend the use of SFCs to foods.[20] That is, the agency expressed a willingness to go beyond the permitted "calcium builds strong bones" type of claim and allow SFCs that characterize the means by which substances act to maintain such structure or function; for example, "fiber maintains bowel regularity" or the claim may describe general well-being from consumption of a nutrient or dietary ingredient. In this notice, the FDA also indicated it would allow SFCs to describe a benefit related to a nutrient deficiency disease (like vitamin C and scurvy), "as long as the statement also tells how widespread such disease is in the U.S....[and] such claims may not explicitly or implicitly link the relationship to a disease or health-related condition." Further, the FDA insists that "the claims must derive from the nutritional value[21] of the product,"[22] (albeit the fact that the few, even in the agency, would admit that fiber is added for nutritional value). Howsoever, this statement, if read in the context of 21 CFR 172.5 (i.e., "The quantity of the substance added to food does not exceed the amount reasonably required to accomplish its intended physical, nutritive, or other technical effect in food") will likely be construed such that the SFC must be achieved at the level at which provides the original effect (of taste, aroma, or nutritive value; or any of the technical effects recited in 21 CFR 170.3(o)). For example, if a flavor ingredient provided the needed flavor at 10 ppm, but supports a structure–function effect at a minimum of 50 ppm, but no change in flavor, is the manufacturer limited to only 10 ppm? Therefore, if an effective dose is higher, clearly, there will have to be a review of safety at this higher level (i.e., GRAS or food additive petition), but the manufacturer is still without an "approved" effect.

6.4.2 The exclusionary clause and section 912

Of particular interest to dietary supplement manufacturers should be the *exclusionary clause* of DHSEA (Table 6.9). This clause simply indicates that a dietary supplement cannot have had prior use as a drug or have been under investigation as a drug with clinical studies having been instituted. This provision theoretically prohibits the redirection of substances once intended for use as drugs to the dietary

---

[20]Guidance for Industry and FDA: Dear Manufacturer Letter Regarding Food Labeling (January 2007) (http://www.fda.gov/Food/GuidanceRegulation/GuidanceDocumentsRegulatoryInformation/LabelingNutrition/ucm053425.htm) (site accessed 27 May 13).
[21]Presumably, "nutritional value" includes "taste" and "aroma" as per *Nutrilab v. Schweiker*, 713 F.2d 335 (CA 7, 1983); although according to 21 CFR 101.14(b)(3), a substance identified as the basis of a health claim, may be any of these three or "or any other technical effect listed in §170.3(o) of this chapter, to the food and must retain that attribute when consumed at levels that are necessary to justify a claim." Because Hutt [10] asserts the limitation that an SFC must directly relate to the nutritive value of the food has doubtful enforceability, the FDA will likely extend this inclusion of technical effects to SFCs as well.
[22]Nutritive value means a value in sustaining human existence by such processes as promoting growth, replacing loss of essential nutrients, or providing energy (21 CFR 101.14(a)(3)).

> **Table 6.9** Exclusionary Clause (FFDCA 201(ff)(3)(B))
>
> [The definition of a dietary supplement does] not include
>
> "(i) an article that is approved as a new drug, . . .certified as an antibiotic. . . or licensed as a biologic . . . or (ii) an article authorized for investigation as a new drug, antibiotic, or biological for which substantial clinical investigations have been instituted and for which the existence of such investigations has been made public, which was not before such approval, certification, licensing, or authorization marketed as a dietary supplement or as a food. . ."

supplement market or the marketing of dietary supplement version of a drug (for example, as a less potent form, such as might be used for an over-the-counter drug). A case in point was the marketing of red yeast rice (Cholestin); although having been used as a food in China for centuries and in the United States for decades, it was known to contain HMG-CoA reductase inhibitors (which lower blood cholesterol). These inhibitors, although natural to the food, are indistinguishable from the synthetic lovastatin, the active ingredient in the prescription drug Mevacor by Merck. Based on this similarity and the fact that the manufacturer had tweaked the manufacturing process to maximize the level of HMG-CoA reductase inhibitors, Cholestin was determined to be a "drug" by the FDA and could not be marketed as a dietary supplement [9]. Therefore, there are three take-home lessons:

1. A manufacturer should not use a process designed to heighten the concentration of a constituent.
2. There should be no promotion of a particular constituent of a supplement, unless that constituent was previously marketed as a dietary supplement.
3. The mere presence of a substance in the food supply may not be sufficient to meet the standard as having been previously marketed.

In a case similar to the Cholestin/lovastatin case, a citizen petition to the FDA by BioStratum, Inc. states, among other things, that it is the manufacturer of Pyridorin (pyridoxamine dihydrochloride), which is the subject of an investigational new drug (IND) that was filed with the FDA in July 1999, and that Pyridorin was being tested for use as a potential therapeutic agent to slow or prevent the progression of diabetic nephropathy in patients with type 1 and 2 diabetes. The petition further states that substantial clinical trials have been conducted for this drug and that the existence of those studies has been made public. In addition, the petition states that pyridoxamine was not marketed as a dietary supplement or as a food prior to Pyridorin's authorization for investigation as a new drug under an IND.[23] The FDA is awaiting comments regarding the prior use of

---

[23]Federal Register: November 18, 2005 (Volume 70, Number 222), page 69976−69977, Request for Comment on the Status of Pyridoxamine. Letter from Michael A. Chappell, Acting Associate Commissioner of Regulatory Affairs, FDA, to Kathleen M. Sanzo, Morgan, Lewis & Bockius LLP, responding to Citizen Petition 2005P-0259 from BioStratum, Inc. (Jan. 12, 2009). Docket No. FDA-2005-P-0259 [Document ID: FDA-2005-P-0259-0004].

this substance as a dietary supplement before it makes a final decision. An ironic twist of fate may result if the FDA determines that pyridoxamine is, in fact, covered by the exclusionary clause, and if BioStratum, Inc. decides not to market Pyridorin; the end result may be a loss to the consumer because pyridoxamine would be forever excluded from being marketed as a dietary supplement.

A similar situation exists for food ingredients, a product of §912 of the FDA Additives Amendment of 2007. At first glance, §912 appears to be no more than a formalization of a "rule of thumb" enforced in food approvals for many years; that is, the use of a drug in a food (or a cosmetic, for that matter) was considered a trivial use and prohibited. For example, while addition of antibiotics to meat or underarm deodorants may provide for longer shelf-life in the former and greater efficacy in the latter, the addition of the antibiotic was considered a trivial use and may contribute to antibiotic resistance. In addition to antibiotics, other drugs are forbidden as well, because the intention of food is to nourish, not medicate, the consumer. There are always exceptions to the rule, and an exception here is a drug called Cytellin, the primary constituent of which was β-sitosterol, which blocks cholesterol absorption[24] and was marketed from 1954 to 1982. β-Sitosterol, along with other phytosterols, are now added to margarine and marketed for essentially the same purpose.

A second look at the exact language of §912 reveals some wiggle room for interpretation by the FDA. That is, while there is a reference to *approved* drugs and biological products (D&BP), there is also an "or," which indicates the law is also referring to D&BPs that have not been approved (i.e., approved by the FDA). Therefore, D&BPs have been reduced to a generic term akin to the term "article" (used in DSHEA), to indicate any substance for which substantial clinical investigations have been undertaken. The key paragraph is as follows:

> *[Section 912(a) prohibits the sale] of any food to which has been added [an approved] drug . . ., a biological product. . ., or a drug or a biological product for which substantial clinical investigations have been instituted and for which the existence of such investigations has been made public, unless—*

Therefore, if "drug or biological product" is interpreted in a generic sense, this has the effect of not only codifying the old rubric of not adding drugs to food, but casts a wider net to capture any substance for which a clinical investigation has been undertaken. This wider net would have a chilling effect on conduct of any studies in pursuit of a claim. For example, the FDA mandates that any substance submitted for a health claim must have substantive (e.g., double-blind, placebo controlled) interventional studies demonstrating efficacy. Further, these studies are made public during the FDA vetting process. Therefore, in order to meet the FDA's requirement for a health claim, the law is violated. Additionally, there is no room for exemption in the law; although there is an exception where the Secretary

---

[24]http://www.fda.gov/ohrms/DOCKETS/DOCKETS/95s0316/95s-0316-rpt000343-039-appx-E-Ref-27-GRAS-vol268.pdf (site accessed 20Aug08).

can grant a waiver through regulation, but this is only if the substance enhances the safety of the food, not for a health claim. Strict implementation of the law could have even more insidious effects; for example, the FDA may demand clinical studies be conducted for fiber (for tolerance), a cooking oil substitute (such as Olestra), or a non-nutritive sweetener (such as aspartame) and, under a strict interpretation of §912, these substances would not be permitted for addition to food.

6.4.3 The standard for safety

A cause of controversy with DSHEA is the setting of a different standard for safety than that for food ingredients. The new standard is the concept of *reasonable expectation of no harm*, although articulated in the FFDCA as [no] "significant or unreasonable risk of illness or injury" (Table 6.10).[25]

The basis for this rationale is that consumption of a dietary supplement is by choice, not involuntary as for a food and, because there is choice, there is an assumption of some risk on the part of the consumer. In many respects, DSHEA was a safety valve, venting consumer discontent with the high degree of restriction placed upon health claims. A tacit bargain between Congress and the consumers was struck, whereby Congress granted continued access by the public to dietary supplements by (1) providing for a lower threshold of evidence for safety, (2) changing the role of FDA from gatekeeper to policeman (i.e., abandoning pre-market approval), and (3) allowing a type of claim (i.e., SFCs, not health claims) because supplements are foods. The consumer's concessions were that (1) supplements could not be added to food (because of the lower threshold for safety for the supplement); (2) consumption will always remain the product of an overt, voluntary act on the part of the consumer (a dietary supplement can never be

---

Table 6.10 Definition of an Adulterated Food

SEC. 402. [342] A food shall be deemed to be adulterate

(f)(1) If it is a dietary supplement or contains a dietary ingredient that

(A) presents a significant or unreasonable risk[a] of illness or injury under

- (i) conditions of use recommended or suggested in labeling, or
- (ii) if no conditions of use are suggested or recommended in the labeling, under ordinary conditions of use;

(B) is a new dietary ingredient for which there is inadequate information to provide reasonable assurance that such ingredient does not present a significant or unreasonable risk of illness or injury;

(C) the Secretary declares to pose an imminent hazard to public health or safety…

[a]Importantly, this was changed from "significant and unreasonable risk" in the original bill.

---

[25]Also articulated in the FFDCA (§413) as "reasonably be expected to be safe." (http://www.fda.gov/opacom/laws/fdcact/fdcact4.htm) (site accessed 27 May13).

represented as a food); and (3) because the recommended daily dose is presented on the supplement, the consumer will assume at least some risk[26] from consumption (articulated by the standard of reasonable *expectation* of no harm (21 CFR 190.6(a)[27]). Because no system is perfect, Congress empowered the Secretary of Health and Human Services (not the Commissioner of the FDA), to take action through the "imminent hazard" clause of the regulation.

Therefore, keys to an SFC are the following: (1) the substance must maintain a physiologic function and (2) the standard of safety of *reasonable expectation of no harm.*

The results so far on DSHEA are uneven. One provision of DSHEA was to allow continuation of the sale of dietary supplements marketed prior to October 1994; for substances not marketed prior to this date, a New Dietary Ingredient Notification (NDIN) to the FDA was required. However, the exact criteria for what constitutes marketing of a substance prior to October 1994 are unclear, especially as it relates to what sort of marketing is claimed and to whom. What is very clear, is that the FDA intends to use the NDIN provision as a mechanism for pre-market approval of new dietary supplements; that is, out of the 715 substances submitted from November 1999 through November 2011, approximately 70% have been rejected.[28] The FDA has cited several reasons for objecting to these submissions, including (1) mismatches between the marketed and tested ingredient; (2) use of therapeutic language, i.e., ingredient is represented for treatment or mitigation of a disease state; (3) use of food language, i.e., ingredient is represented as a traditional food; and (4) clinical studies have been done via non-oral routes, but most often (5) a lack of sufficient safety data and/or that the substance fails to meet the safety standard.

Despite the fact that 18 years have elapsed since passage of DSHEA, the FDA has yet to recognize the gift given to it by Congress and embodied within the Act—a new standard by which supplement safety could be judged—"a reasonable *expectation* of no harm." Instead, the FDA has created its own dilemma by implying that safety has only one standard, that the reasonable *expectation* of no harm for supplements and reasonable *certainty* of no harm for food ingredients is a distinction without a difference. This has led to the continued high rejection rate of NDINs. Further, the Division of Dietary Supplement Programs (DDSP)[29] has raised the bar for safety even further with the July 2011 proposed guidelines with

---

[26]The FDA has stated that it would use a "reasonable consumer" standard in determining whether a claim is misleading. The reasonable consumer standard replaced the standard of "the ignorant, the unthinking, and the credulous" consumer, used by courts at the request of the FDA in the past [11].
[27]Referred in the FFDCA (§402(f)(1)(A)) as "significant or unreasonable risk of illness or injury." (http://www.fda.gov/opacom/laws/fdcact/fdcact4.htm) (site accessed 27 May 13).
[28]§402 (f)(1) "If it is a dietary supplement or contains a dietary ingredient that—(B) is a new dietary ingredient for which there is inadequate information to provide reasonable assurance that such ingredient does not present a significant or unreasonable risk of illness or injury…"
[29]DDSP is the division within the Office of Nutrition Product Labeling and Dietary Supplements charged with reviewing New Dietary Ingredient Notifications.

a more daunting rule set than suggested for food additives in the Redbook and, further, the only concession for an automatic NDIN approval would be a successful GRAS notification. This concession has not been the sharpest rebuke of Congressional will on record, but certainly one that has not escaped the notice of Congress in the form of a letter from Senators Hatch and Harkin to the Commissioner.[30] Still, the guidelines remain on the FDA Website and communications from DDSP have included demands for unquantified safety factors, extensive and unnecessary studies, and aggregate consumption estimates to include concomitant use in supplements and food (the latter of which is generally unavailable from food manufacturers). All of the above proceeds even in light of Congress' stated findings that: "although the Federal Government should take swift action against products that are unsafe or adulterated, the Federal Government should not take any actions to impose unreasonable regulatory barriers limiting or slowing the flow of safe products and accurate information to consumers."[31]

Still, the FDA maintains that it protects public health and safety by maintaining a 77% rejection rate of NDINs—granted, many rejections are the fault of the notifier by including foreign language documents in the notification, inadequate identification of a botanical, incomprehensible labeling instructions, or obvious drug claims. Notwithstanding the foregoing, however, a healthy percentage of rejections are based on DDSP's conclusion of an inadequate demonstration of safety. Is it any wonder then why, with less than two hundred NDIN "no objections," there are by the agency's own estimate, approximately 55,000 supplements on the market. Manufacturers, many of whom are undercapitalized and unprepared for the costs of extensive testing, may well have decided that a warning letter or even a recall is simply a cost of doing business, especially in the face of a high probability of an NDIN rejection letter. Going to market with an unnotified supplement is simply a roll of the dice for some manufacturers, but it is an even greater roll of the dice for consumers who have been placed in harm's way by the agency indirectly encouraging bad behavior by manufacturers by setting the bar too high for NDIN acceptance. The bottom line—the high threshold for approval ostensibly in place to protect the consumer—has had the opposite effect by exposing consumers to unvetted and potentially adulterated substances.

6.5 The immediate future and the path forward for FDA
6.5.1 The immediate future

For the immediate future, without a comprehensive change in the statute or movement within the Agency, it is likely that (unqualified) health claims will remain

---

[30]December 22, 2011

[31]Dietary Supplement Health and Education Act of 1994, Pub. L. No. 103-417, 108 Stat. 4325 (1994) §2 Congressional findings. (http://www.fda.gov/RegulatoryInformation/Legislation/FederalFoodDrugandCosmeticActFDCAct/SignificantAmendmentstotheFDCAct/ucm148003.htm).

as now positioned: approved under SSA in response to overwhelming evidence of health benefit and no proprietary protection of data.

However, QHCs is the area where the greatest opportunity for change could occur. QHCs will likely experience at least one more round in the Courts, addressing the issue of freedom of commercial speech with two significant changes: (1) an abandonment of the various grades of acceptance, settling on a binary system of "yes" or "no" for a claim, rather than leaving the public baffled with confusing statements and (2) abandonment of the strictly "in-house" review system (which currently includes various "evidence-based" practice centers), leveling the playing field and including an option for review by independent panels of experts, much on the model of the GRAS system and similar to that employed by the FTC (i.e., "competent and reliable scientific evidence...based on the expertise of professionals in the relevant area").

The pressure for this change is mounting and will come rather soon, in response to new and different types of claims, many of which will be generated as the result of progress in nutrigenomics, proteomics, and metabolomics. For example, for substances such as sulforanes (as detoxicants, promoting enzymes for type II detoxication reactions), sulfur compounds in garlic (which decrease gastric nitrosamine formation), proanthocyanidins in cranberries, and health claims for specific subpopulations (e.g., higher amounts of biotin are required by pregnant women of Mexican descent than women of Western European heritage, and for which precedent has been established by setting a standard for choline intake, a substance not required so much for women, as it is for men). These changes for QHCs will also be a response to safeguard against a proliferation of inclusions into the regulatory categories foods for special dietary use and medical foods, two categories for which the framework for admission has not been well established.

It is likely also that future QHCs will be directed toward specific enhancements and specific population subsets, rather than addressing the general population. This will require rethinking how clinical studies will be conducted and the general caveats that although a substance will be safe for anyone to consume (within guidelines) and it may have particular benefit for only a specific population subset.

6.5.2 Four things the FDA must do to resolve the impasse

Recalling Figure 6.1, there exists an impasse between the interested parties in functional foods, dietary supplements, and the like, between the FDA, Industry, and Consumers. However, much of the foregoing text has described what amounts to failed attempts by the FDA to respond to the situation. These palliative attempts have resulted only in a stalemate that, as mentioned earlier, the Court may be forced to resolve, but the power of immediate resolution rests with the FDA, with only four actions, all of which are within the regulatory powers of the Agency (Table 6.11).

Table 6.11 Four things FDA Must do to Resolve the Impasse
1. Create a new category of functional claims
2. Promote the use of independent expert determinations
3. Initiate a notification system
4. Provide a term of exclusivity and ROI for manufacturers

6.5.2.1 *Create a new category of "functional claims"*

Functional claims should be defined only as "providing a health benefit beyond basic nutrition." The FDA needs to make the switch in the mindset from "health" claims to functional claims. This category should be more relaxed in its requirements than those for health claims or QHCs to something that is more realistic and attainable.

At this point in time, we have already found the "no brainer" nutrients such as the vitamins and minerals, for which a deficiency was clinically obvious. Among these more obvious nutrients were vitamin C, a deficiency of which resulted in scurvy, thiamine (beriberi), iodine (goiter), and calcium (rickets). Likewise, we have also found those dietary elements whose absence (or excess) extracted a terrible cost to the public and at great expense have been incorporated into regulation as health claims (e.g., calcium and osteoporosis, sodium and hypertension, and, dietary fiber and cancer; see also Table 6.4). As indicated earlier, the cost for clinical studies resulting in a health claim is prohibitive. The cost for such studies basically shuts out most companies, especially when no return on investment can be expected.

To move forward with this new category of functional claims, the FDA must (1) disconnect the health–disease relationship and restore the substance–claim relationship mandated by Congress, (2) allow claims based on changes in biomarkers, (3) allow claims for specific population subsets, and (4) most of all, remain flexible.

For the FDA to make the "disconnect" between substance and disease is essential, and the FDA must acknowledge that not all substances will have the same effect on all or even a majority of consumers. The FDA must sever this mandated connection because the concept is intrinsically hobbled by three factors: (1) for something to be a "disease," it must meet criteria, which often require decades of debate, and the debate is often distorted by political and economic interests; (2) the substance may only effect a "symptom" or a biomarker, which when so affected, may be a desired outcome; and (3) not all or even a majority of people can be expected to experience the same effect from a substance.

Agreement on what constitutes a disease is an arduous process. Many diseases are still unknown or undefined to the satisfaction of a consensus among mainstream scientists—such was the case with what we now recognize as a folate deficiency. What we may now judge to be the result of an aging process, an adverse reaction to a food or drug, and an unexplained event or worse, wrongful attribution; is all too often taking place. How long was "early-onset" senility diagnosed before Alzheimer's disease became known? How many died from unexplained cardiovascular and neurological deterioration in middle age before obstructive sleep apnea was agreed upon? How long was Crohn's disease or gluten sensitivity characterized as "indigestion of unknown origin"?

Some "conditions" if identified as diseases, may have undesirable political or economic outcomes. For example, recognizing a new disease may then require coverage by health insurance, disability insurance, or pension plans. Also, some charitable organizations and researchers have a vested interest in preserving unsolved problems and may find it difficult to make the transition to something else once the *raison d'être* no longer exists. For example, while the March of Dimes[32] has successfully made the transition from combating polio to now "...preventing birth defects, premature birth, and infant mortality,"[33] many foundations might not be able to make such a successful transition.

A change in a biomarker should be sufficient basis to make a claim. The second compelling reason for the "disconnect" between a substance and a disease is that mitigation of a biomarker may be as important as mitigation of a disease. For example, while we know that abatement of hypertriglyceridemia or hypercholesterolemia has a statistically significant association in the reduction of risk of coronary heart disease, might there be other beneficial effects as well? Under the current system, the effect on each possible end disease would have to be tested (see also below). While the debate continues in the scientific community about the beneficial effects of decreased homocysteine levels, many educated consumers and clinical practitioners are already convinced of the need to lower these levels; further, because consumers and clinicians are convinced that the argument among scientists only has to do with efficacy and not safety, why not allow consumers a chance to decrease their homocysteine blood level? Why not allow a truthful statement such as *Substance X will lower homocysteine blood levels* as a claim? Consumers will rarely agree that the agency or mainstream science moves fast enough to respond to scientific developments, so if the association between the biomarker and the substance is demonstrated, then the consumer should be allowed to make the decision for him or herself.

Not all consumers can be expected to benefit equally from a treatment. It has long been known that not all drugs will have the same quantitative or qualitative effects in all patients. For example, it is recognized that the most successful antihypertensive therapy in African-Americans is different than for Caucasians. Also, the FDA Center for Drug Evaluation and Research does not approve drugs on the basis of imputed mechanism of action, but on empirical result, e.g., did the drug lower blood pressure? In the past few years we have learned that efficacy is not only controlled by availability of receptor sites and metabolism, but the "-omics" of the

---

[32]©2007 March of Dimes Foundation

[33]Excerpted from the March of Dimes Foundation mission statement. (http://www.marchofdimes.com/; site accessed 03 Feb 07).

individual (metabolomics, influencing the metabolism of the drug) and even "nutra-genomics" (when certain constituents of foods, or lack of these constituents), may have a profound effect on the individual (e.g., tyramine in fish and monoamine oxidase inhibitors) and/or the ability of the individual to "up-" or "downregulate" specific genes. Could it be that some of the several hundred "orphan diseases" are actually as yet unidentified nutritional or micronutrient deficiencies in those victims with the disease? The FDA cannot expect that all subjects will respond equally or even at all; the criterion for confirmation should be simply that the biomarker exhibited change in a susceptible population when treated with the substance.

Now is the time to make the tough choices and allow claims for which a consensus may not exist, but for which persuasive clinical and mechanistic data do exist. Now is the time to allow changes in biomarkers and persuasive data from the new sciences of proteomics and metabolomics and to allow special subpopulations to experience a benefit. However, it should be clear that while a functional food may benefit only a small subset of the population, all functional foods must be safe for all consumers.[34]

6.5.2.2 Promote the use of independent expert determinations

The second thing the FDA must do is to accept input from independent expert panels, a concept for which there is ample precedent. These outside experts include GRAS determinations for food ingredients; the now unused provision for experts determining Generally Recognized As Safe and Effective, but which has since morphed into the over-the-counter drug reviews and; the numerous FDA advisory panels. Use of independent expert panels could alleviate much of the staffing problems for review of dietary supplements as well as claims for functional foods. Further, there are adequate provisions under the Food Safety Modernization Act for the FDA to demand the evidence on which a claim is made and to sanction those whose determinations are not supported by an adequate amount of evidence.

6.5.2.3 Initiate a notification system

Experts making claims for safety and efficacy will submit a confidential dossier on the substance, describing the safety, efficacy, and rationale for both (Figure 6.4).

The agency can make a determination on the credibility of the experts and the credibility of the supporting data (i.e., "credible evidence" as mandated by the court, not "weight of evidence" as now observed by the FDA). If the decision is that the dossier is not persuasive, the dossier will be returned to the submitter, without prejudice, for a possible resubmission. If, on the other hand, the document is found to be persuasive, the FDA would inform consumers through a

---

[34]Interestingly, this concept is very nearly already addressed in the regulations at 21 CFR 172.5(c) "The existence of any regulation prescribing safe conditions of use for a nutrient substance does not constitute a finding that the substance is useful or required as a supplement to the diet of humans."

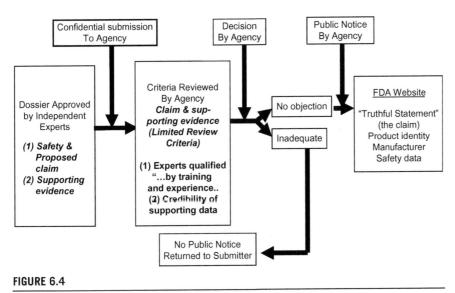

FIGURE 6.4

Proposed notification system.

posting on the FDA website with the product name, name of the manufacturer, and the safety data. The efficacy data would remain confidential and placed in a food master file. While all safety data would be made public and could be relied upon by other submitters for their dossiers, any additional substances making a claim must submit their own efficacy data—there could be no piggy-backing on efficacy claims.

This system would respond to demands for consumer empowerment, demands for commercial free speech, and would relieve a potential log jam of petitions.

6.5.2.4 *Provide a term of exclusivity and return on investment for manufacturers*

Keeping efficacy data proprietary is the essential fuel for driving research. Without the possibility for return on investment (ROI), there is no incentive for research. Making the efficacy data public has effectively killed the value that might otherwise have been derived from health claims or QHCs. Again, safety data should be made public, but efficacy data should remain secret, at least for some period in which the investment can be recaptured (Figure 6.5).

As depicted in Figure 6.5, the solid lines in the schematic illustrate the proposed method, i.e., manufacturers investing in research and upon a no objection notice from the Agency, the manufacturing, marketing, and distribution chains can be started up to serve the public and return profit to the manufacturer. However, if the no objection notice by the Agency also reveals the efficacy data (see dashed lines), then pirates (for lack of a better term), can also start up their manufacturing, marketing, and distribution chains and *without the cost of*

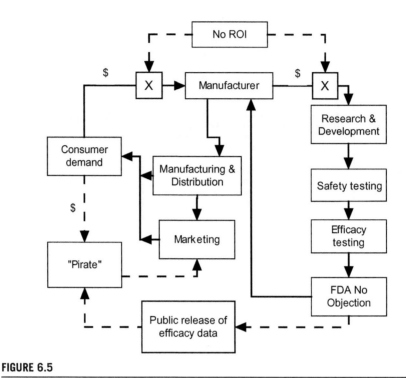

FIGURE 6.5

The proprietary nature of efficacy data is the essential fuel for driving research.

research, can sell the product at a much lower price. This selling at a lower price has the predictable effect of bleeding off profits for the ROI needed to fuel more research and product innovation stops.

This requirement for disclosure of the basis for a claim of efficacy is a disincentive for claims generation. If the FDA truly wants to contribute to public health, only disclosure of safety data should be required; efficacy data should be sequestered in a food "master file" much as that for a drug. If another company wants to make the same claim, it should either generate its own data or negotiate with the first company for access to the master file. The bottom line is that the concept of a claim is the same for a drug or functional food; it is the ability to make the claim that sells the product and sales of the product drive more research. If it were not for drug master files, would we have many new drugs today?

6.6 Discussion

At a time when nearly half of the American population spend over $30 billion on functional foods, nutraceuticals, and dietary supplements, the FDA is doing little to ensure that all the safe and efficacious products that could come to the market are allowed to do so. FDA has only responded slowly and reluctantly to Congressional action and to mandates from the Courts to implement the law. Additionally, the FDA had set the bar too high for health claims and was forced by the Courts to implement a more reasonable standard, but the response, QHCs, has failed to gain the confidence of the public because of the confusing wording of the claims demanded by the FDA. Further, disclosure of efficacy data has prevented ROI and development of new products. If the same principle of disclosure of efficacy data were applied to the drug side of the FDA, there would not be many new drugs today. While the FDA has a duty to protect the public from unsafe substances and fraud, it also has an obligation to take affirmative steps to improve public health by working with manufacturers to bring beneficial substances to market.

Congressional efforts to assure consumer access to dietary supplements have been met with similar resistance from the FDA. DSHEA was the product of a compromise with a lower threshold for demonstration of safety (reasonable *expectation* of no harm) that would be met by consumer self-policing and assumption of some risk. The FDA has thwarted this effort by raising the bar for NDINs to what appears to be the higher threshold for the safety of food ingredients (reasonable *certainty* of no harm); the FDA apparently sees these two safety thresholds as a distinction without a difference. As a result, increasing numbers of dietary supplement manufacturers, unwilling to gamble the future of their products to a system that provides little hope for the FDA's response of no objection, have simply gone to market with their products, hoping to stay off of the FDA's radar.

The pressure on the FDA and Congress for change is again building with increased dissatisfaction among consumers as the result of confusing labels and among manufacturers saddled with unreasonable requests for data with no expectation for return on their investment. As if the confusion about health claims and the claims for dietary supplements were not enough, there is now public clamor for resolution on claims for functional foods. The FDA has within its power to resolve the conundrum of so many claims borne of fine technical differences and presumptions of the ability of the consumer to distinguish the Byzantine wording of one type of claim or grade from another. The FDA needs to consolidate the claims into three categories: health claims for foods, functional foods, and dietary supplements. Health claims should remain the gold standard and require the highest level of proof of a health effect. The new category of functional foods would be those substances falling somewhat short of the threshold of a health claim and be substantiated by criteria such as biomarkers and mechanistic studies, not requiring the evidence of disease incidence reduction. Dietary supplements would remain relegated to structure function claims and not permitted to be added to food as they only satisfied the lower threshold of *reasonable expectation* of safety (Table 6.12).

Table 6.12 Resolution to the Impasse

	Health Claims	Functional Food	Dietary Supplements
Safety standard	Reasonable certainty	Reasonable certainty	Reasonable expectation
Addition to food	Yes	Yes	No
Type of claim	Health and disease reduction	Reduction of symptom, change in biomarker or clinical finding	Structure/function
Level of evidence supporting efficacy	Gold standard interventional studies	[a]Meta-analysis of studies, biomarker change, epidemiological studies, mechanistic evidence, smaller clinical trials	One or more persuasive studies related to maintenance of healthy function or structure
Deciding body	As presently exists (FDA and regulation promulgated)	[b]Expert Panel with Notification to FDA (no regulation promulgated)	As presently exists
Public disclosure	Safety information, product identification and manufacturer. Efficacy information held in confidential master file for >10 years	Safety information, product identification and manufacturer. Efficacy information held in confidential master file for >10 years	As presently exists

[a]Generally considers two or more studies with consistent results within the more developed category shown in Figure 6.2 as "emerging evidence." Discrepancies in data must be resolved to the satisfaction of the Expert Panel.
[b]The agency notification review limited to (1) credibility of the experts (modeled on the description of experts in §201(s) of the FFDCA) and (2) credibility of submitted supporting evidence.
[c]SFCs are permitted for food ingredients.

6.7 Conclusion

The FDA has the power and the duty to resolve the conflict and relieve the burden that has been created for industry, consumers, and for itself. With the addition of the functional food category, use of expert panels, and maintenance of proprietary information (on efficacy), the FDA can be seen as the champion of consumers and encouraging industry research to come up with new and better solutions for an aging and demanding customer base. There is precedent for all of these suggested actions and compelling reasons to take the suggested route. The public sees at least the food side of the FDA as monolithic and all the good works on protection of the food supply from bioterrorism, spoiled goods, adulteration, and mislabeling may be lost. The time for the FDA to step up to the plate is now, before the credibility of the Agency is compromised.

Acknowledgments

We thank Dr. James Griffiths for his editorial comments on the first edition and, to Arlene Morales and Silvia Ulm for their technical support for this second edition. Also, thanks to Carra Carabin for her patience and wise counsel.

References

[1] Burdock GA. Commentary: FDA must overcome skepticism toward health claims. The Tan Sheet 2013;21(16):11—2.

[2] Rulis, A. Food Safety and Nutritional Risk, Bioactive Food Components. CSL/ JIFSAN Symposium on Food Safety and Nutrition. July, 2005.

[3] Emord J, Schwitters B. FDLI's food and drug policy forum: do qualified health claims deceive when they are not misleading? perspectives from the European Union and United States. FDLI 2012;2(12):12.

[4] Carabin IG. The clinical aspects of claim substantiation: clinical trial costs. FDLI Update 2004a;39—43 Issue 4.

[5] Carabin, IG The clinical aspects of claim substantiation: inclusion/exclusion criteria, screening and baseline evaluations in clinical trials. FDLI Update 2004b, Issue 5, p. 41—45.

[6] International Food Information Council (IFIC). Qualified Health Claims Consumer Research Project Executive Summary <http://www.ific.org/research/qualhealthclaimsres. cfm>; 2005 [site visited 25 Dec 05]

[7] Burdock GA. Dietary supplements and lessons to be learned from GRAS. Journal of Regulatory Toxicology and Pharmacology 2000;31:68—76.

[8] Council for Agricultural Science and Technology (CAST). Issue Paper No. 24, nutraceuticals for health promotion and disease prevention. Camire, ME (chair). CAST, Ames, IA; 2003, 16 pages.

[9] Kracov DA, Rubin PD, Dwyer LM. Dietary Supplements and Drug Constituents: The *Pharmanex v. Shalala* Case and Implications for the Pharmaceutical and Dietary Supplement Industries. In: Hasler C, editor. Regulation of Functional Foods and Nutraceuticals. Ames, IA: Blackwell Publishing; 2005. p. 137—48.

[10] Hutt PB. U.S. government regulations of food with claims for special physiological value. In: Schmidl K, Labuza T, editors. Essentials of Functional Foods. Gaithersburg, MD: Aspen Publishers, Inc.; 2000. p. 339—62.

[11] Walsh EM, Leitzan EK, Hutt PB. The importance of the court decision in *Pearson v. Shalala* to the marketing of conventional food and dietary supplements in the United States. In: Hasler C, editor. Regulation of Functional Foods and Nutraceuticals. Ames, IA: Blackwell Publishing; 2005. p. 109—35.

[12] Rowe, SB. Round Table Forum, Moderators Comments. In: What is a Nutrient? Defining the Food-Drug Continuum. Georgetown University; March 30, 1999.

NSF International's Role in the Dietary Supplements and Nutraceuticals Industries

7

Greta Houlahan and Edward Wyszumiala

NSF International, Ann Arbor Michigan

7.1 A look at the market

From 2009 to 2010, almost half of all U.S. adults consumed dietary supplements—including vitamins, minerals, and herbals—according to the Centers for Disease Control and Prevention, and dietary supplement sales surpassed $30 billion in 2011 [1].

According to estimates from the *Nutrition Business Journal* [2], U.S. consumer sales of dietary supplements reached $28.1 billion in 2010, a 4.4% growth over 2009 sales. The top supplement categories included: multivitamins ($4.9 billion), sports nutrition powders and formulas ($2.8 billion), B vitamins ($1.3 billion), calcium ($1.3 billion), and fish/animal oil ($1.1 billion) [3].

7.2 Brief history of NSF International

NSF International is a global independent public health organization that writes standards and tests and certifies products for the food, water, and consumer goods industries to minimize adverse health effects and protect the environment (www.nsf.org).

NSF International helps protect and improve public health and the environment by offering services in several different areas including food, water, health sciences, consumer products, sustainability, and management systems registration.

7.2.1 History in standards development and independent third-party certification

NSF International was founded in 1944 within the University of Michigan's School of Public Health as the National Sanitation Foundation to standardize sanitation and food safety requirements. The transparent, consensus-based process

Nutraceutical and Functional Food Regulations in the United States and Around the World.
DOI: http://dx.doi.org/10.1016/B978-0-12-405870-5.00007-4

established to develop NSF International's first standard on the sanitation of soda fountain and luncheonette equipment became the process by which NSF International developed nearly a hundred other public health and safety standards, including the NSF American National Standard for dietary supplements (NSF/ANSI Standard 173). NSF/ANSI Standards are developed through involvement of those who are directly and materially affected by the scope of the standard. The process ensures balanced participation from industry representatives, public health/regulatory officials, and users/consumer representatives.

7.2.2 Organizational growth

Since its founding in 1944, NSF International has grown significantly in terms of staff, locations, and services offered. Today, NSF has more than 1200 employees operating in 150 countries worldwide offering testing, auditing, and certification services to industries such as water, food, and consumer goods, as well as dietary supplements.

7.2.3 NSF testing and certification

Millions of products worldwide are authorized to bear the NSF Certification Mark (Figure 7.1). The NSF Certification Mark indicates that the product meets quality and safety standards. For instance, dietary supplement products that have been authorized to bear the NSF Mark must comply with numerous requirements including label, testing, and auditing requirements. Refer to Table 7.1 for more information on these requirements.

7.2.4 Why did NSF International start a certification program for dietary supplements?

NSF entered the dietary supplement market when industry came to them. In 2000, dietary supplement companies started approaching NSF to create and draft

FIGURE 7.1

NSF dietary supplement certification mark.

Table 7.1 NSF Certification Requirements

NSF developed NSF/ANSI Standard 173 to help protect consumers by testing for contaminants and certifying that NSF Certified supplements contain the ingredients listed on the label and nothing else. The certification process requires:

- Label claim review: to certify what is on the label is in the bottle
- Toxicology review: to certify product formulation
- Contaminant review: to ensure there are no undeclared ingredients present or unacceptable levels of contaminants in the product
- Facility audits: GMP audits of the plant annually
- Ongoing monitoring: to certify compliance through periodic auditing and testing.

---

standards for the industry. The goal was to come together with the various stakeholders to devise a national standard that would also promote public health. See Table 7.2 for more information on dietary supplement safety. In response, NSF formed a working group comprised of key regulatory officials, trade associations, and industry representatives. The result was the first and only national standard for dietary supplements.

NSF helped write the accredited American National Standard (NSF/ANSI 173), which verifies the purity and safety of dietary supplements and also tests and certifies products to this standard. The standard is used to evaluate and analyze dietary supplements to ensure they do not contain undeclared ingredients or unsafe levels of contaminants such as pesticides and heavy metals as well as verify that the nutrient levels of the products are accurately labeled. Adding an extra layer of credibility, the standard is accredited and verified by the ANSI, a private, non-profit organization that administers and coordinates the U.S. voluntary standardization and conformity assessment system.

A list of NSF Certified dietary supplements can be found on the NSF's Web site. Go to www.nsf.org and click "Search certified products."

7.3 What are good manufacturing practices?

Good Manufacturing Practices (GMPs) describe the methods, equipment, facilities, and controls for producing dietary supplement products. Only through rigorous compliance with GMPs can a company be confident in the products it produces.

7.3.1 GMP regulations

First to provide some background on GMPs, the Food and Drug Administration (FDA) introduced GMP regulations in 2007 to establish quality control standards for U.S. nutritional products. They took effect on June 25, 2008, for companies who employ more than 500 employees. Businesses that employ 20−500 had until

Table 7.2 Supplement Safety Q&A
Who Ensures the Safety of Dietary Supplements?
By law, manufacturers are responsible for ensuring their supplements are safe before they are marketed. Unlike drug products, dietary supplements are not reviewed by the government before being made available to the consumer.
What Information Needs to be on the Label?
Federal laws require the labels on dietary supplement products to contain the following information: • Statement of identity • Net quantity of contents • Directions for use • Supplement Facts Panel, listing serving size, amount, and active ingredient(s) • Other ingredients in descending order of predominance • Name and place of business of manufacturer, packer or distributor
Is It Necessary to Check with a Health Care Professional before Using Supplements?
Dietary supplements may not be totally risk free under all circumstances. Some supplements can interact with over-the-counter or prescription medications or have unwanted effects during surgery, while others may contain active ingredients that can cause adverse reactions in some users. Check with a health care provider prior to taking a dietary supplement.
Why Should I Purchase NSF Certified Supplements?
With studies showing that not all supplement products contain the ingredients or quantities shown on the label, purchasing supplements that are NSF Certified is important for today's health-conscious consumers. A list of NSF certified supplements can be found at www.nsf.org.
Why Aren't More Dietary Supplements NSF Certified?
Since product certification is voluntary, not all companies will pursue independent testing and certification if their customers do not demand it. While financial reasons are a barrier for some companies, for many others certification is not obtainable either because the manufacturing facility has difficulty passing the GMP audit or there is a problem with the actual product. Not all products are able to pass the label claim testing portion of the certification process, meaning the stated nutrient content on the product label and what is actually in the product do not match. Products can also fail if they have heavy metal or microbial contamination or, in the case of sports supplements, if the product does not pass the athletic banned substance portion of the testing.

June 25, 2009, and businesses that employ fewer than 20 employees had until June 25, 2010, to comply with GMPs.

The FDA defines these safety and quality standards or GMPs in the Code of Federal Regulations Title 21 (21 CFR111). Dietary supplement manufacturers, packagers, and distributors in the United States are all required by law to comply with these regulations, and the cost of noncompliance is steep [4]. This also affects any supplements produced outside of the United States that are manufactured for distribution in the United States.

Failing to comply can result in a number of undesirable outcomes: public warning letters, products labeled as adulterated, seizure by authorities, injunction from manufacturing, and damage to brand and company reputation. Unfortunately, evident by the alarming number of FDA-issued warning letters in recent years, many firms still struggle to comply with GMPs. According to the FDA, the agency issued warning letters to one in four dietary supplement facilities it inspected.

When GMP regulations first took effect, seven GMP inspections occurred in 2008, which increased to 34 in 2009 and 84 in 2010 [5]. During fiscal 2012, there was an all-time high of 341 dietary supplement facility inspections by the FDA [6]. What does this mean for dietary supplement manufacturers? It means that GMP compliance is essential (see Figure 7.2).

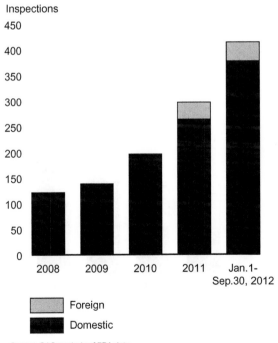

Source: GAO analysis of FDA data.

FIGURE 7.2

Number of foreign and domestic inspections of dietary supplement firms the FDA or its state partners conducted, January 1, 2008 through September 30, 2012 [8].

According to FDA officials, the key factors underlying this increase in inspections were the full implementation of dietary supplement current GMP (cGMP) regulations in 2010 (i.e., describing the conditions under which supplements must be manufactured, packed, and held) and an increase in field investigators available to conduct inspections [7].

As the cGMPs were being phased in, the FDA identified problems or concerns during inspections, such as a manufacturer not maintaining, cleaning, or sanitizing equipment [8]. The percentage of inspections where the FDA identified problems or concerns increased from 51% in 2008 to 73% in 2011, largely resulting from cGMP inspections. See Figure 7.3 for the proportion of dietary supplement inspections where the FDA identified problems or concerns from 2008 to 2011.

With the increasing number of inspections, the number of warning letters issued by the FDA also continues to increase. As mentioned previously, the NSF American National Standard 173—Dietary Supplements contains a section on GMP requirements that references the new FDA requirements for both dietary supplement GMPs as well as the Adverse Event Reporting (AER) requirements (see Table 7.3).

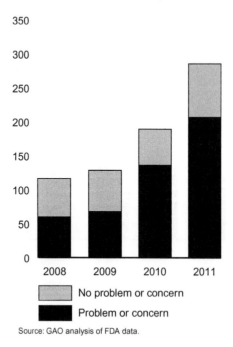

Source: GAO analysis of FDA data.

FIGURE 7.3

Proportion of dietary supplement inspections for which the FDA identified problems or concerns, 2008 through 2011 [7].

Table 7.3 What Is the Difference between GMP Registration and Product Certification?

NSF International provides three different services to dietary supplement manufacturers:

1) **GMP registration.** This registration is for a facility and involves on-site audits of the entire production operation to verify the facility is observing GMPs established for the industry. Products produced at the facility would be NSF registered, not be NSF Certified, and products could not bear the NSF Certification Mark.

2) **Product certification.** In addition to earning GMP registration, companies may also have their products evaluated to determine whether they meet NSF/ANSI Standard 173–Dietary Supplements. This American National Standard limits the amount of impurities that can be present and requires that the ingredients and quantities shown on the label match what is present in the product. In addition, no unlisted ingredients are permitted to be present in products certified to this standard. Products that are NSF certified to NSF/ANSI 173 are permitted to display the NSF Mark on the product label.

3) **Certified for Sport.** Products that are specifically targeted for use by athletes can be evaluated by the NSF Certified for Sport program. In addition to meeting GMP requirements and the requirements of the NSF American National Standard for dietary supplements, products also are reviewed on a lot-by-lot basis to ensure they do not contain athletic substances banned by major sports organizations including: stimulants, narcotics, steroids, diuretics, β-2-agonists, beta blockers, masking agents, and other substances.

7.3.2 GMPs and compliance issues

Inadequate testing is the most common GMP infraction. Testing is broken down into three components: identity testing, supplier qualification testing, and finished-product testing.

7.3.2.1 Identity testing

According to FDA reports and warning letters, many firms are still not spending enough time developing appropriate test methods to validate the identity of incoming ingredients. To satisfy this GMP requirement, many companies use in-house verification technology, such as Fourier transform infrared (FTIR) and near-infrared (NIR) spectrometers [9]. NSF auditors find these technologies are commonly used, as they are relatively inexpensive and quick; however, they may not be appropriate for identifying all incoming materials, such as blends or botanicals. When ingredients are mixed with other compounds or are derived from plant material, if the proper validated methods are not being used, these verification techniques could potentially misidentify the botanical species or fail to detect an adulterant.

Also, FTIR and NIR spectrometers create a "fingerprint" or data pattern of the ingredient: the more complex the ingredient, the more complex the fingerprint. The most important aspect of using this method—and where many companies continue to struggle—is establishing a reference standard of what a sample fingerprint of good, high-quality material should look like. Companies need these reference standards with which to compare incoming ingredients. An FTIR and NIR spectrometer will produce a scan for virtually any ingredient, but without an established standard to compare an ingredient against, it is hard to tell whether the incoming ingredient meets the manufacturer's specification.

To properly develop these reference standards, companies should have several samples (six or more, preferably) that they have qualified using other techniques to develop a library of good ingredient datasets. NIR instruments in particular are sensitive enough to pick up very subtle differences in various ingredient lots, which may be particularly critical when dealing with botanicals. Factors such as moisture level, particle size within the sample, and even the time of year the ingredient was harvested can affect the resulting fingerprint or data pattern. For this reason, having multiple reference samples to compare against will accommodate these very miniscule variations and still provide a means of identifying the correct match to the ingredient specification.

Properly identifying incoming ingredients also enables a company to protect itself against bad players in the supply chain. For instance, NSF auditors have observed cases in which chondroitin sulfate has been diluted with a cheaper substitute called carrageenan. If misidentified ingredients make their way into a finished product, they compromise the integrity and accuracy of the formulation and label claims. Again, testing is crucial.

7.3.2.2 Supplier qualification testing

GMPs require that dietary supplement companies qualify their suppliers. While the GMP regulations do not spell out what "qualify" means exactly, they do say that companies must verify the Certificate of Analysis (CofA) of incoming ingredients. However, ingredients and their CofAs pass through many hands before making their way to a company's facility for processing, providing plenty of opportunity for contamination or fraud along the way.

Depending on the nuances of a company's supply chain, it may not be enough to just compare a CofA against specifications. CofAs may not always be a reliable testament to the quality of the ingredient. Often, CofAs for a particular ingredient are a reproduction created by the ingredient distributor rather than the true ingredient CofA created by the manufacturer of the ingredient. NSF auditors have seen synthetic vitamin E sold and represented as the more expensive natural vitamin E, as well as instances in which a company sourced a high-quality version of glucosamine, but after testing found that it was a much cheaper form that did not match the CofA. Proper testing to verify the source eliminates this type of fraud. This is why the leading dietary supplement companies have heeded the guidance of the FDA and verify CofAs through proper testing.

Additionally, NSF auditors continue to see facilities struggle with properly identifying their suppliers. Supply chains are usually long and complex, making it hard to determine who originally manufactured the ingredient. Known cases of adulterated raw materials making their way into the United States have even sparked new regulations. The Food Safety Modernization Act (FSMA) in particular keys in on suppliers as a control point for preventing contaminated products from entering U.S. facilities.

NSF auditors recommend keeping the paper trail as clean and clear as possible. To ensure continued compliance to specifications and to GMPs, it is required that companies perform periodic requalification of suppliers and testing to verify CofAs. To help, trade organizations such as the Council for Responsible Nutrition (CRN), Consumer Healthcare Products Association (CHPA), and United Natural Products Alliance (UNPA), as well as many of their member organizations, have started a collaboration known as Standardized Information on Dietary Ingredients (SIDI), with the goal of developing guidelines for industry on certificates of analysis and supplier qualification.

7.3.2.3 Finished product testing

Consumers buying a product place a great deal of trust in that brand and expect product labels to be accurate. Testing finished products to verify the accuracy of their labels, as well as the absence of harmful levels of contaminants such as lead, is paramount to GMP compliance; however, testing complex finished products requires high-tech instrumentation, test methods, and expertise that many companies do not have in-house.

Some finished products have upward of 20−50 different ingredients and label claims. This becomes even more complex as testing and verification involves trace amounts—micrograms and milligrams in dosage quantities. The challenge can sometimes lie in selecting the appropriate test methods and instruments. High-performance liquid chromatography is often most suitable for the analysis of various phytochemicals. Inductively coupled plasma (ICP) instruments are ideal for confirming mineral levels such as calcium, while ICP mass spectrometry is needed for measuring trace heavy metal contaminants. For products containing fatty acids, such as fish oil, gas chromatography instruments are required.

Companies that manufacture complex proprietary blends have an even greater challenge, as there are no compendial standards to compare their finished products against (as opposed to a product containing a single ingredient). This is why many companies seek the expertise of outside, third-party accredited laboratories to help review their formulas and develop test methods, or even perform the testing for them.

Separating the different compounds in the finished products and then controlling the cost of the various testing required can also be difficult. To help control the cost of testing, perform a majority of the contaminant testing on incoming ingredients and then perform label claim verification testing on finished products. For complex finished products, this can still add up to a lot of tests. By always ensuring the quality of the ingredients and monitoring the ingredient inputs during the manufacturing process, the focus in the finished-product testing can be on those parameters that are most sensitive to change or degradation. Building quality into the system can save on overall test cost and goes a long way toward achieving the ultimate GMP goal of producing and releasing only high-quality products.

> *Perform testing at the beginning*: Sufficiently qualify suppliers by either sourcing ingredients from other third-party GMP-registered facilities and/or perform the appropriate identity testing and validation of the CofAs *before* ingredients go into the manufacturing process. Then, validate the blending process to ensure that the proper equipment and systems are in place to produce a consistent homogenous blend and finished product.
> *Perform testing at the end*: In addition to testing ingredients prior to manufacturing, companies should perform testing post-manufacturing to validate relevant marker compounds in the finished products. Thoroughly test finished products to verify formulation and label claims and to ensure that there are no unacceptable levels of harmful contaminants.

7.4 Developing master manufacturing records and batch records

Other areas in which companies struggle to comply with GMP regulations include developing master manufacturing records (MMRs) and batch records (BRs) that include all 14 requirements outlined in the Code of Federal Regulations. MMRs should identify specifications for the control points, steps, or stages in the manufacturing process for which control is necessary to ensure the quality of the dietary supplement. This includes all processing steps, from production through packaging and labeling. Companies may see these as just a lot of paperwork, but without traceable, accurate records, product consistency can be a real challenge.

Finally, the ensured recording and review of customer complaints is also an area of GMPs. Serious adverse events are vastly under-reported, according to the FDA. Recording customer complaints and adverse events is just the beginning; companies must then investigate the root cause of the deficiency and develop appropriate corrective actions to be implemented. Without this process, continuous improvement cannot be achieved (see Table 7.4).

Table 7.4 Top Five Areas of Concern

1. Formulas different from labels
2. Inadequate level of testing
3. Lack of knowledge and training
4. Poor plant design
5. Deficiencies in batch documentation

According to Mollie Kober, NSF International Auditor, "While improvements are needed, companies are working to comply. Understanding the regulations and where a company is deficient is the first step." Kober adds, "the regulations are a change in the way business has been conducted, but many of the GMP requirements point to a smarter business approach."

- Companies may find minor facility changes, things that can be done over a weekend with little business impact, can create vast improvements in the product flow and efficiency of the operation.
- Defining the product specifications up front leads to fewer mix-ups and rework situations on the floor. This reduces cost.
- Companies need to develop MMR and BR paperwork around the process first and then ensure that it also meets regulatory requirements. The paperwork needs to be adaptable to the changing business needs and customer requirements.

Companies should understand that while the regulations may make their businesses function differently, the end result does not have to mean significantly higher manufacturing costs.

The companies, which embrace the regulations and use that information to create a smarter more efficient business, are the ones that will succeed long term in this market place.

After reviewing the top five GMP non-conformances, the following questions probably come to mind:

- How do you qualify suppliers?
- Do you have established standard operating procedures in place?
- Do you test, audit, and train on an ongoing basis?

Asking these key questions, knowing what to expect and being prepared will help companies stay ahead and help prevent warning letters.

7.4.1 Importance of GMP registration and GMP registration process

GMP registration demonstrates that a company has been audited and found compliant with the FDA GMPs. To ensure continued compliance, NSF GMP-registered facilities are audited twice annually.

> *Growth in NSF GMP registrations*: Since NSF International started in GMP registration program in 2002, nearly 400 companies have earned GMP registration to demonstrate compliance with GMP requirements.
> *NSF GMP registered companies*: These companies can be found at the following Web site: http://www.nsf.org/Certified/GMP/Listings.asp? PlantType = .

7.5 Importance of independent, third-party certification

As discussed earlier, independent, third-party product certification of dietary supplements demonstrates that the product complies with NSF/ANSI Standard 173: Dietary Supplements requirements.

This American national standard specifies the amount of impurities that can be present and requires that the ingredients and quantities shown on the label match what is present in the product. In addition, no unlisted ingredients are permitted to be present in products certified to this standard. Products that are NSF certified to NSF/ANSI 173 are authorized to use the NSF Mark on the product label.

7.5.1 Increase in dietary supplement product certification

Since NSF International launched its dietary supplement certification program in 2001, NSF has tested and certified hundreds of products. With the seemingly continual media coverage about the latest dietary supplement allegedly containing unsafe levels of contaminants or pharmaceutical ingredients, more and more consumers are starting to look for products that have earned NSF certification.

According to an independent survey conducted on behalf of NSF, 51% of the general population expressed concern about the safety and quality of their nutritional supplements, including vitamins and herbs. That same survey revealed that 73% prefer products that are certified or have a seal of approval.

7.6 Future outlook

GMP regulations are just starting to be enforced and will be an ongoing issue for industry to deal with in the coming years. These regulations, if properly adopted by industry, will lead to higher quality products on the shelves for consumers, which will help to build more consumer confidence and credibility for supplement users.

In addition to GMPs, dietary supplement product certification and FSMA also will help with consumer reassurance, as they call for more transparency for both food and dietary supplements.

The educated consumer, increased enforcement, and the importance of accountability will all impact where the dietary supplement industry goes in the next 5 years. The future looks promising as long as companies meet GMP requirements and products continue to be tested and certified to ensure that what is on the label matches what is in the bottle and there are no unsafe levels of contaminants. For educational resources for consumers, see Table 7.5 and for educational resources for athletes, see Table 7.6.

Table 7.5 Educational Resources for Consumers

NSF International offers to help answer many common questions consumers have about supplements:

- What should I look for on a dietary supplement label?
- What kinds of claims can manufacturers really make?
- How does certification protect consumers?
- What special considerations exist for student and professional athletes who take supplements?

For the answers, consumers can visit www.nsf.org/consumer.

Table 7.6 Educational Resources for Athletes

NSF International offers answers to help athletes at all levels choose safer, verified supplements:

- Online Resource for Consumers, Coaches, and Athletes (www.nsfsport.com)
- Guide to Sports Supplements Testing (http://www.nsf.org/consumer/newsroom/fact_abs.asp)
- NSF for Sport App (http://nsfsport.com/sport_app.asp)

References

[1] Supplement Usage, Consumer Confidence Remain Steady According to New Annual Survey from CRN at, <http://www.crnusa.org/CRNPR12-ConsumerSurvey100412.html>.

[2] Nutrition Business Journal-NBJ's Supplement Business Report 2012 at, <http://newhope360.com/site-files/newhope360.com/files/uploads/2013/04/TOC_SUMM120928.supp%20report%20FINAL%20standard.pdf>.

[3] Dietary Supplements: Onward...and Upward? at, <http://www.nutraceuticalsworld.com/issues/2012-04/view_features/dietary-supplements-onwardand-upward/>.

[4] Ingredient Suppliers Need Their Own GMP Audit Program at, <http://www.naturalproductsinsider.com/articles/2013/03/ingredient-suppliers-need-their-own-gmp-audit-pro.aspx>.

[5] AERs, GMP Inspections Can Be Double Trouble for Product Liability Applicants at, <http://www.naturalproductsinsider.com/articles/2011/11/aers-gmp-inspections-can-be-double-trouble-for-product-liability-applicants.aspx>.

[6] The Tan Sheet, January 21, 2013.

[7] United States Government Accountability Office Dietary Supplements: FDA May Have Opportunities to Expand Its Use of Reported Health Problems to Oversee Products at, <http://www.gao.gov/assets/660/653113.pdf> (pg. 28).

[8] United States Government Accountability Office Dietary Supplements: FDA May Have Opportunities to Expand Its Use of Reported Health Problems to Oversee Products at, <http://www.gao.gov/assets/660/653113.pdf> (pg. 26).

[9] Dietary Supplement GMP Weak Spots: An Auditor's Perspective at, <http://www.nutritionaloutlook.com/article/dietary-supplement-gmp-weak-spots-auditor%E2%80%99s-perspective-3-10994>.

Regulations Around the World

FDA Perspectives on Food Label Claims in the United States

James E. Hoadley[*] **and J. Craig Rowlands**[†]

EAS Consulting Group, Alexandria, Virginia [†]*The Dow Chemical Company, Toxicology and Environmental Research and Consulting, Midland, Michigan*

8.1 Introduction

Products, other than tobacco products and alcoholic beverages, marketed in the United States and intended for oral consumption, are categorized for regulatory purposes as either foods or drugs. Within the context of the U.S. Federal Food, Drug, and Cosmetic Act (FFD&C Act), dietary supplements are considered to be foods. Nutraceuticals and functional foods are food marketing concepts rather than regulatory categories and there are no U.S. regulatory definitions to accommodate them separately from other foods. The U.S. Food and Drug Administration (FDA) regulates the labeling of foods (other than the U.S. Department of Agriculture (USDA)-regulated meat products). The advertising of food falls under the jurisdiction of the Federal Trade Commission (FTC). Because the FDA and FTC enforce different laws, the regulations and policies pertaining to advertising and labeling of food differ. This chapter pertains to FDA regulation of food label claims and, as such, will not address advertising. Current U.S. laws pertaining to food labeling are able, with some restrictions, to accommodate "functional foods." The statutory definition of a "food" in the FFD&C Act is: "an article used for food or drink for man or for other animals, chewing gum, and articles used for components of any such article." The definition of a "drug" in the FFD&C Act is, in part: "articles intended for use in the diagnosis, cure, mitigation, treatment, or prevention of disease in man or other animals; and articles, other than food, intended to affect the structure or any function of the body of man or other animals." Consistent with these definitions, the labeling[1] of food may not include any information about the usefulness of a food to cure, mitigate, treat, or prevent a disease, but food labels can present information about

---

[1]There is a legal distinction between the "label" (on the immediate product container) and "labeling" (material accompanying the product). In this chapter, we use the terms label and labeling interchangeably to describe the written, printed, or graphic matter on any food, its containers or wrappers, or other material accompanying the food.

how a food may affect a structure or function of the body. The FFD&C Act also provides for the use of food label claims that describe how a food or food component may affect disease risk (i.e., health claims). The presence of an authorized health claim on a food label does not cause the intended use of the product to be a drug use. However, the use on a food label of a health claim not authorized by the FDA will cause the product to be subject to regulation as a drug. The statutory distinction between what is a food and what is a drug is based to a large extent on how the product is represented for intended use. It is possible for a substance to have both food and drug uses. An intended drug use does not necessarily preclude a product from also having a food use, or vice versa. Baking soda, for example, is useful both as an ingredient in baking (food use) and as a treatment of acid indigestion (drug use). A key phrase in the FFD&C Act drug definition is "intended for use." Product labeling is one obvious way to show intended use of a product. A baking soda label stating that baking soda may be used for relief of acid indigestion shows that the product is intended to be used as a drug.[2] Any intended drug use must be in accordance with the pre-market review and approval requirements for drugs, including efficacy and safety standards applicable to drugs. A product label claiming that the product is effective in the mitigation of a disease symptom (e.g., relief from osteoarthritis pain) shows the intended use of the product to be as a drug and therefore would be regulated as a drug, not as a food. A product labeled as a food may not bear a claim pertaining to use of the product as a drug (i.e., use in the cure, mitigation, treatment, or prevention of any disease), irrespective of how good the evidence is to substantiate the effectiveness of the product for that particular use, unless the product has also been approved for use as a drug.

8.2 Legal basis for U.S. regulation of food label claims

Most U.S. laws pertaining to the labeling of food are contained within the FFD&C Act. The FDA implements these laws through food labeling regulations found in title 21, Chapter 1, subchapter B, of the Code of Federal Regulations. The FFD&C Act was amended by the U.S. Congress several times in the 1990s to make significant changes in how food label claims are regulated. These amendments include: the 1990 Nutrition Labeling and Education Act (NLEA) [1], the 1994 Dietary Supplement Health and Education Act (DSHEA) [2], and the 1997 Food and Drug Administration Modernization Act (FDAMA) [3].

NLEA amended the FFD&C Act in a number of important ways. Two chief aspects of the NLEA amendments to the FFD&C Act are that they established

---

[2]Bicarbonate ion-containing active ingredients, such as sodium bicarbonate (baking soda), are an approved antacid over-the-counter drug for relief of heartburn and acid indigestion (21 CFR 331). There are brands of baking soda marketed in the United States with the labeling required for intended use both as a food and as a drug.

mandatory nutrition labeling for nearly all packaged foods sold in the United States and that they established the FDA's authority to regulate food label statements that characterize levels of nutrients in a food or that characterize relationships of food substances to a disease or health-related condition. In general, information provided by manufacturers on food labels is restricted only to the extent that the information may not be false or misleading. For nutrient content and health-related information, however, NLEA stipulated that statements characterizing the level of a nutrient in a food (nutrient content claims) and statements that characterize a relationship of a nutrient to a disease or health-related condition (health claims) may only be used in food labeling in accordance with relevant statutory provisions and FDA regulations. Although label statements such as "high fiber" (a nutrient content claim) and "fiber may reduce the risk of heart disease" (a health claim) may be perceived by many consumers as equivalent health messages, NLEA mandated that the FDA regulate nutrient content claims and health claims as separate regulatory categories of food label statements. Although the FFD&C Act has always recognized that foods may be used to affect the structure or any function of the body, DSHEA added new provisions to the FFD&C Act defining and prescribing the use of structure/function claims in dietary supplement labeling. Structure/function claims describe a role of a nutrient or other dietary supplement ingredient in affecting the structure or function of the body (e.g., *calcium builds strong bones*).

The process for substantiation and authorization of food label health claims in the United States has evolved from that initially established in 1990 under NLEA. The NLEA amendments require that the FDA issues regulations authorizing use of a health claim about a substance/disease relationship only after having evaluated the scientific evidence for the substance/disease relationship and determining that the relationship is substantiated by an appropriate level of evidence. The validity requirement for the substantiation of health claims mandated by the NLEA amendments require an FDA assessment of "significant scientific agreement" (SSA) among qualified experts that the totality of the scientific evidence supports the dietary substance/disease relationship. In 1997, FDAMA amended the FFD&C Act to provide an alternative to the FDA's process for SSA assessment and regulation of health claims. The FDAMA amendments provided for a process whereby the substantiation of a health claim is based on authoritative statements about a dietary substance/disease risk reduction relationship from any U.S. federal scientific agency with official responsibility for public health protection or human nutrition research, or by the National Academy of Sciences or any of its subdivisions, rather than being based on an FDA review of the supporting scientific evidence. The FDAMA amendments also provided for the authorization of new authoritative statement-based nutrient content claims. The FDAMA provision for authoritative statement-based health claims and nutrient content claims was intended to expedite the process by which new claims are authorized.

The basis for the use of health claims further evolved following a series of court decisions that led to the 2003 FDA Consumer Health Information for Better

Nutrition Initiative [4] under which certain "qualified health claims" supported by credible, but less than conclusive scientific evidence, can be used in food labeling based on the FDA's exercise of regulatory discretion to allow the use of certain qualified health claims without an authorizing FDA regulation. The FDA uses the term qualified health claim to refer to health claims for which the scientific evidence does not meet the SSA standard, that include a disclaimer that explains (or qualifies) the level of scientific evidence supporting the claim, and that are the subject of a current FDA regulatory discretion enforcement decision.

8.3 Nutrient content claims

A food label statement that directly or by implication characterizes the level of a nutrient in the food is a nutrient content claim. The NLEA amendments restrict the use of nutrient content claims to those statements made in accordance with the FDA's authorizing regulations for nutrient content claims. The FDA's nutrient content claim regulations define how descriptive terms can be used to characterize the level of a nutrient in a food. The level of a nutrient in a food may be characterized using descriptive terms such as *free*, *high*, and *low*, or the level of a nutrient in one food may be compared to that in another food using descriptive terms such as *more*, *reduced*, and *lite*. The requirements governing the use of nutrient content claims help ensure that descriptive terms are used consistently for all types of food products and are thus meaningful to consumers. Nutrient content claims include both explicit claims (e.g., *low fat*) and implied claims (e.g., *healthy*). With a few exceptions, only nutrients with an established daily value for nutrition labeling may be the subject of a nutrient content claim. The general principles for the use of nutrient content claims are listed in the Code of Federal Regulations, title 21 part 101, section 13 (21 CFR 101.13).

Nutrient content claims have been defined only for situations where they will be nutritionally relevant. The terms *good source* (defined as 10−19% of daily value per reference amount customarily consumed; RACC[3]), *high* (defined as at least 20% of daily value per RACC), and *more* (defined as at least 10% of daily value more than that in a reference food) have been defined for nutrients that have a daily value. The requirements for the use of *good source*, *high*, and *more* content claims are listed in 21 CFR 101.54. The terms *free*, *low*, and *reduced* have been defined for use only with calories, total fat, saturated fat, cholesterol, sodium, and, in some cases, sugars and salt. These descriptive terms are defined individually for each of the listed nutrients and there are a number of additional criteria that apply as well. For example, the criteria for *sugar free* include a sugar

---

[3]Reference Amounts Customarily Consumed per Eating Occasion (RACC) are used as a reference point for determining the serving sizes used for nutrition labeling and as a uniform reference amount for basing nutrient content claims. RACCs for 139 food product categories are listed in 21 CFR 101.12.

content limit (0.5 g per RACC), a requirement that the label shows the calorie profile (e.g., *low calorie*) and trigger levels for other nutrients (such as total fat) that require a disclosure statement. Consequently, the regulations governing these claims should be read carefully before using them on a food label. Authorized nutrient content claims for sugar include *free*, *reduced*, and *no added sugar*; but a *low sugar* nutrient content claim is not permitted. Light (or lite) is a nutrient content claim that has been defined for certain foods reduced in fat and/or calories, or certain foods reduced in sodium content. The term light may also be used outside of a nutrient content claim context in describing organoleptic properties such as color, e.g., "light brown sugar." Requirements for the use of light/lite nutrient content claims are found in 21 CFR 101.56. Requirements for the use of content claims about calories and sugar are found in 21 CFR 101.60. Requirements for the use of content claims about sodium and salt are found in 21 CFR 101.61. Requirements for the use of content claims about fat, saturated fat, and cholesterol are found in 21 CFR 101.62.

One exception to the rule that only nutrients with established daily values can be the subject of nutrient content claims is a provision that permits the use of quantitative statements (e.g., 100 mg of lycopene). A quantitative statement may be used in food labeling to describe the amount of a nutrient present in a food, whether or not a daily value has been established for the nutrient, provided that the statement does not use language that characterizes the amount. A statement such as "packed with 100 mg of lycopene" characterizes the amount of lycopene in the food by suggesting this to be a significant amount of lycopene. Because there is not an established daily value for lycopene, the "packed with 100 mg of lycopene" statement would not be permitted. The terms *contains* and *provides* are defined as nutrient content claim descriptors synonymous with good source. However, when contains or provides are used with a quantitative statement about a nutrient for which there is no established daily value, the term is considered customary English usage, not a descriptive term nutrient content claim, thus "contains 100 mg lycopene" may be used in the same manner as "100 mg lycopene." When a quantitative statement is made about a nutrient for which there is an established daily value, the food must either meet the criteria of an appropriate nutrient content claim or carry a statement disclosing that the food does not comply with an authorized content claim. For example, "low sodium" is defined as less than 140 mg sodium per RACC; therefore a quantitative statement referencing 200 mg of sodium would need to be qualified such as: "200 mg sodium. Not low in sodium."

Nutrient content claims include both explicit and implied claims. Implied nutrient content claims describe a food or ingredient in a manner that suggests that a nutrient is either present in a certain amount, or is absent. For example, a food labeled "high in oat bran" implies that the food is high in dietary fiber and therefore the food should contain at least 20% of the daily value per RACC of dietary fiber. The term "healthy," and any derivation of healthy, when used in association with an explicit statement about a nutrient, e.g., "healthy low sodium," is considered to be an implied nutrient content claim characterizing the levels not only of the indicated nutrient, but also characterizing the amounts of total fat, saturated fat, cholesterol, and sodium in the food as all being at "healthy" levels. The FDA has set criteria to ensure that healthy is used consistently and in a meaningful way. There are five criteria to qualify as a healthy food. For most packaged foods these criteria include:

1. Low in total fat
2. Low in saturated fat
3. Cholesterol content less than 60 mg per RACC and per 50 g of food if the RACC is under 30 g or 2 tablespoons
4. Sodium content less than 480 mg per RACC and per 50 g of food if the RACC is under 30 g or 2 tablespoons
5. Contain at least 10% of the daily value per RACC for one or more of vitamin A, vitamin C, calcium, iron, protein, or dietary fiber

The fifth criterion (occasionally called the jelly bean rule) is included to prevent foods with little intrinsic nutritional value (such as candy) from being characterized as healthy. There are some variations in these criteria depending upon whether the labeled food is a fruit or vegetable, a grain product, a seafood or game meat product, or meal-type product. The rules for labeling a food with a healthy nutrient content claim are explained in the regulation 21 CFR 101.65(d).

Two nutrient content claim regulations of potential relevance to functional foods are regulations for the use of the terms "high potency" and "antioxidant." The level of vitamins and minerals in a food may be characterized as high potency when the individual vitamins or minerals characterized by the claim are present at 100% or more of the daily value per RACC. A high potency claim must identify the vitamin(s) and/or mineral(s) that are characterized as high potency unless the product is a single nutrient dietary supplement. Alternatively, a multivitamin/mineral product may be characterized as high potency without specifying the individual nutrients when at least two-thirds of all the vitamins and minerals listed in the product's nutrition information are present at 100% or more of the daily value. Nutrient content claims using the term antioxidant may be used for nutrients that have an established daily value and that have recognized bioavailable antioxidant activity when ingested. The nutrients that meet both of these criteria are vitamin C, vitamin E, and selenium. β-Carotene may also be the subject of an antioxidant nutrient content claim when the amount of vitamin A activity in the food, present as β-carotene, is sufficient to qualify for the claim. Vitamin A itself is not an antioxidant and the bioavailability of β-carotene's antioxidant potential is questionable [5]. Nevertheless, under current nutrient content claim regulations, β-carotene still may be the subject of an antioxidant nutrient content claim. To use the claim "high in antioxidant β-carotene," the food would need to contain at least 1000 International Units (IU) of vitamin A activity per RACC from β-carotene (20% of the vitamin A daily value is 1000 IU). At the time the FDA's nutrition labeling regulations were published, the accepted β-carotene/vitamin A conversion factor was 1.8 μg of β-carotene as equivalent to 1 IU of vitamin A activity. Thus, high in antioxidant β-carotene requires a minimum of 1.8 mg β-carotene per RACC. It is possible to craft a claim that is not an antioxidant content claim by describing a substance's antioxidant function rather than characterizing the amount of the substance. However, for such a claim to be truthful there must be evidence that the substance functions as an antioxidant within the human body after being ingested in food. *In vitro* measures of antioxidant potential, such as ORAC, are not evidence for *in vivo* antioxidant function.

A summary of the rules for the use of nutrient content claims can be found in A Food Labeling Guide [6]; examples of authorized nutrient content claims are provided in Appendices A and B of this guide. The FDA *A Food Labeling Guide* is available as a guidance document on the FDA webpage, www.FDA.gov.

8.4 **Health claims**

8.4.1 **Validity standard of health claims**

NLEA mandated that health claims may not be used in food labeling prior to FDA evaluation of supporting evidence and issuance of a regulation to authorize the health claim. NLEA allows the FDA to authorize a health claim only after the agency has determined that there is SSA among qualified scientific experts that the claim is supported by the totality of publicly available scientific evidence. The SSA standard for health claims has two components:

1. There is publicly available supporting scientific evidence from well-designed studies conducted in a manner consistent with generally recognized scientific procedures and principles.
2. There is SSA among qualified experts that the substance/disease relationship is supported by the totality of the publicly available scientific evidence.

An explanation of the FDA's use of the SSA standard to evaluate the supporting scientific evidence for a health claim is available on the FDA's Internet Website [7]. The SSA standard is intended to be a strong but flexible standard that provides a high level of confidence in the validity of a substance/disease relationship [7]. NLEA did not require the FDA to apply the SSA standard to dietary supplement health claims, rather the agency was given discretion to establish a validity standard it considered appropriate for dietary supplements. The FDA considered a "level playing field" approach to be preferable to multiple health claim standards; therefore, the FDA concluded that the same standard to which conventional food health claims were held (i.e., SSA) was also an appropriate standard for dietary supplement health claims.

8.4.2 General health claim regulation

The FDA General Requirements for Health Claims regulation (21 CFR 101.14) sets out general requirements applicable to all food label health claims. The FDA also issues specific authorizing regulations with additional criteria specific to each individual health claim meeting the SSA validity criteria. The general health claim regulation defines a health claim as a statement made on a food label that expressly, or by implication, characterizes the relationship of any substance to a disease or health-related condition. Implied health claims include any material, such as third-party references, written statements (e.g., a brand name including the word "heart"), and symbols (e.g., a heart symbol) presented in food labeling that suggests, within the context in which it is presented, that a relationship exists between a substance in the food and a disease or health-related condition.

There are two components to a health claim: the substance and the disease or health-related condition. A food label statement that does not include both components would not be a health claim. The FFD&C Act identifies the substance component of a health claim as any "nutrient which is of the type" required to be listed in the food label nutrition information. However, in that the NLEA also included dietary supplements of herbs or other similar nutritional substances in the health claim provisions of the Act, it is clear that the substance of a health claim was not intended to be limited strictly to "nutrients." The legislative history of NLEA shows that Congress intended that individual foods also could be the subject of health claims. The legislative history of NLEA also makes it clear that a health claim about a food must, at least by implication, be a claim about a substance in the food. When a consumer could reasonably interpret a claim about the relationship of a food to a disease or health-related condition to be an implied claim about a specific substance in the food, that claim would satisfy the "substance" component of a health claim. The FDA therefore has defined "substance" as meaning any specific food or a component of a food (e.g., nutrients). Defining substance broadly allows authorization of health claims for a wider variety of relationships (e.g., health claims for phytosterols) than would a narrower definition of substance encompassing nutrients only. Health claims about consuming a substance at decreased levels (e.g., saturated fat, cholesterol, and heart disease), however, are restricted to the nutrients that are required to be listed in the nutrition information on the food label.

The FDA's general health claim regulation includes a requirement that the substance in a health claim must contribute taste, aroma, nutritive value, or any other recognized technical effect of a food additive. This requirement impacts on the potential for use of health claims for food ingredients promoted for functional attributes other than the traditional food ingredient attributes of taste, aroma, and nutrition. Because the FDA recognizes that health benefits may be derived from food ingredients through a variety of processes, the FDA has provided for flexibility in determining whether a food substance has nutritional value. "Nutritive value" has been defined in the FDA's general health claim regulation as value in sustaining human existence by processes such as promoting growth, replacing loss of essential nutrients, and providing energy and includes assisting in the efficient function of classical nutritional processes and of other metabolic processes necessary for normal maintenance of human existence.

The second component of a health claim, the "disease or health-related condition," means damage to an organ, part, structure, or system of the body such that it does not function properly (e.g., cardiovascular disease), or a state of health leading to such dysfunction (e.g., hypertension). The disease component of a health claim must be one for which the general U.S. population, or an identified population subgroup (e.g., the elderly), is at risk. Diseases resulting from essential nutrient deficiencies (e.g., scurvy or pellagra) are excluded from the diseases that may be the subject of a health claim. A food label statement about a nutrient deficiency disease would not be regulated as a health claim. A food label statement about a nutrient deficiency disease should identify the prevalence of the nutrient deficiency disease in the United States so as not to be misleading.

The nature of the substance/disease relationships addressed in health claims is that of disease risk reduction. Whenever the intended use of a product is to cure, mitigate, treat, or prevent a disease, the product is considered to be a drug, rather than a food. Health claims cannot describe an effect of the substance to cure, mitigate, treat, or prevent a disease because health claims apply only to foods. For example, a label statement about use of a food or dietary supplement for relief from arthritis pain (e.g., mitigation of a disease symptom) is not a health claim because it infers intent that the product be used as a drug. This principle was upheld in the 2004 *Whitaker v. Thompson* court decision (353 F.3d 947 (D.C. Cir), cert. denied, 125 S Ct. 310 (2004)), which held that it was constitutionally permissible for the FDA to deny a health claim about use of saw palmetto extract dietary supplements for relief of urinary symptoms of benign prostatic hyperplasia on the basis that the claim constituted a drug use for the product.

NLEA prohibited the FDA from authorizing the use of health claims for foods that contain any nutrient in an amount that increases the risk of a diet-related disease. As such, the general health claim regulation has set threshold levels for several nutrients above which a food is disqualified from using health claims. The disqualifying nutrient levels for individual foods are 13 g of total fat, 4 g of saturated fat, 60 mg of cholesterol, and 480 mg of sodium per RACC. Where the actual serving size of a food is greater than the RACC, the disqualifying nutrient levels are determined on the labeled serving size, and the disqualifying nutrient levels for foods with a small RACC (defined as 30 g or less or 2 tablespoons or less) are determined on a per 50 g of food basis. Any food that exceeds one or more nutrient disqualifying levels is prohibited from using any health claim, even though that food may otherwise meet other requirements of a specific health claim. The FDA has authority to waive disqualifying nutrient levels in instances where the agency finds that allowing a health claim will assist consumers in maintaining healthy dietary practices in spite of a high level of a disqualifying nutrient, provided that the label contains a disclosure statement about the nutrient (e.g., "See nutrition information for fat content").

The general health claim regulation also restricts the use of health claims on labels of foods with minimal nutritional value. Except for dietary supplements, a food may not use a health claim unless the food contains at least 10% of the daily value for one or more of six core nutrients (dietary fiber, protein, vitamin A, vitamin C, calcium, or iron) per RACC. The food must be able to meet the 10% daily value minimum nutrient content requirement prior to any nutrient addition. This provision has come to be known as the jelly bean rule, because it was intended to restrict the use of health claims by foods with little inherent nutritional value, such as jelly beans. The FDA has made exceptions to the jelly bean rule in some health claims when a public health benefit warranted.

8.4.3 Authorized health claims

The FDA was directed initially to evaluate the available scientific evidence for 10 nutrient/disease relationships that Congress had considered to be probable health claim topics (Table 8.1). The FDA concluded there was SSA on 5 of the initial 10 nutrient/disease relationships and health claims for these 5 substance/disease relationships were authorized. The FDA also concluded that the available evidence did not warrant authorizing claims for two other nutrient/disease relationships. It concluded that the evidence for the remaining three nutrient/disease relationships (i.e., dietary fiber and cardiovascular disease, dietary fiber and cancer, and antioxidant vitamins and cancer) was inconclusive but that there was SSA for a relationship between diets characterized by the types of foods that are rich sources of these nutrients (i.e., fruits, vegetables, and grain products) and reduced disease risk. Therefore, the FDA authorized three alternative health claims for fruits, vegetables, and grain products that are rich in antioxidant vitamins or in soluble fiber (see Table 8.1).

The NLEA amendments directed the FDA to establish a petition procedure through which interested parties could request the FDA to authorize additional health claims. The NLEA amendments permitted the FDA to authorize a petitioned health claim only when the agency has determined that the claim meets general requirements for health claims established by the agency and that the SSA standard has been satisfied. NLEA specified that health claims authorizing regulations issued by the FDA will do the following:

1. Describe the substance/disease relationship of the health claim and the significance of the substance in affecting the disease.
2. Require that the claim be stated in a manner that is an accurate representation of the substance/disease relationship.

Table 8.1 Authorized Health Claims from the 1990 Nutrition Labeling and Education Act

Relationship	Initial 10 NLEA Health Claims
Calcium, vitamin D, and osteoporosis	Foods and supplements that are high in calcium, or high in both calcium and vitamin D, and reduced risk of osteoporosis,[a] 21 CFR 101.72[b]
Dietary fat and cancer	Foods that are low in total fat,[c] or extra lean fish and game meat and reduced risk of cancer, 21 CFR 101.73
Sodium and hypertension	Foods and supplements that are low in sodium and reduced risk of high blood pressure, 21 CFR 101.74
Saturated fat/cholesterol and cardiovascular disease	Foods and supplements that are low in both saturated fat and cholesterol,[c] or extra lean fish and game meat and reduced risk of coronary heart disease, 21 CFR 101.75
Dietary fiber and cancer	Fruit, vegetable, or grain products that are low fat and are a good source of dietary fiber and reduced risk of cancer, 21 CFR 101.76
Dietary fiber and cardiovascular disease	Fruit, vegetable, or grain products that are low in total fat, saturated fat, and cholesterol, and contain soluble dietary fiber and reduced risk of coronary heart disease, 21 CFR 101.77
Antioxidant vitamins and cancer	Fruits or vegetables that are low fat and a good source of vitamin A, vitamin C, or dietary fiber and reduced risk of cancer, 21 CFR 101.78
Folic acid and neural tube birth defects	Foods/supplements that are a good source of folate and reduced risk of neural tube birth defects,[d] 21 CFR 101.79
Zinc and immune function in the elderly	No authorized claim
Omega-3 fatty acids and heart disease	No authorized claim[e]

[a]Initially authorized as a calcium and osteoporosis claim; the authorizing regulation was amended in 2010 to include an optional "calcium and vitamin D" claim.
[b]Food labeling regulations are codified in Title 21 of the Code of Federal Regulations (CFR), Part 101. The regulation authorizing the health claim about the relationship of dietary calcium and osteoporosis risk is in Section 72 of this part, i.e., 21 CFR 101.72.
[c]Although dietary supplements are not explicitly excluded from use of these claims, these claims are not available for many dietary supplement products because the "low" nutrient content claims for total fat, saturated fat, and cholesterol are not defined for "low calorie"/"calorie-free" dietary supplements that are excluded from use.
[d]Authorization of the folate health claim was delayed until 1996 pending resolution of safety questions.
[e]Although there was insufficient scientific evidence to support an authorizing regulation, there is a qualified health claim for omega-3 fatty acids and heart disease.

3. Require that the claim be stated in a manner that enables the public to understand the information provided in the claim and to understand the relative significance of such information in the context of a total daily diet.

The FDA has authorized new health claims in response to a number of health claim petitions over the past decade (Table 8.2). The health claims that have been authorized as the result of health claim petitions include claims about sugar substitutes and dental caries, whole oat and barley β-glucan fiber and heart disease, psyllium husk fiber and heart disease, soy protein and heart disease, and plant sterol/stanol esters and heart disease (see Table 8.2).

The health claim authorization procedure established by NLEA involves FDA evaluation of supporting scientific evidence. Upon a determination of SSA that the available scientific evidence substantiates the substance/disease relationship, the FDA follows "notice and comment" rulemaking to issue an authorizing regulation. The FDAMA provision for authoritative statement-based health claims serves as an alternative to the health claim petition and FDA rulemaking process. Substantiation of an authoritative statement-based health claim derives from a review of scientific evidence conducted by any U.S. federal scientific agency, other than the FDA, with official responsibility for public health protection or human nutrition research or by the National Academy of Sciences or any of its subdivisions, where the result of such review is published as an authoritative statement of that agency about a dietary substance/disease risk reduction

Table 8.2 Authorized Health Claims from Health Claim Petitions

Relationship	Authorized Health Claim
Non-cariogenic carbohydrate sweeteners and dental caries	Sugar-free foods sweetened with certain sugar alcohols and other sugar substitutes: xylitol, sorbitol, mannitol, maltitol, isomalt, lactitol, hydrogenated starch hydrolysates, hydrogenated glucose syrups, erythritol, D-tagatose, and sucralose, 21 CFR 101.80
Soluble fiber from certain foods and coronary heart disease	Foods that contain a barley (whole grain or β-fiber) or oat (oat bran, oatmeal, whole oat flour, or oatrim) ingredient, and contain least 0.75 g β-glucan fiber per reference amount, and are also low in total fat, saturated fat, and cholesterol. Foods that contain psyllium seed husk providing at least 1.7 g soluble fiber per reference amount, and are also low in total fat, saturated fat, and cholesterol, 21 CFR 101.81
Soy protein and coronary heart disease	Foods low in total fat, saturated fat, and cholesterol and that contain at least 6.25 g soy protein per reference amount, 21 CFR 101.82
Phytosterols and coronary heart disease	Foods/supplements that are low in saturated fat and cholesterol and that contain at least 0.5 gram phytosterols per reference amount,[a] 21 CFR 101.83

[a]The authorizing health claim regulation had been amended in 2011; however, enforcement of claim criteria in the amended regulation that are different from the original regulation is being delayed. Check the FDA Website for the current status of this health claim.

relationship. The same SSA validity standard applies in both processes, but they differ in where the deliberative review of the scientific evidence occurs. The process leading to authorization of an authoritative statement-based health claim begins with submission to the FDA of a notice that includes the exact wording to be used in the health claim, the authoritative statement that is the source of the claim, and a balanced representation of the scientific literature relating to the substance/disease relationship to which the claim refers. The new health claim that is the subject of an authoritative statement health claim notification becomes authorized by law 120 days after receipt of the notification by the FDA provided it has taken no regulatory action during that period to prohibit or modify the claim. The FDA uses the 120 day period to determine whether both the notification and the claim meet the requirements established by FDAMA. It may issue regulations to prohibit, or to modify, authoritative statement-based health or nutrient content claims. The FDA guidelines on the procedure for notifying it of intent to use health claims based on an authoritative statement from a scientific body is available on the Internet [8]. Table 8.3 lists the authorized health claims that are based on authoritative statement health claim notifications.

8.4.4 Qualified health claims

In 1999, the U.S. Court of Appeals for the DC Circuit issued its decision in *Pearson v. Shalala* (164 F.3d 650, D.C. Cir. 1999) (*Pearson*). The *Pearson* plaintiffs had challenged the FDA's application of the SSA standard to dietary supplement health claims and the FDA's decision not to authorize health claims for four specific substance/disease relationships: dietary fiber and colorectal cancer, antioxidant vitamins and cancer, omega-3 fatty acids and coronary heart disease, and the claim that 0.8 mg of folic acid in dietary supplement form is more effective in reducing the risk of neural tube birth defects than a lower amount in conventional food form. The FDA had denied these four claims after concluding that these claims did not meet the SSA substantiation standard and the claims therefore would be misleading. The court held in *Pearson* that the Constitutional protection of free speech does not permit the FDA to reject health claims that the agency determines to be potentially misleading unless the agency can also show that adding disclaimer would not eliminate the potential deception. The court stated in *Pearson* that the potential for claims that do not meet the SSA validity standard to mislead the consumer could be remedied by including "qualifying" language in the claim, which would disclose the level of scientific support for the claim (i.e., disclosure of more information is preferential to suppressing information).

The FDA interpreted the *Pearson* court opinion to suggest that where there was more scientific evidence against a specific claim than there was evidence in support of the claim, the claim would be inherently misleading (rather than merely potentially misleading) and could be prohibited. FDA initially applied a "weight of the evidence" standard, as an alternative to the SSA standard, in evaluating one of the first health claim topics to be re-evaluated under *Pearson*

Table 8.3 Authorized Health Claims Based on an Authoritative Statement

Relationship	Authorized Health Claim
Whole grains and heart disease and cancer	Foods with 51% or more whole grain ingredients; dietary fiber content is the measurable marker of whole grain content. There are two versions of the claim:
	Diets rich in whole grain foods and other plant foods and low in total fat, saturated fat and cholesterol may reduce the risk of heart disease and some cancers.
	Diets rich in whole grain foods and other plant foods and low in saturated fat and cholesterol may reduce the risk of heart disease.
	FDA Docket no. 1999P-2209 and 2003Q-0547
Potassium and high blood pressure and stroke	Foods that are good sources of potassium in addition to being low in total fat, saturated fat, cholesterol, and sodium.
	Diets containing foods that are a good source of potassium and that are low in sodium may reduce the risk of high blood pressure and stroke.
	FDA Docket no. 2000Q-1582
Fluoridated water and dental caries	Fluoridated bottled water, except those specifically marketed for use by infants.
	Drinking fluoridated water may reduce the risk of dental caries [or tooth decay]. FDA Docket no. 2006Q-0418
Dietary fats and heart disease risk	Foods that are low in saturated fat and cholesterol, moderate in fat content, and contain less than 0.5 g/RACC trans fat.
	Diets low in saturated fat and cholesterol and as low as possible in trans fat may reduce the risk of heart disease.
	FDA Docket no. 2006Q-0458
Substitution of saturated fat in the diet with unsaturated fatty acids and heart disease risk	Vegetable oil and shortening-containing foods that have a total unsaturated fat content of 80% or more of total fat content.
	Replacing saturated fat with similar amounts of unsaturated fats may reduce the risk of heart disease. To achieve this benefit, total daily calories should not increase.
	FDA Docket No. 2007Q-0192

(a claim for antioxidant vitamin supplements and cancer risk). The FDA concluded that there was more scientific evidence against the validity of this claim than there was to support the claim, thus it would not permit a qualified health claim for antioxidant vitamins and cancer risk because the weight of available scientific evidence did not support this substance/disease relationship [9]. The FDA's interpretation of *Pearson* in the decision to deny a qualified health claim for antioxidant vitamin supplements and cancer was challenged in another lawsuit (*Whitaker v. Thompson*, 248 F. Supp 2d 1 (D.D.C. 2002)) (*Whitaker*). The court's opinion in *Whitaker* was that the FDA's weight of the evidence standard misinterpreted *Pearson* and that qualified health claims must be permitted if there is any credible scientific evidence in support of a claim. Consistent with *Whitaker*, the FDA now uses regulatory discretion to permit qualified health claims where there is any credible scientific evidence to support the claim.

To develop a framework to help consumers obtain accurate, up-to-date, and science-based information about conventional foods and dietary supplements, the FDA convened the Consumer Health Information for Better Nutrition Initiative Task Force to establish guidelines for health claims that do not meet the SSA standard [4]. Guidance setting out the procedures that the FDA will use for qualified health claims in conventional foods and dietary supplements was explained by the Guidance for Industry and FDA: Interim Evidence-Based Ranking System for Scientific Data [10].

To provide for the use of qualified health claims as directed by *Pearson*, the FDA uses regulatory discretion in enforcing the requirement that the use of health claims on food labels must be in accordance with authorizing regulations. The FDA's qualified health claim enforcement discretion decisions are posted on the FDA Internet Website rather than codifying the decisions in a regulation. The agency letters announcing an enforcement discretion decision include an explanation of the rationale for the decision and specify the conditions under which the agency will consider the exercise of its enforcement discretion (e.g., the exact wording of the qualifying language). The enforcement discretion decision letters caution that since the qualified health claim topics are areas where the evidence for the claim is incomplete and subject to change, the FDA intends to evaluate new information as it becomes available to determine whether the new evidence necessitates a change in the decision. Table 8.4 presents a listing of qualified health claims that have been evaluated by the FDA.

8.5 Structure/function claims

Food and supplement labels may use claims about the products' effects on structures or functions of the body. The FFD&C Act recognizes explicitly that foods, as well as drugs, can be used to affect the structure or any function of the body.[4] The DSHEA amended the FFD&C Act to clarify the difference between health claims and of structure/function claims made for dietary supplement products. The FDA uses the phrase "structure/function claim" to refer to food label statements that describe the role of a nutrient or other dietary supplement ingredient in maintaining normal structure or function in humans (e.g., calcium builds strong bones) or to promote general well-being. Structure/function claims can also characterize the means by which a nutrient or other dietary ingredient acts to maintain a body structure or function (e.g., antioxidants maintain cell integrity or fiber

---

[4]Drugs are defined, in part, as "... articles (other than food) intended to affect the structure or any function of the body of man ..." FFD&C Act section 201(g)(1)(C).

Table 8.4 Qualified Health Claims Considered under the FDA's Exercise of Enforcement Discretion

Enforcement discretion considered[a]
Atopic dermatitis
100% whey-protein partially hydrolyzed infant formula and reduced risk of atopic dermatitis
Cancer
Antioxidant vitamins C and E and reduction in the risk of site-specific cancers Tomatoes and prostate, ovarian, gastric, and pancreatic cancers Tomatoes and prostate cancer Calcium and colon/rectal cancer and calcium and colon/rectal polyps [6] Green tea and risk of breast cancer and prostate cancer Selenium and certain cancers Antioxidant vitamins and risk of certain cancers
Cardiovascular disease
Folic acid, vitamin B6, and vitamin B12 and vascular disease Nuts and coronary heart disease Walnuts and coronary heart disease Omega-3 fatty acids and reduced risk of coronary heart disease Corn oil and corn oil-containing products and a reduced risk of heart disease Unsaturated fatty acids from canola oil and reduced risk of coronary heart disease Monounsaturated fatty acids from olive oil and coronary heart disease
Cognitive function
Phosphatidylserine and cognitive dysfunction and dementia
Diabetes
Chromium picolinate and a reduced risk of insulin resistance, type 2 diabetes
Hypertension
Calcium and hypertension, pregnancy-induced hypertension, and preeclampsia
Neural Tube Defects
Folic acid and neural tube defects

The list of qualified health claims permitted and the claim criteria frequently changes. A current listing of qualified health claim petitions considered by FDA is posted at http://www.fda.gov/Food/IngredientsPackagingLabeling/LabelingNutrition/ucm072756.htm.
[a]*The enforcement discretion letters describe the specific conditions under which the claim can be used, including the qualifying language that characterizes the strength of the scientific evidence behind the claim.*

maintains bowel regularity). Health claims and structure/function claims differ in that there are two elements to a health claim: a substance and a disease or health-related condition, whereas, structure/function claims have the first element (a substance) but cannot imply any relationship of the substance to a disease or health-related condition. The context of a statement within the entire label is considered in determining whether a label statement implies a relationship to a disease or a health-related condition. For example, a statement such as "supports the cardiovascular system" does not of itself imply a disease connection. However, this same statement on the label of a product called "CardioCure" would constitute a disease claim within the context of the entire label. A food label statement that makes, or implies, a relationship with a disease or health-related condition is either a health claim or a drug claim, not a structure/function claim. An exception to the prohibition of structure/function claims from referencing a disease is that structure/function claims may describe a benefit related to a classical nutrient deficiency disease (e.g., vitamin C and scurvy) when the statement also tells how widespread such a disease is in the United States. The distinction between a disease claim and a dietary supplement structure/function claim is described in 21 CFR 101.93(g). Structure/function claims, like all food labeling statements, must be truthful and not misleading. Unlike health claims and nutrient content claims, there are no FFD&C Act provisions that provide for the FDA's prior review or approval of structure/function claims before their use. Structure/function claims are not regulated by the FDA to the extent of health claims and nutrient content claims, but DSHEA set three conditions for the use of structure/function claims in the labeling of dietary supplements. The first condition is that the manufacturer of a dietary supplement using a structure/function claim has substantiation that the claim is truthful and not misleading. The second condition is that the following statement must be displayed prominently and in boldface type on the label with the structure/function claim:

> *This statement has not been evaluated by the Food and Drug Administration.*
> *This product is not intended to diagnose, treat, cure, or prevent any disease.*

The third condition is that the dietary supplement manufacturer notify the FDA, within 30 days of first marketing a dietary supplement label bearing a structure/function claim, that the statement is being made. The FDA examines the dietary supplement structure/function claim notifications only to determine that the claims being made are not claims for an effect related to a disease. It does not evaluate the substantiation for, or approve, the structure/function claims in these notifications.

DSHEA's structure/function claim conditions apply specifically to dietary supplement labeling. Structure/function claims, whether used on conventional food labels or dietary supplement labels, need to be adequately substantiated to satisfy the "not false or misleading" criterion that applies to all label information. However, there are no provisions in the FFD&C Act requiring manufacturers using structure/function claims on conventional food labels to include the disclaimer statement or to notify the FDA of their structure/function claims. Unlike dietary supplements, structure/function claims made for conventional foods must be about effects that derive from the taste, aroma, or nutritive value of the food or food ingredient that is the subject of the claim. This limitation on use of structure/function claims for conventional foods is based on a federal court ruling that structure/function claims on a conventional food label cause the product to be a drug if the claim promotes the product for an effect that is unrelated to the product's "food attributes" of taste, aroma, and nutritive value (*Nutrilab v. Schweiker*, 713 F. 2d 335 (7th Cir. 1983)). Table 8.5 compares regulatory differences between health claims and structure/function claims used on conventional foods and dietary supplements.

Because structure/function claims for dietary supplements must be truthful and not misleading, the FDA recommends that manufacturers possess adequate substantiation for all reasonable interpretations of the claims made. The FDA recommends applying a standard that is consistent with the FTC standard of "competent and reliable scientific evidence" to substantiate a claim [11]. The FDA considers

Table 8.5 Requirements for Health Claims and Structure/Function Claims

	Health Claims: Conventional Food and Dietary Supplements	Structure/ Function Claims: Dietary Supplements	Structure/ Function Claims: Conventional Food
Petition FDA for new claims	Yes	No	No
Requires FDA review and approval	Yes	No	No
Must not be false or misleading	Yes	Yes	Yes
Notify FDA within 30 days of marketing product labeled with the claim	No	Yes	No
Include a disclaimer stating the claim has not been evaluated by FDA	No	Yes	No
Each manufacturer using the claim must possess scientific evidence to substantiate the claim is true for their product	No	Yes	Yes
Product subject to disqualifying nutrient levels and minimum nutrient levels	Yes	No	No
Substance must have food attributes such as nutritive value	Yes	No	Yes

the following factors important to establish whether information would constitute competent and reliable scientific evidence:

- Does each study or piece of evidence bear a relationship to the specific claim?
- What are the strengths and weaknesses of the individual study or evidence? Consider the type of study, the design of the study, analysis of the results, and peer review.
- If multiple studies exist, do the studies that have the most reliable methodologies suggest a particular outcome?
- If multiple studies exist, what do most studies suggest or find? Does the totality of the evidence agree with the claims?

The FDA has recently provided guidelines on the amount, type, and quality of evidence that FDA recommends a manufacturer have to substantiate a structure/function claim [12].

References

[1] Nutrition Labeling and Education Act of 1990. Public Law 101−535. November 8, 1990. Summary available at, <http://thomas.loc.gov/cgi-bin/bdquery/z?d101:HR03562: @@@D&summ1&TOM:/bss/d101query.html>.

[2] Dietary Supplement Health and Education Act of 1994. Public Law 103−417. October 25, 1994. Available at, <http://www.fda.gov/RegulatoryInformation/Legislation/FederalFood DrugandCosmeticActFDCAct/SignificantAmendmentstotheFDCAct/ucm148003.htm>.

[3] Food and Drug Administration Modernization Act of 1997. Public Law 105−115. November 21, 1997. Available at, <http://www.fda.gov/RegulatoryInformation/ Legislation/FederalFoodDrugandCosmeticActFDCAct/SignificantAmendmentstotheFDC Act/ FDAMA/default.htm>.

[4] Food and Drug Administration. Consumer Health Information for Better Nutrition Initiative. Task Force Final Report. July 2003. Available at, <http://www.fda.gov/ Food/IngredientsPackagingLabeling/LabelingNutrition/ucm096010.htm>.

[5] Food and Nutrition Board, Institute of Medicine. Vitamin A. Dietary reference intakes for vitamin A, vitamin K, arsenic, boron, chromium, copper, iodine, iron, nickel, silicon, vanadium, and zinc. Washington, DC: The National Academy Press; 2001.

[6] Food and Drug Administration. A Food Labeling Guide. September 1994. Available at, <http://www.fda.gov/Food/GuidanceRegulation/GuidanceDocumentsRegulatoryInformation/ LabelingNutrition/ucm2006828.htm>.

[7] Food and Drug Administration. Guidance for Industry: Evidence-Based Review System for the Scientific Evaluation of Health Claims. January 2009. Available at, <http://www. fda.gov/Food/GuidanceRegulation/GuidanceDocumentsRegulatoryInformation/Labeling Nutrition/ucm073332.htm>.

[8] Food and Drug Administration. Guidance for Industry: Notification of a Health Claim or a Nutrient Content Claim Based on an Authoritative Statement of a Scientific Body, June 1998. Available at, <http://www.fda.gov/Food/GuidanceRegulation/ GuidanceDocumentsRegulatoryInformation/LabelingNutrition/ucm056975.htm>.

[9] Letter to Jonathan Emord (Emord & Associates) from Christine Lewis (FDA, CFSAN). Re: Petition for Health Claim: Antioxidants and Cancer (Docket No. 91N-0101). May 4, 2001.

[10] Food and Drug Administration. Guidance for Industry and FDA: Interim Evidence-Based Ranking System for Scientific Data. July 10, 2003. Available at, <http://www.fda.gov/Food/GuidanceRegulation/GuidanceDocumentsRegulatoryInformation/LabelingNutrition/ucm053832.htm>.

[11] Bureau of Consumer Protection, Federal Trade Commission. Dietary Supplements: An Advertising Guide for Industry. April 2001. Available at, <http://business.ftc.gov/documents/bus09-dietary-supplements-advertising-guide-industry>.

[12] Food and Drug Administration. Guidance for Industry: Substantiation for Dietary Supplement Claims Made Under Section 403(r)(6) of the Federal Food Drug, and Cosmetic Act, November 2004. Available at, <http://www.fda.gov/Food/GuidanceRegulation/GuidanceDocumentsRegulatoryInformation/DietarySupplements/ucm073200.htm>.

Nutrition and Health-Related Labeling Claims for Functional Foods and Dietary Supplements in the United States

Sanjiv Agarwal*, Stein Hordvik[†] and Sandra Morar**

*NutriScience LLC East Norriton, Pennsylvania [†]Hordvik's Consulting, Elkhorn, Nebraska
**McGrath North Mullin & Kratz PC LLO, Omaha, Nebraska

9.1 Introduction

A diet and health relationship was initially proposed in the fourth century BC by Hippocrates. Although, there is significant scientific agreement that diet plays an important role in health, many Americans make food choices that do not follow healthy dietary pattern [1], are not meeting the dietary guidelines recommendations, and not consuming enough nutrients including vitamins and minerals from foods [1,2]. Food labels are important tools to educate consumers about the healthfulness of specific foods and the benefits of following a nutritious diet. In addition to the mandatory nutrition labeling information (Nutrition Facts or Supplement Facts), current Food and Drug Administration (FDA) regulations [3] provide several provisions to communicate the healthfulness of the food products to the consumers.

9.2 Nutrient content claims

Nutrient content claims are regulated by the FDA under the Nutrition Labeling and Education Act (NLEA) [4]. They explicitly or implicitly characterize the level of a nutrient in a food. These claims are made to describe the nutrient levels using such terms as "free," "low," and "high" as well as to compare the nutrient levels using terms "more," "reduced," and "light":

Nutraceutical and Functional Food Regulations in the United States and Around the World.
DOI: http://dx.doi.org/10.1016/B978-0-12-405870-5.00009-8

1. Characterizing the nutrient levels
   - "Free" identifies food that contains a nutrient at inconsequential levels.
   - "Low" identifies food that is distinctly low in a nutrient compared to a daily value (DV).
   - "Good Source" or "Excellent Source" identifies foods that contain higher levels and contribute significantly toward the DV.
2. Comparing the nutrient levels
   - "Reduced," "More," or "Light" identifies nutritionally meaningful differences from a reference product.
3. Implied nutrient claim: implied nutrient content claims are claims about a food or an ingredient in a food suggesting that a nutrient or ingredient is present in certain amounts.
   - "Healthy" is an implied nutrient content claim.
   - "Lean" and "Extra Lean" are terms that can be used to describe the fat content of meat, poultry, seafood, and game meats.

Table 9.1 provides a list of nutrient content claims and their specific requirements (as of March 2013). These claims are, by and large, allowed for those nutrients for which the FDA has established a daily value. Nutrient content claims

Table 9.1 Nutrient Content Claims (Numbers are Rounded) as of March 2013

Nutrients CFR Reference	Claims
Calories 21 CFR 101.60(b)	Free: Less than 5 calories per reference amount (RACC) and serving
	Low: 40 calories or less per RACC or per 50 g if RACC is small; 120 calories or less for meals and main dishes
	Reduced or less: At least 25% less calories per RACC (per 100 g for meals and main dishes) than an appropriate reference food
Total fat 21 CFR 101.62(b)	Free: Less than 0.5 g fat per RACC and serving
	Low: 3 g or less fat per RACC or per 50 g if RACC is small; 3 g or less fat per 100 g and not more than 30% calories from fat for meals and main dishes
	Reduced or less: At least 25% less fat per RACC (per 100 g for meals and main dishes) than an appropriate reference food
Saturated fat 21 CFR 101.62(c)	Free: Less than 0.5 g saturated fat and less than 0.5 g trans fat per RACC and serving
	Low: 1 g or less saturated fat per RACC and less than 15% calories from saturated fat; 1 g or less fat per 100 g and not more than 10% calories from fat for meals and main dishes
	Reduced or less: At least 25% less saturated fat per RACC (per 100 g for meals and main dishes) than an appropriate reference food

(Continued)

Table 9.1 (Continued)	
Nutrients CFR Reference	**Claims**
Cholesterol 21 CFR 101.62(d)	Free: Less than 2 mg cholesterol per RACC and serving or per serving for meals and main dishes
	Low: 20 mg or less cholesterol per RACC or per 50 g if RACC is small; 20 mg or less cholesterol per 100 g for meals and main dishes
	Reduced or less: At least 25% less cholesterol per RACC (per 100 g for meals and main dishes) than an appropriate reference food
	(Cholesterol claims only allowed when food contains 2 g or less saturated fat)
Sodium 21 CFR 101.61	Free: Less than 5 mg sodium per RACC and serving or per serving for meals and main dishes
	Low: 140 mg (35 mg for "Very Low") or less sodium per RACC or per 50 g if RACC is small; 140 mg or less sodium per 100 g for meals and main dishes
	Reduced or less: At least 25% less sodium per RACC (per 100 g for meals and main dishes) than an appropriate reference food
Sugar 21 CFR 101.60(c)	Free: Less than 0.5 g sugar per RACC and serving or per serving for meals and main dishes
	Low: Claim is not defined and therefore not permitted
	Reduced or less:– At least 25% less sugar per RACC (per 100 g for meals and main dishes) than an appropriate reference food
Vitamins, minerals, fiber and protein 21 CFR 101.54	Good source: 10–19% DV per RACC Excellent source: 20% or more DV per RACC

CFR, Code of Federal Regulations; small RACC, RACC of 30 g or less.

are currently permitted for calories, fat, saturated fat, cholesterol, sodium, sugar, vitamins, minerals, fiber, and protein. Nutrient content claims for total carbohydrate (except for sugar and fiber) are currently not defined by the FDA and therefore are prohibited.

For nutrients/food components, which do not have a DV, a factual statement specifying the nutrient quantity present in the food, e.g., "... mg of ... per serving" can be used on a food label. However, statements that characterize the amount of the nutrient as being high or low, e.g., "only ... mg of ... per serving" are not permissible and would be considered to be implied nutrient content claims. Also, quantitative statements must include the unit of measure.

The general principles for making these claims are provided in 21 CFR 101.13, 101.54–101.69 (6). Additional requirements for nutrient content claims

Table 9.2 Implied Nutrient Content Claims as of March 2013

Claim	Requirements
Healthy 21 CFR 101.65(d)(2)	Low fat, Low saturated fat: Cholesterol below 60 mg (90 mg for meals and main dishes) and sodium below 480 mg (600 mg for meals/main dishes); at least 10% of the DV for one (individual foods), two (main dishes), or three (meals) of the six nutrients (vitamin A, vitamin C, iron, calcium, protein. or fiber)
Lean 21 CFR 101.62(e)(1)–(3)	Less than 10 g total fat, 4.5 g saturated fat, and 95 mg of cholesterol per RACC and labeled serving (per 100 g and per serving for meals/main dishes); less than 8 g fat, 3.5 g saturated fat, and 80 mg cholesterol per RACC for mixed dishes not measurable with a cup
Extra Lean 21 CFR 101.62(e)(4)&(5)	Less than 5 g total fat, 2 g saturated fat, and 95 mg of cholesterol per RACC and labeled serving (per 100 g and per serving for meals/main dishes)

CFR, Code of Federal Regulations.

include: (1) for foods with a Reference Amount Customarily Consumed (RACC) less than 30 g (small RACC), nutrient content claims must be determined as if there were 50 g of product and (2) a disclosure statement, "See nutrition information for … content," is also required as part of the claim when levels of fat, saturated fat, cholesterol, and/or sodium exceed 13 g, 4 g, 60 mg, and 480 mg, respectively, to call the consumer's attention to one or more nutrients in the food that may increase the risk of a disease or health-related condition that is diet related. Table 9.2 provides a list of implied nutrient-content claims and their requirements (as of March 2013).

9.3 Structure/function claims

Structure/function claims describe the role of a nutrient or a substance in food or a food supplement in affecting the normal structure or function in the body. These also refer to a change in, the support of, maintenance, or functions of the body and may also describe the mechanism of action. These, however, do not relate to a disease or a health-related condition. For example, "calcium helps build strong bones" is a structure/function claim, whereas "calcium reduces the risk of osteoporosis" is a health claim. The key components of structure/function claims are safety, scientific basis, and the nutritive value. Table 9.3 provides some examples of structure/function claims used in the market. Some of these claims may not have strong scientific support and therefore the regulatory agencies may consider them as "misleading to consumers." Pre-market approval for structure/function claims is not required. The FDA requires that these claims be truthful, non-misleading, and substantiated by the appropriate scientific data and the food/

Table 9.3 Common Structure/Function Claims Currently in the Marketplace

Structure/Function Claims	Nutrient
Calcium is important for both men and women to help build strong bones	Calcium
Fiber promotes digestive health	Fiber
Protein for muscle strength	Protein
Grape juice may promote healthy arteries	Flavonoids
Antioxidant vitamins A, C, and E help the body's natural defenses	Vitamin A, C, and E
Antioxidants may help protect against the damaging effects of free radicals	Lycopene, vitamin A, C, and E, selenium, flavonoids
Antioxidants help provide support for your body's natural defenses	Lycopene, vitamin A, C, and E, selenium, flavonoids
For healthy eyes	Lutein, vitamin A

supplement manufacturer must ensure their accuracy and truthfulness. In the case of dietary supplements, these claims are also accompanied by a disclaimer stating that the "FDA has not evaluated this claim and this product is not intended to diagnose, treat, cure, or prevent any disease." Moreover, the FDA has to be notified about the claim, including the text, within 30 days of marketing a dietary supplement bearing a new structure/function claim. The FDA has provided a guidance document to distinguish between structure/function claims and disease claims [5].

9.4 Health claims

Health claims describe the relationship between a substance (food, food component, or dietary supplement) and a disease or a health-related condition. There are two important constituents of a health claim: a substance and a disease or health-related condition. Claims relating to a dietary pattern instead of a substance are not considered as health claims. Unlike structure/function claims, health claims must be pre-approved by the FDA. The FDA conducts an evidence-based review to ascertain the scientific validity of the claim. It reviews and authorizes the health claims by three means [6]:

1. *Claims based on significant scientific agreement*: NLEA and the Dietary Supplement Health and Education Act (DSHEA) of 1994 allow health claims on food or supplement labels describing the role of a substance in disease risk reduction. The FDA authorizes these claims based on the totality of publicly available scientific evidence and using significant scientific agreement (SSA) criteria to determine the validity of the substance/disease relationship. A list of current (as of March 2013) FDA-approved SSA health claims is presented in Table 9.4.

Table 9.4 Approved NLEA Health Claims with SSA as of March 2013

Health Claim	CFR Reference
Calcium and osteoporosis	21 CFR 101.72
Dietary lipids (fat) and cancer	21 CFR 101.73
Sodium and hypertension	21 CFR 101.74
Dietary saturated fat and cholesterol and risk of coronary heart disease	21 CFR 101.75
Fiber-containing grain products, fruits and vegetables, and cancer	21 CFR 101.76
Fruits, vegetables, and grain products that contain fiber, particularly soluble fiber, and risk of coronary heart disease	21 CFR 101.77
Fruits and vegetables and cancer	21 CFR 101.78
Folic acid and neural tube defects	21 CFR 101.79
Dietary non-cariogenic carbohydrate sweeteners and dental caries	21 CFR 101.80
Soluble fiber from certain foods and risk of coronary heart disease	21 CFR 101.81
Soy protein and risk of coronary heart disease	21 CFR 101.82
Stanols/sterols and risk of coronary heart disease	21 CFR 101.83

CFR, Code of Federal Regulations.

Table 9.5 Authorized FDAMA Health Claims as of March 2013

Health Claim	Docket Reference
Whole grain foods and the risk of heart disease and certain cancers	1999P-2209 03Q-0574 2008Q-0270
Potassium and the risk of high blood pressure and stroke	2000Q-1582
Nutrient content claim (good source) of choline	01Q-0352
Fluoridated water and reduced risk of dental caries	2006Q-0418
Saturated fat, cholesterol and trans fat and the risk of heart disease	2006Q-0458 2007Q-0192

2. *Claims based on authoritative statement*: The FDA Modernization Act (FDAMA) of 1997 also allows health claims based on an authoritative statement issued by a scientific body of the U.S. Government bearing a public health protection responsibility, such as The National Institutes of Health, The Centers for Disease Control, or The National Academy of Science. However, FDAMA claims are not available for dietary supplements. Table 9.5 lists the health claims that are currently (as of March 2013) approved under the FDAMA process.

3. *Qualified health claims*: Consumer Health Information for Better Nutrition Initiative of FDA (2003) provides the use of health claims (qualified health

claims; QHC) for foods or dietary supplements where the scientific evidence to support a substance/disease relationship is still emerging and has not developed enough to meet the SSA standard. These claims have to include qualifying language as part of the claim indicating that the evidence supporting the claim is limited. In July 2003, the FDA provided interim guidelines outlining the petition process and the evidence-based ranking system to evaluate the scientific data concerning the claim. Since July 2003, the FDA has issued letters of enforcement discretion for (authorized) several qualified health claims for foods and dietary supplements. Table 9.6 shows all the approved QHCs to date (as of March 2013).

Table 9.6 Qualified Health Claims as of March 2013

Permitted QHCs	Docket Reference
Folic acid (0.8 mg) and the risk of neural tube birth defects[a]	91N-100H
B vitamins (B6, B12 and folic acid) and the risk of vascular disease[a]	99P-3029
Selenium and the risk of cancer[a]	02P-0457 2008-Q-0323
Phosphatidylserine and the risk of cognitive dysfunction and dementia[a]	02P-0413
Antioxidant vitamins (C and E) and the risk of cancer[a]	91N-0101 2008-Q-0299
Nuts and the risk of heart disease	02P-0505
Walnuts and the risk of heart disease	02P-0292
Omega-3 fatty acids (EPA and DHA) and the risk of coronary heart disease	2003Q-0401
Monounsaturated fatty acids from olive oil and the risk of coronary heart disease	2003Q-0559
Green tea and the risk of certain cancer	2004-Q-0427
Calcium and the risk of colorectal cancer[a]	2004Q-0097
Calcium and the risk of hypertension[a]	2004Q-0098
Chromium picolinate and the risk of insulin resistance or type 2 diabetes[a]	2004Q-0144
Tomatoes and/or tomato sauce and the risk of prostate, ovarian, gastric, and pancreatic cancer	2004Q-0201
Unsaturated fatty acids from canola oil and the risk of coronary heart disease	2006Q-0091
Corn oil and corn oil-containing products and a reduced risk of heart disease	2006P-0243
Whey protein and reduced risk of atopic dermatitis	2009-Q-0301

[a]Approved only for the dietary supplements.

Health claims in general relate to disease risk reduction, but do not quantify the degree of risk reduction. Health claims language always uses "may" or "might" to express the substance and disease relationship. All FDA-approved health claims are generic and not for the exclusive use of the petitioner. These claims are available to any conventional food or dietary supplement product that meets the SSA or FDAMA claims criteria or the QHC enforcement discretion conditions specified by the FDA. Additionally, the foods bearing the health claims must contain 10% or more of the DV for one or more of the six nutrients (vitamin A, vitamin C, iron, calcium, protein, or fiber) without fortification. Dietary supplements are exempt from this requirement. Foods bearing the health claims must also contain less than the 13 g, 4 g, 60 mg, 480 mg of fat, saturated fat, cholesterol, and sodium, respectively, the disqualifying nutrients 21 CFR 101.14 (6).

9.5 Dietary guidance statements

The dietary guidance statements address the role of dietary patterns or of general categories of foods (e.g., fruits and vegetables) in health and are not considered health claims. These statements describe the health effects of a broad category of foods rather than specific nutrients, e.g., "Diets rich in fruits and vegetables may reduce the risk of some types of cancer and other chronic diseases." Although dietary guidance statements are not subject to FDA review and authorization, these must be truthful and non-misleading.

9.6 Factual statements

There are several other claims/factual statements that are often used by manufacturers. Some of the common examples are provided in the following list. Many of these may not have strong scientific support or regulatory guidance and therefore the regulatory agencies may view them as misleading to consumers. Legal counsel should be consulted prior to making any claims not clearly defined in the regulations.

1. *Net carbs*: This is a way of representing the amount of carbohydrate not contributing significantly toward energy or blood sugar. The net carbs number is often calculated by subtracting fiber and/or sugar alcohol from total carbohydrate. The net carbs calculation is generally presented next to the claim. The FDA has not provided any guidance on how to calculate or make net carb claims.
2. *"Zero" g trans fat or 0 g trans fat*: This is a factual statement. In the absence of any guidance from the FDA about "trans fat free," this statement is becoming a common claim of expressing that the product does not have any significant amount of trans fat.

3. *Whole grain goodness*: This is a statement indicating that the product contains whole grain and therefore may provide the health benefits that are associated with whole grain.
4. *Glycemic index (GI)*: This is measured as a rise in the blood sugar following ingestion of a food and is representative of the quality of carbohydrate in the food. Usually foods with a GI of 55 or less are considered as low GI foods.
5. *Other common factual statements*: "As much calcium as a glass of milk," "contains same amount of vitamin C as a glass of orange juice," and "as much fiber as in an apple."

9.7 Nutritional claims display on packages

Manufacturers use a variety of ways to communicate the nutritional attributes and health advantages of their products:

1. Attribute on the principle display panel as a burst
2. List of attributes in a check box style on the side panel
3. Goodness corner to highlight the attributes
4. Complete back panel to display the nutritional attributes and their associated functional/health benefits
5. Company/brand's Website
6. Sales literature and other materials available at the point of purchase (grocery store/health food store)

Nutritional rating systems and front of pack (FOP) labeling have also been commonly used by the food industry and stakeholders with the intention of helping consumers make healthier choices [7]. However, many of these systems were considered controversial as they used self-defined nutritional criteria, which is often not transparent and not validated for their efficacy. The Institute of Medicine's (IOM) recent report on FOP labeling [8] highlighted the need for a simple, universal FOP symbol system aligned with current dietary guidance and designed to encourage the consumers to make healthier food choices. The IOM Committee recommended that the FDA and U.S. Department of Agriculture should develop, test, and implement a single, universal, and standardized FOP system.

According to a Food Label Package Survey conducted by the FDA [9], 49.7% of the FDA regulated products sold during 2000−2001 had nutrient content claims, 6.2% had structure/function claims, and 4.4% had health claims on food labels.

In summary, a wide range of nutritional and health claims are available under NLEA that intend to provide scientifically substantiated nutrition information to the consumer to help them make healthier food choices.

References

[1] U.S. Department of Agriculture, U.S. Department of Health and Human Services. Dietary guidelines for Americans. 7th ed. Washington, DC: U.S. Government Printing Office; 2010, December 2010.

[2] Moshfegh A, Goldman J, Cleveland L. What we eat in America, NHANES 2001−2002: usual nutrient intakes from food compared to dietary reference intakes. USDA, ARS; 2005.

[3] FDA. Code of Federal Regulation, Title 21, Part 101 Food Labeling, <http://www.accessdata.fda.gov/scripts/cdrh/cfdocs/cfcfr/CFRSearch.cfm?CFRPart = 101&show FR = 1>; 2012 [accessed 11.03.13].

[4] Nutrition Labeling and Education Act of 1990. Public Law 101−535, 1990; 1990.

[5] FDA. Guidance for Industry: Structure Function Claims − Small Entity Compliance Guide, <http://www.fda.gov/Food/GuidanceComplianceRegulatoryInformation/Guidance Documents/DietarySupplements/ucm103340.htm>; 2002 [accessed 11.03.13].

[6] FDA. Claims that can be made for conventional foods and dietary supplements, <http://www.fda.gov/Food/LabelingNutrition/LabelClaims/ucm111447.htm>; 2003 [accessed 11.03.13].

[7] IOM (Institute of Medicine). Examination of front-of-package nutrition rating systems and symbols: phase I report. Washington, DC: The National Academies Press; 2010.

[8] IOM (Institute of Medicine). Front-of-package nutrition rating systems and symbols: promoting healthier choices. Washington, DC: The National Academies Press; 2012.

[9] LeGault L, Brandt MB, McCabe N, Adler C, Brown A-M, Brecher S. 2000−2001 Food label and package survey: an update on prevalence of nutrition labeling and claims on processed, packaged foods. J Am Diet Assoc 2004;104:952−8.

Assessment of Safety and Quality Assurance of Herbal Dietary Supplements

10

Peter P. Fu and Qingsu Xia

National Center for Toxicological Research, Jefferson, Arkansas

10.1 Introduction

Botanicals have been used as herbal medicines for over 4000 years [1]. In China, traditional Chinese medicine and its theoretical concepts have been recorded for about 2000 years [1]. Production of modern Western pharmaceuticals, mainly by chemical synthesis, started soon after the advent of the petroleum industry, about 100 years ago. Western medicine is highly effective in healing human illness and improving quality of life. Besides, the therapeutic rationale of botanical/herbal medicine is difficult to understand in terms of modern science. Consequently, botanical herbal medicine is not viewed favorably in the United States and Western countries. It is currently called "alternative medicine" and is often considered inferior to modern Western conventional medicine.

Ironically, during the last several decades, use of botanicals/herbal products for improving the quality of life and/or for purported medicinal purposes has been increasing dramatically worldwide [2]. During the period of 2003–2006, about 20% of American adults used botanical supplements [3]. In 2004, the American Herbal Products Association estimated that there are about 3000 plant species used in as many as 50,000 different products sold as herbal dietary supplements in the United States [4]. Although herbal medicinal plants and herbal products are natural materials and are often considered safe for use, numerous cases of toxic effects in humans have been reported [2,5–7]. To date, many acute and chronic adverse effects of herbs and herbal products, such as herbal dietary supplements, including genotoxicity and carcinogenicity, have not been systematically examined. We previously published a review article titled "Quality Assurance and Safety of Herbal Dietary Supplements" in 2009 [2] and a book chapter titled "Quality Assurance and Safety Protection of Traditional Chinese Herbs as Dietary Supplements" in 2011 [8]. As a continuation of this work, this review presents an overview of the following: (1) the current understanding of quality and safety

Nutraceutical and Functional Food Regulations in the United States and Around the World.
DOI: http://dx.doi.org/10.1016/B978-0-12-405870-5.00010-4

assurance of herbal dietary supplements, (2) the regulatory decisions of the U.S. federal agencies, (3) the current toxicological studies conducted by the U.S. federal agencies, and (4) the safety assurance of raw Chinese medicinal herbs used for the preparation of herbal dietary supplements that are sold worldwide. More precisely, this review is focused on the toxicity of herbs and herbal dietary supplements, especially on the safety of Chinese medicinal herbs.

10.2 Quality control of herbal dietary supplements

Quality control is a fundamental aspect of the manufacture of products. Both Good Agriculture Practice (GAP) and Good Manufacturing Practice (GMP) are guidelines designed to standardize procedures to achieve safety and quality assurance of manufactured products [8]. GMP guides all the practices, from the raw material to delivery to the customer. For raw medicinal herbal plants, GAP mandates procedures for the entire production process, including planning, sowing, cultivating, and producing the raw material.

While a Western drug in most cases contains only one or several effective therapeutic constituents, an herbal medicine or herbal product contains hundreds of chemical components. In addition, the quality of herbal plants is markedly affected by external factors such as genetic variations of the plant species and different environmental and growing conditions [8]. These external variables are not easy to control because they vary among crops, geographic locations of growth, and growers. Consequently, raw herbal plants produced at different locations and times may not possess the same quality and properties. These factors result in the lack of science-based conventions to define and standardize the quality of herbal plants. The lack of comprehensive toxicological data is another concern. These combined problems result in the difficulty in establishing GAP for medicinal herbs. Consequently, although GMP guidelines have been well established in manufacturing processes, the GAP for efficacy assurance and safety of Chinese herbal plants is difficult to establish; as a result, the GAP for efficacy assurance and safety of Chinese herbal plants is still in a development stage.

10.2.1 Species authentication and standardization

Identical herbal plant species grown at different locations and under different environmental and cultivation conditions may have different profiles and different quantities of chemical constituents. Herbal plants may contain different chemical constituents in leaves, stems, and/or roots. These variations result in different beneficial and/or toxicological effects. Therefore, although rational selection of species samples used as standards for authentication is required, in practice it is very difficult.

There are several established methodologies adopted for authentication, standardization, and quality assurance of traditional herbs in China. These methods include plant taxonomic identification, morphological and microscopic examination, fingerprint chromatographic identification, and DNA molecular marker characterization. A near-infrared (NIR) technique is also used at times [8].

10.2.2 **Selection of chemical markers**

For quality control of herbal materials in manufacture and pharmacological research, there are seven types of chemical markers: (1) "active principle," (2) "active marker," (3) "group marker," (4) "chemical fingerprint," (5) "analytical marker," (6) "phantom marker," and (7) "toxic marker" (negative marker) [9]; these various markers are described below.

An active principle is a chemical constituent that possesses functional/ medicinal activity in an herbal plant. Active principles are ideal chemical markers for quality assurance of the herb and its extract. For example, ginkgolides and bilobalide are the principal active constituents of *Ginkgo biloba* leaves that produce the pharmacological activities of *G. biloba*, and are thus ideal active principles [10].

An active marker is a chemical component that has pharmacological activity but may or may not contribute to the functional/medicinal effects of the herb. In most cases, the active principles are not easily determined and the use of an active marker is considered appropriate. For example, ferulic acid is the appropriate active marker for the quality assessment and pharmacological evaluation of *Ligusticum chuanxiong* [11].

A group marker is a group of chemical constituents possessing similar chemical structures and/or physical properties. Since the functional activity of an herbal plant is in most cases attributed to more than one chemical constituent, use of a group marker is practical, although it has limited utility for quality control. For example, the total amount of polysaccharides is used as the group marker for *Ganoderma* deriving from *Ganoderma lucidum* and *G. japonicum* [12].

A chemical fingerprint is the spectroscopic pattern and/or chromatographic profile of an herb. The chemical fingerprint is useful as a marker for quality assurance because it represents the whole group of chemical constituents that are diagnostic for the herb. Chemical fingerprints are often used to compare the similarity of chemical profiles between a herb of interest and a reference herbal material [13]. Since chemical fingerprints are routinely used in many countries, this method will be discussed separately in the following section.

An analytical marker is a chemical constituent that may not possess any biological activities [9], and a phantom marker is a chemical constituent that possesses known pharmacological actions but is present in very low quantity. Consequently both analytical markers and phantom markers are not suitable for quality control practices.

A toxic marker or a negative marker [9] is a chemical constituent that is toxic.

10.2.3 **The fingerprinting technique**

Chemical fingerprinting is a widely accepted method for quality control of medicinal herbs, including in China [8]. Many chemical analysis techniques have been applied successfully for the chemical fingerprinting of traditional Chinese herbs and herbal products. The most commonly used chromatographic techniques are high performance liquid chromatography (HPLC), thin layer chromatography, and gas chromatography; and the routinely used spectroscopic techniques are Fourier transform infrared, NIR, and nuclear magnetic resonance (NMR) spectroscopy, and mass spectrometric (MS) techniques. Because of the complexity of herbs and herbal products, comprehensive compositional analysis of their chemical constituents is highly challenging and time-consuming. Because in most cases, the biological, functional, and medicinal activities of herbs and herbal products arise from the combined actions of a group of chemical constituents, the analysis of marker compounds by themselves can simplify the analytical process, but the results can be misleading. However, fingerprinting analysis can avoid such problems and is, therefore, an effective method.

Since fingerprinting analysis is an effective and more reliable approach than other methods, this method has been recommended for use in China and the United States [8], and was also recommended for use by the World Health Organization (WHO) in 2003 [14].

10.3 Safety assurance of herbal dietary supplements

10.3.1 Contamination by microbes, heavy metals, and pesticides

Following GAP and GMP guidelines, herbal plants and herbal products must be free from contamination by microbes, toxic heavy metals, and pesticide residues. Compared with many other problems that are difficult to solve, such as product standardization, the problem of contamination is manageable. Unfortunately, contamination problems still persistently occur in the United States and other countries. For example, dietary supplements labeled as ephedra free and sold in the San Francisco bay area in 2003 were found by the Food and Drug Laboratory of the California Department of Health Services to contain significant concentrations of lead, arsenic, cadmium, and mercury [15]. Organochlorine pesticide residues have also been found in a number of Chinese herbal plants cultivated in China and sold in Hong Kong [16]. These examples show that the safety of Chinese medicinal products must be critically assessed before the products are placed on the market.

10.3.2 Adulteration

Adulteration of herbal products frequently occurs, making quality assurance difficult to accomplish. Adulteration of traditional Chinese medicines frequently involves incorporating Western drugs of known pharmacological activity, such as steroids, with the intention of increasing therapeutic efficacy. An effective ingredient being replaced by a much cheaper chemical component is another commonly employed adulteration practice. The adulteration of G. biloba dietary supplements is a good example. The food labeling practice requires that the commercial product must contain 24% *Ginkgo* flavonoids and 6% terpene lactones (ginkgolides and bilobalide). However, the requirement is based upon quantifying the total flavonoid and terpene content, not the individual flavonoids. Mustafa et al. [17] found that the content of flavones and terpene in the *G. biloba* dietary supplements commercially sold in the United States was significantly different between different batches. Thus, *G. biloba* can be adulterated to have higher total flavonol content, in order to meet the standard, by adding rutin or quercetin to the extract. Another type of adulteration is that, based on the Federal Food, Drug, and Cosmetic Act (FFD&C Act), a food product is unsafe and adulterated if the product is contaminated with pesticide for which a tolerance level has not been defined [18].

10.3.3 Toxicity bioassays of herbal dietary supplements

For safety assurance, toxicological assessments of herbal products are a necessity. Currently, the toxicological effects of herbs and herbal products, including cytotoxicity, mutagenicity, carcinogenicity, teratogenicity, reproductive toxicity, neurotoxicity, immunotoxicity, and cardiovascular toxicity, are generally unknown, and the mechanisms of toxicity are not well understood. Because herbal plants and herbal products contain hundreds of chemical constituents, the study of mechanism of action is highly challenging. Since 1999, the U.S. National Toxicology Program (NTP) has been conducting long-term tumorigenicity bioassays on a large number of herbal plant extracts, herbal dietary supplements, and their functionally active chemical constituents nominated by the public and federal agencies. The nominated herbs and active ingredients are among the most commonly sold products in the United States and are suspected to possess a high potential for toxic activity. The studies focus on determining potential adverse health effects, including reproductive toxicity, neurotoxicity, immunotoxicity, and tumorigenicity. Safety assessment (toxicological) studies by the NTP and the National Center for Toxicological Research (NCTR) will be discussed later. One of the major concerns about herbs and herbal products is that they may cause alteration of metabolizing enzymes, leading to herb–drug and herb–herb interactions [19].

10.3.4 Herb–drug and herb–herb interaction and alteration of metabolizing enzymes

Both herbs and Western drugs require metabolizing enzymes for metabolism. Herbs and herbal supplements are often used in combination with therapeutic drugs, and raise the potential of herb–drug interactions that may result in adverse effects. It has been reported that a number of herbal extracts and herbal constituents can significantly alter the activity of phase I and phase II metabolizing enzymes, which may lead to herb–herb and herb–drug interactions [2,20–28]. The herbs that have been studied include kava extract and its kavalactones [2,24–26,28–30], *Panax ginseng* [20,23], St. John's Wort [23,28], *G. biloba*

[23,31], goldenseal (*Hydrastis canadensis*) [23], comfrey [32], and tumorigenic pyrrolizidine alkaloids [21,22,27,33−35]. Among these, Gurley et al. [23] determined that *P. ginseng* significantly inhibited CYP2D6, although the magnitude of inhibition did not appear to be clinically relevant.

G. biloba leaf extract is one of the most studied herbal extracts. It has been shown to inhibit CYP450 enzyme activity and thus produced CYP-mediated herb−drug interactions [10,23,31,36]. In a 2 year toxicology study, *G. biloba* administered to B6C3F1 mice resulted in gene expression changes in liver, with significant numbers of drug-metabolizing genes altered [31]. *G. biloba* leaf extract is often taken in combination with prescribed Western medicines and thus may potentially lead to herb−drug interactions [20,25,37−39].

It has been demonstrated that the biologically inactive or toxic constituents can alter enzyme activity and induce herb−drug interaction as well. For example, retrorsine, a tumorigenic pyrrolizidine alkaloid, increased the expression of hepatic CYP1A1, 1A2, 2E1, and 2B1/2 enzymes in rats [21]. Dwivedi et al. [40] reported that, in rats, monocrotaline increased the activities of hepatic succinate dehydrogenase, acid ribonuclease, acid phosphatase, γ-glutamyl transpeptidase, and 5′-nucleotidase, and reduced activities of glucose-6-phosphatase and CYP enzymes. The expression of phase I and II drug-metabolizing genes was significantly altered in livers of female Big Blue transgenic rats gavaged with riddelliine [27]. For example, four CYP genes, Cyp2c12, Cyp2e1, Cyp3a9, and Cyp26, were significantly upregulated. In addition, the phase II drug-metabolizing enzymes GST genes (Gsta3) and ATP-binding cassette transporter genes (Abcb1a and Abcc3) were also significantly upregulated.

Metabolizing enzymes are required for the metabolism, elimination, and detoxification of xenobiotics. Among the phase I enzymes, the CYP1, CYP2, and CYP3 subfamilies play a major role in the metabolism of drugs and as well as in the metabolic activation of toxic and carcinogenic xenobiotics [41,42]. The main role of phase II enzymes is to eliminate the phase I metabolites by forming aqueous soluble conjugates that are more easily excreted.

10.4 Regulatory activities concerning botanical/herbal dietary supplements

China produces traditional Chinese medicines (herbs) that are used as drugs throughout Asia. Traditional Chinese medicines are also the raw materials used for preparing herbal products, including herbal dietary supplements, which have recently become popular in Western countries. The Chinese government has imposed a number of regulations, including GAP, GMP, and Acts for safe control of the herbal products. In the United States, a series of regulations has been enacted for manufacturers to follow. The regulations set by China and the United States will be addressed separately in the following sections. Besides China, the concept of GAP has also been initiated in Europe and Japan. In 2000, the European Agency for the Evaluation of Medicinal Products announced official GAP guidelines for botanical drugs and herbal products. In 2003, WHO published guidelines on good agricultural and collection practices for herbal plants [14].

10.4.1 **The U.S. Food and Drug Administration**

The U.S. Food and Drug Administration (FDA) initially regulated food, drugs, and cosmetics, including botanicals, by the provisions of the FFD&C Act), a set of laws passed by the U.S. Congress in 1938. After the production of herbal dietary supplements in the United States increased drastically, beginning around 1980, the U.S. Congress, in 1994, enacted the Dietary Supplement Health and Education Act (DSHEA) that amended the FFD&C Act and created a new regulatory category, safety standard, and other rules for the FDA to regulate dietary supplements. According to DSHEA, a dietary supplement is defined as a product taken by mouth that contains a "dietary ingredient" intended to supplement the diet. The "dietary ingredients" include herbs and other botanicals. Dietary supplements are specifically classified under the general umbrella of "foods," not drugs, and DSHEA requires that every supplement be labeled a dietary supplement. According to DSHEA, a dietary supplement is considered unsafe only if it presents a significant or unreasonable risk of illness or injury under conditions of use recommended or suggested in the labeling, or, if no conditions of use are suggested or recommended in the labeling, under ordinary conditions of use [43].

Since, according to the DSHEA, manufacturing companies are not required to submit safety reports on (herbal) dietary supplement products to the FDA, the FDA has difficulty determining whether or not their botanical ingredients are safe for use. Accordingly, the FDA must determine information about the safety and the efficacy from other sources, including scientific literature and reports from the media. Under such circumstances, consumers who take (herbal) dietary supplements may not be adequately protected. Accordingly, to ensure quality control of herbal dietary supplement products, the FDA subsequently promulgated a series of regulations. In 1997 the initial GMP regulations were issued. In 1999, the FDA Food Labeling Act was enacted. In 2003, the FDA issued the final GMP regulations that specify detailed conditions for preparation, packing, and storing of dietary supplements, and required that dietary supplements be unadulterated and accurately labeled to fully meet safety and sanitation standards. Also in 2003, the FDA established current good manufacturing practice requirements (CGMPs) for dietary supplements and dietary supplement ingredients. A final dietary supplement CGMPs and an Interim Final Rule were enacted in 2007 [44].

Furthermore, the Dietary Supplement and Nonprescription Drug Consumer Protection Act (Public Law 109-462, effective in December 2007) was issued to require that serious adverse events related to dietary supplements and nonprescription drugs be reported. This Act requires that the manufacturers, distributors, and retailers of dietary supplements collect all adverse event reports and submit them to the FDA; that the firms maintain records of reports of all adverse events for inspection by the FDA; and that dietary supplement labels bear information to facilitate the reporting of serious adverse events associated with the use of dietary supplements by consumers. According to this Act, the definition of serious adverse events includes death, a life-threatening experience, inpatient hospitalization, a persistent or significant disability or incapacity, or a congenital anomaly or birth defect. This Act requires records retention, specifying that "The responsible person must maintain records of all adverse reports it receives, whether serious or not, for 6 years." This Act also mandates that manufacturers provide a domestic telephone number or a domestic address on product labels so that consumers can contact them. This Act enables the FDA to better fulfill its public health mission and to monitor more effectively the herbal medicines and dietary supplements it regulates.

10.4.2 Regulation of Chinese herbs in China

During the last two decades, the Chinese government promulgated regulations on herbal medicines. In China, Chinese herbal products are governed by the State Food and Drug Administration (SFDA) and are registered either as functional foods or as drugs. Thus there are two types of drugs in China: Western chemical and biological synthetic drugs and traditional medicines. The China SFDA initiated GMP guidelines in 1982, enacted good laboratory practice guidelines for experimentation on animals in 1988, and approved Chinese herbal medicines in 1992. Several related regulations were issued in 2000, 2002, and 2003 [8]. The Drug Administration Law of China, enacted in 2001, is the basic law governing drug administration in China to ensure drug quality and safety for humans and to protect their health. The SFDA for Chinese Crude Drugs (Interim) implemented the GAP in June 2002. The SFDA also stated in 1998 that all manufacturers comply with the GMP by April 2004, and farms producing raw herbal ingredients had until 2007 to meet the guidelines specified in GAP.

10.5 Toxicological study of herbal dietary supplements by the NTP and the NCTR

10.5.1 Toxicological study by the NTP

The use of herbal medicines has recently been greatly increasing in the United States. To date, there are about 3000 herbal plants used for herbal dietary supplements sold in the United States. The safety risk that these products pose to humans has received much public attention. In response, in 1998, the NTP started working with the Office of Dietary Supplements of the National Institutes of Health (NIH), the FDA, and the academic community to nominate chemical candidates for toxicological studies [19]. Since then, long-term chronic bioassays of a series of the most commonly sold herbal dietary supplements and their active ingredients have been conducted. According to an NTP fact sheet on medicinal herbs published in 2006, the NTP studies of medicinal herbs are designed to "focus on the characterization of potential adverse health effects, including reproductive toxicity, neurotoxicity, and immunotoxicity, as well as those effects associated with short-term high-dose exposure or long-term exposure to lower doses" [19]. Special attention has also been given to studies of herb—herb and herb—drug interactions and to the adverse effects in sensitive subpopulations. A number of herbal plants and their chemical constituents that have been found to be potentially tumorigenic have been studied by the NTP. To date, these include echinacea, golden seal, ginseng, kava extract, *G. biloba* extract, *Aloe vera* extract, berberine, milk thistle extract, pulegone, thujone, quercetin, green tea extract, resveratrol, D-carvone, furfural, and *Usnea* lichen.

Pyrrolizidine alkaloids are a class of toxic plant secondary metabolites that have been detected in various herbal plants widespread in the world and some of them have been studied. Pyrrolizidine alkaloid-containing plants are probably the most common poisonous plants affecting livestock, wildlife, and humans [33,45,46]. In view of their widespread occurrence, two tumorigenic pyrrolizidine alkaloids, lasiocarpine [47] and riddelliine [48], have been studied for tumorigenicity by the NTP.

10.5.2 **Toxicological studies by the NCTR**

The NCTR is a research center of the FDA that plays a critical role in supporting FDA regulatory decision making. In recent years, the NCTR has been working together with the NTP on long-term tumorigenicity bioassays and mechanistic studies of a large number of herbal dietary supplements, including *Aloe vera* [49—52], aristolochic acid [53—55], riddelliine [5,6,33,56—62], comfrey [32], kava [29,30,63,64], *G. biloba* leaf extract [31], usnic acid [65], and *P. ginseng* [66]. The toxicity studies include: cytotoxicity [64], genotoxicity [32,53—55], and tumorigenicity [5,6,33,50,57—62], DNA adduct formation [5,6,33,57—62,67], induction of herb—drug interactions, and alteration of cytochrome P450 metabolizing isozymes [29—31].

An example of the long-term carcinogenicity bioassays conducted at the NCTR is the study of the photo-co-carcinogenicity and carcinogenicity of *Aloe vera* by Boudreau et al. [50,68]. It was determined that topical application of creams containing *Aloe vera* plant extracts, combined with exposure to simulated solar light for 1 year, resulted in the development of a significantly increased multiplicity of squamous cell neoplasms in B6C3F1 mice [68]. A 2 year long-term toxicity study of *Aloe vera* whole leaf extract administered in drinking water to F344/N rats resulted in increased incidences of adenomas and carcinomas in the large intestine [50,69]. These results provide important information to the FDA for regulatory decision making.

A good example of studying the mechanism of tumor induction by an herbal constituent at the NCTR is the study of riddelliine, a representative pyrrolizidine alkaloid, which was determined to induce liver tumors in rats and mice [56]. Since this study represents one of the best mechanistic studies on herbal dietary supplements so far reported, it is described below in detail.

10.5.3 Case study: mechanism of liver tumor induction by riddelliine

Pyrrolizidine alkaloids are heterocyclic compounds that are common constituents of hundreds of plant species around the world [5,6,33,45,46,59,70,71] and are the most widespread phytochemicals, having been identified in over 6000 plants, and are present in about 3% of the world's flowering plants [72,73]. It has been recognized since the eighteenth century that many pyrrolizidine alkaloids are highly toxic, with about half of them exhibiting hepatotoxic activity [46,74]. Pyrrolizidine alkaloids exhibit a variety of toxicities and genotoxicities, including acute toxicity, mutagenicity, chromosomal aberrations, DNA cross-linking, DNA−protein cross-linking, inhibition of colony formation, and megalocytosis [2,33]. More recently, the toxic effects of pyrrolizidine alkaloids gained attention when experimental animals dosed with these compounds developed liver tumors and pulmonary lesions [46−48,75]. It became even more serious when human poisoning caused by pyrrolizidine alkaloids was reported [33].

Pyrrolizidine alkaloids were among the first naturally occurring carcinogens identified in plants. In 1954, a pure pyrrolizidine alkaloid, retrorsine, was found to induce liver tumors in experimental animals [75]. Subsequently, a series of pyrrolizidine alkaloids was found to induce liver tumors in experimental animals [47,48]. Riddelliine is a representative genotoxic and tumorigenic pyrrolizidine alkaloid. Due to its potential for human exposure and genotoxicity, riddelliine was nominated by the FDA to the NTP for genotoxicity and 2 year carcinogenicity bioassays. Riddelliine induced liver hemangiosarcomas in male and female F344 rats and male B6C3F$_1$ mice [56]. At the same time, Fu and coworkers [57,59,61,72] at NCTR determined the mechanism of tumor formation induced by riddelliine. The mechanistic study involved characterization of the metabolic activation pathway, identification of activated metabolites, and development of a ^{32}P-postlabeling/HPLC method [67] for the identification and quantification of the riddelliine-derived DNA adducts in livers of F344 rats treated with riddelliine [61]. Riddelliine was shown to induce liver tumors through a genotoxic mechanism mediated by 6,7-dihydro-7-hydroxy-1-hydroxymethyl-5H-pyrrolizine (DHP)-derived DNA adduct formation [61]. The levels of DNA adduct formation correlated closely with the tumorigenic potencies of different doses of riddelliine fed to rats [61]. Furthermore, the levels of DHP-derived DNA adduct in F344 rat and B6C3F1 mouse liver endothelial cells, the cells of origin for the hemangiosarcomas, were significantly greater than those in the parenchymal cells [57], showing that the levels of riddelliine-induced DNA adducts in specific populations of liver cells correlate with the preferential induction of liver hemangiosarcomas by riddelliine. Subsequent studies with human liver microsomes showed that the metabolic pattern and DNA adduct profiles obtained from riddelliine metabolism with human liver microsomes were similar to those formed in rat liver *in vitro* and *in vivo*. Thus, the results of *in vivo* and *in vitro* mechanistic studies with experimental rodents are highly relevant to humans [60]. The fact that riddelliine induces liver tumors in rats and mice and that DHP-derived DNA adducts are responsible for tumor induction suggest that riddelliine may be highly genotoxic to humans, mediated by DHP-derived DNA adduct formation [60].

A major limitation associated with the ^{32}P-postlabeling/HPLC method for DNA adduct analysis is the lack of structural information since the method does not show how the reactive metabolites bind to cellular DNA, and the molecular structures of the resulting DNA adducts cannot be determined. Subsequently, an HPLC-ES-MS/MS method was developed for the identification and quantitation of DHP-derived DNA adducts *in vivo* and *in vitro* [59]. MS and NMR spectroscopic analysis showed that four DNA adducts were formed in rats dosed with riddelliine. The complete structures of the four DNA adducts were elucidated: there are two DHP-dG adducts (DHP-dG-3 and DHP-dG-4), which are a pair of epimers of 7-hydroxy-9-(deoxyguanosin-N^2-yl)dehydrosupinidine and there are two DHP-dA adducts (DHP-dA-3 and DHP-dA-4), which are a pair of epimers of 7-hydroxy-9-(deoxyadenosin-N^6-yl)dehydrosupinidine [62].

With the structures of the DNA adducts unequivocally elucidated, it is now clear that cellular DNA preferentially binds dehydroriddelliine, the reactive metabolite, at the C9 position of the necine base, rather than at the C7 position as suggested previously. This study is the first report with detailed structural assignments of the DNA adducts that are responsible for liver tumor formation induced by a pyrrolizidine alkaloid. Thus these studies have elucidated the mechanism of tumor initiation by a pyrrolizidine alkaloid (riddelliine) at the molecular level [62].

10.5.4 Official actions by the FDA and NTP on herbal dietary supplements

In a 2001 alert [76], the FDA "strongly recommended that firms marketing a product containing comfrey or another source of pyrrolizidine alkaloids remove the product from the market and alert its customers to immediately stop using the product." The FDA also advised that "it was prepared to use its authority and resources to remove products from the market that appeared to violate the Federal Food, Drug, and Cosmetic Act" [77]. In 2004, under the FFD&C Act, the FDA issued a guidance for industry, ruling that dietary supplements containing ephedrine alkaloid be removed from the market because they present an unreasonable health risk [78].

The Report on Carcinogens, published by the NTP, is a congressionally mandated document listing chemical substances that may potentially put people at increased risk for cancer. In 2011, the NTP classified aristolochic acids as "known to be human carcinogen" and riddelliine as "reasonably anticipated to be human carcinogen" in the NTP 12th Report on Carcinogens [77].

10.6 Difficulties in safety assurance of herbs and herbal dietary supplements

10.6.1 Intrinsic problems: herbs and herbal products contain multiple chemical constituents

While production of high-quality and contaminant-free raw herbal plants, under GAP principles, is a prerequisite for quality and safety assurance of medicinal herbs and herbal products, there are hundreds of chemical constituents present in a single herbal plant species, at concentrations that can differ drastically when grown at different locations and in different seasons. This makes standardization and quality control of medicinal herbs nearly impossible.

The large number of chemical constituents in an herbal plant also makes it difficult to conduct risk assessments on herbs and herbal products. Although there are methods that can be used to determine the mechanism by which a pure chemical induces toxicity and tumorigenicity, these methods are not applicable for assessing the risk posed by chemical mixtures, such as herbal plants, herbal dietary supplements, tobacco smoke condensates, and environmental pollution mixtures. As a general practice, safety assessment of a purported therapeutic constituent is commonly conducted as the initial study. However, this approach cannot determine the additive, synergistic, or antagonistic effects exerted by other constituents in the herb. As a consequence, it is not known how much each single constituent contributes to the total genotoxicity of the herb, even the toxicity of this herbal component has been determined. Thus, to elucidate mechanisms by which herbal dietary supplements exert toxicity and tumorigenicity, additional practical and reliable methods for toxicological assays have to be developed.

10.6.2 Understanding the theory of traditional Chinese medicine

A key question is whether traditional Chinese medicine can help us understand how herbal supplements work and how they can be used safely. The problem seen by Western *scientists* is that there is scant *scientific* evidence for the effectiveness of traditional medicine and so published papers on the toxic effects of traditional Chinese herbs and/or herbal dietary supplements do not cite information from the Chinese literature and/or Chinese medicinal books. This situation is compounded because the Chinese literature is not readily accessible to the Western scientists. Most of the toxicological results compiled in classical Chinese herbal medicinal books were obtained by clinical observations and may be valuable in directing future toxicological studies and for validating the functional and therapeutic effects that the Chinese believe in, even though these results were not obtained by scientific (e.g., well-controlled) studies. For example, the Chinese medicinal book *Compendium of Materia Medica* (本草纲目) compiled by Li Shizhen more than 400 years ago states that drinking too much tea can damage health, and the severity of the damage depends on the sex, health, and age of the drinker; the toxic effect is higher when drinking tea during the winter, early in the morning, and on an empty stomach. This example demonstrates that traditional Chinese medicine can help us take herbal dietary supplements more efficiently and safely if the "science" behind such findings can be explained.

10.7 Alternative approaches for safety assurance of herbal dietary supplements

A number of reports have indicated that in some cases, a dietary supplement believed to exhibit beneficial effects to human health turns out to be harmful. A prominent example is the result of nutritional intervention trial of the antioxidant β-carotene indicating that β-carotene does not act as an antioxidant but exerts harmful health effects [79,80]. Consequently, for safety assurance of herbal dietary supplements, it is important and practical to bear this fact in mind.

One of the purported beneficial functional activities of herbal dietary supplements is antioxidant activity. However, under certain conditions, some dietary antioxidants display pro-oxidant effects that promote oxidative damage leading to the formation of reactive oxygen species and lipid peroxidation [81]. This phenomenon has recently been stressed by Watson [80] who stated "free-radical-destroying antioxidative nutritional supplements may have caused more cancers than they have prevented." Watson also stated that "in light of the recent data strongly hinting that much of late-stage cancer's untreatability may arise from its possession of too many antioxidants, the time has come to seriously ask whether antioxidant use much more likely causes than prevents cancer." Since a number of herbal dietary supplements, including *G. biloba*, ginseng, and tea, exhibit antioxidative activity [10,43], it is prudent to ascertain whether or not herbal dietary supplements with antioxidant activity may also display pro-oxidative activity.

10.8 The role of the Chinese government in safety assurance of Chinese herbal medicine

The Chinese people have long experience in practicing Chinese herbal medicine and in cultivating Chinese herbs for use, and the majority of herbal dietary supplements sold worldwide are manufactured by using raw herbal plants produced in China. Thus, to participate in current research on quality control and safety assurance of herbal dietary supplements, the Chinese government and Chinese research scientists can and should play an important role in the following: (1) developing practical approaches for achieving standardization and safety assurance of Chinese medicinal herbs and herbal products, (2) promoting public acceptance of and education about the principles of Chinese herbal medicine, and (3) investigating Chinese medicine-based alternatives to current toxicity bioassays.

10.9 Perspectives

Herbal dietary supplements are popularly used around the world. For human health protection, quality control and safety assurance of raw herbal plants and herbal dietary supplements are a necessity. Herbal plants consist of multiple chemical constituents; we have difficulties in the standardization of raw medicinal herbs and often lack scientific methods to study the mechanisms by which a chemical mixture induces toxic effects. We need to determine the best procedures for manufacturing herbal dietary supplements with minimal toxic effects, but the above-described problems hamper our pursuit of this task. A global endeavor to address this task is necessary. The FDA has promulgated a series of regulations to ensure safety of dietary supplements. The NTP and the NCTR as well as other federal agencies have been conducting long-term research programs to determine the toxicity and tumorigenicity of herbal dietary supplements. The NIH Office of Dietary Supplements has also been establishing analytical methods and reference materials for ensuring safe use of herbal dietary supplements. Information described in this review illustrates actions taken by the U.S. federal agencies on this subject. The Chinese government and research community have simultaneously undertaken similar endeavors to assess the safety of the raw medicinal herbs. It is anticipated that this worldwide effort will improve quality control and safety assurance of medicinal herbs and herbal dietary supplements.

Acknowledgments

We thank Drs. William Melchior and Daniel Doerge for critical review of this manuscript. This article is not an official guidance or policy statement of U.S. Food and Drug Administration (FDA). No official support or endorsement by the U.S. FDA is intended or should be inferred.

References

[1] Huang KC, editor. The pharmacology of Chinese herbs. Boca Raton, FL: CRC Press; 1998.

[2] Fu PP, Chiang H-M, Xia Q, Chen T, Chen BH, Yin J-J, et al. Quality assurance and safety of herbal dietary supplements. J Environ Sci Health C 2009;27:91−119.

[3] Bailey RL, Gahche JJ, Lentino CV, Dwyer JT, Engel JS, Thomas PR, et al. Dietary supplement use in the United States, 2003-2006. J Nutr 2011;141:261−6.

[4] Zurer P, Hanson D. Chemistry puts herbal supplements to the test. Chem Eng News 2004;82:16.

[5] Fu PP, Xia Q, Chou MW, Lin G. Detection, hepatotoxicity, and tumorigenicity of pyrrolizidine alkaloids in Chinese herbal plants and herbal dietary supplements. J Food Drug Anal 2007;15:400−15.

[6] Fu PP, Yang Y-C, Xia Q, Chou MW, Cui Y, Lin G. Pyrrolizidine alkaloids - tumorigenic components in Chinese herbal medicines and dietary supplements. J Food Drug Anal 2002;10:198–211.

[7] Li N, Xia Q, Ruan J, Fu PP, Lin G. Hepatotoxicity and tumorigenicity induced by metabolic activation of pyrrolizidine alkaloids in herbs. Curr Drug Metab 2011;12:823–34.

[8] Lee FSC, Wang X, Fu PP. Quality assurance and safety protection of traditional Chinese herbs as dietary supplements. In: Shi J, Ho C-T, Shahidi F, editors. Nutraceutical science and technology. Series 10, functional foods of the east. New York, NY: CRC Press, Taylor and Francis Group; 2011. p. 18.

[9] Chan SS, Jiang Y, Jiang ZH, Lin G. Pitfalls of the selection of chemical markers for the quality control of medicinal herbs. J Food Drug Anal 2007;15:365–71.

[10] Chan P-C, Xia Q, Fu PP. Ginkgo biloba leave extract: biological, medicinal, and toxicological effects. J Environ Sci Health C 2007;25:211–44.

[11] Shu M, Hou DQ, Hou CJ, Zhang W, Xie G. Microwave extraction of ferulic acid from Ligusticum chuanxiong. Zhongchengyao 2007;29:908–9.

[12] Chan SS, Cheng TY, Lin G. Relaxation effects of ligustilide and senkyunolide A, two main constituents of Ligusticum chuanxiong, in rat isolated aorta. J Ethnopharmacol 2007;111:677–80.

[13] Chen J, Lee FS, Li L, Yang B, Wang X. Standardized extracts of Chinese medicinal herbs: case study of Danshen (Salvia miltiorrhiza Bunge). J Food Drug Anal 2007;15:347.

[14] WHO. Guidelines on Good Agricultural and Collection Practices for Medicinal Plants, <http://www.who.int/medicines/library/trm/medicinalplants/agricultural.shtml>; 2003.

[15] Tam JW, Dennehy CE, Ko R, Tsourounis C. Analysis of ephedra-free labeled dietary supplements sold in the San Francisco Bay area in 2003. J Herb Pharmacother 2006;6:1–19.

[16] Leung KS, Chan K, Chan CL, Lu GH. Systematic evaluation of organochlorine pesticide residues in Chinese *Materia Medica*. Phytother Res 2005;19:514–8.

[17] Mustafa O, Brendan M, Pei C. Comparison of the terpene lactones and flavonols. J Food Drug Anal 2007;15:55–62.

[18] FDA. FDA. Warning on imported ginseng. FDA Consum 2005;39:.

[19] NTP. Fact Sheet, Medicine herbs; 2006.

[20] Bressler R. Herb-drug interactions. Geriatrics 2005;60:.

[21] Gordon GJ, Coleman WB, Grisham JW. Induction of cytochrome P450 enzymes in the livers of rats treated with the pyrrolizidine alkaloid retrorsine. Exp Mol Pathol 2000;69:17–26.

[22] Guengerich FP. Separation and purification of multiple forms of microsomal cytochrome P-450. Activities of different forms of cytochrome P-450 towards several compounds of environmental interest. J Biol Chem 1977;252:3970–9.

[23] Gurley BJ, Gardner SF, Hubbard MA, Williams DK, Gentry WB, Cui Y, et al. Clinical assessment of effects of botanical supplementation on cytochrome P450 phenotypes in the elderly. Drugs Aging 2005;22:525–39.

[24] Gurley BJ, Swain A, Barone GW, Williams DK, Breen P, Yates CR, et al. Effect of goldenseal (Hydrastis canadensis) and kava kava (Piper methysticum) supplementation on digoxin pharmacokinetics in humans. Drug Metabol Dispos 2007;35:240–5.

[25] Hu Z, Yang X, Ho PCL, Chan SY, Heng PWS, Chan E, et al. Herb-drug interactions. Drugs 2005;65:1239–82.

[26] Mathews JM, Etheridge AS, Black SR. Inhibition of human cytochrome P450 activities by kava extract and kavalactones. Drug Metabol Dispos 2002;30:1153−7.

[27] Mei N, Guo L, Liu R, Fuscoe JC, Chen T. Gene expression changes induced by the tumorigenic pyrrolizidine alkaloid riddelliine in liver of Big Blue rats. BMC Bioinformatics 2007;8(Suppl. 7):S4.

[28] Singh YN. Potential for interaction of kava and St. John's Wort with drugs. J Ethnopharmacol 2005;100:108−13.

[29] Guo L, Li Q, Xia Q, Dial S, Chan P-C, Fu P. Analysis of gene expression changes of drug metabolizing enzymes in the livers of F344 rats following oral treatment with kava extract. Food Chem Toxicol 2009;47:433−42.

[30] Guo L, Shi Q, Dial S, Xia Q, Mei N, Li Q-z, et al. Gene expression profiling in male B6C3F1 mouse livers exposed to kava identifies changes in drug metabolizing genes and potential mechanisms linked to kava toxicity. Food Chem Toxicol 2010;48:686−96.

[31] Guo L, Mei N, Liao W, Chan P-C, Fu PP. Ginkgo biloba extract induces gene expression changes in xenobiotics metabolism and the Myc-centered network. OMICS 2010;14:75−90.

[32] Mei N, Guo L, Fu PP, Fuscoe JC, Luan Y, Chen T. Metabolism, genotoxicity, and carcinogenicity of comfrey. J Toxicol Environ Health B Crit Rev 2010;13:509−26.

[33] Fu PP, Xia Q, Lin G, Chou MW. Pyrrolizidine alkaloids--genotoxicity, metabolism enzymes, metabolic activation, and mechanisms. Drug Metab Rev 2004;36:1−55.

[34] Mei N, Chou MW, Fu PP, Heflich RH, Chen T. Differential mutagenicity of riddelliine in liver endothelial and parenchymal cells of transgenic big blue rats. Cancer Lett 2004;215:151−8.

[35] Mei N, Guo L, Fu PP, Heflich RH, Chen T. Mutagenicity of comfrey (Symphytum officinale) in rat liver. Br J Cancer 2005;92:873−5.

[36] Zou L, Harkey MR, Henderson GL. Effects of herbal components on cDNA-expressed cytochrome P450 enzyme catalytic activity. Life Sci 2002;71:1579−89.

[37] Messina BA. Herbal supplements: Facts and myths--talking to your patients about herbal supplements. J Perianesth Nurs 2006;21:268.

[38] Williamson EM. Interactions between herbal and conventional medicines. Expert Opin Drug Saf 2005;4:355−78.

[39] Wold RS, Lopez ST, Yau CL, Butler LM, Pareo-Tubbeh SL, Waters DL, et al. Increasing trends in elderly persons'' use of nonvitamin, nonmineral dietary supplements and concurrent use of medications. J Am Diet Assoc 2005;105:54−63.

[40] Dwivedi Y, Rastogi R, Sharma SK, Mehrotra R, Garg NK, Dhawan BN. Picroliv protects against monocrotaline-induced hepatic damage in rats. Pharmacol Res 1991;23:399−407.

[41] Gonzalez FJ, Gelboin HV. Role of human cytochromes P450 in the metabolic activation of chemical carcinogens and toxins. Drug Metab Rev 1994;26:165−83.

[42] Gonzalez FJ, Yu AM. Cytochrome P450 and xenobiotic receptor humanized mice. Annu Rev Pharmacol Toxicol 2006;46:41−64.

[43] Abdel-Rahman A, Anyangwe N, Carlacci L, Casper S, Danam RP, Enongene E, et al. The safety and regulation of natural products used as foods and food ingredients. Toxicol Sci 2011;123:333−48.

[44] FDA Public Health News. Dietary Supplement Current Good Manufacturing Practices (CGMPs) and Interim Final Rule (IFR) Facts; 2007.

[45] (IPCS), I. P. o. C. S. Health and safety criteria guides 26., Geneva, Switzerland; 1989.

[46] Mattocks AR. Chemistry and toxicology of pyrrolizidine alkaloids. London, NY: Academic Press; 1986.

[47] NTP. Bioassay of lasiocarpine for possible carcinogenicity. Natl Cancer Inst Carcinog Tech Rep Ser 1978;39:1−66.

[48] NTP. Toxicology and carcinogenesis studies of riddelliine (CAS No. 23246-96-0) in F344/N rats and B6C3F1 mice (gavage studies). Natl Toxicol Program Tech Rep Ser 2003;:1−280.

[49] Boudreau MD, Beland FA. An evaluation of the biological and toxicological properties of Aloe barbadensis (miller), Aloe vera. J Environ Sci Health C Environ Carcinog Ecotoxicol Rev 2006;24:103−54.

[50] Boudreau MD, Mellick PW, Olson GR, Felton RP, Thorn BT, Beland FA. Clear evidence of carcinogenic activity by a whole-leaf extract of Aloe barbadensis miller (aloe vera) in F344/N rats. Toxicol Sci 2013;131:26−39.

[51] Xia Q, Boudreau MD, Zhou YT, Yin JJ, Fu PP. UVB Photoirradiation of Aloe vera by UVA-formation of free radicals, singlet oxygen, superoxide, and induction of lipid peroxidation. J Food Drug Anal 2011;19:396−402.

[52] Xia Q, Yin JJ, Fu PP, Boudreau MD. Photo-irradiation of aloe vera by UVA-formation of free radicals, singlet oxygen, superoxide, and induction of lipid peroxidation. Toxicol Lett 2007;168:165−75.

[53] Chen L, Mei N, Yao L, Chen T. Mutations induced by carcinogenic doses of aristolochic acid in kidney of Big Blue transgenic rats. Toxicol Lett 2006;165:250−6.

[54] Chen T. Genotoxicity of aristolochic acid: a review. J Food Drug Anal 2007;15:387−99.

[55] Chen T, Mei N, Fu PP. Genotoxicity of pyrrolizidine alkaloids. J Appl Toxicol 2010;30:183−96.

[56] Chan PC, Haseman JK, Prejean JD, Nyska A. Toxicity and carcinogenicity of riddelliine in rats and mice. Toxicol Lett 2003;144:295−311.

[57] Chou MW, Yan J, Nichols J, Xia Q, Beland FA, Chan PC, et al. Correlation of DNA adduct formation and riddelliine-induced liver tumorigenesis in F344 rats and B6C3F (1) mice. Cancer Lett 2003;193:119−25.

[58] Chou MW, Yan J, Williams L, Xia Q, Churchwell M, Doerge DR, et al. Identification of DNA adducts derived from riddelliine, a carcinogenic pyrrolizidine alkaloid, in vitro and in vivo. Chem Res Toxicol 2003;16:1130−7.

[59] Fu PP, Chou MW, Churchwell M, Wang Y, Zhao Y, Xia Q, et al. High-performance liquid chromatography electrospray ionization tandem mass spectrometry for the detection and quantitation of pyrrolizidine alkaloid-derived DNA adducts in vitro and in vivo. Chem Res Toxicol 2010;23:637−52.

[60] Xia Q, Chou MW, Kadlubar FF, Chan PC, Fu PP. Human liver microsomal metabolism and DNA adduct formation of the tumorigenic pyrrolizidine alkaloid, riddelliine. Chem Res Toxicol 2003;16:66−73.

[61] Yang YC, Yan J, Doerge DR, Chan PC, Fu PP, Chou MW. Metabolic activation of the tumorigenic pyrrolizidine alkaloid, riddelliine, leading to DNA adduct formation in vivo. Chem Res Toxicol 2001;14:101−9.

[62] Zhao Y, Xia Q, Gamboa da Costa G, Yu H, Cai L, Fu PP. Full structure assignments of pyrrolizidine alkaloid DNA adducts and mechanism of tumor initiation. Chem Res Toxicol 2012;25:1985−96.

[63] Fu PP, Xia Q, Guo L, Yu H, Chan PC. Toxicity of kava kava. J Environ Sci Health C Environ Carcinog Ecotoxicol Rev 2008;26:89−112.

[64] Xia Q, Chiang H-M, Zhou Y-T, Yin J-J, Liu F, Wang C, et al. Phototoxicity of Kava-formation of reactive oxygen species leading to lipid peroxidation and DNA damage. Am J Chin Med 2012;40:1271−88.

[65] Guo L, Shi Q, Fang J-L, Mei N, Ali AA, Lewis SM, et al. Review of usnic acid and Usnea barbata toxicity. J Environm Sci Health C 2008;26:317−38.

[66] Chan P-C, Peckham JC, Malarkey DE, Kissling GE, Travlos GS, Fu PP. Two-year toxicity and carcinogenicity studies of Panax ginseng in Fischer 344 rats and B6C3F1 mice. Am J Chin Med 2011;39:779−88.

[67] Yang Y, Yan J, Churchwell M, Beger R, Chan P, Doerge DR, et al. Development of a (32)P-postlabeling/HPLC method for detection of dehydroretronecine-derived DNA adducts in vivo and in vitro. Chem Res Toxicol 2001;14:91−100.

[68] NTP. Photocarcinogenesis study of Aloe vera in SKH-1 mice, Vol. <http://ntp.niehs. nih.gov/ntp/htdocs/lt_rpts/tr553.pdf>; 2008.

[69] NTP. Toxicology and carcinogenesis Studies of a nondecolorized whole leaf extract of Aloe Vera in F344 rats and B6C3F1 mice, pp. <http://ntp.niehs.nih.gov/Ntp/ About_Ntp/.../2011/April/DraftTR577.pdf>; 2011.

[70] Roeder E. Medicinal plants in Europe containing pyrrolizidine alkaloids. Pharmazie 1995;50:83−98.

[71] Roeder E. Medicinal plants in China containing pyrrolizidine alkaloids. Pharmazie 2000;55:711−26.

[72] Fu PP, Chou MW, Xia Q, Yang YC, Yan J, Doerge DR, et al. Genotoxic pyrrolizidine alkaloids and pyrrolizidine alkaloid N-oxides - mechanisms leading to DNA adduct formation and tumorigenicity. Environ Carcinogen Ecotoxicol Rev 2001;19:353−86.

[73] Smith LW, Culvenor CC. Plant sources of hepatotoxic pyrrolizidine alkaloids. J Nat Prod 1981;44:129−52.

[74] Stegelmeier BL, Edgar JA, Colegate SM, Gardner DR, Schoch TK, Coulombe RA, et al. Pyrrolizidine alkaloid plants, metabolism and toxicity. J Nat Toxins 1999;8:95−116.

[75] Schoental R, head MA, Peacock PR. Senecio alkaloids: Primary liver tumours in rats as a result of treatment with (1) a mixture of alkaloids from S. jacobaea lin.; (2) retrorsine; (3) isatidine. Br J Cancer 1954;8:458−65.

[76] FDA. FDA Advises Dietary Supplement Manufacturers to Remove Comfrey Products From the Market. U.S. Food and Drug Administration. U.S. Food and Drug Administration, <http://www.fda.gov/Food/DietarySupplements/Alerts/ucm111219. htm>; 2001.

[77] NTP. NTP 12th report on carcinogens. 12, iii-499; 2011.

[78] FDA. FDA Acts to Remove Ephedra-Containing Dietary Supplements From Market; 2004.

[79] Bjelakovic G, Nikolova D, Gluud LL, Simonetti RG, Gluud C. Mortality in randomized trials of antioxidant supplements for primary and secondary prevention: systematic review and meta-analysis. JAMA 2007;297:842−57.

[80] Watson J. Oxidants, antioxidants and the current incurability of metastatic cancers. Open Biol 2013;3:.

[81] Yin J-J, Fu PP, Lutterodt H, Zhou Y-T, Antholine WE, Wamer W. Dual role of selected antioxidants found in dietary supplements: crossover between anti-and prooxidant activities in the presence of copper. J Agri Food Chem 2012;60:2554−61.

Understanding Medical Foods under FDA Regulations

11

Claudia A. Lewis and Michelle C. Jackson

Venable LLP, Washington, DC

11.1 History of medical foods

It took the Food and Drug Administration (FDA) 50 years after the passage of the Federal Food, Drug, and Cosmetic Act (FDCA) to formally define "medical foods." Prior to 1972, what we now would consider medical foods were regulated as prescription drugs under section 201(g)(1)(B) of the FDCA because of their role in mitigating serious adverse effects of diseases.[1] Furthermore, prior to 1972, to market new products, manufacturers of medical foods were subject to onerous requirements, such as conducting complete drug trials, Investigational New Drug license applications, and New Drug Applications. Extremely time-consuming and cost-restrictive, such requirements choked the life out of medical food product innovation.

In 1972, the FDA reassessed its position on medical foods. This action was prompted by the agency's interest in fostering innovation in the development of medical foods and ensuring that such products were available to the public at a reasonable cost. However, due to safety concerns, the agency still sought to differentiate medical foods from general use foods. For example, the FDA reasoned that Lofenalac, an infant product designed for use in the dietary management of a rare genetic condition known as phenylketonuria (PKU), would be hazardous for healthy infants since it would be nutritionally inadequate for them. Therefore, the agency reclassified medical foods provided *enterally* (i.e., ingested via the digestive tract) as "foods for special dietary use," but injectable medical foods remained classified as drugs

---

[1] Advanced Notice of Proposed Rulemaking, Regulation of Medical Foods, 61 Fed. Reg. 60661, 60662 (Nov. 29, 1996).

Nutraceutical and Functional Food Regulations in the United States and Around the World.
DOI: http://dx.doi.org/10.1016/B978-0-12-405870-5.00011-6

subject to the FDA's Drug Efficacy Study (DESI) program.[2] In short, enterally administered nutrition was transferred to the food category while parenteral nutrition (i.e., injected into the body) retained its drug status.

Just one year later, when the agency made nutrition labeling mandatory for certain foods, it exempted certain types of foods for special dietary use from this requirement.[3] In the preamble of the final rule, the FDA noted that nutrition labeling developed for foods intended for consumption by the general population was not well suited for some food products. Two foods for special dietary use were exempted from the nutrition labeling required for other food: (1) any food represented for use as the sole item of the diet and (2) foods represented for use solely under medical supervision in the dietary management of specific diseases and disorders.

A statutory definition of medical foods was finally promulgated in the Orphan Drug Amendments of 1988, Section 5b, Orphan Drug Act.[4] A medical food was defined as "a food which is formulated to be consumed or administered enterally under the supervision of a physician and which is intended for the specific dietary management of a disease or condition for which distinctive nutritional requirements, based on recognized scientific principles, are established by medical evaluation."[5] This statutory definition remains unchanged. Unfortunately, the legislative history of the amendments does not discuss the statutory definition of medical foods, thus failing to provide any additional information regarding the types of products Congress intended the definition to cover.[6]

Soon after the Orphan Drug Act Amendments, the FDA formally launched its initiative to improve the content and format of food labels with the publication of an Advanced Notice of Proposed Rulemaking (ANPR).[7] As part of this overall initiative, the agency sought to resolve consumer confusion about food labels, aid consumers in health food decisions, and encourage product innovations so that manufacturers were

---

[2]The FDA recognized foods for "special dietary use" as early as 1941. Per regulation, the FDA stated that the term "special dietary uses" as applied to food for man, meant, among other things, "uses for supplying particular dietary needs which exist by reason of a physical, physiological, pathological or other condition, including but not limited to the conditions of disease, convalescence, pregnancy, lactation, allergic hypersensitivity to food, underweight, and overweight." Amendment to the General Regulations, Regulations for the Enforcement of the Federal Food, Drug, and Cosmetic Act, 6 Fed. Reg. 5921 (Nov. 22, 1941). This part of the regulation remains unchanged in the Code of Federal Regulations. See 21 C.F.R. § 105.3(a)(1) (September 17, 2013).
[3]8 Fed. Reg. 2124, 2126 (January 19, 1973).
[4]21 U.S.C. § 360ee(b)(3). Foods for special dietary use were often referred to as "orphan" because they were developed for the treatment of rare disorders that affect fewer than 200,000 persons in the United States. T.P. Labuza, *Food Laws and Regulations: The Impact on Food Research*, 36 Food Drug Cosmetic L.J. 293 (1981).
[5]21 U.S.C. § 360ee(b)(3). The amendments also introduced a subcategory called "orphan medical foods" to be used in the management of "...any disease or condition that occurs so infrequently in the United States that there is no reasonable expectation that a medical food for such a disease or condition will be developed without assistance." *Id.*
[6]61 Fed. Reg. 60662.
[7]54 Fed. Reg. 32610 (August 8, 1989).

given an incentive to improve the quality of the food and provide consumers with more healthy food choices.[8] In the ANPR, the FDA asked the industry for guidance on a wide range of food-labeling issues to assist the agency in determining what, if any, changes to food labeling requirements were necessary, and it was quickly followed by four public hearings. With obvious public support for a thorough modernization of food labeling, the FDA published proposed regulations on July 19, 1990.[9]

It was during the comment period for the proposed regulations that Congress passed the Nutrition Labeling and Education Act of 1990 (NLEA), and on November 8, 1990, the legislation was signed into law by President George H. Bush. Not only did the NLEA affirm the FDA's authority to mandate nutrition labeling on most foods and clarify the agency's role in regulating nutrient content claims and health claims on food labels, it also incorporated the definition of medical foods contained in the Orphan Drug Amendments of 1988 into Section 403(q)(5)(A)(iv) of the FDCA and exempted medical foods from the nutrition labeling, health claim, and nutrient content claim requirements applicable to foods generally.[10]

Quickly thereafter, the FDA published a proposal to implement the mandatory nutrition labeling provisions of the NLEA, focusing specifically on the statutory exemption for medical foods.[11] The proposal advised that the agency considered the statutory definition of medical foods to "narrowly constrain the types of products that can be considered to fall within this exemption,"[12] a sentiment that the FDA has since reiterated time and again. Further, the FDA explained how medical foods are distinguished from the broader category of foods for special dietary use and from foods that make health claims. In the FDA's opinion, "under the supervision of a physician" within the NLEA meant "that the intended use of a medical food is for the dietary management of a patient receiving active and ongoing medical supervisions (e.g., in a health care facility or as an outpatient). The physician determines the food that is necessary to the patient's overall medical care,"[13] and the patient visits the doctor for instructions on the use of the medical food. In its closing remarks on medical foods, the FDA stressed the vital public health interest in proper labeling of the nutrient content and purported uses of medical foods, which it noted may require a different manner and more detail than more traditional foods, adequate and appropriate directions for use, and product quality assurance. Thus, the agency declared that it intended to develop regulations covering these aspects "in the near future."[14]

---

[8]Virginia Wilkening, *The Nutrition Labeling and Education Act of 1990* (November 27, 1991), http://www.nutrientdataconf.org/PastConf/NDBC17/8-2_Wilkening.pdf (last visited September 17, 2013).
[9]55 Fed. Reg. 29456 (July 19, 1990).
[10]Nutrition Labeling and Education Act § 2(a)(1990).
[11]56 Fed. Reg. 60366 (November 27, 1991).
[12]*Id.* at 60377.
[13]*Id.*
[14]*Id.*, at 60378.

In the Federal Register of January 6, 1993, the FDA published the final rule on mandatory nutrition labeling, which exempted medical foods from the nutrition labeling requirements and incorporated the statutory definition of medical foods into the agency's regulations at Section 101.9(j)(8).[15] In the regulation, the FDA enumerated criteria intended to clarify the characteristics of medical foods. Accordingly, a food was defined as a medical food and, thus, not subject to the nutrition labeling requirements only if:

1. It is a specially formulated and processed product (as opposed to a naturally occurring foodstuff used in its natural state) for the partial or exclusive feeding of a patient by means of oral intake or enteral feeding tube.
2. It is intended for the dietary management of a patient who, because of therapeutic or chronic medical needs, has limited or impaired capacity to ingest, digest, absorb, or metabolize ordinary foodstuffs or certain nutrients, or who has other special medically determined nutrient requirements, the dietary management of which cannot be achieved by the modification of the normal diet alone.
3. It provides nutritional support specifically modified for the management of the unique nutrient needs that result from the specific disease or condition, as determined by medical evaluation.
4. It is intended to be used under medical supervision.
5. It is intended only for a patient receiving active and ongoing medical supervision wherein the patient requires medical care on a recurring basis for, among other things, instructions on the use of the medical food.[16]

This definition remains unchanged. In addition, the agency acknowledged that further clarification on the specific types of products the FDA considers medical foods would be helpful. Accordingly, it expressed its intention to address the issue in the future, but also noted its objective to develop much-needed medical food-labeling regulations.[17]

Citing the enactment of a statutory definition of medical food, the rapid increase in the variety and number of products marketed as medical foods, safety problems associated with the manufacture and quality control of these products, and the potential proliferation of fraudulent claims not supported by sound science, the FDA issued an ANPR on the "Regulation of Medical Foods" in 1996.[18] The agency also sought to clarify the distinct differences between medical foods and foods for special dietary purpose. Though this ANPR was withdrawn in 2004,

---

[15] 58 Fed. Reg. 2079, 2151 (January 6, 1993).
[16] 21 C.F.R. § 101.9(j)(8).
[17] 58 Fed. Reg. 2151.
[18] 61 Fed. Reg. 60662.

it largely remains the guiding force for industry understanding of the agency's views of medical food regulation.[19]

In the 1996 ANPR, the FDA acknowledged that the universe of products purporting to be medical foods had surpassed the statutory definition of a medical food to include foods that would more appropriately be categorized as foods for special dietary use. Looking to statutory language, the FDA sought to outline distinctions between these two types of foods, beginning with the meaning of "distinctive nutritional requirements" in the FDCA's definition of medical food. Pursuant to the statute, distinctive nutritional requirements must be based on recognized scientific principles and established by medical evaluation. Unfortunately, as the FDA noted in the ANPRM, the law does not define distinctive nutritional requirements. As a result, the agency proposed two possible interpretations of the phrase: (1) physiological interpretation and (2) alternative interpretation.

In the physiological interpretation, the FDA advised that distinctive nutritional requirements could be understood as referring to the body's need for specific amounts of nutrients to maintain homeostasis and sustain life. Under this interpretation, medical foods are

> foods that are formulated to aid in the dietary management of a specific disease or health-related condition that causes distinctive nutritional requirements that are different from the nutritional requirements of healthy people. Foods for special dietary use, on the other hand, are foods that are specially formulated to meet a special dietary need, such as a food allergy or difficulty in swallowing, but that provide nutrients intended to meet ordinary nutritional requirements. The special dietary needs addressed by these foods do not reflect a nutritional problem per se; that is, the physiological requirements for nutrients necessary to maintain life or homeostasis addressed by foods for special dietary use are the same as those of normal, healthy

---

[19]69 Fed. Reg. 68834 (November 26, 2004). The FDA declared that:

Because of competing priorities that have tied up FDA's limited resources, the agency has been unable to consider, in a timely manner, the issues raised by comments on the ANPRM, and does not foresee having sufficient resources in the near term to do so. Therefore, the agency is withdrawing this ANPRM. However, FDA believes that the basic principles described in the ANPRM provide an appropriate framework for understanding the regulatory paradigm governing medical foods. Therefore, FDA advises that it will continue to refer to the basic principles described in the ANPRM and in FDA's Medical Foods Compliance Program (CP 7321.002) when evaluating medical foods. With regard to the specific points made in the comment regarding regulation of medical foods, the comment is correct that the act exempts medical foods from the nutrition labeling, health claim and nutrient content claim requirements that are applicable to most other foods. However, all statements on food labels (including medical foods) must be truthful and not misleading (see section 403(a)(1) of the [A]ct). FDA advises that medical foods with false and misleading labeling are subject to enforcement action. The agency also advises that withdrawal of this ANPRM does not change the requirement that all ingredients used in medical foods must be approved food additive, GRAS, or otherwise exempt from the food additive definition. Medical foods that do not comply with this requirement are subject to enforcement action.

persons. These foods are formulated in such a way that only the ingredients or physical form of the diet is different.[20]

On the other hand, the agency stated in the alternative interpretation that distinctive nutritional requirement may be construed to encompass physical and physiological limitations in a person's ability to ingest or digest conventional foods, as well as distinctive physiological nutrient requirements.[21] Similarly, the Life Sciences Research Office of the Federation of American Societies for Experimental Biology (LSRO/FASEB Panel) noted in its 1990 *Guidelines for the Scientific Review of Enteral Food Products for Special Medical Purposes* that medical foods are for "patients with limited or impaired capacity to ingest, digest, absorb, or metabolize ordinary foodstuffs or certain nutrients contained therein, or (who) have other specialized medically determined nutrient requirements."[22] After quoting this purpose, the FDA stated that the definition of distinctive nutritional requirement would include

foods intended for persons not able to ingest foods in certain physical forms (e.g., solid food), foods intended for persons who need a concentrated form of nutrition because of reduced appetite as a result of disease or convalescence, or foods intended for persons who may have other physical limitations on the amount or composition of food that they can consume. Although these types of conditions do not necessarily result in nutrient needs different from those of healthy persons, they represent a situation where it may be necessary that the food be formulated and manufactured within very narrow tolerances to ensure that the food provides most or all of the essential nutrients, as the person for whom the food is intended may not be able to eat a variety of foods to ensure that they meet their nutritional requirements.[23]

The second element that the FDA recognized as a distinguishing attribute of medical foods is the statutory requirement that a medical food be "formulated to be consumed or administered enterally under the supervision of a physician." As a general requirement, the patient must be receiving short- or long-term "active and ongoing" medical supervision (e.g., in a health care facility or as

---

[20]61 Fed. Reg. at 60667. The agency provided an example of a person possessing a special dietary need for a food that is in liquid form due to problems swallowing, noting that this special dietary need does not change his or her physiologic nutrient requirements. Along the same lines, a person allergic to gluten may need foods specially formulated, but the food would still provide the same amount of amino acids as needed by the general population because the quantitative and qualitative amount of overall protein required by the body is similar in both healthy and protein-sensitive individuals. *Id.*

[21]*Id.*

[22]J.M. Talbot, *Guidelines for the Scientific Review of Enteral Food Products for Special Medical Purposes*, Life Sciences Research Office, Federation of American Societies for Experimental Biology (1990).

[23]61 Fed. Reg. at 60668.

an outpatient).[24] Unlike foods for special dietary purposes, the FDA views medical foods as an integral component of the patient's clinical management. Medical foods are not just simply recommended by a physician as a "part of an overall diet designed to reduce the risk of a disease or medical condition, to lose or maintain weight, or to ensure the consumption of a healthy diet."[25]

The final fundamental element of the definition of medical food addressed in the 1996 ANPR is the statutory requirement that a medical food be intended for the "specific dietary management" of a disease or condition. The FDA advised that the term "specific dietary management...evidences that Congress intended [medical] foods to be an integral part of the clinical treatment of patients."[26] The agency also cited the LSRO/FASEB Panel's conclusion that the objective of incorporating the use of medical foods into patient management was, in part, to "ameliorate clinical manifestations of the disease," "favorably influence the disease process," and "positively influence morbidity and mortality (patient outcomes)."

Axona is an example of a medical food widely prescribed by physicians today. Axona, developed by Accera, is a medical food to provide the necessary nutrients for patients with Alzheimer's disease (AD). It has been clinically shown to improve cognitive function in some patients with AD, the leading cause of dementia, and does not increase metabolism. AD is a neurodegenerative disease characterized by a decline in the ability of the brain to metabolize glucose, even in its early stages. Axona is made from caprylic triglyceride and other medium chain triglycerides, which are converted to ketone bodies by the liver, an alternative energy source for cerebral neurons.

11.2 **FDA guidance**

In May 2007, the FDA published its first draft guidance (Draft Guidance) for the industry of medical foods.[27] This guidance reiterated the statutory and regulatory provisions addressing medical foods, as well as the FDA's long-standing interpretations of those authorities. However, the guidance included no new statements of policy. Instead, the agency noted that the Draft Guidance was intended to be a convenient place for the industry to find answers to common medical food questions.

Just recently, in August 2013, the FDA released a revised version of its Draft Guidance (Revised Draft Guidance).[28] The Revised Draft Guidance was published the very same day the agency issued its second Warning Letter regarding medical

---

[24]*Id.* (*citing* 56 Fed. Reg. 60377).

[25]*Id.*, at 60668.

[26]*Id.*

[27]FDA, Draft Guidance for Industry, Frequently Asked Questions About Medical Foods (May 2007).

[28]FDA, Draft Guidance for Industry: Frequently Asked Questions About Medical Foods (2d ed. August 13, 2013), http://www.fda.gov/downloads/Food/GuidanceRegulation/GuidanceDocuments RegulatoryInformation/MedicalFoods/UCM362995.pdf (last visited September 17, 2013) (hereinafter, "Revised Draft Guidance").

foods in 2013,[29] discussed in the enforcement section below. This Revised Draft Guidance both amended and expanded upon the original draft guidance published in 2007 by incorporating 15 new questions and answers. Specifically, the agency addressed medical food labeling, physician supervision, and the scope of permissible diseases or conditions that medical foods may be labeled or marketed to manage.

The Revised Draft Guidance first addressed the FDA's understanding of the medical food definition. The 2007 Draft Guidance described medical foods as foods that are "specially formulated and processed (as opposed to a naturally occurring foodstuff used in a natural state) for the patient who is seriously ill or who requires the use of the product as a major treatment modality." The Revised Draft Guidance replaced the technical term "treatment modality" with the phrase "component of a disease or condition's specific dietary management," signifying the agency's growing emphasis on the idea that a medical food must be designed for dietary management of a disease condition, rather than generalized treatment.

The agency also reiterated that medical foods cannot be labeled or marketed for a disease or condition that can be *managed solely by a normal diet alone.* Such conditions or diseases discussed specifically in the Draft Guidance include inborn errors of metabolism (IEMs), pregnancy, diabetes mellitus (types 1 and 2), and nutrient deficiency diseases.

11.2.1 IEMs

IEMs, which include inherited biochemical disorders in which a specific enzyme defect interferes with the normal metabolism of protein, fat, or carbohydrate, are generally considered to be diseases or conditions that a medical food may be used to manage, according to the Revised Draft Guidance. Some IEMs can be managed solely with modification to the normal diet, but others cannot. For those IEMs that can be managed solely with modification to the normal diet (e.g., reduction of galactose and lactose for galactosemia), the FDA indicated that it would not be appropriate for a medical food to be labeled or marketed for that condition. For those IEMs that cannot be managed solely with modification to the normal diet, a medical food is required in addition to a specific dietary modification (e.g., reduced total protein/phenylalanine for PKU). A non-exclusive list of specific IEMs that medical foods could be used to manage is included in the 2013 Draft Guidance.

> *Pregnancy.* The FDA does not consider pregnancy a disease, instead agreeing with the Institute of Medicine that it is a "life stage." The agency also does not consider pregnancy to be a condition for which a medical food could be labeled or marketed. The Revised Drafted Guidance explained that "generally the levels of micronutrients necessary for pregnancy can be achieved by the modification of the normal diet alone."

---

[29]FDA, Warning Letter, Metagenics (August 13, 2013), http://www.fda.gov/iceci/enforcementactions/warningletters/2013/ucm367142.htm (last visited September 17, 2013).

Diabetes Mellitus (DM) Types 1 and 2. The FDA does not generally consider a product labeled and marketed for DM to meet the regulatory criteria for a medical food, based on the theory that "diet therapy is the mainstay of diabetes treatment." In the Revised Draft Guidance, the agency provided that a regular diet can be modified to meet the needs of a person with DM (along with appropriate drug therapy, if necessary).

Nutrient Deficiency Diseases. The agency explained in the Revised Draft Guidance that it does not consider classical nutrient deficiency disease, like scurvy or pellagra, to be diseases for which a medical food could be labeled and marketed. Excluding any permanent physical damage, such diseases can typically be corrected once foods (or dietary supplements) with these essential nutrients are consumed. In short, nutrient deficiency disease can be managed by normal diet alone.

The Revised Draft Guidance also emphasized that no written or oral prescription is necessary for medical foods. However, the FDA reiterated that it does not consider foods that are simply recommended by a physician or other health care professionals as part of an overall diet designed to reduce the risk of a disease or medical condition or to help support weight loss to be medical foods. Rather, the statutory requirements that a medical food be consumed or administered enterally "under the supervision of a physician" mean that "the intended use of a medical food is for the dietary management of a patient receiving active and ongoing medical supervision (e.g., in a health care facility or as an outpatient) of a physician who has determined that the medical food is necessary to the patient's overall medical care." The FDA stated it expects that the patient should generally see the physician on a recurring basis for, among other things, instructions on the use of the medical food.

With regard to labeling of medical foods, the guidance explained that medical foods are misbranded if their labeling bears the symbol "Rx only" and/or National Drug Code (NDC) numbers. However, the FDA does not object to the use of language communicating that the medical food may only be distributed enterally under the supervision of a physician. The FDA provided the following example of a permissible statement: "must be used under the supervision of a physician."

The FDA is accepting comments on the Draft Guidance until October 15, 2013.

11.3 Good manufacturing practices and import/export

Medical foods must comply with all applicable FDA requirements for foods. This includes the regulations pertaining to Current Good Manufacturing Practices (cGMPs),[30] Registration of Food Facilities,[31] and, if applicable, those specific to the

---

[30]21 C.F.R. part 110.
[31]21 C.F.R. part 1, Subpart H.

product formulation and processing.[32] Examples of formula- and processing-specific regulations include those for thermally processed low-acid foods packaged in hermetically sealed containers,[33] acidified foods,[34] and emergency permit control.[35] Even though the level of industry experience in the cGMP and quality control procedures necessary to produce medical foods (i.e., products that contain nutrients within a narrow range of declared label values) increased in the decade directly following the establishment of the statutory definition of medical food, the agency felt it necessary to create compliance programs specifically designed for medical foods.

As noted in the 1996 ANPR, medical foods are complex formulated products requiring sophisticated and exacting technology comparable to that used in the manufacture of infant formulas and drugs.[36] Moreover, the populations that consume such foods are often extremely vulnerable, such as pediatric patients at periods of growth and development or the elderly. For these reasons, the FDA published its Medical Foods Compliance Program for domestic and imported products as part of the agency's Compliance Program Guidance Manual in 1996.[37] This program, which the FDA has explicitly stated is a "high priority" due to the "susceptible population for which the products are intended,"[38] remains in effect today.

The FDA's compliance program for medical foods provides FDA inspectors direction to (1) obtain information regarding the manufacturing/control processes and quality assurance programs employed by domestic manufacturers of medical foods through establishment inspections, (2) collect domestic and import surveillance samples of medical foods for nutrient and microbiological analyses; and (3) recommend action when significant violations of the FDCA and/or related regulations are detected.[39] During an inspection of a medical food facility, agency inspectors review labeling, promotional materials, brochures, and correspondence with physicians. They also collect samples of recent lots for microbiological and nutrient content analyses.[40]

Pursuant to its medical food compliance program, the FDA has also compiled a list of known non-U.S. medical food manufacturers and their products.[41] The inspection of these firms includes collection of medical food products intended for exportation to the United States, but they do not need to be routinely

---

[32]See Revised Draft Guidance at 6.

[33]21 C.F.R. part 113.

[34]21 C.F.R. part 114.

[35]21 C.F.R. part 108.

[36]61 Fed. Reg. at 6066.

[37]FDA, Compliance Program Guidance Manual, Medical Foods Compliance Program § 7321.002 (1996), http://www.fda.gov/downloads/Food/ComplianceEnforcement/ucm073339.pdf (hereinafter, "Compliance Program") (last visited September 17, 2013).

[38]*Id.*, Introduction at 1.

[39]Revised Draft Guidance at 7.

[40]Compliance Program, Part III at 2.

[41]*Id.*, Part III at 3. This list, labeled Attachment A in the Compliance Program, is not for public distribution.

sampled under the compliance program when offered for import.[42] All imported and other shipments of medical foods not on the FDA's list "must be sampled and held" pending test results.[43]

In addition to the compliance program, the FDA has identified certain non-U.S. medical foods to be detained without physical examination under Import Alert #41-03.[44] According to the Alert, firms may be listed on the so-called "Red List" because the most recent FDA inspection conducted revealed that the facility was (1) not following cGMPs or was otherwise preparing, packing, or holding products under insanitary or other conditions that could render the products injurious to health and/or (2) one or more medical foods manufactured at the facility were analyzed by the FDA and classified as violating the FDCA (and/or its corresponding regulations).[45] Of note, a firm may also appear on the Red List if the product label misstates the active amount of an ingredient. An example is if a product is marketed as "low carbohydrate" contains 11 grams of carbohydrate based on agency testing when its label claims it only has 3 grams.[46] Two firms are currently on the Red List: Laboratorio Pisa Sa De Cv (Mexico) and Sunspray Food Ingredients Ltd. (South Africa).

11.4 **FDA enforcement of medical foods**

To our knowledge, the FDA issued its first Warning Letter concentrating on medical foods in 2001.[47] Since then, 11 more Warning Letters have been distributed, the most recent of which was in August 2013,[48] the very same day as the release of the Revised Draft Guidance.

As in its other letters, the FDA stated in the August 2013 letter that it considers the statutory definition of medical food to "narrowly constrain" the types of products that fit within this category. Accordingly, the FDA told the company that:

> [A] medical food must be intended for a patient who has a limited or impaired capacity to ingest, digest, absorb, or metabolize ordinary foodstuffs or certain nutrients, or who has other special medically determined nutrient requirements, the dietary management of which cannot be achieved by the modification of the normal diet alone. . . . [Y]our products do not meet these requirements and therefore do not qualify as medical foods under either the statute or FDA's regulations.

[42]*Id.*

[43]*Id.*

[44]FDA, Import Alert #41-03, *Detention Without Physical Examination of Adulterated and Misbranded Medical Foods* (Oct. 12, 2011), http://www.accessdata.fda.gov/cms_ia/importalert_117.html (last visited September 17, 2013).

[45]*Id.*

[46]*Id.*

[47]FDA, Warning Letter, Bristol-Myers Squibb Company (August 29, 2001), http://www.fda.gov/iceci/enforcementactions/warningletters/2001/ucm178451.htm (last visited September 17, 2013).

[48]FDA, Warning Letter, Metagenics (August 13, 2013), http://www.fda.gov/iceci/enforcementactions/warningletters/2013/ucm367142.htm (last visited September 17, 2013).

Pursuant to this letter, the agency does not consider the following as diseases or conditions eligible for treatment and/or maintenance with medical foods: chronic fatigue syndrome; fibromyalgia; leaky gut syndrome; metabolic syndrome; cardiovascular disease; inflammatory bowel disease and/or conditions; type 2 diabetes, atopic disorders such as eczema, rhinitis, and allergy-responsive asthma; bariatric patients preoperatively and postoperatively; and peripheral artery disease.

Other conditions named within the last decade in Warning Letters as being inappropriate for medical foods include inflammatory conditions, migraines,[49] immune system deficiencies, AD[50]; arthritis; colitis; constipation; lactose intolerance; diarrhea[51]; chronic illnesses; failure to thrive; pre- and post-surgery conditions[52]; and vitamin deficiency throughout pregnancy, postnatal, and the lactating periods.[53] In each of these letters, the agency took one or more of the following positions: (1) there was inadequate evidence that the disease had distinct nutritional requirements, (2) there was inadequate evidence that the particular food at issue would meet the distinct nutritional requirements of the disease, and/or (3) there was no available evidence illustrating that the nutrient levels cannot be achieved by diet modification alone.

Aside from the Warning Letters, the FDA has not actively enforced the medical food regulations. However, the publication of the 2013 Revised Draft Guidance, combined with the issuance of the August 2013 Warning Letter, signals the FDA's heightened commitment in regulating this industry. Additional law enforcement efforts may soon follow in the coming year.

11.5 Looking forward

The use of medical foods has grown steadily since the introduction of Lofenalac® over 40 years ago. Just within the last few years, however, the medical food market has prolifically expanded. In the United States and globally, there have been more than 100 new medical food product launches annually since 2009.[54] However, the actual size of the medical food market is unclear. In light of increasing use in long-term care and the aging baby boomer population,

[49]FDA, Warning Letter, Metagenics (October 1, 2003), http://www.fda.gov/ICECI/EnforcementActions/WarningLetters/2003/ucm147751.htm (last visited September 17, 2013).

[50]FDA, Warning Letter, Neuroscience, Inc. (December 19, 2011), http://www.fda.gov/ICECI/EnforcementActions/WarningLetters/2011/ucm284391.htm (last visited September 17, 2013).

[51]FDA, Warning Letter, Ganeden Biotech Inc. (December. 8, 2006), http://www.fda.gov/ICECI/EnforcementActions/WarningLetters/2006/ucm076208.htm (last visited September 17, 2013).

[52]FDA, Warning Letter, Nestle Healthcare Nutrition (December. 3, 2009), http://www.fda.gov/ICECI/EnforcementActions/WarningLetters/2009/ucm194121.htm (last visited September 17, 2013).

[53]FDA, Warning Letter, Pan American Laboratories, Inc. (November 20, 2009), http://www.fda.gov/ICECI/EnforcementActions/WarningLetters/2009/ucm191841.htm (last visited September 17, 2013).

[54]Gregory Stephens, *Convergence of the Health Practitioner Channel & Medical Foods*, Nutraceuticals World (June 3, 2013), http://www.nutraceuticalsworld.com/issues/2013-06/view_columns/convergence-of-the-health-practitioner-channel-medical-foods/ (last visited September 17, 2013).

continued strong growth is likely. In 2011, global sales were projected at just less than $9 billion,[55] but the lack of industry association and scarcity of public data have made it difficult to estimate medical food revenue in the United States. The best estimate is $2.1 billion for 2011 with a growing rate of approximately 10%.[56]

The primary drivers identified for the medical food industry include (1) the aforementioned rise in the aging population, (2) a shift to enteral nutrition, and (3) the demand for personalized medicine. As life expectancy has steadily increased over the last few decades, the older population has, in turn, grown at a significantly higher rate than the total population. In fact, over the next half century, the proportion of older persons is projected to more than double.[57] Consequently, health care services are already strained, and surgical areas of cardiothoracic, ophthalmology, and urology are particularly expected to be overwhelmingly burdened in the near future.[58] This has prompted medical providers to ponder the question of how to keep people healthy for longer without abusing medical industry resources. Valuable solutions suggested have been medical foods and other nutritional substances, especially when long-term effects of medication are important.[59]

There has also been a shift to using enteral nutrition (i.e., absorption through the gastrointestinal (GI) tract) rather than total parenteral nutrition (i.e., absorption not through the GI tract, e.g., intravenous administration). Reasons for this change include the fact that enteral administration is safer since there is less risk of infection and it offers physiological benefits such as the maintenance of small intestine mass and pancreatic function. Moreover, technological advances have helped enteral administration become more cost-effective and led to improved feeding devices. For example, pumps are becoming lightweight, constructed from materials that alleviate cracking in high stress applications, and they are now easy to use and clean. In addition, nutrients can now be delivered in various formats, such as tablets and capsules, as opposed to sterile liquids and rehydratable powders.[60]

An outstanding issue of medical treatment continues to be that people respond differently. Medical foods, however, have the ability to cater to a person's individualized needs. Drugs often have severe side effects, thus developing nutritional

---

[55]Dilip Ghosh, *Medical Foods: Opportunities In An Emerging Market*, Nutraceuticals World (April 1, 2013), http://www.nutraceuticalsworld.com/issues/2013-04/view_features/medical-foods-opportunities-in-an-emerging-market/ (last visited September 17, 2013).

[56]*Id.*

[57]United Nations, *World Population Aging 1950−2050* (May 23, 2013), http://www.un.org/esa/population/publications/worldageing19502050/pdf/62executivesummary_english.pdf (last visited September 17, 2013).

[58]D.A. Etzioni, J.H. Liu, M.A. Maggard & C.Y. Ko, *The Aging Population and Its Impact on the Surgery Workforce*, 238 Annals of Surgery 170 (2003).

[59]I. Siro, E. Kapolna, B. Kapolna & A. Lugasi, *Functional Food Product Development, Marketing and Consumer Acceptance: A Review*, 51 Appetite 456 (2008).

[60]See Rebecca Pullon, *An Introduction to the Medical Foods Industry*, Healthcare Innovation Centre for Doctoral Training (May 24, 2013), http://www.colleripmanagement.com/downloads/Medical%20Foods%20Report.pdf (last visited September 17, 2013).

alternatives to treat the underlying cause of a condition or disease is an effective way to not only personalize medicine, but hopefully avoid risks associated with treatment and/or maintenance of the condition or disease.

Since medical foods are relatively lightly regulated, especially in comparison to drugs, the barriers to market entry are lower than for other medical products. The recent publication of the FDA's Revised Draft Guidance, coupled with the issuance of two Warning Letters in 2013 regarding the definition of medical food, indicates that the agency may be acknowledging the need for stricter enforcement of its medical food regulations. However, since Warning Letters appear to be the only real FDA enforcement efforts to date, only time will tell if the agency truly intends to intensify its regulation of medical foods.

Current Canadian Regulatory Initiatives and Policies for Natural Health Products (Dietary Supplements)

12

John R. Harrison* and Earle R. Nestmann[†]

*JRH Toxicology, Ontario, Canada †Health Science Consultants Inc., Ontario, Canada

12.1 Introduction

In the United States the public recognizes products that include vitamins, minerals, and herbs as dietary supplements and in Canada these products are called natural health products (NHP), regulated under the Canadian Food and Drugs Act (the Act) as a specific regulation incorporated under the Act. The Natural Health Products Regulations (NHPRs) take into account the unique nature and properties of these products and, since 2004, Health Canada has authorized more than 61,000 NHPs for sale in Canada (as of February 27, 2013) as well as 1250 manufacturing sites for these products. We have described the Canadian NHP regulations in an earlier review [1], and subsequently have provided an interim update [2] on emerging issues confronting the NHP Directorate (NHPD). The purpose of this article is to continue this series with another up-to-date summary of significant developments initiated by NHPD as well as to provide the latest performance statistics on the status of product applications that have been submitted.

According to Global Industry Analysts Inc., the global market for functional foods and drinks is forecast to exceed U.S. $130 billion by 2015. Canada's functional food and NHP industry is a leading contributor to global innovation and growth. Canada has more than 680 specialist functional food and NHP companies with revenues of $3.7 billion [3]. NHP firms had total revenue from all sources of $2.5 billion. Sales of NHP and services accounted for 68% of this revenue [4].

Nutraceutical and Functional Food Regulations in the United States and Around the World.
DOI: http://dx.doi.org/10.1016/B978-0-12-405870-5.00012-8

12.2 What is an NHP?

NHP is the term used in Canada to refer to a group of regulated health products that include:

- Vitamin and mineral supplements
- Herbal and other plant-based health products
- Traditional medicines, such as traditional Chinese medicines
- Homeopathic medicines
- Probiotics and enzymes
- Certain personal care products, e.g., toothpastes, that contain natural ingredients

NHPs are regulated according to the NHPRs, which came into effect on January 1, 2004. Once these regulations came into effect, the anticipated flood of submissions created tens of thousands of unprocessed applications. Health Canada has been actively making changes to its administrative processes to ensure that applications are processed with increasing efficiency and to clear the backlog by February 2013, which was their stated goal for pending applications.

In 2012 Health Canada published their stated principles for setting out a new approach or modernization of the regulation of NHPs in Canada as achieving a balance that ensures product safety and consumer access while enabling NHP industry innovation and growth [5]. Put another way, the NHPD is going in the direction of cutting bureaucratic red tape while ensuring that consumers have access to many safe NHPs.

In May 2012 Health Canada initiated a consultation with stakeholders by publishing revised industrial guidance documents, holding many open meetings for all stakeholders toward the fall and holding "road show meetings" in the late fall of 2012. All input from stakeholders on the documents, surveys, and meetings were to be folded into revised guidance documents before year end 2012 (with the exception of some segments of regulatory activities [e.g., compliance], which will have further stakeholder input). All these initiatives are characterized by well-thought-out plans, good communications, and careful listening to the NHP industry. This high level of communication with and listening to stakeholders has been the hallmark of NHPD and the Office of Natural Health Products, the prior organization that initiated the development and drafting of the NHP Regulations in 1999.

Health Canada commissioned a study by Ipsos-Reid in 2011 [6] on NHPs to gauge current awareness, attitudes, knowledge, and behaviors among consumers. The study reported that while the incidence of using NHPs held steady compared to a study conducted by Ipsos-Reid in 2005 (73 vs 71%), frequency of usage has shifted such that fewer people surveyed are using them on a daily basis (32%, down from 38%) and more people surveyed are only using them during certain seasons (41%, up from 37%).

Vitamins and minerals (53%, down from 57%) continue to be the NHPs used most often by Canadians, although usage has declined compared to 2005. Some changes have taken place in terms of the NHPs used compared to 2005. Most notably, mentions of omega 3/essential fatty acids (18%, up from 3%) and herbal teas (11%, up from 0), have increased significantly, while usage of echinacea has dropped off considerably (7%, down from 15%). The exception to this is residents of Quebec who maintain a relatively high usage of echinacea [6].

There was a significant proportion of Canadians who seriously question the safety and quality of NHPs as indicated by their responses to a number of related questions; 4 in 10 (39%) indicate that they are concerned about the safety of NHPs, about one-third (34%) agree with the statement "I think natural health products can be harmful to use," 4 in 10 (42%) indicate that they question the quality of NHPs, and just under one-third (32%) indicate that they do not trust the information on the labels of NHPs. Public sentiment that NHPs are safe because they are made from natural substances has declined compared to the previous study, with fewer (42%, down from 52% in 2005) who agree with this statement. Moreover, only one in five agree that if a NHP is made of natural substances, there is no risk associated with its use [6].

12.3 Current statistics on the NHPD regulatory products: what do the numbers show?

Health Canada's NHPD, the federal Directorate responsible for the regulation of NHPs sold in Canada, produces quarterly reports on the status of industry applications sent to the NHPD. The purpose of each public report is to provide statistical data on the product license, site license, and clinical trial applications received and processed by NHPD. As a general comment, the NHPD has been operating at a greater level of efficiency in the past few years, in part due to the maturing nature of the organization as they gain more experience and insight into these complicated regulatory areas.

The first statistics relate to product license applications (PLAs): NHPD received 73,669 PLAs from January 1, 2004 to December 31, 2012. Of this total, 70,078 (95%) PLAs were completed, including the issuance of 41,255 product licenses representing 59,382 products. The remaining completed applications were either refused by NHPD or withdrawn by the industry applicant. In total, 2045 companies have received a product license to date [7].

The NHPD created a special class of PLAs in 2010 to essentially catch up on their heavy backlog. Natural Health Products (Unprocessed Product License Applications) Regulations are referred to as UPLARs or NHP-UPLARs and were registered on August 4, 2010, and were repealed 30 months following that date on February 4, 2013. These Regulations allowed such NHPs to be sold legally and to be issued exemption numbers by Health Canada. To qualify, products must have had a complete PLA submitted, they must not have been withdrawn, must still be with Health Canada for a final licensing decision, and, in addition, not fall within certain risk criteria.

Pre-UPLAR PLAs consist of PLAs received prior to the publication of the Natural Health Products (Unprocessed Product License Applications) Regulations on August 5, 2010, which were not yet complete as of that date. For the 10,885 PLAs and amendments received before August 5, 2010 (pre-UPLAR PLAs), and which had not been completed prior to that date, 10,271 (94%) were completed by December 31, 2012. NHPD was committed to completing the remaining 614 applications before NHP-UPLAR was repealed in February 2013 [7]. As of February 27, 2013, the percentage of pre-UPLAR applications completed was at 99.8%.

Post-UPLAR PLAs are those PLAs received on or after August 5, 2010. These applications are subject to the performance targets set out in the NHPD Application Management Policy, which outlines the way in which PLAs for NHPs are to be submitted. The policy also outlines the responsibilities and expectations for NHP industry applicants before and throughout the application review process. The policy also outlines the proposed performance targets for the management of all types of PLAs.

For the 23,138 applications and amendments received on and after August 5, 2010 (post-UPLAR PLAs), 46% fully met NHPD's pre-cleared information (PCI; 60 day review target) and the remaining 54% required full assessment (180 day review target). During Quarter 3 ending on December 31, 2012, NHPD met performance targets as follows: 30 day initial assessment at 95%, 60 day PCI at 99%, and the 180 day full assessment at 62% [7]. See Section 12.5.1 for further information on these three product classes.

The NHPD received 2027 Site License Applications (SLAs) from January 1, 2004 to December 31, 2012. Of this total, 2016 (99%) SLAs have been completed, including the issuance of 1386 site licenses. The remaining SLAs were either refused by NHPD or withdrawn by the industry applicant. A total of 2773 Site License Renewals have been received for issued site licenses and 98% of all Site License Renewals have been completed. During fiscal year Quarter 3, April 1 to December 31, 2012, the NHPD received 20 SLAs and 128 Site License Renewals and completed 36 SLAs and 134 Site License Renewals [7].

For the period January 1, 2004 to December 31, 2012, NHPD received 405 clinical trial applications (CTAs). Of this total, 398 (98%) CTAs have been completed, including the issuance of 340 Notices of Authorization. The remaining completed applications were either refused by NHPD or withdrawn by the applicant. A total of 364 CTA Amendments and Notifications have been received for issued Notices of Authorization, and 98% of all CTA Amendments and Notifications have been completed. During Quarter 3 ending on December 31, 2012, NHPD received 8 CTAs and 1 CTA Amendments and Notifications and completed 9 CTAs and 0 CTA Amendments and Notifications [7].

12.4 **NHPs transitioning to food products: a regulatory transition process**

Health Canada recognized the confusion around foods that were marketed as NHP as more and more foods incorporating NHP ingredients entered the marketplace over the years. This is all part of the spectrum of regulated products that exists between foods and NHPs, which could be termed nutraceuticals or functional foods. What makes things even more complicated is that some people consider food as NHPs.

Since May 2010, Health Canada has worked with manufacturers to safely transition food products that were marketed as NHPs to the food regulatory framework while attempting to minimize disruption to the marketplace. Health Canada first announced in October 2011 its intent to transition caffeinated energy drinks to the food regulatory framework [8]. Additional categories of foods being marketed as NHPs in 2011 began to transition in April 2012. Health Canada stated that it was aiming to complete the transition process by December 2012.

The goal of this regulatory transition process is to make sure that products that look like foods are used and regulated as foods. Canadians will be able to make better informed choices due to consistent nutrition information and labeling requirements on these food products. For the food industry, this transition will level the playing field, reduce red tape, and continue to help support innovation [9].

The transition of eligible food products marketed as NHPs to the food regulatory framework was carried out through the process of issuing a Temporary Marketing Authorization Letter (TMAL). Manufacturers were required to collect and provide data to Health Canada to fill the knowledge gaps needed to finalize the regulatory requirements for these products (Section B.01.054 of the Food and Drug Regulations). Manufacturers also had to comply with the additional requirements as stipulated in the TMAL, including:

- Conducting research according to a protocol mutually agreed upon by the manufacturer and Health Canada
- Collecting and reporting data on consumption incidents when deemed necessary by Health Canada
- Withdrawing the product if requested

12.5 **A "new approach" to regulating NHPs**

Since their initial development, the Canadian NHP regulations have been seen as comprising a new and innovative model for regulating NHPs, or dietary supplements as they are known in the United States and elsewhere in the world [1]. Recently, the NHPD has been developing a new approach to regulating these NHPs.

Largely, the carefully thought-out modifications of the way the NHPD does business appear to be excellent. They are based, in part, on the regulatory experience of Health Canada and input of many stakeholders since the regulations came into force in 2004 as well as other extensive work dating back to 1999 with the considerable stakeholder input that produced the Transition Team's Final Report to the Minister of Health in 2000.

The stakeholder discussions and stakeholder input, with many new ideas and approaches from all sides, have been carefully considered and used by the NHPD to develop the latest new approach to the regulation of these products [5]. An earlier publication [2] had set out in detail the basic framework of the NHP regulations and the emerging policies and practices under these regulations at that time.

A process aimed at developing more efficient, flexible ways of regulating NHPs by Health Canada was initiated by the NHPD, holding stakeholders meetings and consultations prior to and then in June 2012, and publishing a document that addressed the goals of providing the public an increased access to products while maintaining the important responsibility of ensuring product safety to consumers. It was intended to repeal UPLAR in February 2013. Implementation of new approaches ensures that UPLAR is no longer required to regulate NHPs. NHPD is well on track to clear the backlog of industry submissions. With the new approach, there will be a predictable and relatively stable regulatory environment for the efficient licensing of NHPs in Canada.

An overview of the new approach to regulating NHPs consists of the following five major components: a three-class system of NHPs based on risk, requirements and pathways for licensing NHPs, site licensing modifications, quality guidance redevelopment for NHPs, and compliance transition over time [5]. These major components are described in more detail in the following sections.

12.5.1 Three-class system

A three-class system of NHPs is based on risk. Products about which there is most knowledge and certainty regarding safety will be reviewed in the shortest amount of time, while complex applications will be the focus of more detailed evaluation efforts. This three-class system builds on the knowledge of ingredients, claims, and combinations of NHPs that the NHPD has seen and evaluated over the years. The overall goal in this three-class system of NHPs is to speed up the NHP submission review process:

12.5.1.1 Class 1

The first class are those products about which the NHPD holds the highest certainty (i.e., how much is known about the product), comprise 75% of all NHPs, and are supported by PCI from prior NHPD decisions. Examples of this class are vitamins, minerals, and various herbal products. An initiative that took years to build was the development of NHPD monographs for specific NHPs for inclusion in the Compendium of Monographs. In the past year, 30 additional monographs, representing hundreds of ingredients, were completed based on PCI and thus add to the Compendial NHP review system. The current performance target for this class is 60 days and the proposed new target is 10 days.

12.5.1.2 Class 2
The second class holds moderate certainty, has 20–24% of products, with at least one claim and/or ingredients supported by PCI for these NHPs. An example for this second class would be an existing authorized product with a new claim related to the product use. The current performance target for this class is 180 days and the proposed new target is 30 days.

12.5.1.3 Class 3
The third class holds the lowest certainty, comprises 1–5% of products, and no ingredients or claims supported by PCI. An example would be a new product claiming to help prevent rheumatoid arthritis. Clinical trial evidence with full pre-market assessment is the level of review needed. Companies could revise their claims for the NHP to meet the PCI to reduce review time. The current performance target for this class is 180 days and it is proposed to remain as 180 days.

An important aspect of classifying an NHP, therefore, is its risk profile. When evaluating the health information on an NHP, Health Canada must consider the inherent risk of the product, which includes its toxicity and the level of certainty or degree to which NHPD is familiar with the product. For instance, ephedrine has a high risk profile with potential serious adverse reactions (ARs); however, with an NHPD monograph, combined with risk mitigation strategies that are well known and documented by the NHPD over the years to produce high certainty, such an NHP could be placed in the first class with a 10 day review period.

12.5.2 Guidance document 2012
The second of the five components to the new approach to regulation is a document produced by the NHPD covering requirements and pathways for licensing NHPs, having involved a prior 90 day public consultation. The consultation on this guidance document was scheduled to end August 21, 2012, and finalized by the end of December 2012. The guidance document for requirements and pathways for licensing NHPs is composed of three subsections including: (1) evidence requirements based on risk; (2) a general health claims annex; and (3) a combination ingredients annex, which can also be called a subtherapeutic ingredients annex.

12.5.2.1 Evidence requirements based on risk
For the first subsection there would be a proportionate oversight based on the gradation of high, medium, and low risk to consumers, as discussed previously.

12.5.2.2 Health claims annex

For the second subcomponent, general health claims, an annex would be developed whose purposes would be to facilitate access to low-risk NHPs and improve consumer confidence by creating a gradient of health claims and providing consumers with more transparency regarding product effects. The principles for these general health claims would cover the concepts that the NHP is for health maintenance, for self-limiting, or minimally bothersome conditions and/or for purposes that will cause little or no harm if the NHP is ineffective for the consumer. In terms of general claims for health maintenance, this area would be reserved for essential nutrients, include the highest levels of scientific evidence, and be based on an ingredient or constituent of a product. One example would be "Strawberry extract provides vitamin C for the maintenance of good health." Specific claims can be made based on constituents with clinically demonstrated benefits and one example is "Cod liver oil is a source of vitamin A to help maintain immune function." For manufacturers making statements relating to microorganisms as the "source of" a constituent in a product such as "source of digestive enzymes" or "source of probiotics," these statements will be considered by NHPD as long as they do not imply health benefit (or as long as the organisms are measured by assay or input). Claims that are effective when used in conjunction with other therapies are another area, and clinical evidence must demonstrate efficacy when used in conjunction with an activity or other constituents. An example is "Helps in weight maintenance when used in conjunction with adequate exercise and a calorie reduced diet."

There is also the area of general claims for the relief, resolution, and/or risk reduction of low therapeutic impact conditions. These would be for self-diagnosable, self-treatable, and/or self-resolving conditions in which the consumer can easily tell if a product is effective. Regulatory decisions for these types of claims can be based on strong evidence, a totality of weak evidence, or trends in evidence. Other general claims that will be considered by NHPD include the use of "used in aromatherapy" (e.g., based on two aromatherapy textbook references), "used in herbal medicine" (e.g., based on two references), or for traditional products (e.g., modification from original formula/manufacturer or practitioner formula). On the other hand, a high-level claim (e.g., used to treat diabetes symptoms) would require a high level of evidence such as two phase III or IV clinical trials. Low-level claims (e.g., soothes a sore throat) would require lower levels of evidence such as published peer-reviewed narrative reviews.

12.5.2.3 Ingredient combinations annex

The third subsection for requirements and pathways for licensing NHPs is a combination ingredients annex. NHPD permits combinations of substances, which are listed in the Schedule 1, NHP Regulations as long as:

- There is no increased health risk (e.g., additive risk, overmedication, altered bioavailability, or pharmacological activity) that cannot be mitigated

- There is no decrease in how the combination product works (e.g., antagonistic effects)
- There are no contradictions in the recommended conditions of use (e.g., contradictory claims, durations of use, risk information)

A subtherapeutic ingredient should be listed as non-medicinal when it is added to confer suitable consistency or form to the medicinal ingredients as per the definition of a non-medicinal ingredient. When an additive effect cannot be demonstrated and it does not meet the definition of a non-medicinal ingredient, such subtherapeutic ingredients should be listed as medicinal. Currently the NHPD is reviewing its subtherapeutic ingredients policy.

12.5.3 **Site licenses**

Site licensing is the third of the five major components of the new approach to regulating NHPs. It was an important NHP regulatory area for stakeholders' sessions in 2012 to discuss a proposed new approach to site licensing, focused on providing greater quality assurance for products. Stakeholders were invited to provide feedback on key elements of the proposed concept. A site license from Health Canada is required for NHP manufacturing, packaging, labeling, and importation for sale. To obtain a site license, a company submits a Quality Assurance Report demonstrating that buildings, equipment, procedures, etc., comply with the Good Manufacturing Practices (GMPs) outlined in the Natural Health Product Regulations (NHPR). For the current site licensing model, a company's quality assurance person demonstrates compliance with GMPs by providing supporting documentation and specific information to Health Canada. For the proposed site licensing model, there are two main aspects including: first, there is a required on-site inspection for companies demonstrating critical noncompliance and, second, there is an optional on-site inspection by a recognized third party to obtain a "seal of approval" for exportation and marketing purposes.

Currently there is no on-site verification for NHPs' compliance with quality standards, unlike other health products and food. The 2011 Risk-Based Approach to Site Licensing proposal included three points: (1) support for independent quality assurance, (2) less support for proposed regulatory changes related to GMP standards, and (c) concern about the costs to small- and medium-sized businesses.

NHPD conceived the idea, after extensive stakeholder input, of a "New Approach to Site Licensing" pilot. NHP companies would continue to apply for and renew site licenses through paper-based assessment but could opt to provide a voluntary third-party audit within existing regulations as evidence of compliance. Where critical deficiencies are noted, Health Canada would request a company to undertake a third-party audit. The expected benefits for this new pilot approach would include the following: creating market-driven demand for products with independent quality assurance, facilitating ability to export to international markets, increasing license renewal interval, and it could help expedite licensing in subsequent renewal periods. The consultations for this proposed pilot occurred through to February/March 2013 at which time there was a posting of a proposed pilot for this new site licensing approach, and the launch of the pilot was aimed for the late summer of 2013.

12.5.4 Quality

Quality of NHPs is another major regulatory component under modernization, as existing quality guidance is considered too prescriptive and not clearly indicative of who is responsible for what. Stakeholders indicated that the existing guidance document did not account for industry experience and expertise. Revised guidance on quality for NHPs was introduced in 2012 and it incorporates the flexibility required to keep up with advances in product and manufacturing technology and knowledge. It is also aimed at reducing regulatory red tape and is in alignment with international bodies. This particular quality document was posted for 90 day consultation until November 30, 2012, and was discussed in stakeholders sessions prior to that date.

This new draft Quality of NHPs guide states that the license holder is responsible for ensuring NHP quality. It is the responsibility of applicants to ensure completeness and accuracy of information submitted. The document incorporates experience gained through review of applications and stakeholders' communications, and to date there has been positive stakeholder reaction. This draft Quality guidance has a number of good aspects as it:

1. Reduces the regulatory burden on industry
2. Harmonizes with international bodies (e.g., World Health Organization)
3. Leverages national and international sources for quality standards (e.g., U.S. Pharmacopoeia)
4. Recognizes that industry and manufacturers are knowledgeable and expert regarding their own manufactured products
5. Allows the use of judgment and discretion to ensure the production of high-quality NHPs
6. Recognizes that testing requirements can vary greatly between ingredients and NHPs
7. Provides a clearer pathway to assure that high-quality NHPs are produced while allowing industry alternative routes to ensure the same outcome for these high-quality NHPs

In addition the draft Quality guidance has less emphasis on finished product testing and more emphasis on the need for appropriate quality control systems that follow GMPs. It complements the specifications section in NHPD Monographs and supplements the use of ingredient-specific information contained in the Natural Health Products Ingredients Database (NHPID) for ingredient-specific requirements, test methods, and tolerance limits. Within the context of the draft Quality guidance, certain aspects of site licensing are set out. First, a valid site license is the most important component in ensuring high-quality NHPs as applicants are required to demonstrate compliance with GMPs by ensuring batches meet established specifications; establishing shelf-life to support stability; and maintaining Standard Operating Procedures, records, and samples. A proposed future approach to site licensing includes on-site verification as evidence of GMP compliance. The time lines for this draft Quality guidance include: November 30, 2012, as the end of consultation period; end of December 2012 for a planned review of stakeholder feedback and a revision of the guidance document as necessary, and; by March 2013, NHPD is aiming to have the finalized quality guidance posted.

12.5.5 Compliance transition

Compliance Transition is the final major component of this new approach to regulating NHPs. After the repeal of UPLAR, compliance activities will continue to be risk based, and Health Canada will continue compliance promotion to assist companies in understanding their regulatory responsibilities. The goal is to highlight to Canadians, when purchasing NHPs, the importance of NHPs having natural product numbers. The most advantageous ways of managing the transition period to avoid any market disruptions for safe products will be covered in stakeholder discussions.

In 2009 NHPD workshops were held across Canada and the NHP Program Advisory Committee established a working group on Compliance & Enforcement (C&E) to develop key areas of a new policy. In the summer of 2010, the C&E Policy & Annex was published and by the fall of 2010, an implementation of the C&E Policy was to be preceded by a 6 month compliance promotion period. In the winter of 2010 the promotion period was extended, and as of December 2012, the compliance promotion period was still in effect. The NHP-UPLAR program was designed to be repealed on February 4, 2013, and then the annex to the NHP C&E Policy for Exempt NHPs will no longer apply. After this time the NHP C&E Policy will remain in effect and there will be the need for a compliance transition period. NHPD has been asking for feedback on these time lines through the stakeholder discussions and workshops during the fall of 2012. The repeal of NHP-UPLAR does not signal the immediate start of a new or different approach to C&E for NHPs. The aim of the NHPD is to have this New Approach to NHPs produce a quicker and clearer market authorization. Health Canada has been working to establish a transition time line that leads to the expectation that all products have a valid market authorization. The transition period was designed to enable companies to make adjustments to business practices and to phase out noncompliant products. The goal of the NHPD is to clear the NHP submission backlog by February 2013 and thus the UPLAR will not be required at that time. The NHPD has developed new pathways for the licensing of NHPs, and a new site licensing approach is underway. Overall, the aim of the New Approach to NHPs is to allow for a new, predictable, and stable approach to licensing NHPs. Throughout this transition, the NHPD has been and will engage stakeholders actively and listen to their concerns with a view to ensuring a successful transition toward compliance of NHPs.

An important goal of this new approach to regulating NHPs is to avoid market disruption for safe products. It appears that with the careful steps of regulatory streamlining stakeholder involvement and input, the probability of reaching this goal is high. For this overall compliance transition there will be three periods, namely, promotion, transition, and implementation.

The NHPD compliance promotion period is slated for March 1, 2013 to December 1, 2013, and the two goals for this period are first, to enhance aware-ness of the NHPRs, to ensure that companies have what they need to understand and comply with the Regulations and second, that companies may use this time to plan and prepare for the additional phases, by beginning to plan for the phasing out of the distribution for non-market authorized products. NHPD will also use all the various communication avenues to send information to all stakeholders, with the goal that all imported or manufactured NHPs should have market authoriza-tion before importation and/or distribution. The phasing out of non-market autho-rized product will continue at the retail level.

The implementation period starts September 1, 2014, and goes from there into the future [5]. The goal for this period is to ensure the completion of the transition period and that all products should have authorization before coming onto the market.

12.6 Recent modifications by NHPD facilitating business and regulatory processes

1. The NHPD informed stakeholders of a pilot plan to streamline CTA review activities: beginning August 1, 2012, the NHPD will no longer review CTAs for NHPs where the use of the product is not appropriate for self-care. This phased approach began with those applications to be reviewed by the Therapeutic Products Directorate. All applications and amendments are to be sent to the NHPD to be routed to the appropriate Directorate in Health Canada that then notifies the sponsor as to which Directorate will review the file [10].

2. NHPD monographs in the Compendium of Monographs provide a preapproved basis for NHP claims, and there is continued development of NHPD monographs to which companies can reference their products to facilitate regulatory licensing. In December 2012, the NHPD sought feedback on eleven newly proposed monographs and three revised monograph for ingredients [11], namely:

   a. New PCI covering: Flower Essences, Gemmotherapy, Lithotherapy, Oligotherapy, Oligotherapy—Plant Sourced, Organotherapy, Homeopathy, Organotherapy and Homeopathy, Hydrolyzed collagen, Marshmallow Leaf (*Althaea officinalis*), and Marshmallow Root (*Althaea officinalis*)

   b. Revised PCI covering: Counterirritants, Dandelion (*Taraxacum officinale*), and Dandelion Juice (*Taraxacum officinale*).
3. The establishment of the Traditional Chinese Medicines (TCM) Advisory Council in 2012 follows the federal Minister of Health's call, in the fall of 2011, for the establishment of a single window for the TCM community to interact with Health Canada. The mandate of the Council is to provide advice on current and emerging issues related to TCM, including:
   a. Importation, sale, and use of TCM in Canada
   b. Practice of TCM in Canada, noting that this is subject to provincial jurisdiction
   c. Novel TCM [12].
4. A Natural Health Product License Application Form User Manual [13] was developed by NHPD for industry use, to explain the terminology and appropriate information to be included in an NHP license application. This guidance document also outlines the application requirements and submission process for each NHP application type. The Electronic Product License Application (ePLA) form has been developed for NHP License Applications and for post-license changes. The ePLA is designed to be completed on the applicant's workstation where it can be saved and retrieved at any time and the NHPD user manual provides the information on how to successfully complete the ePLA form. Through an active Internet connection, the form can automatically access the NHPID to provide drop-down selections of medicinal and non-medicinal ingredients and related information.
5. Health Canada [14] published a third version in 2012 of a document intended to provide a summary of information on how to interact electronically with the NHPD of Health Canada. A key objective of the Natural Health Products Online Solution (NHPOLS) is to provide an end-to-end process for the electronic exchange of protected information between NHPD and applicants. By taking advantage of the electronic tools available through the NHPOLS, industry applicants will experience an efficient, secure, and cost-effective way to submit applications and deal with NHPD electronically. The benefits include quick and cost-effective means of communication, reduced frequency of application errors, receipt confirmations, and tracking of documentation. All correspondence is sent in an electronic pdf format, resulting in less paper.

12.7 **The larger picture: mutual recognition agreements**

Mutual recognition agreements (MRAs) are just that—agreements between countries to facilitate trade. Ideally one would consider that MRAs would assist, for instance, the Canadian importer and the U.S. manufacturer, and vice versa. MRAs do not always exist to help the flow of goods such as NHPs or self-care products between Canada and other countries. In fact, there is no MRA between Canada and the United States for NHPs.

Consider the situation in which a company wants to import a consumer health product such as an NHP from the United States to Canada. Unfortunately, there are many regulatory inconsistencies, even for low-risk health products, that become barriers to trade when a consumer or NHP company wants to import into Canada. Canadian NHP regulations under the Canadian Food and Drug Act are not harmonized for similar U.S. products and, depending on the type of product the company wishes to import to Canada, there are numerous barriers. Consumers are denied product choices and industry is denied the opportunity to offer these choices to the Canadian public. If one considers that an MRA is a useful approach to increase trade and harmonize regulations, then there are a number of steps that can be taken to develop a specific MRA. As one step toward MRAs, the NHPD has reviewed the NHP laws and regulations for Canada's primary NHP trading partners—Australia, China, France, Germany, Hong Kong, India, United States, UK, and the European Union—and compared them to Canada's NHP regulations [15].

This comparison of regulatory frameworks shows that certain countries such as Australia have a regulatory regime very comparable to Canada's regime, whereas China mainly exports only one type of natural health product, i.e., Chinese herbal medicines. For Canada's European trading partners NHPs are categorized as drugs or medical products and are treated, from a regulatory point of view, as drugs. Germany has the most rigorous set of regulations and this is reflected in the way NHPs are dispensed to the consumer: through pharmacies only, except for herbal products with German names. Canada's major trading partner, the United States, classifies some NHPs as dietary substances, which require prior governmental approval before they are sold to consumers. Other NHPs are under the U.S. Food and Drug Administration's control as nonprescription medicine and homeopathic medicines, which are classified as drugs. In the past few years, the regulatory environment in the United States has been moving toward a more regulated system [15].

Strategies for prioritizing policy research needs in the area of international regulations of NHPs are under consideration by the NHPD. The priority areas for policy research fall into two main categories: the safety of the products being manufactured and sold to Canadians and processes to facilitate trade between countries. The overall aims for trade development are to look for methods and means of further harmonization with trading partners and to develop MRAs [15].

Health Canada presently has a Memorandum of Understanding (MOU) related to GMPs with the following countries: Australia, France, Sweden, Switzerland, and the UK. This MOU stipulates that an official inspection report from the regulatory authority of one of these five countries would be considered an appropriate basis to establish that a drug or NHP has been fabricated, packaged, labeled, and/or tested against equivalent standards in the GMP requirements set out in Part 3 of the NHPRs.

Consider the situation of health risks within the context of trade and NHPRs. One of the guiding principles that Health Canada uses in the regulation of NHPs and foods under the Canadian Food and Drug Act is a risk-based approach in protecting health and safety. Low-risk products face fewer regulatory requirements than high-risk products and the NHPD has acknowledged that there are many low-risk NHPs as evidenced from their analysis of NHP information concerning the recent modernization of the NHP regulatory approach. Information required by Canadian authorities but not required in the United States creates barriers to getting low health risk NHP products imported into Canada. There are options that Health Canada may consider to solve these challenges.

The U.S. regulatory authorities appear to be reluctant to move to an MRA and Canada does not appear to be making this a regulatory priority, despite the fact that this is a multibillion dollar industry. New approaches and initiatives could be developed mutually by these governments. Canadian stakeholder surveys could be conducted to develop ideas and examine how consumers and industry could be assisted by getting low health risk products to market. Intergovernmental cooperation on this topic could result in plans, priorities, and eventual pilot projects to slowly build confidence in each other's regulatory systems to then aim for an MRA covering NHPs.

12.8 Post-market activities following product approval
12.8.1 AR reporting

One condition of NHP approval (Section 24 of the NHPRs) is to inform Health Canada of any serious ARs attributed to that product or to its ingredients. Each license holder must report any domestic serious ARs, and any foreign, unexpected serious ARs to the NHPD within 15 days of learning of their occurrence. In addition, each licensee must prepare and maintain a summary report, with critical analysis, of all non-serious and serious ARs that occur in Canada, as well as all foreign, serious unexpected ARs. This report must be prepared on an annual basis. Upon request from Health Canada, an interim report or an annual report must be supplied within 30 days.

In addition to required reporting by industry (i.e., product license holders), consumers and health professionals (e.g., physicians, naturopaths, nurses, dentists, pharmacists) are encouraged to report any suspected ARs. The Spontaneous AR Monitoring Program and Database was developed by Health Canada in 1965. ARs can be reported in writing, fax, or by phone, for possible reactions to drugs or other health products, and to facilitate such reporting, more recently Health Canada launched the MedEffect Canada website (www.healthcanada.gc.ca/medeffect; www.santecanada.gc.ca/medeffet, in French) in 2005, so that ARs can also be reported online. The MedEffect Canada website also is a source of information, such as, advisories, warnings, and recalls; present and past issues of the Canadian Adverse Reaction Newsletter; offerings through the Learning Centre; regulatory guidelines and guidance documents; summaries of stakeholders' consultations; activities of the Expert Advisory Committee; MedEffect e-notices; and reports and other publications [16]. Other useful documents produced by Health Canada include the Guidance Document for Industry—Reporting Adverse Reactions to Marketed Health Products [17] and the Adverse Reaction Reporting for NHPs [18].

The activities of monitoring AR reports, and of determining whether or not action is required, are discussed in the following section.

12.8.2 Post-market surveillance

In addition to the NHPD, responsible for implementation of the NHPRs, another agency within Health Canada is the Marketed Health Products Directorate (MHPD), which is responsible for (1) post-market surveillance of products once they reach the public, (2) risk communication, and (3) oversight of advertising.

In the first instance, MHPD monitors AR reports submitted to the Canada Vigilance Program. These reports are submitted by consumers, health care practitioners, and also by the manufacturers, as part of their regulatory requirements (see Section 7.1). The activities of regulatory agencies in other countries are also monitored. It is the responsibility of MHPD to determine the strength of any signal that these reports provide. For example, are the reports random, or is there some consistency about reports for certain products or for certain ingredients? The under-reporting of ARs is recognized, and the MHPD engages in education and outreach activities to stimulate the reporting of such reactions. A recent paper [19] lays out some of the challenges and opportunities in assessing ARs to NHPs.

Follow-up of the ARs is often required to request additional information and to obtain samples of suspect products that may be obtained for analysis, if required, when available. It is important to determine whether the suspected AR might be due to possible contamination of the product by a chemical impurity or microbial agent. It is also important to know if the product has been adulterated in some way, e.g., addition of drug substances that are not permitted in NHPs. Other possibilities could be mislabeling, misidentification of herbal ingredients, incorrect dosage, incorrect consumption, or interaction with other medications or NHPs. An example of how post-market activities can impact on safety issues of NHPs such as Black cohosh products has been given in detail by Health Canada [20].

These investigations may remain incomplete or equivocal because of lack of consistent or conclusive information. In other cases, it may be determined that the product itself may not be safe and becomes subject to regulatory action such as recalls, warnings, advisories, press releases, etc., as described on the MedEffect Canada website. Health Canada has noted that despite the new approach to licensing natural health products, the responsibilities of manufacturers to report serious ARs under the NHPRs remain. The monitoring of the safety of products will continue under the new licensing paradigm.

12.9 Conclusion

Canadian regulatory initiatives and policies for NHPs or dietary supplements, as they are known in the United States, are undergoing considerable changes; yet the main stated goal of the NHPD remains to provide consumers the informed choice for those self-care products that are safe, efficacious, and produced under good manufacturing processes. Currently, there are over 55,000 products that are in the category of an NHP in Canada. New approaches to modernizing regulatory areas have been developed or are in the last stages of completion, using extensive stakeholder consultations [21,22]. This modernization of NHP regulatory areas consists of the following five major components:

1. A three-class system of NHPs based on risk
2. Requirements and pathways for licensing NHPs
3. Site licensing modifications
4. Quality guidance redevelopment for NHPs
5. A compliance transition over time

The major modernization process still underway at the end of 2012 was the compliance transition. In addition to the overall transitions within the new approaches to modernizing regulatory areas, there was also, concurrently, a transition of specific food products such as caffeinated energy drinks from NHPs to the food regulatory framework, which was completed by December 2012. The NHPD has initiated numerous excellent modifications by facilitating business and regulatory processes with a view to streamline and modernize the business interactions with industry. One area that still presents a challenge for the NHPD is the development of MRAs to develop agreements between countries to facilitate trade. Unfortunately, there is no existing MRA between Canada and the United States for NHPs and dietary supplements. Finally, the post-market activities conducted by Health Canada will continue to be an important aspect of detecting and assessing safety-related issues pertaining to NHPs available to Canadian consumers.

References

[1] Nestmann ER, Harwood M, Martyres S. An innovative model for regulating supplement products: Natural health products in Canada. Toxicology 2006;221(1):50−8.

[2] Martyres S, Harwood M, Nestmann ER. Emerging policies and practices under the Canadian Natural Health Product Regulations. In: Bagchi D, editor. Nutraceutical and Functional Food Regulations in the United States and Around the World. (Food Science and Technology International Series). New York (NY)/Toronto (ON): Elsevier; 2008. p. 159−72.

[3] Foreign Affairs and International Trade. Invest in canada. functional foods and natural health products: canada's competitive advantages. <http://investincanada.gc.ca/download/1607.pdf>; 2012.

[4] Statistics Canada. Results from the functional food and natural health products survey. <http://www.statcan.gc.ca/pub/88f0006x/88f0006x2009001-eng.pdf>; 2009.

[5] Health Canada. A new approach to natural health products. <http://www.hc-sc.gc.ca/dhp-mps/prodnatur/nhp-new-nouvelle-psn-eng.php>; 2012.

[6] Health Canada. Natural Health Products. Ipsos-Reid Survey. <http://www.hc-sc.gc.ca/dhp-mps/prodnatur/index-eng.php>; 2012.

[7] Health Canada. Status of applications quarterly report: Quarter 3 (October 1, 2012 to December 31, 2012). <http://www.hc-sc.gc.ca/dhp-mps/prodnatur/report-rapport/index-eng.php>; 2013.

[8] Health Canada. Health Canada's Proposed Management Approach in Response to the Expert Panel on Caffeinated Energy Drinks. <http://www.hc-sc.gc.ca/fn-an/securit/addit/caf/ced-response-bec-eng.php>; 2011.

[9] Health Canada. Transition process for foods marketed as Natural Health Products. http://www.hc-sc.gc.ca/fn-an/prodnatur/transit-process-food-aliment-eng.php>; 2011.

[10] Health Canada. Stakeholders of clinical trials for natural health products. <http://www.hc-sc.gc.ca/dhp-mps/prodnatur/applications/clini/nhp-psn-eng.php>; 2012.

[11] Health Canada. Consultation - Message to natural health product stakeholders regarding development of pre-cleared information. <http://www.hc-sc.gc.ca/dhp-mps/consultation/natur/pre-cleared_information_preautorisee-041212-eng.php>; 2012.

[12] Health Canada. Advisory Council on Traditional Chinese Medicines. <http://www.hc-sc.gc.ca/dhp-mps/prodnatur/activit/com/tcm-mtc/index-eng.php>; 2012.

[13] Health Canada. Natural Health Product Licence Application Form User Manual. <http://www.hc-sc.gc.ca/dhp-mps/pubs/natur/eplaguide-eng.php>; 2012.

[14] Health Canada. Guidance document on how to interact with the Natural Health Products Directorate electronically. <http://www.hc-sc.gc.ca/dhp-mps/pubs/natur/trading_part_commerce-eng.php>; 2012.

[15] Harrison JR. "*International Regulation of Natural Health Products.*". Boca Raton: Universal Publishers; 2008.

[16] Jordan SA. Natural health products: regulation and product vigilance activities at health canada. Lecture in Course NSF 1212. University of Toronto; 2012, October 4.

[17] Health Canada. Guidance Document for Industry-Reporting Adverse Reactions to Marketed Health Products <http://www.hc-sc.gc.ca/dhp-mps/pubs/medeff/_guide/2011-guidance-directrice_reporting-notification/index-eng.php>; 2013.

[18] Health Canada. Adverse Reaction Reporting for Natural Health Products. <http://www.hc-sc.gc.ca/dhp-mps/alt_formats/pdf/pubs/medeff/fs-if/2011-reporting-declaration_nhp-psn/2011_reporting-declaration_nhp-psn-eng.pdf>; 2013.

[19] Jordan SA, Cunningham DG, Marles RJ. Assessment of herbal medicinal products: Challenges, and opportunities to increase the knowledge base for safety assessment. Toxicology and Applied Pharmacology 2010;243:198−216.

[20] Health Canada. Black cohosh products and liver toxicity: update. Canadian Adverse Reaction Newsletter, 20 (1). <http://www.hc-sc.gc.ca/dhp-mps/medeff/bulletin/carn-bcei_v20n1-eng.php#_Black_cohosh_products>; 2010.

[21] Health Canada. Pathway for Licensing Natural Health Products used as Traditional Medicines. <http://www.hc-sc.gc.ca/dhp-mps/prodnatur/legislation/docs/tradit-eng.php>; 2013.

[22] Health Canada. Pathway for Licensing Natural Health Products Making Modern Health Claims. <http://www.hc-sc.gc.ca/dhp-mps/prodnatur/legislation/docs/modern-eng.php>; 2013.

European Regulations on Food Supplements, Fortified Foods, Dietetic Foods, and Health Claims

13

Patrick Coppens and Simon Pettman

EAS Strategic Advice, Brussels, Belgium

13.1 Introduction

This article reviews the various rules that may be of importance in the marketing of "functional foods" in the European Union (EU). It will take the reader through the basic principles of EU food law as applied through the General Food Law Regulation 178/2002 [1]. It will also cover the most recent legislation established or under way in the area of Food Supplements (Directive 2002/46) [2], Fortified Foods (Regulation 1925/2006) [3], Dietetic Foods (Directive 2009/39) [4], and Nutrition and Health Claims (Regulation 1924/2006) [5]. Finally it will illustrate that political considerations have become more significant in recent years in the development of legislation having a fundamental impact on the legal framework or resulting in some important initiatives being placed on hold, in particular relating to the use of botanicals and to dietetic and novel foods.

13.2 The General Food Law Regulation 178/2002

The 1997 Green Paper on Food Law [6], preceding the major food scares of the late 1990s, gave a new impetus to the foundation of European Food Law. It laid down for discussion a number of important principles for the revision of EU Food Law and was followed by the 2000 White Paper on Food Safety [7] announcing some 80 proposals for new and improved legislation in this field, covering a legal framework for foods with functional properties.

In the EU, nutraceuticals and functional foods have no specific legal status. However, most of these products are covered by legislation on food supplements, fortified foods, and dietetic foods and to the extent that health claims are made by the EU legislation covering nutrition and health claims. As such products belong

Nutraceutical and Functional Food Regulations in the United States and Around the World.
DOI: http://dx.doi.org/10.1016/B978-0-12-405870-5.00013-X

to the area of foods; they also need to be in compliance with the requirements of the General Food Law regulation and all relevant legislation covering safety, hygiene, additives, residues, and contaminants.

The "General Food Law Regulation" 178/2002 contains the foundation of European Food Law. Its aim was to provide the basis for the assurance of a high level of protection of human health and consumer interest in relation to food, taking into account in particular diversity in the supply of food including traditional products, while ensuring the effective functioning of the internal market. It establishes common principles and responsibilities, the means to provide a strong science base, efficient organizational arrangements, and procedures to underpin decision making in matters of food and feed safety.

The Regulation applies to all foodstuffs and defines "food" (or "foodstuff") as "any substance or product, whether processed, partially processed or unprocessed, intended to be, or reasonably expected to be ingested by humans." "Food" includes drink, chewing gum, and any substance, including water, intentionally incorporated into the food during its manufacture, preparation, or treatment. It includes water but does not include feed, live animals unless they are prepared for placing on the market for human consumption, plants prior to harvesting, medicinal products, cosmetics, tobacco and tobacco products, narcotic or psychotropic substances, and residues and contaminants.

The fact that this definition excludes medicinal products means that the decisive step to judge whether medicinal law or food law applies to a specific product will have to be sought in medicinal law. This is not a theoretical issue as products with added functions with regard to health may closely resemble medicinal products given their composition, claims, and presentation and the Court of Justice of the European Union (CJEU) has been called upon to judge on this borderline on several occasions.

Medicinal products are defined by Directive 2001/83 [8] on the Community code relating to medicinal products for human use as:

Any substance or combination of substances presented as having properties for treating or preventing disease in human beings; or any substance or combination of substance which may be used in or administered to human beings either with a view to restoring, correcting or modifying physiological functions by exerting a pharmacological, immunological or metabolic action, or to making a medical diagnosis.

Furthermore, Article 2.2. of the said Directive also states that in cases of doubt, where, taking into account all its characteristics, a product may fall within the definition of a "medicinal product" and within the definition of a product covered by other Community legislation, the provisions of this Directive will apply. This provision puts into legal text the jurisprudence that has been established over the years by the CJEU on borderline cases.

Article 6 of Regulation 178/2002 also firmly establishes scientific risk analysis as the basis of the decision-making process in matters of food law. It defines this process through a number of definitions and lays down the responsibilities for each of the respective stages of the risk analysis process.

In the area of "nutraceuticals," relevant European legislation where the risk assessment–risk management process is well developed includes the areas of food supplements and fortified foods.

13.3 **The Food Supplements Directive 2002/46**

In the absence of EU harmonized legislation, the principle of mutual recognition applies. Embedded in the Treaty, this principle was taken as the basis for a new Food Law approach by the 1985 White Paper on the Completion of the Internal Market [9]. This required that harmonization would only be undertaken for the elements that are essential to protect consumer health and provide correct information to consumers.

The principle says that a product, lawfully manufactured in one Member State, should be able to move within the European Union without restrictions. The only exceptions to this principle are those justified on grounds of public morality, public policy, or public security; the protection of health and life of humans, animals, or plants; the protection of national treasures possessing artistic, historic, or archaeological value; or the protection of industrial and commercial property. Such prohibitions or restrictions must not constitute a means of arbitrary discrimination or a disguised restriction on trade between Member States.

Since 2008, the principles of mutual recognition and the procedures to follow by Member States in cases of denial of mutual recognition are covered by the provisions of Regulation 764/2008 [10]. The practicalities of the application of this legislation have been reviewed elsewhere [11].

Food supplements have remained under national law until 2002 subject to this principle of mutual recognition, which proved rather ineffective to ensure free movement of such products. Since 2002 the EU has tried to develop harmonized legislation for food supplements so the same rules apply in all Member States in the same way.

In Article 2 of the Food Supplements Directive 2002/46, food supplements are defined as

> *"Foodstuffs the purpose of which is to supplement the normal diet and which are concentrated sources of nutrients or other substances with a nutritional or physiological effect, alone or in combination, marketed in dose form, namely forms such as capsules, pastilles, tablets, pills and other similar forms, sachets of powder, ampoules of liquids, drop dispensing bottles, and other similar forms of liquids and powders designed to be taken in measured small unit quantities."*

A number of elements from that definition are important to ascertain the products that fall under the scope. An important element is that food supplements are considered as foodstuffs. This means that regulations that cover foodstuffs are applicable to food supplements except where specifically provided for in the food supplements legislation itself. This does not completely resolve the issue of the borderline between medicinal products and food supplements, given that specific compositional provisions are still missing, but it does indicate that when they are in line with the definition, such products should follow food law.

A second important element is that food supplements may contain substances with a nutritional or physiological effect. Until that point it was not legally recognized that substances possessing physiological, but non-medicinal properties could be regulated outside medicinal law, although this was the practical reality in most Member States. Since 2006 this principle is also firmly established through specific legislation covering nutrition and health claims.

The third element is the fact that in order to fall under the food supplements law the product must be in a concentrated form, presented in measured small unit quantities.

The Directive represents a first step in the harmonization process as it focuses only on vitamins and minerals. And it has proven to be particularly difficult to carry out initiatives for further harmonization because of the differences in appreciation of the various Member States as illustrated by two examples:

1. For the range of other substances used in food supplements, the Commission was required to submit, not later than 12 July 2007, to the European Parliament (EP) and the Council a report on the advisability of establishing specific rules, including, where appropriate, positive lists, on categories of nutrients or of substances with a nutritional or physiological effect, accompanied by any proposals for amendment to this Directive, which the Commission deems necessary. This report was published in 2008 [12]. Its main conclusion was that laying down specific rules applicable to substances other than vitamins and minerals for use in food supplements was not justified, because it was judged not necessary in the short term and also the feasibility of such a measure was questioned, because of the different way these substances had been addressed by the Member States, the complexity of the matter and the scientific and methodological difficulties which would have to be overcome. The Commission was of the opinion that existing instruments, in particular the then recently adopted legislation on fortified foods (Regulation 1925/2006), on nutrition and health claims (Regulation 1924/2006), on novel foods (Regulation 258/1997 [13], legislation that was intended to be revised at that time) and on mutual recognition (Regulation 764/2008) would already constitute useful tools to harmonize specific aspects of the products concerned. The Commission referred in particular to the procedure laid down in Article 8 to Regulation 1925/2006, which allows a substance to be placed under scrutiny for a given period if the available scientific information is insufficient. The Commission was then of the opinion that this type of procedure was particularly well suited to plants and plant extracts, for which sufficient and appropriate scientific data

are not always available, and for which the safety assessment methodology was still being developed.

In 2013, however, the environment had evolved quite differently, with Article 8 of Regulation 1924/2006 only applied to two substances, the novel foods revision having failed, the nutrition and health claims legislation having resulted in most claims for non-vitamin and mineral substances being rejected and the issue of how to regulate botanicals still not resolved.

2. In addition, the Food Supplements Directive foresaw that maximum amounts of vitamins and minerals present in food supplements would be set, taking into account the upper safe levels of vitamins and minerals established by scientific risk assessment based on generally accepted scientific data, taking into account, as appropriate, the varying degrees of sensitivity of different consumer groups, and intake of vitamins and minerals from other dietary sources. It specified that when the maximum levels would be set, due account should also be taken of reference intakes of vitamins and minerals for the population. Similar principles had been introduced in the fortification Regulation 1925/2006 for the setting of maximum levels for the addition of vitamins and minerals to regular foods. Although the European Commission was invited to initiate work on this by January 2009, by the end of 2013 no proposals had been put forward. The European Commission had not come further than the development of two discussion papers [14,15].

The principles for the setting of maximum levels for vitamins and minerals in foods and food supplements proposed in these papers are a good example of the application of the Risk Analysis process. In a first phase the Upper Tolerable Safe Intake Levels (UL) of the vitamins and minerals are set by the Scientific Committee on Food (SCF) and later by its successor, EFSA [16]. The UL is the intake level that can be consumed daily over a lifetime without being likely to pose any risk to health according to available evidence. It guides the Risk Assessor in characterizing the risk associated with each individual vitamin and mineral.

In a second phase the risk manager considers the risk assessment. He will have to determine what maximum levels will be established for the use in food supplements based on the established UL and the intake of the vitamins and minerals from other food sources including water. The European Responsible Nutrition Alliance (ERNA) has developed a risk management model as an example of how such maximum levels can be set, taking into consideration the risk characterization of the vitamins and nutrients and European intake data, based on the population safety index of each individual nutrient [17−19]. The International Life Sciences Institute developed a similar model for determining the maximum amounts of vitamins and minerals for addition to regular foods [20]. Other models have also been proposed [21−24]. The model developed by ERNA is the model that was retained in the latest discussion document of the European Commission. However, given the differences in appreciation of the Member States on how maximum levels should be set and other priorities, this work has not proceeded since then.

Another example in the Food Supplements Directive for managing risk through scientific risk assessment is the establishment of a positive list of nutritional substances that are allowed in food supplements. The list, initially included in the proposed text, was the list of Directive 2001/15/EC [25] on substances that may be added for specific nutritional purposes in foods for particular nutritional uses. The argumentation was that the substances included in that Directive had been evaluated by the SCF and thereby had been considered safe and appropriate for use in dietetic foods. The Food Supplements Directive foresaw an authorization procedure for those sources of vitamins and minerals currently in use after a favorable EFSA opinion and a period during which Member States can grant derogations to those substances that were lawfully marketed in the European Union until EFSA gives a favorable opinion. However, this aspect has been very much criticized. In fact, in the UK, the High Court of Justice of England and Wales brought it before the CJEU for a preliminary ruling concerning the validity of the Food Supplements Directive.

The Court Decision of 12/07/05 [26], however, rejected the action and judged that the approach chosen, including the establishment of positive lists and subsequent authorization procedure, in this case was rightful and not disproportionate for regulating vitamin and mineral substances to be used as sources of nutrients in food supplements. A number of nutritional substances have since been added to the list after an assessment of their safety by EFSA, based on submissions of applications by food businesses.

13.4 Regulation 1924/2006 on the addition of vitamins and minerals and other substances to food (fortified foods)

Another example of the application of the risk analysis principle can be found in the Regulation on the addition of vitamins and minerals and other substances to foods (Regulation 1925/2006). As indicated before, the setting of maximum levels, as is the case with food supplements, is governed by similar principles: Such levels will be set taking into account the upper safe levels of vitamins and minerals established by scientific risk assessment based on generally acceptable scientific data, taking into account, as appropriate, the varying degrees of sensitivity of different groups of consumers and the intakes of vitamins and minerals from other dietary sources. In this process, due account will also be taken of reference intakes of vitamins and minerals for the population. In addition, when the maximum amounts are set for vitamins and minerals whose reference intakes for the population are close to the upper safe levels, the contribution of individual products to the overall diet of the population in general or of subgroups of the population and the nutrient profile of the product established as provided for by Regulation 1924/2006 on claims will also be taken into account, as necessary.

However, the same political considerations of the Member States apply as in the case of food supplements and therefore this work has not progressed since the adoption of the Regulation.

Also as is the case with food supplements, this legislation includes a list of nutritional substances that are allowed to be added as sources of vitamins and minerals to foods. This list largely contains the same substances and has also been established based on the results of a risk assessment by the European risk assessment bodies (SCF and EFSA).

The most important element that is of relevance for safety is Article 8 of the Regulation. It deals with the concern of Member States with regard to substances that are added to foods or used in the manufacturing of foods in conditions that are likely to result in the ingestion of amounts of this substance greatly exceeding those reasonably expected to be ingested under normal conditions of consumption in a balanced and varied diet.

It foresees two possibilities of dealing with such risk:

- In the case that, following an assessment of available information by EFSA, a harmful effect on health resulting from such use has been identified, the use of the substance and/or the ingredient containing the substance, could either be prohibited or only allowed under specific conditions.
- In the case that, following an assessment of available information by EFSA, the possibility of harmful effects on health resulting from such use is identified but scientific uncertainty persists, the substance will be placed on a list. Food business operators, or any other interested parties, may at any time submit, for evaluation to EFSA, a file containing the scientific data demonstrating the safety of a substance listed on the list, under the conditions of its use in a food or in a category of foods and explaining the purpose of that use. Within four years from the date a substance has been listed, a decision would be taken, taking into account the opinion of EFSA on any files submitted for evaluation to generally allow the use of a substance, prohibit it, or only allow it under specific conditions.

In all cases, the decision is taken through the Standing Committee on the Food Chain and Animal Health, after having obtained a scientific risk assessment opinion by EFSA. The Standing Committee, which is composed of experts from the Member States, assists the European Commission in the risk management process and is endowed with the powers to decide on measures. When approved, this legislation can have considerable consequences for those substances in use today, when used at higher levels. In such cases, the burden of proof is reversed. It is no longer the enforcing authority that must prove the harmfulness of the substances. It is the food business operator that must prove its safety. Although liability law and general food law impose the safety criterion to foodstuffs, in this case scientific proof of safety will be required. By the end of 2013 only two substances had been introduced into this process (*Pausinystalia yohimbe* and Ephedra spp.) and no decisions taken.

13.5 Regulation 1924/2006 on nutrition and health claims made on foods

All foods making nutrition or health claims are subject to specific legislation since 2006: Regulation 1924/2006. The system put in place is based on pre-marketing approval of all claims. In addition, the scope of the regulation is broad, since it not only covers the labeling of foodstuffs but also the presentation and advertising of the food.

The regulation covers all possible claims relating to health. It covers nutrition claims (claims which state, suggest, or imply that a food has particular nutrition properties due to the energy (caloric value) it provides, provides at a reduced or increased rate, or does not provide, and/or the nutrients or other substances it contains, contains in reduced or increased proportions, or does not contain), health claims (claims that state, suggest, or imply that a relationship exists between a food category, a food, or one of its constituents and health), reduction of disease risk claims (claims that state, suggest, or imply that the consumption of a food category, a food, or one of its constituents significantly reduces a risk factor in the development of a human disease), and claims referring to the development and health of children. This last category was a late addition in the course of the final negotiations between the EU institutions. It was not defined and clarification followed in December 2007 by means of a guidance document issued by the European Commission [27].

The most criticized aspect of this legislation is the so-called concept of nutrient profiles. This was included very late, after the public consultation stage in the text, and became the political focus of much of the discussions. Article 4 foresees that nutrient profiles would be set for foods or categories of foods based on their nutritional composition and more specifically the content of saturated fatty acids, sugar, and salt. The main argument in favor is that it would not be appropriate to allow nutrition and health claims on foodstuffs with an unfavorable nutrient profile since the consumption of such products should not be encouraged by giving them a positive health image. The main criticism of the concept was that it would be very difficult to determine in an objective scientific manner nutrient profiles for foods, since their effect is not determined so much by their composition but rather by their use. The EP decided to delete nutrient profiling in the first reading of the draft regulation. The Council, representing the Member States in their political agreement, chose to approve them unanimously. In the end the concept remained in the text and a later amendment to delete the profiles in the context of new labeling legislation again showed how divided the ideas on this concept are. The EC started work to develop nutrient profiles carefully with a discussion paper on the principles and a number of draft proposals containing detailed simulations [28]. It also asked scientific advice of EFSA [29]. However, because of the contentious nature of the concept itself and fierce opposition against the proposed values by many food sectors, the work was stopped by a political decision of the EC in 2009 to allow further reflection. By the end of 2013 no further action had been taken.

The way the legislator has chosen to regulate claims is through authorization prior to marketing. Furthermore, a number of conditions, restrictions, and prohibitions are included in the text. It is this approach that was met with great opposition, especially in the EP because it was judged disproportionate with regard to the intended purpose.

Regulation 1924/2006 foresees three ways for making Nutrition and Health Claims:

- Nutrition claims would only be possible if they are included in the annex to the text. This annex was considered to be too restrictive, but a number of nutrition claims, e.g., for essential fatty acids have been added at later stages
- Health Claims based on generally accepted scientific data can only be allowed when included on a list. It was feared that Member States would considerably limit the number of claims to be included in this list, especially those regarding other substances and botanicals. After the implementation of the legislation the scientific principles applied by EFSA turned out to be the limiting factor. The establishment of the list suffered considerable delay and a first list was only adopted and published mid-2012 (Regulation 432/2012) [30], with many claims still remaining on hold.
- Other Health Claims, Reduction of Disease Risk Claims, and Health Claims referring to Children's Development and Health can only be allowed after submission of a file to EFSA and approval through the Standing Committee. It was feared that the requirements for such files and the conditions for their approval would only be possible for companies with substantial financial and other resources, which turned out to be the case. In addition, the principles for scientific assessment as they have been established by EFSA are very strict and do not include evidence grading (i.e., the concept based on WHO work that takes into consideration the amount of evidence that is available for a specific claim and the corresponding qualifying language for the claim's wording) [31].

It is noted that the EP chose in first reading to replace the authorization procedure by a notification procedure, arguing that control on nutrition and health claims should not be submitted to the most extreme measure, which is a priori approval but that leaving them under the responsibility of food operators and acting only in case of inappropriate use of claims would be more proportionate. It was feared that the burdensome procedure would create the same disadvantages as the novel foods procedure (i.e., be time-consuming, expensive, unpredictable, inaccessible for small and medium enterprises) and would make illegal many of the claims currently lawfully made in many Member States. The EP thereby asked the rightful question: Is the chosen risk management route the most proportionate, least burdensome way of achieving the desired objective of consumer protection? And if so, have the consequences for economic operators, public authorities, and other stakeholders been properly studied by appropriate impact assessments? In the end, the legislation retained the pre-marketing authorization process. However, the consequences raised by the EP have all become reality. It goes beyond the scope of this paper to go deeper into the consequences and impact the claims legislation has had for businesses since its application in 2007 and the measures that have been needed to correct the many flaws and inconsistencies that have marked the implementation process. A comprehensive overview of these aspects was published by the European Responsible Nutrition Alliance in 2011 [32].

Since 2007 the application of the nutrition and health claims Regulation is covered by numerous guidance documents and communications from the European Commission, EFSA, and national authorities. Comprehensive guidance on how to apply this legislation is published elsewhere [33,34].

13.6 Novel Foods Regulation 258/97

The novel foods legislation is seen as a cornerstone of the safety management of the EU. Regulation 258/97 concerning novel foods and novel food ingredients lays down an authorization procedure for a number of foods or food ingredients that have not been consumed to a significant degree in the EU. In order to protect public health, it was deemed necessary to ensure that such novel foods and novel food ingredients are subject to a safety assessment before they are placed on the market. In contrast to the procedures applicable to claims and food improvement agents (e.g., additives) by virtue of Regulation 1331/2008 [35], which are full centralized procedures, the novel foods authorization procedure is, in essence, a procedure managed by the Member States. But in cases where concern is raised by Member States because of public health or food safety concern (which is more the rule than the exception), the procedure is continued at the EU level in the same way as if it were a centralized procedure.

The novel foods legislation applies to the following categories of foods:

- Foods and food ingredients with a new or intentionally modified primary molecular structure
- Foods and food ingredients consisting of or isolated from microorganisms, fungi, or algae
- Foods and food ingredients consisting of or isolated from plants and food ingredients isolated from animals, except for foods and food ingredients obtained by traditional propagating or breeding practices and having a history of safe food use
- Foods and food ingredients to which has been applied a production process not currently used, where that process gives rise to significant changes in the composition or structure of the foods or food ingredients, which affect their nutritional value, metabolism, or level of undesirable substances

If a food or food ingredient falls within one or more of the abovementioned food categories, the decisive factor on whether it will be regarded as novel is whether it has already been sold under food law on the EU market before 1997. The EU Novel Foods Regulation stipulates that a food/food ingredient is novel if it has not been used to a significant degree for human consumption in the EU before May 15, 1997.

In order to prove the non-novel food status of a food/food ingredient, proof is required of "significant" EU use under food law before May 1997. The key aspects that need to be considered and complications that may arise to provide such proof are described elsewhere [36]. The European Commission published guidance on this aspect only in 2012 [37].

The Novel Foods Regulation provides two different procedures for authorization:

- The simplified procedure is based on a notification and can only be followed if the applicant can provide evidence that the novel food/food ingredient is substantially equivalent to an existing food ingredient. A notification providing evidence of its substantial equivalence needs to be submitted to the European Commission. All notifications made so far have relied upon opinions of the competent food assessment bodies of the Member States. Guidance on this aspect was adopted only in 2013 [38].
- The full authorization procedure applies if no evidence of substantial equivalence can be provided and to all foods and food ingredients with a new or intentionally modified primary molecular structure and foods and food ingredients to which has been applied a new production process. An application needs to be submitted to one of the Member State authorities for initial assessment. The other Member States can and usually do object to the initial assessment opinion. In that case, a decision is then taken on the basis of a risk assessment opinion by EFSA.

All in all, the novel food procedure is not a very clear and transparent one. The number of applications is low, the costs for the data needed considerable, decisions take years, and incentive for companies is low as competitors can catch up quickly by using the simplified notification procedure. Further hampering factors include the extensive dossier requirements, the uncertainty of success, and the objections of various nature raised by the Member States.

Due to many of the above reasons the European Commission initiated a revision of the novels foods Regulation in 2008 [39]. The proposal aimed to revise the definition of novel food; abolish the simplified notification procedure; make all authorizations generic, except where protection of proprietary data is requested; and provide for a centralized authorization procedure and a simplified procedure for novel foods with a tradition of safe use in third countries. During the legislative process, the status of nanotechnology and cloned animals raised most of the attention and the fact that the Council and EP continued to fundamentally differ in opinion on how to cover the offspring of cloned animals ultimately led to the proposal being rejected at the end of the process. Nevertheless, during the discussions, some aspects of the original proposal were fundamentally changed so the original revision towards a new framework was gradually watered down. A new proposal is expected once clarity and agreement is reached on the cloning issue.

13.7 Foods for particular nutritional uses (dietetic foods)

There is one category of products that has had a firmly established legal framework for the last 30 years in the EU, which is "foods for particular nutritional uses" (also called PARNUTS or dietetic foods). It offers a good example of legislation that balances scientific risk assessment, consumer protection, manufacturer responsibility, and market innovation in a rather proportionate way. Despite this by mid-2013 considerable progress has been made to abolish most of this legislation, resulting in a shift of many products into the category of regular foods and thus subject only to the requirements of the nutrition and health claims Regulation [40].

Harmonization efforts in this field were among the first within the EU in the food sector, but given the complexity and diversity of these products this exercise has never been completely finalized. The so-called PARNUTS framework Directive 39/2009/EC (previously Directive 89/398/EEC) defines the scope of PARNUTS and creates the framework for a number of vertical Directives with specific provisions for a number of subcategories. Originally the European legislator foresaw nine subcategories, which was later reduced to five of them having been finalized over time:

- Infant formulae and follow-on Formulae (Directive 2006/141/EC [40], previously Directive 91/321/EEC) (originally two different Directives were foreseen)
- Processed cereal-based foods and baby foods for infants and young children (Directive 2006/125/EC [41], previously Directive 96/5/EC)
- Foods intended for use in energy-restricted diets for weight reduction (Directive 96/8/EC) [42]
- Foods for Special Medical Purposes (Directive 1999/21/EC) [43].

A fundamental revision of dietetic food legislation initiated in 2012 will abolish the framework of Directive 39/2009/EC and only leave infant and baby foods, foods for special medical purposes, and certain slimming products (whole diet replacers) under specific legislation. After the finalization of this revision, it is foreseen that the legislation covering these products already in existence (Directive 2006/141, Directive 125/2006, Directive 98/8 and Directive 1999/21) will be revised and updated. The future for milks for toddlers and sports foods will depend on the outcome of an opinion by EFSA on whether these groups of people need specific nutritional requirements. This is all covered by Regulation 609/2013 that was published on 26 June 2013 and will apply from 20 July 2016 [44].

According to Directive 39/2009/EC, PARNUTS foods should have a special composition or manufacturing process, must be clearly distinguishable from foodstuffs for normal consumption, and must be suitable for their claimed nutritional purposes and marketed in such a way as to indicate such suitability. With "particular nutritional use" is understood that the product must fulfill the particular nutritional requirements of certain categories of persons whose digestive processes or metabolism are disturbed, or of certain categories of persons who are in a special physiological condition and who are therefore able to obtain special benefit from controlled consumption of certain substances in foodstuffs. Given these requirements, it is clear that only products that are destined to be used in small target populations suffering from disturbed metabolism or digestion could fall under this legislation. Products destined for the normal general population would clearly not be covered.

However, some differences in interpretation are seen between Member States as to products specifically destined for pregnant and lactating women and even low fat polyunsaturated fat spreads, for example. In addition in recent years more and more products are being positioned as dietetic foods under the so-called Article 11 notification, covering dietetic foods for which no specific compositional rules have been established. The main reason is that the intended use of the product is not subject to pre-marketing authorization whereas a nutrition or health claim is. Also in this case major differences were being observed between the Member States, as described in an EC report [45]. The revision of dietetic food legislation clearly has the intention to cover all such foods under the rules for nutrition and health claims.

One category of PARNUTS that will remain and is of particular relevance because of its functional properties is foods for special medical purposes. Directive 1999/21/EC defines such products as a category of PARNUTS specially processed or formulated and intended for the dietary management of patients and to be used under medical supervision. They are intended for the exclusive or partial feeding of patients with a limited, impaired, or disturbed capacity to take, digest, absorb, metabolize, or excrete ordinary foodstuffs or certain nutrients contained therein or metabolites, or with other medically determined nutrient requirements, whose dietary management cannot be achieved only by modification of the normal diet, by other foods for particular nutritional uses, or by a combination of the two. Such products have to bear the statement "For the dietary management of (diseases, disorders or medical conditions for which the product is intended)" and a description of the properties and/or characteristics that make the product useful in particular, as the case may be, relating to the nutrients that have been increased, reduced, eliminated, or otherwise modified and the rationale of the use of the product.

The procedure to put such products on the market is notification, which means that the manufacturers must notify the competent authority of the Member State at the time of placing the product on the market. The manufacturers should have at hand the necessary documentation to demonstrate that the product is suitable for its intended use as a food for special medical purposes. Member States may request to submit such evidence for verification.

This is seen as a proportionate measure enabling new products benefiting the needs of patients to be quickly put on the market without needing prior assessment of the efficacy and appropriateness for the intended use. Such assessment can and in some Member States is carried out on the basis of the notified material.

Together with the General Food Law requirements that put primary responsibility for food safety with the food business operator, that require them to have in place a system for traceability of the foodstuff and adequate procedures to withdraw the foodstuff from the market if they consider or have reason to believe that it is not in compliance with the food safety requirements, and that impose a notification duty (i.e., to immediately inform the competent authorities if they consider or have reason to believe that a food which they have placed on the market may be injurious to human health), this would seem to be a balanced way for creating a regulatory framework that fosters both consumer protection and quick innovation as a result of scientific developments.

Finally, under the so-called "safeguard clauses," included in most food-related legislation, Member States have the possibility to temporarily suspend or restrict the application of certain provisions of legislation within their territory where there are serious grounds for considering that a product endangers human health despite complying with EU legislation.

It remains to be seen if under the revision of Directive 1999/21 the notification procedure will be maintained or be replaced by a pre-marketing authorization.

13.8 The future of botanicals

Plants and plant preparations have been used for decades in the EU as ingredients in a variety of products, including foods, food supplements, cosmetics, and medicinal products. EU and national legislation is in place to ensure that the use of these ingredients is safe.

Although the use of botanicals for nutritional or physiological purposes in food supplements has been covered under food law since 2002 by the food supplements Directive, specific requirements for the use of botanicals in food supplements only exist at the national level. Many EU Member States have regulated the use of botanicals in food supplements by means of positive and/or negative lists and have procedures in place for the notification and assessment of these products on their territory. Products that are lawfully marketed in one Member State are in principle free to be sold in all other Member States following the principle of mutual recognition (Regulation 764/2008).

The use of botanicals for therapeutic or preventive purposes in medicinal products has been harmonized in the EU by the Traditional Herbal Medicinal

Product Directive (THMPD 2004/24) [46]. The use of such products to cure or prevent diseases remains under medicinal legislation, but a simplified registration procedure has been put in place, based on the principle of traditional use.

Both legal frameworks coexist. Therefore the same botanical can be used as both a component of a food supplement and of a medicinal product, depending on the intended use of the product and its modalities of use. This coexistence has caused issues in certain Member States and the CJEU has been called upon in a number of cases to judge on this borderline. In summary its main conclusions are:

- Member States have the competence to determine whether a certain product is a medicinal product or not, but have to base that decision on a *case-by-case assessment* of all of the product's characteristics, particularly its composition, its pharmacological properties as they may be ascertained in the current state of scientific knowledge, the way in which it is used, the extent to which it is sold, its familiarity to the consumer, and the risks that its use might entail. This means that a decision cannot be taken solely on the basis of the composition, the form or the nature of the ingredients of a product, but must be based on all of its characteristics.
- All products that are *presented* as having therapeutic or preventative effects in relation to diseases should be subject to medicinal law in order to be able to ascertain the efficacy of the product in relation to its claimed effects on the basis of the clinical studies performed. It is the aim of medicinal law that if efficacy cannot be established, a marketing license can effectively be refused.
- However, medicinal law is not intended to cover products that have an effect on the body or health but are not presented for the treatment, prevention, or cure of diseases. The concept of a physiological effect is not specific to medicinal products only but is also among the criteria used for the definition of food supplements. Products having an effect on the human body, but which do not significantly affect the metabolism and thus do not strictly modify the way in which it functions, cannot be considered as medicinal products by *function*. This is the case with many botanicals and botanical preparations.
- The fact that a risk to health may be present is not sufficient to classify a product as medicinal by function. The legal framework of food (Regulation 178/2002) and national legislation in place contain sufficient provisions to ensure the safety of any food, including botanical food supplements.
- The fact that similar products are registered as medicinal products is also not a determining factor to consider all similar products as medicinal products.

With the publication of the EU nutrition and health claims Regulation (Regulation 1924/2006) a new challenge has arisen. Under this legislation, all nutrition and health claims made on foods (including food supplements) can only be used when approved after an assessment of the scientific data available by EFSA. The key element is the requirement of scientific studies that demonstrate cause-and-effect relationships between intake of a compound and a health benefit.

For botanicals most of the knowledge has been gained through use and experience over time and communication of the effects to consumers has never required confirmation of the acquired knowledge by studies. This was explicitly recognized in the development of the THMP legislation and was the reason for establishing a simplified procedure for products that had been marketed for 30 years, 15 of which were in the EU, and for which the long tradition of use makes it possible to reduce the need for clinical trials in so far as the efficacy of the medicinal product is plausible on the basis of long-standing use and experience.

This is, however, not the case for nutrition and health claims made on foodstuffs. This has resulted in a situation where the requirements for demonstrating simple health effects for food are more demanding than for therapeutic effects for medicinal products. When EFSA started its assessment of the scientific evidence and published its first opinions in 2009 it quickly became clear that this would result in negative opinions for all claims for botanicals. This situation was judged sufficiently serious for the EC to stop the process of claims assessment for botanicals in September 2010, pending further reflection.

In August 2012 the EC relaunched this discussion with the publication of a discussion document addressed to the Member States [47]. This document asks for input on one of two options:

> **Option 1**: Proposes asking EFSA to continue the assessment of claims for botanicals as was originally foreseen. It notes that this is likely to ultimately result in all health claims for botanicals being rejected, while claims for medicinal products could continue to exist based on traditional use, but it would ensure that all foods and food ingredients were treated in an equal way.
>
> **Option 2**: Proposes addressing further legislation to specifically regulate the use of botanicals in food. This new legislation would enable the Member States to address the specificities of botanicals by taking into consideration traditional use and could even go beyond the scope of health claims and include appropriate provisions relating to safety and quality of these products.

Choosing the first option would legitimate the unequal treatment of food and medicinal products as an appropriate risk management measure but not solve any of the problems, while the second option would offer the possibility of developing further harmonization in this area. The responses by the Member States have shown the deep divide that exists between the Member States, and a marked and vigorous opposition to option 2 has been expressed by many pharmaceutical stakeholders, largely defending the use of botanicals under medicinal legislation only.

Until a decision is taken, both product categories, THMPs and food supplements, can continue to be marketed in parallel. It is clear, however, that both legislations have resulted in considerable overlap: On the one hand THMP indications described in the monographs elaborated by the European Medicines Agency are similar to health effects accepted for foods in opinions published by EFSA, and on the other hand health claims are allowed to refer to the reduction of disease risk. The acceptance of health claims for a number of substances that are considered to be restricted to medicinal products by many Member States (e.g., melatonin, red rice yeast, lactulose) is a further element that is likely to shift the legal borderline between medicinal products and food.

In the end, the risk management decision lies with the EC, an institution that is obliged to observe the basic principles as established by the CJEU in terms of necessity and proportionality.

References

[1] Regulation (EC) no 178/2002 of the European Parliament and of the Council of 28 January 2002 laying down the general principles and requirements of food law, establishing the European Food Safety Authority and laying down procedures in matters of food safety. 01/02/2002. Official Journal of the European Union: L31/1.

[2] Directive 2002/46/EC of the European Parliament and of the Council of 10 June 2002 on the approximation of the laws of the Member States relating to food supplements. 12/06/2002. Official Journal of the European Union: L136/85.

[3] Regulation (EC) No 1925/2006 of the European Parliament and of the Council of 20 December 2006 on the addition of vitamins and minerals and of certain other substances to foods. 30/12/2006. Official Journal of the European Union: L404/26.

[4] Directive 2009/39/EC of the European Parliament and of the Council of 6 May 2009 on foodstuffs intended for particular nutritional uses. 20/05/2009. Official Journal of the European Union: L124/21.

[5] Corrigendum to Regulation (EC) No 1924/2006 of the European Parliament and of the Council of 20 December 2006 on nutrition and health claims made on foods. 18/01/2007. Official Journal of the European Union: L12/3.

[6] The General principles of food law in the European Union. Commission Green Paper. 30/04/1997. Com(97) 176 final.

[7] White paper on food safety (COM (1999) 719 final; 12/01/2000).

[8] Directive 2001/83/EC of the European Parliament and of the Council of 6 November 2001 on the Community code relating to medicinal products for human use. 28/11/2001. Official Journal of the European Union: L311/67.

[9] Commission of the European Communities. White Paper on the Completion of the Internal Market. 14/06/1985 COM(85) 310 final.

[10] Regulation (EC) No 764/2008 of the European Parliament and of the Council of 9 July 2008 laying down procedures relating to the application of certain national technical rules to products lawfully marketed in another Member State and repealing Decision No 3052/95/EC. 13/08/08. Official Journal of the European Union: L218/21.

[11] EAS Strategic Advice. How to apply mutual recognition for the free trade of food products across the EU, 2009. www.eas.eu

[12] Commission of the European Communities. Report from the Commission to the Council and the European Parliament on the use of substances other than vitamins and minerals in food supplements. 05/12/2008. COM(2008) 824 final.

[13] Regulation (EC) No 258/97 of the European Parliament and of the Council of 27 January 1997 concerning novel foods and novel food ingredients. 14/02/1997. Official Journal of the European Union: L043/1.

[14] European Commission: Discussion paper on the setting of maximum and minimum amounts for vitamins and minerals in foodstuffs; 2006.

[15] European Commission: Orientation paper on the setting of maximum and minimum amounts for vitamins and minerals in foodstuffs. July 2007.

[16] European Food Safety Authority: Compilation of the Scientific Opinions on Tolerable Upper Intake Levels for Vitamins and Minerals; 2006.

[17] European Responsible Nutrition Alliance (ERNA): Vitamin and Mineral Supplements: a risk management model; 2005.

[18] Richardson David P. Risk management of vitamins and minerals: a risk categorization model for the setting of maximum levels in food supplements and fortified foods. Food Sci Technol Bull Funct Foods 2007;4(6):51−66.

[19] Richardson David P. Risk management of vitamins and minerals in Europe: quantitative and qualitative approaches for setting maximum levels in food supplements for children. Food Sci Technol Bull Funct Foods 2010;7(6):77−101.

[20] Flynn A, Moreiras O, Stehle P, Fletcher RJ, Müller DJ, Rolland V. Vitamins and minerals: a model for safe addition to foods. Eur J Nutr 2003;42(2):118−30.

[21] Bundesinstituts für Risikobewertung (Federal Institute for Risk assessment - BfR): Verwendung von Vitaminen in Lebensmitteln (Use of Vitamins in Foods); 2005.

[22] Bundesinstituts für Risikobewertung (Federal Institute for Risk assessment - BfR): Verwendung von Mineralstoffen in Lebensmitteln; 2005.

[23] Danmarks Fødevareforskning (Danish Institute for Food and Veterinary Research−DFVF). A safe strategy for addition of vitamins and minerals to foods. Eur J Nutr 2005; Available from: http://dx.doi.org/10.1007/s00394-005-0580-9.

[24] Kloosterman J, Fransen HP, de Stoppelaar J, Verhagen H, Rompelberg C. Safe addition of vitamins and minerals to foods: setting maximum levels for fortification in the Netherlands. Eur J Nutr 2007.

[25] Commission Directive 2001/15/EC of 15 February 2001 on substances that may be added for specific nutritional purposes in foods for particular nutritional uses. 22/02/2001. Official Journal of the European Union: L52/19.

[26] Court of Justice of the European Union. Joined Cases C-154/04 and C-155/04. Judgment of the Court of 12 July 2005.

[27] Guidance on the implementation of regulation no 1924/2006 on nutrition and health claims made on foods conclusions of the standing committee on the food chain and animal health. 14/12/2007.

[28] European Commission. Working document on the setting of nutrient profiles. 02/06/2008.

[29] European Food Safety Authority. The setting of nutrient profiles for foods bearing nutrition and health claims pursuant to article 4 of the regulation (EC) No 1924/2006. EFSA J 2008;644:1−44.

[30] Commission Regulation (EU) No 432/2012 of 16 May 2012 establishing a list of permitted health claims made on foods, other than those referring to the reduction of disease risk and to children's development and health. 25/05/2012. Official Journal of the European Union: L136/1.

[31] Richardson David P. The scientific substantiation of health claims with particular reference to the grading of evidence. Eur J Nutr 2005;44:319−24.

[32] European Responsible Nutrition Alliance (ERNA), Nutrition and Health Claims in the EU. Regulation (EC) No 1924/2006. A review of the consequences of implementation.

[33] EAS Strategic Advice. How to apply the Nutrition and Health Claims Regulation, 2010. www.eas.eu.

[34] Food Supplements Europe. The application of the Nutrition and Health Claims Regulation 1924/2006. Guidance for food operators, 2013.

[35] Regulation (EC) No 1331/2008 of the European Parliament and of the Council of 16 December 2008 establishing a common authorisation procedure for food additives, food enzymes and food flavourings. 31/12/2008. Official Journal of the European Union L354/1.

[36] EAS Strategic Advice. Novel Foods in the European Union: Developing Regulatory Strategies; 2003.

[37] European Commission. Human Consumption to a Significant Degree. Information and Guidance Document; 2012.

[38] European Commission. EU Guidelines for the presentation of data to demonstrate substantial equivalence between a novel food or food ingredient and an existing counterpart; 2013.

[39] Commission of the European Communities. Proposal for a Regulation of the European Parliament and of the Council on novel foods and amending Regulation (EC) No XXX/XXXX [common procedure]. 14/01/2008. COM(2007) 872 final.

[40] Commission Directive 2006/141/EC of 22 December 2006 on infant formulae and follow-on formulae and amending Directive 1999/21/EC. 30/12/2006. Official Journal of the European Union: L401/1.

[41] Commission Directive 2006/125/EC of 5 December 2006 on processed cereal-based foods and baby foods for infants and young children. 06/12/2006. Official Journal of the European Union: L339/16.

[42] Commission Directive 96/8/EC of 26 February 1996 on foods intended for use in energy-restricted diets for weight reduction. 06/03/1996. Official Journal of the European Union: L55/22.

[43] Commission Directive 1999/21/EC of 25 March 1999 on dietary foods for special medical purposes. 07/04/1999. Official Journal of the European Union: L91/29.

[44] Regulation (EU) No 609/2013 of the European Parliament and of the Council of 12 June 2013 on food intended for infants and young children, food for special medical purposes, and total diet replacement for weight control [. . .]. Official Journal of the European Union: L181/35.

[45] Commission of the European Communities. Report from the Commission to the European Parliament and the Council on the implementation of Article 9 of Council Directive 89/398/EEC on the approximation of the laws of the Member States relating to foodstuffs intended for particular nutritional uses. 27/06/2008. COM(2008) 393 final.

[46] Directive 2004/24/EC of the European Parliament and of the Council of 31 March 2004 amending, as regards traditional herbal medicinal products, Directive 2001/83/EC on the Community code relating to medicinal products for human use. 30/04/2004. Official Journal of the European Union: L136/85.

[47] European Commission. Discussion paper on Health Claims on Botanicals in Food; 2013.

Botanical Nutraceuticals, (Food Supplements, Fortified and Functional Foods) in the European Union with Main Focus on Nutrition And Health Claims Regulation

14

Om P. Gulati*, Peter Berry Ottaway[†] and Patrick Coppens**

*Scientific & Regulatory Affairs, Horphag Research Management S.A., Geneva, Switzerland
[†]Berry Ottaway & Associates Limited, Hereford, United Kingdom **EAS Strategic Advice, Brussels, Belgium

14.1 Introduction

The term "nutraceutical" was originally used by Defelice in 1995 with the definition: "Food or parts of food that provide medical or health benefits, including the prevention and treatment of disease" [1]. As they have defined physiological effects many nutraceuticals do not easily fall into the legal categories of food or drug but inhabit a gray area between the two [2].

Nutraceuticals have currently been receiving international recognition as having potentially beneficial effects on health when consumed as part of a varied diet on a regular basis and at effective levels [3].

The legal categorization of a nutraceutical is, in general, made on the basis of its accepted effects on the body. Thus, if the substance contributes only to the maintenance of healthy tissues and organs its use may be considered to be a food ingredient. If, however, it can be shown to correct, restore, or modify one or more of the body's physiological processes, its use is likely to be considered as medicinal.

Nutraceuticals in a broad sense can be found in functional and fortified foods and food supplements [4]. Functional and fortified foods are those with a similar appearance to their traditional counterparts, while food supplements are

Nutraceutical and Functional Food Regulations in the United States and Around the World.
DOI: http://dx.doi.org/10.1016/B978-0-12-405870-5.00014-1

components that are mostly consumed in unit dose forms such as tablets, capsules, or liquids, and are commonly known as food/dietary supplements.

Botanicals represent a large segment of the nutraceuticals category and present additional problems relating to the complex nature and composition of botanical ingredients. Botanicals in this context are not defined in European Union (EU) food law, but are defined in some detail in European Medicines law as

> *whole, fragmented or cut plants, algae, fungi, lichens, and botanical preparations from these materials involving extraction, distillation, expression, fractionation, purification, concentration and fermentation. Botanical substances are defined by the botanical name of the plant according to the binomial system (genus, species, variety and author) and the part used (e.g. leaf, root and fruit).*

The challenges in the EU for botanical nutraceuticals relate to quality, safety, claims, and the resolution of borderline issues. In many cases there are still a number of unknowns such as the identification of specific active components, their absorption, and metabolism in the human body and the effects of processing on the products containing them.

In this chapter the focus will mainly be on the newly adopted EU legislation governing the enrichment of foods and the list of permitted health claims. It also covers the use of botanicals in functional and fortified foods and food supplements.

Emphasis is particularly given to different approaches adopted by the European Food Safety Authority (EFSA), in evaluation of the "general function" Article 13.1 claims and that adopted for Articles 13.5 and 14 claims, and the issues surrounding their decisions. Traditional use of botanicals was not considered by EFSA, as a result, many of the botanicals failed to pass the stringent assessment procedure adopted by EFSA, particularly for health claims for botanicals under Articles 13.1. This resulted in a pan-European controversy, which the European Commission (EC) attempted to alleviate by halting the assessment of those botanical claim applications that had not already received an EFSA opinion. This step, taken by the EC to put botanicals on hold, has further complicated the situation producing discrimination on claims relating to botanicals by granting transition periods for some and not for the others.

14.2 The nutraceutical concept in the US

Much of the early development of the nutraceutical concept and products was driven from the United States where, since its introduction in 1994, the Dietary Supplement and Health Education Act (DSHEA 1994) has allowed considerable flexibility and blurred the boundaries between foods and medicines found in other parts of the world [5].

Under DSHEA a dietary supplement may contain "a herb or other botanical" or "a concentrate, metabolite, constituent, extract or combination of any ingredient from the other categories." This was subjected to very little qualification and as a consequence a wide variety of botanicals and other substances have been sold as dietary supplement ingredients, including some that are restricted to medicinal use under most regulatory regimes in other parts of the world.

14.3 EU legislation on botanicals as medicines or foods

There is currently no consistency in the legal status of many botanicals across the EU. In all EU countries, botanical products can be sold as foods, food supplements, or incorporated in functional/fortified foods as long as no medicinal claims are made. In some EU Member States, however, certain preparations have been seen as herbal medicines and need to be registered. In some countries, medicinal product status for some botanicals is automatically linked to pharmacy-only status.

1. Although the definition of a medicine is given in EU legislation, (Directive 2001/83/EC, formerly 65/65/EC) as amended by Directive 2004/27/EC, it has been interpreted differently by Member States in the context of products containing botanicals [6−8]. This definition consists of two parts, one relating to the presentation as medicinal product and the other to the function. A product falling into the scope of either part is considered a medicinal product. It is the precise interpretation of the term "pharmacological action" in the second part that has led to the inconsistencies of interpretation between Member States.

2. If medicinal claims are made based on "traditional use" as defined in Directive 2004/24/EC [9], or the herb is considered medicinal by function, the product may be categorized as a "Traditional Herbal Medicinal Product (THMP)," provided the time-related criteria are met. These are 30 years usage, of which 15 years are in the EU.

3. The General Food Law Regulation (EC) No. 178/2002 defines food and lays down the general principles and requirements of food law, establishes EFSA, and lays down procedures in matters of food safety [10]. Its aim was to "provide the basis for the assurance of a high level of protection of human health." The Regulation applies to all foodstuffs; its general principles, therefore, also cover foods with added functional properties (e.g., "functional foods," "nutraceuticals," dietetic foods, and food supplements). "Food" (or "foodstuff") is defined as "any substance or product, whether processed, partially processed or unprocessed, intended to be, or reasonably expected to be ingested by humans." Food includes drink, chewing gum, and any substance, including water, intentionally incorporated into the food during its manufacture, preparation, or treatment. There are a small number of exclusions that include medicinal products.

    The fact that this definition excludes medicinal products means that the decisive step to judge whether medicinal law or food law applies to a specific product has to be sought in the intended use of the product. This is not a theoretical issue as products with added functions with regard to health may closely resemble medicinal products given their composition, claims and presentation.

Article 2.2 of Directive 2004/27/EC states that in cases of doubt, where, taking into account all its characteristics, a product may fall within the definition of a "medicinal product" and within the definition of a product covered by other Community legislation, the provisions of this Directive shall apply [8]. This provision basically puts into legal text the jurisprudence that has been established over the years by the Court of Justice of the European Union on "borderline" cases.

If a botanical is categorized as a food or a food ingredient according to Article 2 of Regulation (EC) No. 178/2002, it must be characterized and further defined as per the following categories of foods [10].

4. Novel Food Regulation (EC) No. 258/97 defines a "Novel Food" as a food or food ingredient, which has not been used to a significant degree for human consumption in the EU before May 15, 1997 (for example, Noni juice). A comprehensive safety evaluation is required before approval for use in foods is given [11].

5. Directive 2009/39/EC defines "Foods for Particular Nutritional Uses" (PARNUTS) as foods, "which, owing to their special composition or manufacturing process are clearly distinguishable from foodstuffs for normal consumption, which are suitable for claimed nutritional purposes, and which are marketed in such a way as to indicate such suitability." This category covers the various types of dietetic foods [12]. Examples are infant formulae, baby foods for infants and young children, slimming foods, foods for special medical purposes, sports foods, food for diabetics, etc.

6. Regulation (EC) No. 1331/2008 defines "Food Additives" as substances that are intentionally added to foods to perform certain technological functions such as to color or to preserve. In the EU, processing aids are not, as yet, regulated as food additives and enzymes and flavorings are the subject of specific legislation [13].

7. Regulation (EU) No. 1334/2008 on flavorings in food contains certain requirements relating to the use of botanicals in foods [14].

8. Directive 2002/46/EC defines "Food Supplements" as "foodstuffs the purpose of which is to supplement the normal diet and which are concentrated sources of nutrients or other substances with a nutritional or physiological effect, alone or in combination, marketed in dose form, namely forms such as capsules, pastilles, tablets, pills and other similar forms, sachets of powder, ampoules of liquids, drops dispensing bottles and other similar forms of liquids and powders designed to be taken in measured small unit quantities" [15].

9. A Regulation on "Nutrition and Health Claims made on Foods" was adopted on October 12, 2006 as Regulation (EC) No. 1924/2006 and came into force in early 2007. This Regulation controls all nutrition and health claims, and only those claims appearing on an approved list of permitted claims will be allowed [16].

10. Regulation (EU) No. 432/2012 establishing a list of "Permitted Health Claims made on Foods," was adopted on May 25, 2012, and came into force in June 2012; this covers claims other than those referring to the reduction of disease risk, and to children's development and health. This list is set out as an annex to the regulation and applied from December 14, 2012 [17].

11. A Regulation on "Food Fortification" [18] with the formal title of "The addition of Vitamins, Minerals and certain other Substances to Foods" was adopted in October 2006 as Regulation (EC) No. 1925/2006 and came into force on January 19, 2007. This Regulation applies to the use of vitamins, minerals, and other substances including botanicals in foods.

    Chapter 3 of this Regulation covers the control of substances, including botanicals, in foods including food supplements. There will be an annex containing a list of prohibited ingredients, a list of ingredients subject to controls on levels or other requirements such as labeling, and a third list of "substances under scrutiny." This third list is intended to contain substances for which safety concerns have been raised and which are required to be evaluated for safety by EFSA.

12. Regulation (EC) No. 1223/2009 on "Cosmetic Products" covers the use of botanicals in cosmetic products, as defined in the legislation [19].

All of the above legislation affects the use of botanicals and botanical preparations in foods and food supplements.

There is also a very considerable amount of EU food legislation covering safety and quality, which impacts significantly on the actual use of botanicals in food products.

14.4 Regulatory status and positioning of botanicals as food supplements, fortified, and functional foods

14.4.1 Food supplement directive

Article 2a of Directive 2002/46/EC defines a Food Supplement, as described earlier in section III, number 8.

As stated in paragraph 6 of its Recitals, Directive 2002/46/EC at present only covers vitamins and minerals. Although the EC may extend the lists to include substances other than vitamins and minerals, work in this area had not commenced by the end of 2013.

The Food Supplement definition specifically makes mention of substances with nutritional or physiological effects such as plant extracts. The Directive, which came in to force in August 2005, has been implemented in the different Member States of the EU. Botanicals are already covered at national level in specific legislation of a number of Member States.

Article 10 of the Food Supplement Directive leaves the choice to EU Member States whether or not to set up a mandatory notification procedure for the first marketing of food supplement products in their country. It is to be noted that all EU Member States, with the exception of Austria, Lithuania, the Netherlands, Sweden, and the UK decided to introduce a mandatory notification procedure. The requirements of food supplement notification procedures vary from country to country. The standard minimum requirement of this procedure is the provision of a copy of the food supplement label.

In countries where notification is required, health claims on the label will therefore be exposed to the immediate attention of national authorities at the time of notification, and authorities could react in cases of non-compliance with the health claim rules. In countries without notification, the risk of challenge for health claims is lower at the start of the product launch. However, post-marketing controls of national control authorities and/or complaints via advertising standard authorities and consumer groups are still possible.

14.5 National control on botanicals in food supplements

In the absence of harmonized EU legislation in the area of botanicals, some Member States have set their own legislation on the use of plants and plant extracts in food supplements. Italy, Belgium, and France, for example, have introduced legal positive lists and, in some cases negative lists, of botanicals that can or cannot be used in food supplements. Examples of these national lists are given below.

Belgium: The Royal Decree of August 29, 1997, relating to the manufacture and commercialization of food products composed of, or containing plants or plant preparations, was amended by Royal Decree of March 19, 2012 in relation to the positive and negative lists, with more detailed conditions of use, restrictions of use, and labeling requirements. This list contains over 650 entries.

France: It has also developed a positive list of botanicals in food supplements. Conditions of use are also in the draft positive list annexed to the Decree. This will come into force on July 1, 2013, according to Article 13 of this draft decree. This list became available in December 2012, as it was notified by the French government to the EC. In contrast to the Belgian and Italian lists it contains only 549 entries.

Ireland: The Irish Medicines Board in April 2011 published a draft list of Herbal Substances that may be acceptable for inclusion in food supplements.

Italy: On July 21, 2012, an Italian Ministry Decree on "the use of plant substances and preparations in food supplements" was published in the Italian *Gazzetta Ufficiale*. The Decree foresees a positive list of plants, with information on the permitted plant parts and, for certain plants, restrictions of use, or warning statements. The list contains over 1200 entries.

Romania: Order no 244/401/2005 of the MAPDR and Health Ministry on processing and marketing of medicinal and aromatic plants.

Slovenia: Slovenia Decree 103/2008 on the classification of medicinal herbs. The list in the Annex of this decree classifies plants into four categories of herbs:

- Herbs permitted for use in foods, including food supplements, provided that no medicinal claims are made = H
- Herbs permitted for use in over-the-counter medicines = Z
- Herbs permitted for use in prescription only medicines = ZR
- Herbs prohibited from use in all types of food and medicinal products = ND

United Kingdom: The Medicines and Healthcare products Regulatory Agency has created a list of herbal ingredients and their reported uses. The list is for information only and has no legal status. However, this list is a useful guide to the probable status of a botanical. The status of a product under medicines legislation is determined on an individual basis.

14.6 Regulation on mutual recognition

In the absence of harmonized EU positive/negative lists of botanicals that are permitted for food supplement use, the EU market is currently still inconsistent with divergent national approaches. This situation can have a positive and/or negative impact on the acceptable market practice for the use of botanicals in food supplements. Depending on the Member State there can be a more liberal or more restrictive approach relating to botanicals sold under food law. A key tool to a successful pan-European market entry for products that contain ingredients not regulated by EU law is the principle of mutual recognition [20].

The so-called principle of mutual recognition derives from the EU principle of "free movement of goods" laid down in Articles 34, 35, and 36 of the Treaty on the Functioning of the European Union (TFEU). It is directly applicable in all EU Member States. The principle of mutual recognition means that each EU Member State is obliged to accept on its territory products lawfully marketed in another EU Member State, in areas that are not subject to EU harmonization, even if such products are not in conformity with the national rules of the Member State of "destination."

Two important areas not yet harmonized in the EU for food supplements include:

- Maximum levels of vitamins and minerals used in food supplements
- The use of ingredients other than vitamins and minerals (including botanicals) for nutritional/physiological purposes, where these ingredients are not regarded as novel foods

There are exceptions to this principle. In accordance with Article 36 TFEU, the Member State of destination may refuse the marketing of a product in its current form, only where it can show that this is strictly necessary for the protection of, for example, human health. In that case, the Member State of destination must also demonstrate that its measure is the least trade-restrictive applicable to the situation. In the area of foodstuffs, the common exception to the principle of mutual recognition is a safety concern (direct or indirect risk to human health).

To make this principle more operational, Regulation (EC) No. 764/2008 of the European Parliament and of the Council of July 9, 2008, was introduced and adopted [20]. This lays down procedures relating to the application of certain national technical rules, to products lawfully marketed in another Member State.

Regulation (EC) No. 764/2008 became applicable in May 2009. It organizes the procedural requirements when a Member State intends to deny the application of mutual recognition for the marketing of a product.

The Regulation applies to administrative decisions addressed to economic operators, on the basis of a technical rule, in respect of any product lawfully marketed in another Member State, where the direct or indirect effect of that decision is the prohibition, modification, additional testing, or withdrawal of the product (Article 2(1)). Any national authority intending to take such a decision must follow the procedural requirements set out in the Regulation.

In practice, however, companies wanting to make use of the principle of mutual recognition for the marketing of food supplement products still face major obstacles. Some substances, in particular certain herbal extracts, are used both in food supplements and in medicinal products, e.g., traditional herbal medicinal products. As a result, there have been borderline cases, which have given rise to situations where a given product is authorized for marketing as a food in some Member States, but as a medicinal product in others.

In most cases, however, classification problems may arise from a case-by-case assessment of the product, taking into account all its characteristics. If this assessment leads a Member State to consider a product as falling under medicinal law, Regulation (EC) No. 764/2008 will no longer apply.

14.7 Other regulations with an impact on functional foods

14.7.1 Traditional herbal medicinal products

Directive 2004/24/EC on THMP introduces a simplified registration procedure, based on "traditional use," ensuring quality and safety, without the need to prove efficacy of the product. Community lists are being prepared for traditional medicinal herbs [9].

To qualify for registration under the THMP Directive, the product must be able to show a history of use of 30 years, of which 15 years are in the EU.

This requirement has had an inhibitory effect on the registrations, as a large number of products have not been able to demonstrate the requirement of 15 years on the market in the EU. Even for products that can meet this requirement, significant changes may have taken place to the product composition.

As a consequence, relatively few Traditional Herbal registrations have been authorized. In parallel, a large number of botanical materials are being used as ingredients in food supplements. As discussed earlier, from the viewpoint of legislation the herbal food supplement category has to be separately identified by carefully interpreting the data on physiological versus pharmacological activity, and health versus therapeutic effects on dose/concentration basis, taking into consideration Directive 2002/46/EC on Food Supplements [15] and Directives 2004/27/EC and 2004/24/EC on medicinal products [8] and THMPs [9], respectively.

This issue has been further complicated due to the stringent criteria for scientific substantiation adopted by EFSA, particularly for health claims for botanicals under Article 13.1 of the Nutrition and Health Claims Regulation (NHCR). As a result, many of the botanicals used in food supplements failed to be accepted onto the list of permitted health claims annexed to the recently adopted regulation (EU) 432/2012.

14.7.2 Botanicals for enrichment of foods and functional foods

When considering the use of botanical ingredients to be added to foods in the EU, there are a number of issues in addition to those already discussed.

The two most important are the "Novel Food status" and whether the ingredient could be judged to be "medicinal by function."

Although the Regulation on Novel Foods and Novel Ingredients came into force in early 1997, it was almost 7 years later that the EC introduced a change in the interpretation of the Regulation. This change was ratified by the EU Standing Committee [21] on the Food Chain and Animal Health in February 2005.

The result of this change is that the pre-1997 use of an ingredient in food supplements only, does not confer exemption from the Regulation for use of that ingredient in other food categories.

This means that if an ingredient can only be demonstrated to have been used in food supplements in the EU before May 1997, any intention to use it in a drink, for example, would require a novel foods application and official approval. The application must include a comprehensive range of toxicity studies. If the ingredient was in use in food supplements only before 1997, there is a high probability that extra data will be required to sustain an application for food use.

In addition to the Novel Food status of an ingredient intended for use in a food, it is important that the type of ingredient and levels of anticipated consumption do not classify it as being "medicinal by function." Such a classification would preclude its use in foods. If the product falls into the classification of a dietetic food (Directive 2009/39/EC) further controls may apply [12].

14.7.2.1 Functional foods

The concept of foods for specified health use (FOSHU) was established [22] in Japan in 1991.

> *Foods that are expected to have certain health benefits, and have been licensed to bear a label claiming that a person using them for a specified health use may expect to obtain the health use through the consumption thereof.*

According to the Japanese Ministry of Health and Welfare FOSHU these include:

- Foods that are expected to have a specific health effect due to relevant constituents, or foods from which allergens have been removed.
- Foods where the effect of such addition or removal has been scientifically evaluated and permission is granted to make claims regarding their specific beneficial effects on health.

According to a working definition adopted in a European Consensus document in 1999:

> *A food can be regarded as "functional" if it satisfactorily demonstrated to affect beneficially one or more target functions in the body, beyond adequate nutritional effects, in a way that is relevant to either an improved state of health and well-being and/or reduction of risk of diseases. Functional foods must remain foods and they must demonstrate their effects in amounts that can normally be expected to be consumed in the diet: they are not pills or capsules, but part of a normal food pattern [23].*

It is in that general context that the EC's concerted action on Functional Food Science in Europe, actively involving a large number of the most prominent European experts in Nutrition and related sciences, were engaged by the International Life Science Institute (ILSI) in Europe [23].

In this context, "target function" refers to genomic, biochemical, physiological, psychological, or behavioral functions that are relevant to the maintenance of a state of well-being and health or to the reduction of the risk of a disease. Modulation of these functions should be quantitatively/objectively evaluated by measuring the biochemical markers (e.g., metabolite, specific proteins, hormone, enzyme, etc.) or physiological parameters (e.g., blood pressure, heart rate, gastrointestinal transit time, etc.) or changes in physical and intellectual performance using objective parameters.

14.7.3 EU legislation and functional foods

There is no legal definition for functional foods in Europe [4].

Current EU legislation does not recognize functional foods as a distinct category of foods, as, for example, in Japan. This means that functional products must comply with all relevant food legislation with respect to composition, labeling, claims etc. Functional foods must comply with the requirements of the Regulation on Nutrition and Health Claims and there are no special concessions for this product category.

In 1995 the United Kingdom Ministry of Agriculture, Fisheries, and Food (MAFF now DEFRA) developed a working definition, which is "a food that has had a component incorporated into it to give a specific medical or physiological benefit, other than a purely nutritional effect." This definition distinguishes functional foods from those that are fortified with vitamins and minerals and from food supplements [24].

According to the ILSI, functional food is defined "as a food which by virtue of physiologically active food components, provides health benefits beyond basic nutrition" [25].

Functional food embraces a wide range of botanical products containing bioactive food components such as soya beans (isoflavones), tomatoes (lycopene), garlic (allicin), and tea (polyphenols, catechins).

Food law always lags behind innovation and developments, sometimes by more than a decade. This is particularly true in the case of functional food, which is still in a state of taking shape in the framework of European legislation. Within Europe there has been increasing recognition of functional foods by national health authorities, particularly in the area of health claims (discussed later in Section 14.12.1).

In January 2008 the EC adopted a proposal on the provision of food information to consumers [26]. This proposal combined consolidating Directive 2000/13/EC and Directive 90/496/EEC on nutrition labeling, presentation, and advertising of foodstuffs, into one instrument. In addition, the proposal simplifies the structure of the horizontal food labeling legislation [27,28].

The Regulation on Food Information to Consumers (Regulation (EU) 1169/2011) contains a long list of labeling requirements, including some specified warnings, all of which must be included on the product label [29].

According to Article 7.3 of this Regulation, it is prohibited to "attribute to any foodstuff the property of preventing treating or curing a human disease, nor refer to such properties." This rule applies also to the presentation of the foodstuffs, and in particular, their shape, appearance or packaging, the packaging materials used, the way in which they are arranged, and the setting in which they are displayed.

14.7.4 **EU legislation on fortified food**

The aim of Regulation (EC) No. 1925/2006 on "The Addition of Vitamins, Minerals and Certain other Substances to Foods [18] is to harmonize divergent national rules concerning the addition of vitamins and minerals and of certain other substances to foods in order to ensure a high level of consumer protection and the free circulation of goods within the Community.

The main focus is on vitamins and minerals; however, the Regulation includes botanicals as "other substances." In the Regulation "other substances" are defined as "a substance other than a vitamin or a mineral that has a nutritional or physiological effect;"

The Regulation has been closely linked to the NHCR (to be described in Section 14.12.1).

The Regulation will not affect:

- The provisions in national legislation that require the addition of vitamins, minerals, or other substances to particular foods (for example, in the UK white and brown bread must contain certain specified quantities of various nutrients).
- The provisions of the Food Supplement Directive already in force.

The Regulation sets out a positive listing of vitamins and minerals and their sources, a prohibited list of other substances, a controlled usage list (with specified conditions), and a scrutiny list. The law required that the positive list must be established within 2 years of the Regulation's entry into force (i.e., 2009). Any vitamins or minerals or their sources not on the first list will not be allowed to be added to foods. This list may be amended from time to time.

The Regulation also requires that maximum levels are assigned for each of the vitamins and minerals, taking into account:

- A scientific risk assessment based on generally acceptable scientific data
- Intakes of vitamins, minerals, or other substances from other dietary sources
- The contribution of individual foods to the overall diet of the population
- Any nutrient profile established for the particular food in accordance with the NHCR.

The Regulation imposes labeling requirements for any fortified foods and requires a fortified product to display the total amounts of vitamins, minerals, or other substance in that food. The NHCR, which requires various conditions to be met before any nutrition or health claim may be made, is also relevant.

The Regulation gives powers to the EC to scrutinize substances other than vitamins and minerals in due course. If necessary, use of such substances may be restricted or forbidden in the future.

The Regulation prohibits the addition of vitamins, minerals, or other substance to unprocessed foodstuffs (including fruit, vegetables, meat, poultry, and fish) and alcoholic drinks, other than fortified tonic wine.

The Regulation provides transition periods, including a transition period for vitamins and minerals not on the approved list, which may continue to be added to foods for 7 years following the date of entry into force of the Regulation (January 19, 2014).

The Regulation also contains:

- Annex I: A list of the vitamins and minerals which may be added to foods in general (excluding food supplements and without prejudice to specific rules

applying to foods for particular nutritional uses). The Regulation sets out a positive list of over 100 substances, which are sources of vitamins and minerals that can be added to foods. (Annex II of the Regulation).
- In addition, on the basis of Article 6 of Regulation (EC) No. 1925/2006, maximum amounts of vitamins and minerals in these products should be set at the EU level. The EC had to submit proposals for the maximum amounts by January 19, 2009. The setting of maximum limits is still on hold at the EU level. As a direct consequence, maximum limits potentially adopted by EU Member States still apply.
- Annex II: A list of the vitamin formulations and mineral substances that may be added to foods.
- Annex III: A list of the substances whose use in foods is prohibited, restricted, or under Community scrutiny. Annex III is currently still empty (under development at the EU level).

Regarding the positive lists of vitamins and minerals and their chemical forms, a transitional period is still running. Member States may provide derogations for vitamins and minerals and their forms currently not listed in Annex I and II of the Regulation until January 19, 2014, provided that certain conditions are met.

Describing the requirements and conditions for the addition of vitamins and minerals to foods is beyond the scope of this chapter. Here we focus on requirements of adding botanicals/bioactive substances to foods. The EU Regulation 1925/2006 introduces a centralized procedure, which makes it possible to put under scrutiny or provide for restrictions or prohibitions on the use of certain substances. Until now, no measures have been taken at the EU level.

In theory, the addition of a herbal extract or another bioactive substance to a food falls under national legislation. It is the responsibility of the person responsible for the marketing of the product to ensure that the food does not present any health risk for the consumer, as generally provided by Articles 14 and 17 of EC Regulation 178/2002 laying down the general principles and requirements of food law.

14.8 **EU legislation and parnuts**

The EU PARNUTS Directive 2009/39/EC was introduced to control products that meet the particular nutritional requirements of certain categories of populations (dietetic foods). The definition of PARNUTS Foods has been given earlier [12].

There are four categories of PARNUTS foods covered by specific legislation:

1. Infant formulae and follow-on formulae [30].
2. Processed cereal-based foods and baby foods for infants and young children [31].
3. Foods for energy-restricted diets (slimming foods) [32].
4. Dietary Foods for Special Medical Purposes (Medical Foods) [33].

A list of permitted PARNUTS nutritional ingredients was attached as Annex 2 of Directive 2001/15/EC [34]. These are: vitamins, minerals, amino acids, carnitine and taurine, and nucleotides, choline and inositol.

In early 2012, the EC, following a general consultation on this legislation, proposed a simplification of the rules.

This revision was aimed at avoiding overlaps with other legislation and closing legal loopholes that had appeared in the existing legislation.

The proposal is to retain only three categories of dietetic foods namely:

- Foods for infants and young children
- Foods for special medical purposes
- Foods for weight control intending to replace the whole of the daily diet

The proposed weight control category removed the previous compositional requirements for meal replacement products.

The proposed amended legislation was adopted by the European Parliament and Council at the end of 2012 and was published as Regulation (EU) No 609/2013 on 29 June 2013 and will be applied from 20 July 2016 [50].

14.9 EU legislation on novel foods

In 1997 the EC introduced Regulation (EC) No. 258/97 concerning novel foods and novel food ingredients [11]. For the definition of a novel food or ingredient see point 4 of the Regulatory Framework in Section 14.3.

A novel food or ingredient is one that has not been used for human consumption in the EU to a significant degree before May 15, 1997. The term "significant" was only defined by the EC in 2012 [35].

There were six categories of foods or food ingredients that fell within the scope of the original Regulation:

1. Containing or consisting of Genetically Modified Organisms (GMOs).
2. Produced from (but not containing) GMOs.
3. Consisting of or isolated from microorganisms, fungi, or algae.
4. Consisting of or isolated from plants, and ingredients isolated from animals except those obtained by traditional propagating or breeding practices and having history of safe food use.
5. Having a new or intentionally modified primary molecular structure.
6. Having been subject to a new production process resulting in significant changes in the composition or structure of foods/food ingredient, which affects the nutritional value, metabolism or level of undesirable substances.

From May 1, 2004, the first two categories were transferred to Regulation (EC) No. 1829/2003 on Genetically Modified Foods [36], leaving 3–6 in the Novel Foods Regulation.

The Novel Food Regulations have well-defined safety evaluation and authorization procedures leading to marketing authorization on a pan-European basis.

The EC on January 14, 2008, adopted a proposal to revise the Novel Foods Regulation with a view to improving the access of new and innovative foods to the EU market, while still maintaining a high level of consumer protection. Under the draft Regulation, novel foods would be subject to a simpler and more efficient authorization procedure, which should enable safe, innovative foods to reach the EU market faster. Moreover, special provisions were made for foods that have not been traditionally sold in the EU, but that have a safe history of use in third-world countries. This proposal was the subject of much discussion, and agreement had not been reached by the end of 2013 [37].

Food supplements or their ingredients including new botanicals/botanical extracts may be considered as "novel foods" or "novel ingredients" within the meaning of Regulation (EC) No. 258/97 [11].

Regulation (EC) No. 258/97 interprets the concept of novel food broadly. Botanical extracts are in particular affected by this Regulation, since a plant extract that was not on the internal EU market, or not being produced, before the date of entry into force of the Regulation could, if it falls under one of the novel food categories and is significantly different from existing counterparts, in principle, be considered a novel ingredient, even though the plant from which it is extracted would not be considered "novel."

Novel foods or novel food ingredients must pass through a case-by-case pre-marketing authorization procedure before their marketing in the EU. Food companies must apply to a EU Member State Authority for authorization, presenting the scientific information and safety assessment report. Following the initial assessment by the Member States, the ingredient may be authorized if no objection by other Member States is raised. If an objection is raised, then the case is evaluated by EFSA. This procedure has proven to be in most cases expensive and long (in average 2–3 years).

Some novel foods or novel food ingredients may fall under the simplified procedure of notification for "substantial equivalence." In this case, data substantiating that the ingredient/food in question is substantially equivalent to existing foods or food ingredients in terms of composition, nutritional value, metabolism, intended use, and the level of undesirable substances must be available. These products only need to be notified to the EC at the time of marketing.

The EC has recently published guidance documents on the concepts of "human consumption to a significant degree" and "significant use" [35,38].

14.10 **Quality aspects**

Botanical products should be well identified and characterized. Identification, standardization and specification are of paramount importance in considering the safety of botanicals [39,40]. A few points to consider for reference are the botanical source, growth conditions, raw material, manufacturing process, botanical preparation, and end product. These are described in detail in the author's earlier publication [2]. The variability of the plant material is due to different climatic conditions, harvesting, and drying and storage conditions. Standardized conditions of cultivation in the form of Good Agriculture Practice are essential.

In mid-2012 the European Botanical Forum issued a quality guide for botanical food supplements. This document is aimed at manufacturers and suppliers who are marketing botanical supplements in the EU and covers all aspects of Good Manufacturing Practice from the selection of appropriate raw materials to the stability of the product in the market place [41].

As botanical products sold in the EU as supplements are subject to all relevant requirements of EU food law, there is more than sufficient legislation in place covering both the quality and safety aspects of botanicals sold as supplements or in functional or fortified foods.

14.11 Safety of botanicals

Botanical ingredients intended for use in food supplements and foods, which are determined to be novel ingredients, are required to undergo a pre-marketing safety assessment as required by the legislation.

The following safety aspects need to be considered when compiling a safety dossier. These safety requirements are given in detail in a document published by the EC's Scientific Committee on Food (SCF) in 2001 [42]. The SCF is the predecessor of EFSA. This document was prepared as guidance for applicants for approval of a novel food or ingredient.

a. *In vitro* Safety Data

These include isolated cells, microorganisms, and subcellular components (enzymes, receptors and DNA). These models are rapid, less expensive, and reveal mechanisms of actions.

The guiding principle is that the *in vitro* study data can serve as signals of potential harmful effects in humans, but not as independent indicators of risk unless an ingredient causes an effect that has been associated with harmful effects in animals or humans, and there is evidence that the ingredient or its metabolites are present in physiological sites where they could cause harm. On its own an *in vitro* study should serve only as a hypothesis generator and indicator of possible mechanisms of harm when the totality of the data from different key factors is considered.

b. *In vivo* Animal Safety Data

Animal studies serve as important signal generators and in some cases, may stand alone as indicators of unreasonable risks. These include acute, subchronic, chronic toxicity, reproduction toxicity, *in vivo* genotoxicity, and safety pharmacology studies. Knowledge of an ingredient's pharmacokinetics and *in vivo* metabolism will allow the most appropriate interpretation of relevancy of the dose used in the *in vitro* tests. All cells react differently to its unique biochemical pathways.

The guiding principle of animal data is that even with the absence of human adverse events, evidence of abnormalities from laboratory animal studies can be indicative of potential harm to humans. This indication may assume greater importance if the route of exposure/administration is similar (e.g., oral), the formulation is similar, and more than one species show the same toxicity.

c. Clinical Safety Data

Vulnerable subpopulations can be defined as groups of individuals who are more likely to experience an adverse event related to the use of a food, food supplement or ingredient, or an individual in whom such events are more likely to be serious in comparison with the general population.

The guiding principle of clinical data is, when data indicate that an identifiable subpopulation may be especially sensitive to adverse effects from certain food or supplement ingredients, then, this higher level of concern should be taken into account when screening the food, food supplement, or ingredient.

EFSA has published a draft guidance document on the procedures to be adopted for the safety assessment of botanicals and botanical preparations intended for use as food supplements, which was adopted on July 22, 2009. This guidance document was finalized and published by EFSA in 2009 [43]. Subsequently, EFSA published a "Compendium of Botanicals Reported to Contain Naturally Occurring Substances of Possible Concern to Human Health When Used in Food and Food Supplement" [44].

These documents need to be taken into consideration when reviewing the safety of a botanical intended for the European food or supplement market.

14.12 **Efficacy of botanicals**

14.12.1 **NHCR**

For many years there had been concerns within governments of the EU Member States and consumer organizations that some claims on the efficacy of foods and ingredients including botanical ingredients, would not stand up to scientific scrutiny.

After over a quarter of a century of discussions and prevarications the EC achieved agreement on a proposal for the scientific assessment and authorization of nutrition and health claims.

The Regulation was adopted on the October 12, 2006, and published in the Official Journal of the EU on December 30, 2006, as the "Nutrition and Health Claims" Regulation (EC) No. 1924/2006 [16]. The Regulation complemented the general principles given in the food labeling Directive 2000/13/EC [27] and lays down specific provisions concerning the use of nutrition and health claims.

The basic principle of the Regulation is the harmonization of provisions laid down by law, regulation, or administrative action relating to nutrition and health claims in the EU member states in labeling, presentation, and advertising of food and related categories of nutraceuticals. This Regulation takes into account the definitions of food provided in Food Regulation 178/2002 [10] and food supplements in Directive 2002/46/EC [15]. Provision is made in the Regulation to support small and medium enterprises to comply with the Regulation.

14.12.1.1 *The conditions and general principles for claims*

According to Article 5 of the Regulation the conditions for a claim are the presence of the active ingredient in sufficient quantity to produce the nutritional or physiological effect claimed, as established by generally accepted scientific evidence. The wording of claims should be understood by the average consumer and should not be misleading in any sense. In the case of food supplements, the nutrition information must be provided in accordance with Article 8 of Directive 2002/46/EC. Special conditions stated in Article 10 prevail, stating the importance of a varied and balanced diet. Appropriate warnings are to be provided for products likely to present a health risk, if consumed in excess.

Article 6 sets out the general principles for substantiation of claims. Health claims would therefore only be approved for use on labeling, presentation, and advertising of foods in the EU market after a scientific evaluation of the highest possible standard.

Fresh food such as fruits, vegetables, and bread are excluded from the regulation.

Article 4 foresees the establishment of nutrient profiles, meaning the maximum amounts of fat, salt, and sugar that products must contain in order to be able to make approved nutrition and health claims. Nutrient profiles need to be based on scientific knowledge about diet and nutrition and their relation to health and had to be established within 2 years of the Regulation coming into force. However, this adoption has suffered considerable delay and no nutrient profiles have been established yet.

14.12.1.2 *Types of claims*

a. Nutrition claims: Approved nutrition claims are given in the Annex to the Regulation, together with specified quantitative conditions. They are also included in the Community Register of approved and rejected claims on the EC website (DG SANCO).

Nutrition claims are banned on products containing alcohol at more than 1.2% v/v, except if these claims refer to reduction in alcohol or calories.

b. Health Claims based on generally accepted scientific evidence: According to Article 13 of the Regulation, applications for health claims had to be routed through the member states, which were required to prepare lists of claims for the EC within 12 months of the Regulation's entry into force. The EC was required to adopt a community list of claims accompanied by conditions applying to them. The list provided by member states for claims already approved at their national level was to be established within 3 years, i.e., by the end of January 2010. Also this process has suffered delays and a partial list was only published in June 2012.

c. Claims that are subject to an application for authorization by an applicant: Health claims based on newly developed scientific evidence and/or containing a request for the protection of proprietary data are subject to an application for authorization under article 18. Also Reduction of Disease Risk Claims (RDRC) and Claims referring to Children's Development and Health, covered by Article 14, can only be approved following an application for authorization as specified in Articles 15–17.

RDRC are subject to additional labeling: "The disease to which a claim refers has multiple risk factors and altering one of those risk factors may or may not have a beneficial effect."

All health claims require specific authorization by the EC before they can be used. All approved claims can be used by any food operator in accordance with the conditions of use, except where protection of proprietary data is granted.

14.12.1.3 Procedures for authorization of claims

a. Application: The two procedures for the application for the authorization of a claim, other than those covered by Article 13, are given in detail in the Regulation. The review of the scientific evidence will be carried out by EFSA who will give their opinion to the EC. The final authorization will come from the EC.

EFSA, on the request of the EC, has published a document [45], "Scientific and Technical Guidance for the Preparation and Presentation of the Application for Authorization of Health Claim," which was adopted on July 6, 2007. The European Commission published Regulation (EC) No. 353/2008 establishing implementing rules for applications for authorizations of health claims [46].

b. EFSA Guidelines for Health Claims Dossier Applications: The guidelines can be found on the EFSA website.

In addition to detailed guidance for preparation and presentation of applications (adopted in 2007), guidance on the general principles employed by the EFSA Panel for scientific substantiation of health claims was developed in 2009 and subsequently updated, with the most recent revision published in 2011. This guidance document covers issues such as the totality of available scientific evidence, selection of pertinent studies for substantiation of health claims, wording of claims, the extent to which a food needs to be characterized for the claimed effect, claimed effects that are considered beneficial physiological effects, definition of a risk factor for the development of a human disease, compliance/eligibility issues for health claims, and procedural aspects, as described above.

Also, between 2011 and 2012 EFSA issued six specific guidance documents on the scientific requirements for health claims related to specific fields of health:

1. Gut and immune function: Adopted on January 28, 2011, and published on April 4, 2011
2. Antioxidants, oxidative damage and cardiovascular health: Adopted on November 24, 2011, and published on December 9, 2011
3. Appetite ratings, weight management, and blood glucose concentrations: Adopted on February 29, 2012, and published on March 21, 2012
4. Bone, joints, skin, and oral health: Adopted on April 25, 2012, and published on May 16, 2012
5. Physical performance: Adopted on June 28, 2012, and published on July 17, 2012
6. Functions of the nervous system, including psychological functions: Adopted on June 28, 2012, and published on July 17, 2012

These guidance documents represent the views of the NDA Panel of EFSA based on the experience gained to date with the evaluation of health claims in these areas. It was not intended that the document would include an exhaustive list of beneficial effects and studies/outcome measures that are acceptable. Rather, it presents examples drawn from evaluations already carried out to illustrate the approach of the NDA Panel, as well as some examples which are currently under consideration within ongoing evaluations.

14.12.1.4 Community register

Article 19 of the Regulation requires that the EC sets up and maintains a register of approved nutritional health claims to which the public can have access.

The salient points include:

i. The Register shall include nutrition and health claims, and any restrictions adopted, the authorized claims and the conditions applying to them, and a list of rejected health claims and the reasons for their rejection. This will allow manufacturers who wish to introduce a product with a particular health claim to simply consult the register in order to know what conditions are to be observed and the wording of the claim without having to go through the authorization process itself.
ii. Health claims authorized on the basis of proprietary data shall be recorded in a separate Annex to the Register together with the following information:
   • The date the EC authorized the health claim and the name of the original applicant that was granted authorization
   • The fact that the EC authorized the health claim on the basis of proprietary data
   • The fact that the health claim is restricted for use, unless a subsequent applicant obtains authorization for the claim without reference to the proprietary data of the original applicant.
iii The Register is in the public domain.

14.12.1.5 Scientific and proprietary data

According to Article 21, the scientific data and other information may not be used for the benefit of a subsequent applicant for a period of 5 years from the date of authorization, unless agreed upon by the prior applicant or until the EC takes the decision whether a claim could be or could have been included in the list provided for in Article 14 or where applicable Article 13 without the submission of these data.

The conditions that apply include:

- The scientific data and other information must have been designated as proprietary data (PD) by the prior applicant at the time of his application
- The prior applicant had exclusive rights of reference to these PD at the time of application
- The health claim could not have been authorized without the submission of the PD

14.12.2 Scientific substantiation of claims

In the early 2000s and prior to the adoption of the NHCR there had been a project under the auspices of ILSI Europe and supported by the EC on the "Process of the Assessment of Scientific Support for Claims (PASSCLAIM)." The objectives of this work included identifying definitions, best practice, and methodology in this area to underpin current and future regulatory development [47].

The essential criteria for the scientific substantiation of claims were developed. These included:

- The food or food component, including the botanical product/ingredient to which the claim is attributed should be well characterized.
- Substantiation of a claim should be based on human data, primarily from well-designed intervention studies taking into consideration target population, appropriate controls, adequate duration of exposure, and a follow-up, to demonstrate the intended effect.
- Generally accepted scientific data must be systematic and objective, balanced and unbiased, strong, consistent, and reproducible, with appropriate statistical analysis.
- The data should be based on validated and predicted biomarkers for
  - Enhanced function
  - Disease risk reduction
- The claim must be evaluated on the totality of the scientific evidence (i.e., both positive and negative evidence).
- The overall assessment should be based on the application of scientific judgment and critical interpretation of the data as a whole.

These principles were largely taken as the basis for the approach by EFSA for the assessment of the scientific evidence underlying Health Claims to ensure that the claims included on the positive list are substantiated by "generally accepted scientific evidence." The person or company placing the product on the market is required to produce all relevant elements and data establishing the science behind the claim.

14.12.3 **Qualified health claims**

The preamble of the EU legislation states that health claims should only be authorized by EFSA after scientific assessment of the highest possible standard. There is no dispute about this, but the major concern is how to accommodate the emerging science in an appropriate grading system. The World Health Organization and the World Cancer Research Fund have established four grades of evidence: "convincing," "probable," "possible," and "insufficient" [48,49]. These correspond to qualified health claims proposed in the United States. The first level "convincing" means convincing scientific evidence. The second level "probable" means that although there is scientific evidence supporting the claim, the evidence is not conclusive. The third level "possible" means "some scientific evidence may support the claim, however, the evidence is limited and not conclusive." The fourth level "insufficient" means "very limited and preliminary scientific research suggests there is little scientific evidence to support the claim."

In their assessment EFSA did not consider the concept of grades of evidence. This has consequences for the value of emerging science and the level of investments industry will be willing to make in scientific research in future. It is crucial to support future scientific initiatives to find an approach where the term "generally accepted scientific data" includes not only generic or well-established linkage between food and food components and health benefits, and to establish standards that should not be revoked or reversed by emerging science at a later stage. The concept of assessing the data in terms of "grades of evidence" was not adopted by EFSA in relation to their assessment of health claims and their focus was only on whether or not a "cause and effect" is established on the basis of Randomized Clinical Trials (RCTs).

As a consequence, the EFSA opinions on the evaluations were on a black or white basis—either accepted or not.

14.13 **Consequences of the NHCR**
14.13.1 **Article 13.1 general function health claims**

A large number of issues have arisen from the adopted procedures for the collation and assessment of the "general function" health claims (Article 13).

The first requirement at the start of the process was that the submission of applications for assessment had to be made by the individual Member States. This meant that all the business operators wishing to retain their claims had to submit an application to the relevant national authorities. This procedure was not well coordinated between Member States, particularly in the context of the amount of data required to support the scientific substantiation of the claim. The Member States authorities were required to collate all the applications made in their country and forward the final list to the EC.

Due to variants of claims in the many national languages, the total number of claims received by the EC exceeded 44,000. The EC then had to undertake the task of rationalization, finally reducing the number submitted to EFSA to 4637. Of these 4637 claims, EFSA was able to complete the evaluation of 2758 by June 2011. A further 331 claims were withdrawn by the applicants.

During the EFSA evaluations a serious controversy arose over the procedures adopted by EFSA for the evaluation of health claims for botanicals. This was temporarily resolved by the EC putting 2085 claims for botanicals "on hold" until agreement could be reached on the way forward. No progress had been made on this issue by the end of 2013.

As a result of the EFSA evaluations carried out between June 2009 and June 2011, 341 opinions were adopted and published in six series.

These opinions have provided the basis for the EC to establish the first list of 222 permitted health claims in 2012, and were just over 8% of the 2758 health claim applications under the Article 13(1) procedure.

In 2012, EFSA also completed further assessments for a number of additional "general function claims." These included 74 claims relating to microorganisms that were considered by the Panel to be insufficiently characterized and 17 claims for which the evidence provided during the initial submission was not sufficient for substantiation. These assessments were based on additional data submitted by Member States and used by the EC and the Member States for amendment of the list of permitted health claims. Only two of these claims were considered to be substantiated in the end, i.e., for prunes and normal bowel function and for α-cyclodextrin and a lower rise in blood glucose after meals.

14.13.2 Article 13.5 and Article 14 claims

In addition to the decisions on the Article 13.1 "general function" health claims, by the end of 2012 EFSA had also given opinions on the following:

- 62 health claims submitted under Article 13.5 (health claims based on newly developed scientific evidence and/or which include a request for the protection of proprietary data)
- 93 health claims submitted under Article 14 (60 on claims for development and health of children and 33 on reduction of disease risk claims)

EFSA's opinions were used as the basis for authorization decisions made by the EC and the Member States (with scrutiny by the European Parliament), the outcomes of which are published in the EU Register of Nutrition and Health Claims made on foods. In the EU Register (July 2012), 19 claims submitted under the individual authorization procedure have been authorized, 77 non-authorized, and the remainder are pending a decision on authorization.

14.14 The article 13.1 list included in Commission Regulation 432/2012

14.14.1 Regulation (EU) 432/2012

Regulation (EU) 432/2012 establishing a list of permitted health claims on foods other than referring to the reduction of disease risk and to children's development and health [17] was published on May 25, 2012, and came into force on June 14, 2012. The list is set out as an annex to this regulation and applied from December 14, 2012.

It is to be noted that this list:

- Includes 222 authorized Article 13.1 non-botanical general function health claims with specific conditions of use.
- Does not include any of the rejected claims. These are published directly in the Community Register of claims.
- Excludes health claims for botanical ingredients at this stage.

The Regulation had a transition period until December 14, 2012, for compliance with the conditions of use/wording for the authorized health claims and for the prohibition of non-authorized claims that were in the Article 13.1 assessment process (only for those that were not put "on hold" by the EC).

14.14.2 Article 13.1 health claims put on hold

The EC has put a number of Article 13.1 health claims on hold to which the transition period of Regulation 432/2012 does not apply. Their approval/prohibition is still under consideration by the EC and Member States, and these claims that have been put on hold can still be used as long as national authorities in the Member States allow their use according to national health claims rules. The application of national rules will stay in place until the EC makes its final decisions and indicates further transition periods for these claims.

At the end of 2012 outstanding Article 13.1 claims on hold included:

- 91 non-botanical claims in the "further Article 13.1 assessment process" (74 on probiotics and 17 other "insufficient evidence" claims)
- 64 non-botanical Article 13.1 claims "under further consideration"
- 2085 Article 13.1 claims for botanical ingredients that had been submitted for evaluation

Relating to the botanical claims, the EC intends to discuss and clarify further with Member States how health claims for botanicals should be evaluated for products sold under food law in the EU. The EFSA opinions may therefore be subject to further discussion (and potential further additional assessment once the criteria have been decided upon by the EC and Member States).

The document listing claim entries, for which the assessment by EFSA or the consideration by the EC is not finalized (on hold), is also published on the EC DG SANCO website.

14.14.3 Commission discussion paper on botanical health claims put on hold

In August 2012 the EC issued a Discussion Paper on botanical health claims providing two options to be considered by the member states on the future of botanical Article 13.1 health claims. It presents two options:

- Option 1: To retain the status quo. This would mean that EFSA will resume its assessment of health claims on botanicals with no change to the approach. This would mean that botanicals would continue to be subject to the same high scientific criteria as used by EFSA for other health claims. From the experience of EFSA opinions, this would result in most botanicals claims being rejected.
- Option 2: A change to the past approach, which is the recognition of the peculiarity of the botanicals case and addressing it through a review of the legislation: This would call for a review of the legislation and recognition given to traditional use. This second option is the one preferred by the food supplement industry overall, and this could also explore other aspects, such as quality and safety. Member States were requested to send their first comments on this Discussion Paper to the EC by September 30, 2012, and the EC and Member States would then continue discussions on a potential approach for botanical claims in the future. The discussions for botanical claims are likely to continue during 2014.

14.14.4 Criteria of evaluation of health claims adopted by EFSA under Article 13 (1) of NHCR

In its methodology, EFSA applies a systematic review of the information provided. It considers the extent to which:

1. The food/constituent is defined and characterized
2. The claimed effect is defined and is a beneficial physiological effect ("beneficial to human health")
3. A cause and effect relationship is established between the consumption of the food/constituent and the claimed effect (for the target group under the proposed conditions of use)

If a cause and effect relationship is considered to be established, the NDA Panel considers whether:

4. The quantity of food/pattern of consumption required to obtain the claimed effect can reasonably be consumed within a balanced diet

5. The proposed wording reflects the scientific evidence
6. The proposed wording complies with the criteria for the use of claims specified in the Regulation
7. The proposed conditions/restrictions of use are appropriate; in the case of Article 13.5 and 14 claims, substantiation may be dependent on data claimed as proprietary by the applicant

14.14.5 Reasons for rejection of health claims by EFSA under article 13 (1) of NHCR

For a judgment of whether these aspects are covered, EFSA only relies on the information as specified in the application. It is therefore not surprising that the main reasons for the rejection of Health Claims as stated by EFSA are as follows:

1. The cause and effect relationship was not established between the dietary intake of the food/ingredient and the claimed effect.
2. The Food constituents, which were the subject of a claim, were not sufficiently characterized.
3. The evidence provided did not establish that this claimed effect was a beneficial physiological effect for the general population.
4. The claim effect is general and nonspecific, and did not refer to any health claim as required by NHCR.
5. The claimed effect was related to prevention or treatment of disease and does not comply with the criteria laid down in NHCR.

14.15 Examples of negative and positive opinions on botanical health claims evaluated by EFSA

The previously mentioned criteria for the evaluation of health claims have been applied to a number of botanical compounds. Some examples of EFSA opinions resulting in acceptance of rejection of the provided scientific justification under Articles 13.1, 13.5, and 14 are presented in Table 14.1. This overview is illustrative and may help understanding of the requirements for positive evaluations, in view of recently issued guidelines and based on experience gained by EFSA.

14.16 Conclusions

The addition of botanical ingredients to food supplements and functional foods encompasses a wide range of issues.

The predominant one is the status of the food or ingredient in relation to EU legislation covering both medicines and foods, as many of the ingredients are very close to the borderline between the two classifications.

Table 14.1 Section 1 Depicts Some Examples of Rejected Health Claims on Botanicals and the Reasoning Provided by EFSA and Section 2 Depicts Some Authorized Health Claims on Botanical with Due Justification

Section 1: Some Examples of Negative Opinions Relating to Botanicals and the Reasoning by EFSA

Claim Entry	Claim Type	Reasons for Rejection
Slimaluma (Extract of *Caralluma fimbriata*) and weight management and appetite [*EFSA Panel on Dietetic Products, Nutrition and Allergies (NDA); Scientific Opinion on the substantiation of a health claim related to ethanol–water extract of Caralluma fimbriata (Slimaluma®) and helps to reduce waist circumference pursuant to Article 13(5) of Regulation (EC) No 1924/2006. EFSA Journal 2010; 8 (5):1602.EFSA Panel on Dietetic Products, Nutrition and Allergies (NDA); Scientific Opinion on the substantiation of a health claim related to ethanol–water extract of Caralluma fimbriata (Slimaluma®) and helps to control hunger/appetite pursuant to Article 13(5) of Regulation (EC) No 1924/2006. EFSA Journal; 8(5):1606.*]	Article 13.5	Cause and effect relationship not established between the consumption of Slimaluma and a reduction of waist circumference because: - While waist circumference was significantly reduced in the experimental group compared to placebo after 60 days of intervention in one of the human studies presented assessing this outcome, no significant effect of the ethanol–water extract of *Caralluma fimbriata* on waist circumference leading to an improvement in adverse health effects associated with an excess abdominal fat was observed when compared to a suitable control. Cause and effect relationship not established between the consumption of Slimaluma and a reduction of appetite because: - While the ratings of hunger were significantly reduced in the experimental group compared to placebo at day 60 in the one human study presented which assessed this outcome, no significant effect of the ethanol-water extract of *Caralluma fimbriata* on energy intake was observed when compared to a suitable control.
Grape OPC Plus (40 mg/day) and reducing the risk of chronic venous insufficiency by increasing microcirculation. [*EFSA Panel on Dietetic Products, Nutrition and Allergies (NDA); Scientific Opinion on the substantiation of a health claim related to OPC PremiumTM and the reduction of blood cholesterol pursuant to Article 14 of Regulation (EC) No 1924/2006. EFSA Journal 2009; 7(10):1356.*]	Article 14	Cause and effect relationship not established between the consumption of OPC (oligomeric procyanidins) from grape seed and reducing the risk of chronic venous insufficiency (CVI) by increasing microcirculation because:

(Continued)

Table 14.1 (Continued)

Section 1: Some Examples of Negative Opinions Relating to Botanicals and the Reasoning by EFSA

Claim Entry	Claim Type	Reasons for Rejection
		- No human intervention studies using the constituent OPC extracted from grape (*Vitis vinifera* L.) seeds have been provided.
		- "Alterations" in the venous microcirculation (i.e., venous microangiopathy) is a consequence rather than a cause of (or a risk factor for) CVI.
		- The evidence provided does not establish that improving the "alterations in the venous microcirculation" is a beneficial physiological effect by reducing the risk of CVI.
Soy proteins (as soy protein isolate and isolated soy proteins) and reduction of blood cholesterol concentrations [EFSA Panel on Dietetic Products, Nutrition and Allergies (NDA); Scientific Opinion on the substantiation of a health claim related to soy protein and reduction of blood cholesterol concentrations pursuant to Article 14 of the Regulation (EC) No 1924/ 2006. EFSA Journal 2010; 8(7):1688.EFSA Panel on Dietetic Products, Nutrition and Allergies (NDA); Scientific Opinion on the substantiation of a health claim related to isolated soy protein and reduction of blood LDL-cholesterol concentrations pursuant to Article 14 of Regulation (EC) No 1924/2006. EFSA Journal 2012; 10(2):2555.]	Article 14	Cause and effect relationship not established between the consumption of soy protein and the reduction of LDL cholesterol concentrations because:
		- The results from four intervention studies identified as being controlled for the macronutrient composition of the test products do not support an effect of the protein component of soy alone) on LDL cholesterol concentrations.
		- The design of the studies on soy protein isolate rated by the applicant as having medium or low quality and the studies on soy foods do not address the effects of the food constituent that is the subject of the health claim (the protein component of soy alone) on LDL cholesterol concentrations.
		- The same argument applies to the nine published and two unpublished meta-analyses provided by the applicant, which aimed to address the relationship between soy protein consumption and blood lipids.

		- Under similar conditions four randomized, control trials (RCTs) reported an effect of isolated soy protein on blood LDL/non-HDL cholesterol concentrations, whereas 14 RCTs did not report such an effect and another RCT showed no consistent effects.
		Most of these RCTs were at high risk of bias.
		Differences in the results obtained between trials appear unrelated to the dose of isolated soy protein used, sample size, or study duration.
		The proposed mechanisms by which the protein component of soy could exert the claimed effect have not been corroborated by available scientific evidence and are therefore not convincing.
Black tea (*Camellia sinensis*) and helps to focus attention [EFSA. Scientific Opinion of the Panel on Dietetic Products, Nutrition and Allergies on a request from Unilever PLC and Unilever NV on the scientific substantiation of a health claim related to black tea from Camellia sinensis and help focus attention. The EFSA Journal (2008) 906, 1–10.]	Article 13.5	Evidence insufficient to establish a cause and effect relationship between the consumption of black tea from *Camellia sinensis* and "helps to focus attention" because:
		- The findings between two studies that were similar in design were inconsistent with only a small difference in cumulative dose for the purported active components (46 vs 36 mg theanine, 100 vs 90 mg caffeine).
		- The cumulative doses of theanine and caffeine used in another study were higher than the cumulative doses in the conditions of use as proposed by the applicant.
Ocean Spray Cranberry Products® and urinary tract infection in women [EFSA Scientific Opinion of the Panel on Dietetic Products, Nutrition and Allergies on a request from Ocean Spray International Services Limited (UK), related to the scientific substantiation of a health claim on Ocean Spray Cranberry Products® and reduced risk of urinary tract infection in women by inhibiting the adhesion of certain bacteria in the urinary tract. The EFSA Journal (2009) 943, 1–16.]	Article 14	Evidence not sufficient to establish a cause and effect relationship between the consumption of Ocean Spray cranberry products® and the reduction of the risk of urinary tract infection in women by inhibiting the adhesion of certain bacteria in the urinary tract because:
		- The studies demonstrate an *in vitro* anti-adherence effect of urine on uropathogenic *Escherichia coli* strains following consumption of cranberry products. However, these studies do

Table 14.1 (Continued)

Section 1: Some Examples of Negative Opinions Relating to Botanicals and the Reasoning by EFSA

Claim Entry	Claim Type	Reasons for Rejection
		not establish the validity of such anti-adherence effects shown *in vitro* to predict the occurrence of a clinically relevant bacterial anti-adherence effect within the urinary tract. - Patients suffering from neurogenic bladder are not representative of the intended normal population. - Intake levels in one study were six times higher than those requested by the applicant. - Many studies have limitation in design, including a small number of subjects, lack of a control group, short duration, high drop-out rate, lack of statistical power, lack of adequate randomization, and intervention matrices (capsules) differing from the foods specified in this application.

Section 2: Some Specific Examples of Positive Opinions on Some Botanicals and Reasoning by EFSA

Claim entry	Claim Type	Reasons for Acceptance
Guar gum and maintenance of normal blood glucose concentrations [EFSA Panel on Dietetic Products, Nutrition and Allergies (NDA); Scientific Opinion on the substantiation of health claims related to guar gum and maintenance of normal blood glucose concentrations (ID 794), increase in satiety (ID 795) and maintenance of normal blood cholesterol concentrations (ID 808) pursuant to Article 13(1) of Regulation (EC) No 1924/2006. EFSA Journal 2010; 8(2):1464.]	Article 13.1	Cause and effect relationship established between the consumption of guar gum and the reduction of blood cholesterol concentrations because: - 18 RCTs conducted in humans and investigating the effects of guar gum on blood cholesterol concentrations have been reviewed in a meta-analysis in healthy (normocholesterolemic), hypercholesterolemic, and diabetic subjects. - In 13 out of the 17 studies, serum total cholesterol concentrations were significantly reduced after the administration of guar gum as compared to the low-fiber control group. - The meta-analysis showed a statistically significant effect of guar gum on serum total and LDL cholesterol at doses of 9–30 g/d. - An inverse (nonlinear) association was found between the dose of guar gum consumed and the changes in total and LDL cholesterol concentrations.
Water-soluble tomato concentrate (WSTC I and II) and platelet aggregation [EFSA. Scientific Opinion of the Panel on Dietetic Products, Nutrition and Allergies on a request from Provexis Natural Products Limited on Water-soluble tomato concentrate (WSTC I and II) and platelet aggregation. The EFSA Journal (2009) 1101, 1–15.]	Article 13.5	Cause and effect relationship established between the consumption of water-soluble tomato concentrate and the reduction in platelet aggregation in humans because: - Taken together the human studies provided consistently show a reduction in platelet aggregation following consumption of WSTC at suboptimal ADP concentrations under the conditions of use proposed by the applicant.

(Continued)

Table 14.1 (Continued)

Section 2: Some Specific Examples of Positive Opinions on Some Botanicals and Reasoning by EFSA

Claim entry	Claim Type	Reasons for Acceptance
		- Possible confounding factors likely to interfere with platelet aggregation have been addressed and the within subject variability accounted for in the studies.
		- Studies show a consistent effect of the supplementation with WSTC on platelet aggregation, which is sustained up to 28 days in subjects that are representative of the target population for which the claim is intended.
		- The biological plausibility of this effect is supported by the presence of 37 identified compounds in aqueous tomato extracts showing different degrees of inhibition of platelet aggregation *in vitro* and by the effects of tomato extract on markers of platelet function in an animal study.
Plant sterols and blood cholesterol [EFSA. Scientific Opinion of the Panel on Dietetic Products Nutrition and Allergies on a request from Unilever PLC/NV on Plant Sterols and lower/reduced blood cholesterol, reduced the risk of (coronary) heart disease. The EFSA Journal (2008) 781, 1–12.]	Article 14	Cause and effect relationship established between the intake of plant sterols added to fat-based foods and low-fat foods such as milk and yogurt and lowering of LDL cholesterol, in a dose-dependent manner because:
		- Many human studies show that phytosterols in the form of supplements or enriched conventional food products can lower blood total and LDL cholesterol concentrations.
		- A meta-analysis of 41 randomized placebo-controlled, double-blind clinical trials (RCT) show that the daily intake of 2–2.4 g/d of plants sterols or stanols added to margarine (or to mayonnaise, olive oil, or butter in 7 trials) reduced on average LDL blood cholesterol levels by 8.9% (95 CI: 7.4–10.5).

Although new legislation covering traditional herbal medicines is now in force, the work on the control of botanically derived substances in food supplements and other foods is focusing on nutrition and health claims and the general requirements of food law to secure safety of the products. Companies wishing to market products in the EU will have to assess their regulatory status in each Member State.

The NHCR and the "Fortified Foods" Regulation came into force in 2007. These were expected to facilitate the market potential of botanically sourced ingredients. However, due to the difficulties on the application of these Regulations, they have not resulted in the anticipated outcome. It has led to discussions on the fundamental nature of botanical ingredients and the way in which future legislation should be drafted. For food business operators, it will become hard to position their products in the right category of foods with appropriate claims or with no claims, depending upon the quality of scientific evidence available to support their products.

References

[1] Defelice SL. The nutraceutical revolution, its impact on food industry research and development. Trends Food Sci Technol 1995;6:*59−61*.

[2] Gulati OP, Berry Ottaway P. Legislation relating to nutraceuticals in the European Union with a particular focus on botanical-sourced products. Toxicology 2006;221:*75−87*.

[3] Mandel S, Packer L, Youdim MBH, Weinreb O. Reviews : current topics- proceeding from the " third international conference on mechanism of action of nutraceuticals". J Nutrition Biochem 2005;16:*513−20*.

[4] Richardson DP. Functional foods-shades of grey: an industry perspective. Nutrition Reviews 1996;54(*11*):*S174−85*.

[5] Dietary Supplement Health Education Act (DSHEA) of 1994, Public Law 103-417 available at FDA Website: <http://www.fda.gov>.

[6] Directive 65/65/EEC of 26 January 1965 on the approximation of provisions laid down by Law, Regulation or Administrative Action relating to proprietary medicinal products. Official Journal 022, 09/02/1965 p. 0369 − 0373.

[7] Directive 2001/83/EC of the European Parliament and of the Council of 6th November, 2001 on the community code relating to medicinal products for human use. Official Journal L-311, p. 0067−0128.

[8] Directive 2004/27/EC of the European Parliament and of the Council of 31st March 2004 Directive 2001/83/EC of the European Parliament and of the Council of 6th November, 2001 on the community code relating to medicinal products for human use Official Journal 136, 30/04/2004, p. 0034−0057.

[9] Directive 2004/24/EC of the European Parliament and of the Council of 31st March 2004 amending, as regards traditional herbal medicinal products, Directive 2001/83/EC on the Community code relating to medicinal products for human use Official Journal L 136, 30/04/2004, p. 0085−0907.

[10] Regulation (EC) No 178/2002 of the European Parliament and of the Council of 28th January 2002 laying down the general principles and requirements of food law, establishing the European Food Safety Authority and laying down procedures in matters of food safety. Official Journal L 31, 01/02/2002, p. 0001−0024.

[11] Regulation (EC) No 258/97 of the European Parliament and of the Council of 27th January 1997 concerning novel and novel ingredients. Official Journal L 043, 14/02/1997, p. 0001−0007.

[12] Directive 2009/39/EC of the European Parliament and of the Council of 6th May 2009 on the approximation of the laws of Member States relating to foodstuffs intended for particular nutritional uses. Official Journal L 124, 20/5/09, p. 00217−0029.

[13] Regulation (EC) No 1331/2008 of the European Parliament and of the Council of 16th December 2008 on the approximation of the laws of Member States concerning food additives, food enzymes and food flavouring stuffs intended for human consumption. Official Journal L 354, 31/12/2008, p. 001−006.

[14] Regulation (EC) No 1334/2008 of the European Parliament and Council on flavourings and certain ingredients with flavouring properties. Official Journal L 354, of 31/12/2008, p. 034−050.

[15] Directive 2002/46/EC of the European Parliament and of the Council of 10th June 2002 on the approximation of the laws of Member States relating to food supplements. Official Journal L 183, 12/07/2002, p. 0051−0057.

[16] Regulation (EC) No 1924/2006 Regulation of the European Parliament and of the Council of 20 December 2006 on nutrition and health claims made on foods. Official Journal L 404, 30/12/2006. p 0009−0025.

[17] Regulation (EC) No 432/2012 of the European Commission establishing a list of permitted health claims made on foods. Official Journal L 136, of 25/5/2012, p. 001−040.

[18] Regulation (EC) No 1925/2006 Regulation of the European Parliament and of the Council of 20 December 2006 on the addition of vitamins and minerals and of certain other substances to foods. Official Journal L 404, 30/12/2006. p 0026−0038.

[19] Regulation (EC) 1223/2009 of the European Parliament and of the Council of 30th November 2009 on the approximation of the laws of Member States relating to cosmetic products. Official Journal L 342, 22/12/2009, p. 059−209.

[20] Regulation (EC) No 764/2008 of the European Parliament and of the Council of 9 July 2008 laying down procedures relating to the application of certain national technical rules to products lawfully marketed in another Member State and repealing Decision No 3052/95/EC. Official Journal L 218, 13/08/2008 p 0021−0029.

[21] Standing Committee on the Food chain and animal health, Proceedings of the committee meeting (Section of Toxicological Safety and section on General Food Law) of 14th February 2005; 2005 p. 6.

[22] The FOSHU system. Nutrition Improvement Law Enforcement Regulations (Ministerial Ordinance No. 41; July 1991.

[23] Diplock AT, Aggett PJ, Ashwell M, Bornet F, Fern EB, Roberfroid MB. Scientific concepts of functional foods in europe - consensus document. Br J Nutr 1999;81 (*Supplement 1*):*S1−S27*.

[24] Ministry of Agriculture, Fisheries and Food (MAFF). Food standards and labelling division discussion paper on functional foods and health claims. MAFF *Publications*; 1995, *August 1995*.

[25] Milner JA. Functional Foods and health: a US perspective. Br J Nutr 2002;88 (*Supplement 2*):*S151−8*.

[26] European Commission. Proposal for a Regulation of the European Parliament and of the Council on the provision of Food Information to Consumers. COM (2008), Brussels of 30.01.08, 2008/028(CODE); (2008)

[27] Directive 2000/13/EC of the European Parliament and of the Council of 20th March 2000 on the approximation of laws of the member states related to labelling, presentation and advertising of food stuffs. Official Journal L 109, 06/05/2000:29.

[28] Directive 90/496/EEC of the European Parliament and of the Council of 24 September 1990 on nutrition labelling for foodstuffs. Official Journal L 276, 6.10.1990, p. 40−44.

[29] Regulation (EC) No 1169/2011 of 25th October 2011 of the European Parliament and of the Council on the Provision of Food Information to Consumers. Official Journal L 304/18, 22/11/2011, p. 018-063.

[30] Directive 2006/141/EC of the European Parliament and of the Council of 22nd December 2006 on infant formulae and follow-on formulae. Official Journal L 401, 30/12/2006, p. 001-033.

[31] Directive 2006/125/EC of 5th December 2006 on processed cereal-based foods and baby foods for infants and young children. Official Journal L 339, 6/12/2006. p. 016−035.

[32] Directive 96/8/EC of 26 February 1996 on foods for use in energy restricted diets for weight reduction. Official Journal L 055 of 06.03.1996. 0022−0026.

[33] Directive 1999/21/EC of 25 March 1999 on dietary foods for special medical purposes Official Journal L 091 of 07.04.1999. 0029−0037.

[34] Directive 2001/15/EC of 15 February 2001 on substances that may be added for specific nutritional purposes in foods for particular nutritional uses- Official Journal L 052 of 22.02.01. 0019−0025.

[35] European Commission. 'Human Consumption to a Significant degree − Information and Guidance Document', Brussels. European Commission Website (Novel Foods); October 2012.

[36] Regulation (EC) No 1829/2003 of the European Parliament and of the Council of 22nd September 2003 on genetically modified food and feed. Official Journal L 268, 18/10/2003. 0001−0023.

[37] European Commission, (2008), Proposal for a Regulation of the European Parliament and of the Council on Novel food amending Regulation EC No 258/97 (Common Procedure) COM (2007), 872 Final, Brussels, of 14.01.08, 2008/028 (COD).

[38] European Commission (2013) Summary of the experience of Novel Food Competent Authorities with the range of products that have been assessed under the simplified procedure (Article 5 of Regulation (EC) No 258/97), Brussels. European Commission Website (Novel Foods) February 2013.

[39] Shilter B, Andersson C, Anton A, Constable A, Kliener, Brien AG, Renwick AG, Korver O, Smit F, Walker R. Guidance for the safety assessment of botanicals and botanical preparations for use in food and food supplements. Food & Chem Toxicol 2003;41:1625−49.

[40] Coppens P, Delmule L, Gulati O, Richardson D, Ruthsatz M, Sievers H, Sidani S. Use of botanicals in Food Supplements- Regulatory scope, scientific risk assessment and claim substantiation. Ann Nutr Metab 2006;50:*538−55*.

[41] European Botanical Forum. Quality Guide for Botanical Food Supplements. Guidance for the manufacture of safe and high quality botanical food supplements across the EU. Available from the European Botanical Forum website <www.botanicalforum.eu>; 2011

[42] Scientific Committee on Food of the EU (2001) 'Guidance on Submissions for Food Additive Evaluations by the Scientific Committee on Food'. SCF/CS/ADD/GEN/26 final of 12 July 2001.

[43] European Food Safety Authority. "Guidance on Safety assessment of botanicals and botanical preparations, intended for use as ingredients in food supplements". EFSA Journal 2009 2009;7(9):1249.

[44] European Food Safety Authority. "Compendium of botanicals reported to contain naturally occurring substances of possible concern for human health when used in food and food supplements". EFSA Journal 2012 2012;10(5):2663.

[45] European Food Safety Authority. "Scientific and technical guidance for the preparation and presentation of the application for authorisation of a health claim". The EFSA Journal (2007) 2007;530:1−44.

[46] Regulation (EC) No 353/2008 of the European Commission implementing rules for applications for authorisations of health claims. Official Journal L 109, of 19/04/2008, p. 011−016.

[47] Aggett PJ, Antoine JM, Asp NG, Bellisle F, Contor L, Cummings JH, et al. PASSCLAIM: consensus on criteria. Eur J Nutr 2005;44(suppl 1):i5−30.

[48] World Health Organisation. Diet, nutrition and the prevention of chronic diseases: report of a joint FAG/WHO expert consultation. WHO Technical Report Series 916. Geneva: World Health Organisation; 2004.

[49] World Cancer Research Fund/American Institute for Cancer Research. Food, nutrition and the prevention of cancer: a global perspective. Washington, DC: World Cancer Research Fund/American Institute for Cancer Research; 1997.

[50] Regulation (EU) No 609/2013 of the European Parliament and of the Council of 12 June 2013 on food intended for infants and young children, food for special medical purposes, and total diet replacement for weight control and repealing Council Directive 92/52/EEC, Commission Directives 96/8/EC, 1999/21/EC, 2006/125/EC and 2006/141/EC, Directive 2009/39/EC of the European Parliament and of the Council and Commission Regulations (EC) No 41/2009 and (EC) No 953/2009.

History and Current Status of Functional Food Regulations in Japan

15

Makoto Shimizu

The University of Tokyo, Bunkyo-ku, Tokyo, Japan

15.1 Introduction

Improving dietary habits would be beneficial in preventing lifestyle-related diseases. This idea does not simply mean better nutrition; it includes a novel concept of consuming special health-promoting food substances or components in our daily life. In the past 35 years, many attempts have been made in Japan to identify those healthy components that eventually result in the creation of the concept of a "functional food."

Research activities, including a large-scale, grant-aided research project on food functions that started in 1984 and lasted for more than 10 years under the sponsorship of the Ministry of Education, Science, and Culture, were begun [1]. A number of researchers from the fields of food science, nutrition, pharmaceutical, and medical sciences participated in this project, and many interesting characteristics of food components in terms of their physiological functions were identified [1–3]. The results from this project led to the world's first policy for legally permitting the commercialization of foods with particular health-promoting functions. This particular food was termed food for specified health use (FoSHU). This concept was introduced in the journal *Nature* in 1993, and since then, the term functional food has been internationally recognized [4].

15.2 FoSHU

A FoSHU system was introduced in 1991 by the Ministry of Health and Welfare, now known as the Ministry of Health, Labor, and Welfare (MHLW), as a regulatory system to approve statements concerning the effects of the food on the human body. In this regulatory system, FoSHU is classified as a special food group placed between medicine and regular food (Figure 15.1).

Nutraceutical and Functional Food Regulations in the United States and Around the World.
DOI: http://dx.doi.org/10.1016/B978-0-12-405870-5.00015-3

Medicine	Food for nutrient function claim (FNFC)	Food with Health Claims (FHC)					So-called health food	Usual food
		Food for Specific Health Use (FoSHU)						
		Ordinary	Standardized	Disease risk reduction claim	Qualified			

FIGURE 15.1

Classification of FHC since 2005.

Table 15.1 FoSHU Categories and Their Corresponding Functional Ingredients

Health Claim Category (Function)	Examples of Functional Ingredients
Promotes gut health	Dietary fiber, oligosaccharide, bacteria
Promotes tooth and gum health	Sugar alcohol, tea polyphenol, milk protein digests, funoran, isoflavone, calcium
Enhances mineral absorption	Casein phosphopeptide, poly-γ-glutamic acid
Promotes bone health and strength	Milk basic protein, isoflavone, vitamin K2
Lowers blood pressure	Food protein-derived peptide, γ-aminobutyrate, acetic acid, chlorogenic acid
Reduces blood glucose level	Indigestible dextrin, wheat albumin, tea polyphenol, fermented soybean extract
Reduces blood cholesterol level	Soybean protein, chitosan, low molecular weight alginate, phytosterol, tea catechin, S-methylcysteine sulfoxide
Reduces blood neutral lipid level and body fat	Polyphenol conjugate, indigestible dextrin, catechin, conglycinin, n-3PUFA

The first FoSHU products were approved in 1993, which included hypoallergenic rice and low phosphorus milk for patients. Since FoSHU health claims must not include medical claims such as to "prevent," "cure," or "treat" human diseases, the previously mentioned products were later transferred from the category of FoSHU to another category termed "food for illness." Despite a few regulation errors in the initial stages, the FoSHU system has been encouraging the Japanese food industry to develop functional food products [5]. Since 2009, the FoSHU system has been under the jurisdiction of The Consumer Commission, Cabinet Office. As of March 2013, 1037 FoSHU products are listed. Table 15.1 shows the categories of FoSHU currently approved; they include foods that (1) improve gastrointestinal health, (2) promote dental and gum health, (3) enhance mineral absorption, (4) promote bone health and strength, (5) reduce blood pressure, (6) lower blood glucose level, (7) lower blood cholesterol level, and (8) lower blood triglyceride levels and reduce body fat accumulation. Underlying mechanisms for

Table 15.2 Examples of Mechanisms of Acton for FoSHU Ingredients

Functional Ingredient	Mechanisms of Action
\<Lower blood pressure\>	
Fish protein hydrolyzate (peptides)	Inhibits angiotensin-converting enzyme
Tochu tea extract (geniposidic acid)	Stimulates the parasympathetic nervous system
γ-Aminobutyric acid	Suppresses the sympathetic nervous system
Chlorogenic acid	Stimulates NO synthesis at endothelial cells
\<Reduce blood neutral lipid level and suppress body fat accumulation\>	
Conjugate tea polyphenol	Inhibits pancreatic lipase action
Indigootiblo doxtrin	Inhibits lipid release from mixed micelles
Coffee mannooligosaccharide	Downregulates fatty acid binding protein expression
Green tea catechin	Regulates the expression of lipid metabolizing enzymes
n-3 polyunsaturated fatty acid	Regulates the expression of lipid metabolizing enzymes
\<Promote bone health and strength\>	
Casein phosphopeptide	Enhances intestinal calcium absorption
Milk basic proteins	Activates osteoblast and suppresses osteoclast activity
Vitamin K2	Activates osteocalcin
Soybean isoflavone	Suppresses osteoclast activity via hormonal regulation

the current FoSHU products are diverse. Even under the same health claim, different functional substances with different mechanisms of action can be included. Some examples are shown in Table 15.2.

A distinguishing feature of FoSHU is that commercially available FoSHU products include a large variety of common food such as beverages, yogurt, steamed rice, noodles, bread, cereals, cracker, margarine, cooking oil, mayonnaise, sausage, and fish paste that can be incorporated into daily meals. In 2011, more than 500 billion yen were spent on FoSHU products, indicating that FoSHU seems to have established itself into the dietary habits of the Japanese.

15.3 Food with nutrient function claim

In 2001, criteria for FoSHU were integrated into the "food with health claim" (FHC) system. FHC is the Japanese regulation system for health foods introduced by the MHLW in 2001 and consists of two categories: FoSHU and "food with nutrient function claim" (FNFC; Figure 15.1) [6–9].

FNFC permits the use of functional claims for nutrients such as vitamins and minerals. Twelve vitamins including vitamin A, B_1, B_2, B_6, B_{12}, C, D, and E, biotin, pantothenic acid, folic acid, and niacin, and five minerals including calcium, iron, magnesium, copper, and zinc are currently permitted for FNFC. Labeling of the nutrient functions for these vitamins and minerals is permitted, because the benefits of taking these nutrients have been internationally recognized based on scientific evidence. The maximum and minimum daily intakes of an individual nutrient have been determined as the standard daily dosage. Therefore, an FNFC product should contain an amount of the nutrient between the designated upper and lower limits [8,9].

15.4 Revision of FoSHU categories

Following the introduction of FNFC to the FHC system in 2001, FoSHU was newly classified into four groups in 2005: "Standardized FoSHU," "FoSHU with Disease Risk Reduction Claims," "Qualified FoSHU," and "Ordinary FoSHU" [8,9].

Standard FoSHU and Qualified FoSHU systems were introduced to relax the frame of FHC. For example, indigestible dextrin and certain oligosaccharides are recognized as functional ingredients that promote gut health, and they are already used in many ($>$100) FoSHU products. In such a case, food containing these ingredients at appropriate concentrations can be approved as Standardized FoSHU, because they are considered to have sufficient scientific evidence to support the claims. The 2005 revision also enabled the manufacturers to apply certain food products with less scientific evidence to the FoSHU system. Food products that do not have sufficient scientific evidence as required for the current licensing examination procedures may be approved as Qualified FoSHU if they have certain efficacy (e.g., $P<0.1$). FoSHU with Disease Risk Reduction Claims was introduced by the MHLW according to the Codex committee under the World Health Organization/Food and Agricultural Organization. Currently, food products containing calcium or folic acid are permitted to have a label identifying a disease-risk reduction, because enough scientific data for their disease preventive roles have been accumulated. Calcium (daily intake, 300−700 mg) may support healthy bones of young women and reduce the risk of osteoporosis. Folic acid (daily intake, 400−1000 μg) may reduce the risk of neural tube defect, supporting healthy fetal development in pregnant women. The FoSHU products of this category should contain calcium or folic acid within their daily intake range.

15.5 Function evaluation of FoSHU

Human testing by a third party is essential for validating the effectiveness or efficacy of FoSHU products. Positive clinical data obtained under reasonable experimental conditions strongly support the effectiveness of these products. The clinical studies should be well designed, using healthy subjects or subjects who are bordering between health and illness. Randomized placebo-controlled, double-blind trials using a sufficient number of subjects to prove significant differences are required. The test should be conducted using the food in question, and not the functional ingredient, over a reasonably long period of time (approximately 12 weeks). The existence of good systematic reviews or meta-analysis data published in scientific journals concerning the functional components may also support the experimental results.

Functional parameters that are internationally authorized, such as blood glucose, blood triglyceride, and blood pressure, should be used to evaluate the food's effectiveness. The parameters analyzed should vary depending on the health claim. For example, in the case of FoSHU products that promote bone health, changes in the bone mass and bone density are acceptable as data to be submitted, but data that only show the blood levels of collagen metabolites or osteocalcin are recognized as "insufficient." An explanation of the mechanism of action is also important to support the food's effectiveness. Mechanism studies on the physiological effect of food substances are generally performed by *in vivo* experiments using animals or by cell-based assays. The experimental data obtained from these studies should have been published in refereed journals. Information found in publications by other research groups is also useful in determining the mechanisms of action.

15.6 Safety evaluation of FoSHU

Japanese consumers are rather sensitive to food safety issues. Therefore, FoSHU products are expected to be particularly safe, and a strict evaluation system is applied to validate the safety [8,9]. A food product using a functional ingredient not yet identified as a FoSHU will be critically reviewed by the Food Safety Commission. A new type of product with a novel health claim will also be carefully examined in terms of its safety. Safety evaluations by the Food Safety Commission are a prerequisite step before the product can be approved as a FoSHU.

The required documentations include the following:

- Information concerning the methods for manufacturing, processing, and quality control of the product
- Identification of the functional component, its mechanism of action, in addition to information on its absorption, metabolism, excretion, and accumulation in the body
- Analytical data on heavy metals, agricultural chemicals, and allergic substances, as the need arises
- Historical diet uses (eating customs) in Japan and foreign countries

- Results of the oral administration tests using animals for acute, subacute (repeated dose for 28 or 90 days), and chronic toxicities (repeated dose for 1 year)
- Results of the tests for antigenicity, allergenicity, breeding toxicity, mutagenicity, and carcinogenicity, as the need arises
- Information on the possibility of plasmid transfer of antibiotic-resistant genes (when the functional ingredient is bacteria)
- Data on human studies where at least three times the minimum effective dosage has been administered
- Information on the safety of the food product when ingested by high-risk groups such as the elderly, infants, babies, and pregnant women
- Information on the safety of the food product when ingested by patients with diabetes, hypertension, hyperlipidemia, etc
- Information concerning the safety of the food product when ingested in combination with drugs

15.7 The future of functional food regulations in Japan

Twenty years have passed since the first FoSHU product was approved. However, the number of FoSHU categories is still limited to eight as shown in Table 15.1. Novel FoSHU categories such as immune-modulation, skin health promotion, anti-fatigue properties, and joint ache relief have not been approved in spite of consumer demand and efforts by the Japanese food industries. This is probably because of the existing laws and regulations concerning food and medicine, difficulties in the objective evaluation of certain food functions (lack of internationally authorized markers), or insufficient knowledge of the mechanisms of action. More basic studies need to be done to solve these problems.

A new project to establish a more reliable evaluation system for food functions was conducted from 2011 to 2013 by the Japan Health and Nutrition Food Association (JHNFA) under the sponsorship of the Japanese Consumer Affairs Agency. To construct an evaluation standard, the JHNFA extensively searched publication databases for parameters useful in validating food functions, and conducted an investigation into foreign evaluation systems as a reference. As a result, a model system to validate the effectiveness and safety of functional food ingredients on humans has been constructed. This model system has been applied to 11 ingredients that are frequently sold as health food or supplements in the Japanese market. One of them, n-3 fatty acid, has been judged to be "convincing" in its reduction in cardiovascular disease risk, lowering blood triacylglycerol levels, and relieving articular rheumatism. Other ingredients have also been suggested to be "probable" in terms of their preventive effect on various diseases and disorders. Based on the results of this project, there is a possibility of constructing a new regulation system for "evidence-based functional food," including supplements and so-called health foods.

References

[1] Arai S. Studies on functional foods in Japan — state of the art. Biosci Biotechnol Biochem 1996;60:9—15.

[2] Arai S, Osawa T, Ohigashi H, Yoshikawa M, Kaminogawa S, Watanabe M, et al. A mainstay of functional food sciences in Japan — history, present status and future outlook. Biosci Biotechnol Biochem 2001;65:1—13.

[3] Arai S, Morinaga Y, Yoshikawa T, Ichiishi E, Kiso Y, Yamazaki M, et al. Recent trends in functional food science and industry in Japan. Biosci Biotechnol Biochem 2002;66:2017—29.

[4] Swinbanks D, O'Brien J. Japan explores the boundary between food and medicine. Nature 1993;362:180.

[5] Shimizu M, Kawakami A. History and scope of functional foods in Japan. In: Losso J, Shahidi F, Bagchi D, editors. Angiogenesis, functional, and medicinal foods. New York: CRC press; 2007. p. 49—68.

[6] Arai S. Global view on functional foods: Asian perspectives. Br J Nutr 2002;88(Suppl. 2):S139—43.

[7] Shimizu T. Health claims on functional foods: Japanese regulations and an international comparison. Nutr Res Rev 2003;16:242—52.

[8] Ohama H, Ikeda H, Moriyama H. Health foods and food with health claims in Japan. Toxicology 2006;221:95—111.

[9] Yamada K, Sato-Mito N, Nagara J, Umegaki K. Health claim evidence requirements in Japan. J Nutr 2008;138:1192S—8S.

Health Foods and Foods with Health Claims in Japan

16

Hirobumi Ohama*, Hideko Ikeda* and Hiroyoshi Moriyama[†]

*Biohealth Research Ltd. [†]The Japanese Institute for Health Food Standards, Bunkyo-ku, Tokyo, Japan

16.1 Introduction

Japanese food is healthy to begin with. The presence of the four distinct seasons in appropriate temperature and humidity in Japan surrounded by sea fostered the Japanese food eating habit and created a distinctive Japanese food culture. Moreover, the availability of agricultural crops, such as soybean, mushroom, rice, wheat, tea, and marine products, which comprise seaweed, algae, bonito, bream, etc., enriched the food resources to establish the Japanese traditional food. The art of fermentation also contributed to constitute Japanese traditional foods, such as "natto" (fermented soybean with *Bacillus subtilis* var. *natto*), "miso" (fermented soybean paste), "katsuobushi" (fermented bonito), and so on utilized not only for preservation of food, but also for adding taste and health benefits.

However, after World War II the lifestyle of the Japanese people drastically changed along with rapid economic growth. Dietary patterns of the Japanese people have also diversified; it is said that the Japanese people live in a "rich food style" or "the age of overabundance." At around this time, the word, "westernization," introduced into the food style among a wide range of aged groups, provided the formation of the westernized modern dietary custom in Japan.

The changes gradually gave rise to the positioning of foods not only for a function such as nutrition, but also functions for sensory/satisfaction and health benefits [1]. From the viewpoint of sensory/satisfaction, in particular, foods based on meats, eggs, milk, butter, etc., of animal origins and processed foods such as so-called "fast food" and "instant food" were prevalent while, in recent years, public health concerns involved excess or insufficient nutritional intake, such as excessive intake of fat or imbalanced diets and also so-called "skipped meals," spreading from school children to working people and even the elderly.

Such derogatory dietary habits are, in large part, responsible for the induction of lifestyle-related diseases (LSRD), such as obesity, diabetes mellitus, high blood pressure, cerebrovascular and cardiovascular diseases, and cancer, which have remarkably increased in recent years [2]. Although causes of LSRD are multifactorial and complex, other social and cultural factors, which might have affected the increase in the incidence of LSRD, are various stresses including anxiety and mental exhaustion, habitual alcohol drinking and smoking, and the lack of exercise and/or sedentary lifestyle. It is obvious that genetic factors play a pivotal role. More recently, the term metabolic syndrome (MS), which is popular and synonymously used as a specific part of LSRD among marketers and consumers, is intensively defined as visceral obesity or excessive fat tissue in and around the abdomen with signs or complications of two or more of hyperglycemia, hyperlipidemia, and hypertension, in Japan.

At the same time, the demographic trend of "shrinking and graying" has progressively become notable and now is becoming an unprecedented population crisis, which generates economic and social impacts such as financial pressure on the government and medical or health care costs for elderly and a decline in the working population. The latest life expectancy in Japan is male and female, 79.4 and 85.90 years old, respectively (http://www.mhlw.go.jp/english/database/db-hw/lifetb11/dl/lifetb11-01. pdf). In 1980 the population of people >65 years of age represented 9.1% against the total population, whereas the population accounted for about 17.4 and 22.7% in 2000 and 2009, respectively (http://www.stat.go.jp/data/topics/topi411.htm). Ironically, the expansion of the aging population has also elevated the rate of population of LSRD associated with an increase in the number of elderly people.

Furthermore, an increase in the population of LSRD has coherently affected the mortality of the Japanese people. In fact, mortality caused by cancer and vascular diseases was expanding, while cancer is the number one cause of death in Japan [2]. According to Sugimura [3], a dietary factor is one of the main causes of cancers, suggesting that the westernization of food eating might have had some negative effect on the balance of nutritional intake of the Japanese people.

In this chapter, we describe the historical development of functional foods (FF) in Japan for health benefits of the Japanese people and how foods (FD), including health foods (HF), so-called health foods (SCHF), and foods with health claims (FHC) are regulated by multiple laws to protect consumers' misuses of FD. We also attempt to explain safety challenges for HF, which are related to safety of raw materials formulated into the products as well as the good manufacturing practice (GMP) guidelines for the products. Much of the information and data in this chapter are referred to and based on the previous work of Ohama et al. [2] with modifications and updates. However, for updating changes in the regulations related to the control of FHC and the protection of consumers, the government established the Consumer Affairs Agency (CAA) in 2011.

16.2 **Historical development of FFs**

In the 1980s, Japanese consumers began to realize the importance of maintaining and improving their health with a gradual increase in the occurrence of LSRD.

Concomitantly, intensive studies were performed on the physiological effects of various foods and their ingredients on the so-called "tertiary function of foods" as described by Arai [1]. In brief, the tertiary function of foods is defined and understood as to be directly involved in the modulation of the human physiological systems such as the immune, endocrine, nerve, and circulatory as well as digestive systems, while the primary and secondary functions are related to nutrition and sensory satisfaction, respectively.

In 1984, the term FF was first assigned in the project initiated by the Ministry of Education (presently The Ministry of Education, Culture, Sports, Science, and Technology), thus crediting Japan for the creation of the FF. The concept of FF evidently attracted the HF industry and health-conscious consumers. The government, however, prohibited the use of the word FF because the word might imply drug-like effects, which would mislead consumers with the expectation of prevention or even cure. Nevertheless, the term HF was extensively used and recognized by consumers, thereby replacing the term FF. In 1991, in the attempt to replace FF, the concept of FF was integrated into the Foods for Specified Health Use (FOSHU) system [4].

More specifically, in 1984, studies on FFs have initially been led by the Special Study Group on the Systematic Analysis and Development of Food Function under the Ministry of Education, as a part of the specified project for the investigation of the FF controlling the physiological function of a living body. The project was carried out for 3 years and then was carried out for an additional 3 years for a project in 1988, elucidating the function of foods controlling physiological activity supported by the Ministry of Education [4].

In 1996, the American Chamber of Commerce of Japan requested the deregulation of an HF system in Japan to remove an important trade barrier of dietary supplements between the United States and Japan. The regulatory system of HF (generally known as dietary supplements in Japan) is unique and rigid to imported products. As an example, orally taken drugs in the form of small round tablets and capsules were prohibited from distribution as HF under this regulation. Therefore, the petition had been accepted by the government and the deregulation was discussed in the special investigative committees on dietary supplements for the following 4 years. Based on the conclusion made by the committees, the Ministry of Health, Labour, and Welfare (MHLW) decided to frame a new regulatory system of HF called FHC and enforced the new system in April 2001. The new system of HF amalgamated with the innovated FOSHU and another new category, foods with nutrient function claims (FNFC). In the FOSHU prior to 2001, only the form of conventional foods had been permitted, where other forms were not allowed.

Although the regulatory system of FHC was successfully introduced, HF that was not regulated by the new system grew and occupied the largest segment of the HF market in Japan. The MHLW decided to reconsider the regulatory system of HF with the request of the Liberal Democratic Party in 2003. As a result of the discussion from the special investigative committee for the reconsideration, the

FOSHU system was relaxed by proposing subcategories such as the qualified FOSHU and standardized FOSHU systems to the existing FOSHU. Concurrently, the disease risk reduction claims were allowed for FOSHU products containing specified nutrients. Two nutrients, calcium and folic acid, are presently allowed for the disease risk reduction claims

It must be emphasized that there is no category or no term for supplements (irrespective of terms dietary or food) according to the regulatory or legal understanding in Japan. While foods having intake forms or shapes such as tablets or capsules are included in the category of HF, discrimination between conventional food and supplement is not legally accepted. In the FOSHU system, all of the products (~1000) are in the form of conventional foods. It is now becoming a practice not to accept FOSHU products in tablet or capsule forms attributed to the misleading of FOSHU to "drug" (DR). The aforementioned situation may often make it difficult to comprehend the Japanese regulatory system of HF for foreign food industries.

According to the investigation conducted by the Consumer Committee using consumer questionnaires ~60% of those who answered take HF to various extents (http://www.cao.go.jp/consumer/iinakaikouhyou/2012/houkoku/20105_report.html).

16.2.1 CAA

The CAA was organized on September 1, 2009, to protect and maximize consumer benefits in a broad array of challenges, including trade and labeling (http://www.caa.go.jp/en/index.html). The integration of regulations related to labeling generated a significant impact in the HF industry. Particularly, controlling and enforcing HF-related regulations under the CAA has tightened the incidence of overexpression of claims and misleading the positioning of HF products.

16.3 HF

It is necessary to summarize the market, definition, and necessity of regulatory systems for HF products in the Japanese marketplace before reviewing various regulations on the HF.

16.3.1 HF and FOSHU markets

With the introduction of the FHC system in 2001, drugs and foods were categorized by the food—drug classification as defined by the Pharmaceutical Affairs Law (Figure 16.1). Since then, the market size of FOSHU rapidly augmented and reached about $4.1 billion dollars in 2001 [2]. By 2007, it is reached about $7.1 billion with the growth rate of 173% in 6 years. In spite of the drastic increase in the number of FOSHU products in the market, SCHF products still

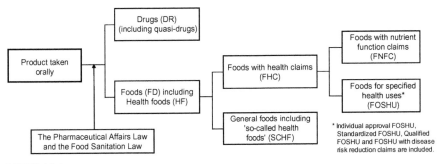

FIGURE 16.1

Positioning and classification of DR and FD, SCHF, and FHC (FOSHU and FNFC).

represent the largest market share. However, in 2006 a decline in the SCHF market size was observed, perceived as a temporary downward trend of the market. The downsizing market was caused by the following:

1. The statement of the MHLW regarding Agaricus (*Agaricus blazei* Murill) as a cancer promoter.
2. Safety concern regarding ingestible dosage of $Q_{10}(CoQ_{10})$ and soybean isoflavones as FD.

16.3.2 Implications of defining HF

According to the proposed definition of HF by the MHLW, the category of HFs is the sum of two categories, i.e., FHC and SCHF as follows:

$$HF = FHC + SCHF$$

Although the proposed definition remained ambiguous, HF has been used by consumers and the HF industry with the implication that these foods are good for health benefits. SCHF comprises a wide range of food products, which include products in tablet or capsule forms as well as other intake forms. However, SCHFs are prohibited to make any health claims, thus causing a risk of overdose, resulting in adverse reactions in this category.

16.3.3 Necessity of regulation for HF

Consequently, the MHLW held a new special investigative committee meeting on the revision of the regulatory system for SCHFs as requested by the Liberal Democratic Party. After discussion by the committee for more than a year, the MHLW announced the implementations of new systems into the previous framework of the FOSHU system. The main objective was for FOSHU to offer better opportunities to the food industry for accepting SCHFs as FOSHU under the new categories such as "Qualified FOSHU" and "Standardized FOSHU."

16.3.4 Labeling and laws

Various laws regulate labeling of HF, such as the Pharmaceutical Affairs Law, the Food Sanitation Law, and the Japan Agricultural Standards Law (Figure 16.2), among which the Pharmaceutical Affairs Law plays an essential role in controlling functional and/or efficacy claims on FD. Functional and/or efficacy claims are not allowed in principle for HF except for FHC because these claims or indications are interpreted as drug related, such as the following:

1. For diagnosis, cure, or prevention of disease in human or animal
2. Items that are intended to affect the structure or functions of the body of human or animal
3. "Hints" for drug efficacy violating the Pharmaceutical Affairs Law; these hints include:
   a. Product naming or promotional statements or phrases suggesting drugs
   b. Description of pharmaceutical ingredients
   c. Description of manufacturing process as suggesting drugs
   d. Description of origin or history suggesting drugs
   e. Reference to articles from newspapers and scientific journals or quotations from interviewing medical doctors or scientists

Claims, however, with expressions such as to maintain or promote the health condition of a healthy individual are acceptable for HF.

16.3.4.1 Use and dosage directions for HF products

Use and dosage directions of HF products must not be stated like drugs, for example:

1. Before meals, after meals or during meals
2. 2−3 times daily
3. 1−2 tablets per dose, twice a day
4. Adult 3−6 tablets a day—adjust according to condition
5. 1−2 tablets before and after meals
6. 1−2 capsules before bedtime

16.3.4.2 Dosage forms of HF products

Dosage forms or shapes, such as tablets, capsules, etc., used for HF products have been deregulated, although several special delivery forms used in drugs are still forbidden. The examples of delivery forms are ampules, sublingual tablets, the product absorbed from mucous membranes, and spray-type products into the oral cavity. On the other hand, the forms allowed for food uses include hard gelatin capsule, soft gelatin capsules, tablets, powder, liquid, or granules.

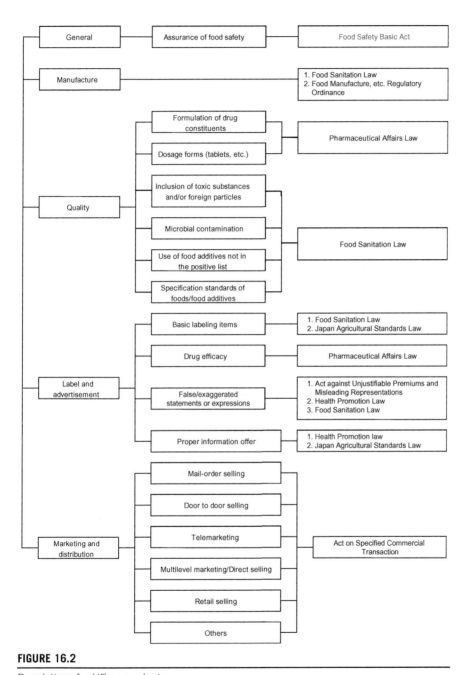

FIGURE 16.2

Regulations for HF as products.

Source: Tokyo Metropolitan Health Bureau and Tokyo Metropolitan Life Culture Bureau, 2010. Health Food Handling Manual. sixth ed, Yakuji-Nipposha, Tokyo. P.6 (in Japanese).

16.4 Regulatory systems on HF

It is preferred to have a unified law to regulate HF as categorized under the FD in safety, quality, manufacturing, labeling, efficacy, advertisement, sales, etc. Therefore, the presence of diverse laws that belong to different ministries may result in the lack of lateral communication and comprehensive control. For example, the label of an HF product carrying a drug-like effect primarily violates the Pharmaceutical Affairs Law, which is under the MHLW. The same product that promoted claims in any of the printed materials without scientific substantiations is subject to a violation ruled by the Health Promotion Law and the unjustifiable premiums and misleading representations under the CAA. However, each law has its own penalty system.

Consequently, it is important to enhance the understanding of each law bound to HF and the claims made by HF to protect consumers from being misled by HF products. The laws related to the FD category affecting the consumers and the HF industries are briefly explained. Some of the regulatory systems have been previously described [2].

16.4.1 Food safety basic act

The main objective of this law is to ensure the safety of foods for consumers by implementing and following:

1. Basic principles for the measurements of consumer health protection
2. Taking appropriate actions on stages of the food supplying process (food chain)
3. Anticipating potential health hazard causing factors based on accessible scientific evidence

The others include statements concerning responsibility of individuals and local authority, responsibility of food industry, and establishing the Food Safety Commission (FSC). The law also states that consumers are recommended to enhance understanding on food safety and to express their opinions regarding policies to ensure food safety.

The FSC plays a key role in achieving the objectives of the law, for example, by conducting risk assessments in response to the concerned minister's consultation or on his discretion, whereas the FSC was established within the Cabinet Office (http://www.fsc.go.jp/english/index.html).

16.4.2 Pharmaceutical affairs law

According to the Food Sanitation Law, any substance ingested orally is defined either as an FD or DR (Figure 16.1). A DR is strictly specified by the Pharmaceutical Affairs Law according to the items:

1. Recognized in the Japanese Pharmacopoeia
2. Other than quasi-drugs, which are intended for use in the diagnosis, cure, or prevention of disease in man or animal, and which are not equipment or instruments

3. Other than quasi-drugs and cosmetics, which are intended to affect the structure or functions of the body of man or animal, and which are not equipment or instruments

Based on the Pharmaceutical Affairs Law, two lists are issued by the MHLW for identifying DR and non-DR:

1. List of ingredients (raw materials) used as exclusively as DR (DR List)
2. List of ingredients (raw materials) used as materials not judged as DR, as long as no DR efficacy claim is made (non-DR List)

These lists are further classified into three subcategories:

1. Substances originated in plants
2. Substances originated from animals
3. Other substances such as chemicals, minerals, and other synthesized or highly purified substances obtained from natural substances (Figure 16.3)

Thus, the substances listed as non-DR may be used mainly as FD including HF, whereas the substances included in the third subcategory (e.g., chemicals) are further included as food additives when the MHLW designated the lists.

When substances or raw materials on the non-DR List are to be processed further to obtain specific ingredient(s) by means of extraction with solvents excluding water and ethanol (and probably carbon dioxide used for the supercritical fluid extraction method), the ingredient(s) obtained will be reinvestigated in the light of the criteria described below to determine whether it should be included in the DR List. This suggests that the solvents excluding water and ethanol are prohibited in the manufacturing of such ingredient(s).

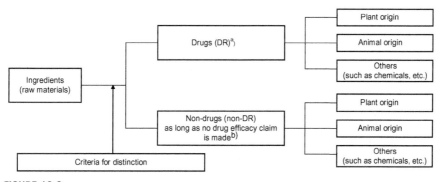

FIGURE 16.3

Distinction between DR and non-DR. [a]List of ingredients (raw materials) used exclusively as drugs and [b]List of ingredients (raw materials) used as materials not judged as DR ("non-DR") as long as efficacy claim is not made [2].

The criteria for identifying substances to be included in the DR List established by the MHLW are briefly summarized in the following:

1. Substances actually used exclusively as DRs (e.g., antipyretic, analgesic, anti-inflammatory agents, hormones, antibiotics, and digestive and other enzymes).
2. Substances such as those of plant or animal origin (including extracts) and synthetic chemicals as poisons or powerful DR (e.g., highly toxic alkaloids or toxic proteins); narcotic, psychotropic DR, or stimulant having DR-like action; and substances that are designated as DR from the standpoint of public health. However, substances that are generally eating or drinking as FD are excluded from the DR List.

Substances such as vitamins, minerals, and 23 amino acids including 4-hydroxyproline and hydroxylysine are excluded from the DR category.

To apply for a new ingredient (raw material) not in the aforementioned two lists for distinction with the intention to be categorized as FD, manufacturers or importers may request its evaluation by submitting data and documents to the Compliance and Narcotics Division, Pharmaceutical and Medical Safety Bureau of the MHLW. The data required for investigation include scientific names of the ingredients (raw materials) expressed in Latin; site of use; pharmacological and/or physiological action; any narcotic, psychotropic, or stimulant DR-like action; and any previous examples of approval as a DR in Japan or overseas to target consumers (Table 16.1). Historical uses as edible materials play a key role for the evaluation. Data and information submitted are reviewed by preferential pharmaceutical and medical authorities of the MHLW.

Furthermore, reclassifying ingredient(s) in the DR List to the non-DR List is also possible. The petition of the reclassification of ingredient(s) is accepted by the aforementioned division of the MHLW and allowed after review by an assigned panel of pharmaceutical and medical authorities.

For example, three prominent ingredients have been reclassified from the category of DR to the non-DR List. They are CoQ_{10} in March 2001, L-carnitine and its organic salts in November 2002, and α-lipoic acid (thioctic acid) in March 2004. It is apparent that some discrepancy arises in the daily doses (Table 16.2). The issue concerning the daily dose as an HF product that exceeds the dose as designed for DR was under dispute in the FSC of the Cabinet Office. Consequently, the FSC and MHLW came to a conclusion on the dosage issue of CoQ_{10}, i.e., the maximum limit is not determined; however, marketing companies are required to perform post-marketing surveillance on adverse reactions to the products.

The other aspect of the law is a role to prohibit FD products from making any drug statements as defined earlier. The law applies to labeling and advertising such as in a package label and inserts; printed materials such as brochures, leaflets, books, and booklets; and mass media such as TV, radio, newspaper, magazines,

Table 16.1 Documentations and Evaluations for New Ingredients (Raw Materials)

1. General name and family	
2. Latin name	
3. Site of use	
4. Active components or contents	
5. Toxicological data	LD_{50} (p.o., etc.; mg/kg) Other chronic toxicological data: Yes, no
6. Narcotic, psychotropic, or stimulant drug-like action	Yes, no
7. Examples of approvals as a drug	Yes, no
a. Previous examples of approvals as a drug in Japan or overseas	If Yes : country Efficacy statement
b. Traditional or historical use as a drug	Yes, no
8. Eating custom in Japan	Product form: eat raw or cooked
9. Eating custom in overseas	Yes, no If Yes: country Product form: eat raw or supplements
10. Refer to herbs, animals, or ingredients in the non-DR List and DR List	

Note: (1) List of such ingredients shall be periodically published (once a year). (2) It will be insufficient if the eating custom is limited to supplements [2].

and the Internet. The law is also extended to direct marketing such as door-to-door, mail order, and telemarketing.

Furthermore, the law restricts the formulation or adulteration of any chemicals or natural substances including herbs in the DR List. A representative example related to this law that developed into a social issue occurred in 1999 and the early 2000s when HFs for body weight control imported from China contained DR such as fenfluramine, *N*-nitroso-fenfluramine, and thyroid hormone in different concentrations, causing hepatic dysfunction and a detrimental effect on thyroid or in some cases, death of the consumers by ingesting the HF. Since then, the HF for weight control has been under the control of the MHLW. The other DRs used for weight control products often were sennosides (A and B) used as a laxative from senna (*Cassia angustifolia* Vahl and *Cassia acutifolia* Delile) leaves. The other DR found to cause adverse reactions were norephedrine and ephedrine in the 1990s and sibutramine hydrochloride monohydrate.

16.4.3 Health promotion law

In 2003 the Health Promotion Law replaced the Nutrition Improvement Law under the control of the MHLW. Its aim is "to improve the health of consumers

Table 16.2 Examples of Reclassifying from DR to non-DR

Substance (Reclassified Date as Food)	Drug	Food[a]
Coenzyme Q_{10} (March 2001)	Dosage: 30 mg/day p.o. Indication: Mild or moderate congestive cardiac failure	Generally 100–200 mg/day
L-Carnitine (November 2002)	Dosage: 30–60 mg/kg/day p.o. (as levocarnitine chloride) Indication: L-carnitine deficiency in patients with propionic acidemia or methyl malonic acidemia	Generally not more than 20 mg/kg/day[b] or 1000 mg/day[c]
α-Lipoic acid (thioctic acid) (March 2004)	Dosage: 10–25 mg/kg/day i.v., i.m., s.c. Indication: (1) supplementation of thioctic acid when highly needed (by hard physical labor), (2) Leigh syndrome, (3) poisoning deafness, noise-induced deafness	Generally 100–200 mg/day

[a]No upper limit has been decided by the MHLW.
[b]Maximum dose determined by the United States. These daily doses are suggested by the MHLW as the maximum doses in Japan.
[c]Maximum dose determined by Switzerland. These daily doses are suggested by the MHLW as the maximum doses in Japan [2].

according to established principles that support and fulfill the necessities including the nutrition improvement from the viewpoint of promoting human well-being." The law considers the rapidly growing awareness of the aging population and LSRD, in particular. The law also governs making regulations for the approval, labeling, quality assurance, ingredients, etc., of Foods for Special Dietary Uses (FSDU) and FHC products. Moreover, under the law, labeling of such products carries nutrition facts, such as energy, protein, fat, carbohydrate, sodium, and other nutrients, if required, with proper units and per amount served in accordance with the nutrition facts standards. The law prohibits any false, exaggerated, and/or any misleading statements on the product label, advertisement, and other means of communicating to consumers such as a catalog. Presently, the CAA manages whether the law functions properly for the benefits of the consumer.

16.4.4 Food sanitation law

The objective of the Food Sanitation Law is to secure the safety of FD by controlling the risk of health hazard risks caused with ingestion of FD. The law establishes the responsibilities of the government, local authorities, and FD industry.

The law also serves diversified roles on HFs as a most crucial regulatory system to ensure the safety of FDs. The law prohibits any product that may contain

FD poisonous or carcinogenic substances, non-healthy and health-hazardous substances, and additives that possess high risk factors for consumer's health to be manufactured, imported, sold, used, displayed, etc. Examples include:

1. Deteriorated or microbial-contaminated FD
2. Toxins from fungi and others, alkaloids of plant origins
3. Foreign particles such as pieces of glasses, metals, etc.

The restrictions on manufacturing and using food additives (FA) are determined by the law. The law as Ministerial Notice enables the establishment of the standards for manufacture, preservation, and use of FD, including genetically recombinant FD or FA and the specifications.

The other function of the law is to monitor whether package labeling of HF products is correctly stated and/or whether any statements are missing from the labels. For example, label statements must include:

1. Designation of the product
2. FA used
3. Preservation method
4. Expiration (consumption or quality limit) date
5. Manufacturer and address
6. Statements of genetically modified organism-containing FD and allergenic FD

The law also controls any false or misleading statements on the labels and advertising as specified by the Pharmaceutical Affairs and Health Promotion Laws. The labeling for FHC products is also controlled under the Food Sanitation Law in addition to the applicable regulations governed by the Health Promotion Law.

Moreover, the law has provisions for risk communication with consumers, positive list on pesticides remaining in FD, etc., and surveillance system for imported FD.

16.4.5 Japan agricultural standards law

The Japanese Agricultural Standards (JAS) Law follows the law concerning the standardization and proper labeling of agricultural and forestry products, stipulating product information requirements such as processed FD and others. Labeling of HF, as of processed FD, should carry information such as designation of product; list of raw materials and FA used; preservation method; manufacturer or importer name; and the address, content (net weight), expiration date, and statements for genetically modified organism-containing FD or organic FD, if applicable, and the country of origin in the case of imported FD products. If the HF products are made from organic materials, the organic JAS symbol is allowed on the labels. The Law is under the control of the Ministry of Agriculture, Forestry, and Fisheries or MAFF (http://www.maff.go.jp/e/index.html).

16.4.6 **Act against unjustifiable premiums and misleading representations and other laws**

The other laws concerning FD products include:

1. Act Against Unjustifiable Premiums and Misleading Representations
2. Act on Specified Commercial Transaction

Both laws protect consumers from the standpoint of sale, while the Act Against Unjustifiable Premiums and Misleading Representations is presently governed under the CAA. The laws extend not only FD products, but also other products and services that consumers can purchase in order to protect consumers from being misled or misinterpreting purchase motivations created by sellers of products and services, resulting in financial loss or any kind of damage caused to consumers.

For example, false and/or exaggerated promotional statements on health benefits of HF products without scientific substantiations directed to the products are strictly regulated by the following:

1. Products are excluded from the marketplace unless any rational evidence is presented within 15 days
2. Involves mostly methods or ways of selling products, for example, door-to-door, mail order, direct-marketing, telemarketing, etc.

Any violation of the Act Against Unjustifiable Premiums and Misleading Representations and/or the Act on Specified Commercial Transaction evokes serious social concerns. Furthermore, illegal methods of selling often results in negative perceptions of HF by consumers, thereby ultimately affecting the entire HF industry. Therefore the Act Against Unjustifiable Premiums and Misleading Representations and the Act on Specified Commercial Transaction play a pivotal role to control correct information and data on HF products.

16.5 **FA**

In Japan the Food Sanitation Law basically regulates uses of FA, which must be approved by the MHLW for the following purposes:

1. To maintain the nutritional quality of FD
2. To provide certain ingredients in FD for consumers who require special diets as not intended for prevention or treatments of certain diseases
3. To enhance the preserved quality and/or stability of FD or to improve the organoleptic properties without altering the nature, substantial characteristics, or quality of the FD for the benefit of consumer
4. To support the manufacture, processing, preparation, treatment, packing, transport, or storage of FD without generating any effects in the use of labile raw materials or of undesired practice or technique

The FA are required to perform rigorous tests for effectiveness, safety, and physicochemical characteristics and specifications.

The effectiveness studies required for FA such as antimicrobial, flavoring, coloring, stabilizing agents, etc., include:

1. Effectiveness and comparison in the effect(s) with other similar FA
2. Effects on nutrients of FD

Information/data or studies for demonstrating safety of FA include:

1. LD_{50} (single administration study)
2. 28 day toxicity study
3. 90 day toxicity study
4. 1 year toxicity study
5. Reproduction study
6. Teratogenicity study
7. Carcinogenicity study
8. Combined 1 year toxicity/carcinogenicity study
9. Antigenicity study
10. Mutagenicity study
11. General pharmacological studies
12. Information concerning metabolism and pharmacokinetic data
13. The daily intake of the FA
14. Information concerning safety in humans

In addition, information on the physicochemical characteristics and specifications of FA generally required are name, formulas, assay, manufacturing methods, identification and purity tests, stability, etc.

Moreover, according to the Food Sanitation Law as described earlier, FA used to prepare HF, regardless of administration, forms such as tablet and capsule or of conventional food forms must be stated on package labels.

16.6 FHC

Conventional FD such as yogurt, cooking oil, and lactic acid drink were originally in the FOSHU category, but intake forms such as tablets or capsules had not been recognized. In 2001, criteria for FOSHU were integrated into the FHC system and then revised to include capsules, tablets, or other dosage forms commonly used in supplements. The FOSHU is allowed to make the structure/function claims on the human body and the disease risk reduction claims with the approval of the CAA. Consequently, health benefits are allowed to be included on the labels of these products. The domain of claims are limited to the contents:

1. Indicating the benefits for maintaining or improving healthy condition that are clearly observed and measured (e.g., to help maintain normal blood pressure)

2. Indicating that the FD product aids to maintain or improve healthy physical condition or tissue functions (e.g., to help maintain the bowel movement)
3. Indicating that the FD product temporarily improves the physical condition, provided that it will not provide continuous or long-term effects and the expectation of such effects (e.g., appropriate to the one who feels tiredness; Figure 16.4).

16.7 FSDU

FSDU regulations were established after the enactment of the Food Sanitation Law in 1947 to fulfill the needs of the Japanese people by improving the unsatisfactory nutritional conditions after World War II. Since then, FSDUs have since been developed and diversified into different diets for subjects meeting the

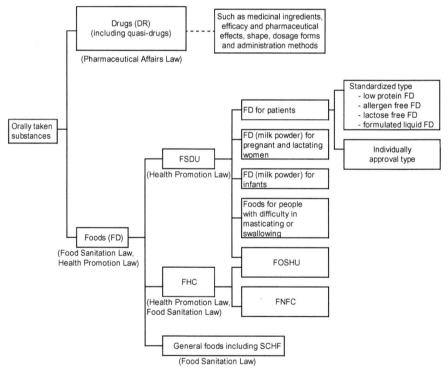

FIGURE 16.4

Positioning of DR and FD including FSDU, FHC (FOSHU and FNFC), and SCHF in accordance with the Pharmaceutical Affairs Law, the Food Sanitation Law, and the Health Promotion Law. (See text for the symbols.)

requirements of their physical conditions. Now FSDUs are based on approvals of the CAA under the Health Promotion Law. FSDUs were categorized as:

1. FD for the patients such as low-protein FD, allergen-free FD, lactose-free FD, formulated liquid FD, etc.
2. FD (milk powder) for pregnant and lactating women
3. FD (milk powder) for infants
4. FD for the people with difficulty masticating or swallowing
5. FOSHU (Figure 16.4).

16.8 FOSHU

FOSHU is a subcategory of FHC as well as of FSDU, and FNFC belongs to the other subcategory of FHC. FOSHU is separately approved by the CAA on the basis of the conclusions from deliberations in the CAA and FSC. The FNFC is another system under the regulations of HF restricted to the specified nutrients (vitamins and minerals) for which nutritional function claims are established in the CAA on the basis of well-confirmed scientific evidence (Figure 16.4).

16.8.1 Approval system for FOSHU

To obtain approvals for FOSHU products, identification of the key or decisive ingredient and elucidation of the mechanistic action of the ingredient are essential. Before a FOSHU product is ready for approval, the following requirements must be considered:

1. Improvement to dietary habits and contribution to maintaining and enhancing the state of health is expected.
2. The decisive ingredient of FD related to the claim for health-related uses should be clearly based on the medical and nutritional evidence demonstrated scientifically from clinical studies.
3. Appropriate doses of FDs or the decisive ingredient can be established on the basis of the medical and nutritional evidence.
4. Safety of FDs or the decisive ingredient can be confirmed by referring published scientific papers that are peer reviewed.
5. The decisive ingredient of FDs should be identified by employing the following methods:
   a. The physical, chemical, and biological characteristics must be clearly identified and test methods for the characteristics must be established.
   b. Quantitative and qualitative analytical methods must be established. However, the above is exempted if reasonable rationale(s) are presented.
6. It must be confirmed that the decisive ingredient does not interfere with the composition of other nutritional ingredients in FD or in the same kind of FD.
7. The ingredient needs to be included in commonly used FDs.
8. FD or the decisive ingredient should not be included in the DR List.

Effectiveness or efficacy is substantiated by elucidating the mechanism of action; establishing the dosage level; performing tests in animal and human subjects conducted under the double-blind, placebo-controlled, randomized clinical trial; and assessing the findings with the appropriate statistical analyses. Safety is confirmed by demonstrating the historical diet usages in both Japan and foreign countries, conducting *in vitro* and *in vivo* toxicity tests that include acute toxicity, 28 and 90 day toxicities, and mutagenicity as minimum requirements to ensure safety. Safety must also be clinically proven by carrying out an overdose (usually 3- to 5-fold of the established dosage level) administration study. Reproduction, chronic toxicity, antigenicity, carcinogenicity or teratogenicity, or general pharmacology studies may be required based on the outcomes of the toxicity studies or if the product contains a new key or decisive ingredient as previously not found in the list of the approved FOSHU products. Physicochemical properties of the decisive ingredient as well as the stability of the product formulated with the decisive ingredient and ingredient itself must be conducted in terms of quality.

All toxicity tests should be performed on the basis of good laboratory practice, whereas clinical studies must be approved by a committee on ethical consideration pertaining to the protection of human rights in accordance with the spirit of the Helsinki Declaration and be carried out on the basis of good clinical practice. Furthermore, the clinical studies should be conducted as randomized, placebo-controlled, double-blind trial and the results must show statistical significance against the control at a P value ≤ 0.05. The subjects for the clinical trial should be healthy subjects or subjects who are on the borderline of being healthy as judged by health testing such as blood cholesterol, triglycerides and glucose, and blood pressure (Table 16.3). The findings from clinical studies are crucial in the determination of an approval for the product as FOSHU.

Accurate determination of an active constituent is essential as part of the requirements for FOSHU approval. An established manufacturing procedure of a FOSHU product is required. Labeling, shelf-life, formulation, dosage, nutritional information, precautions, etc., of the intended product for FOSHU status should be clearly stated.

16.8.2 Number of FOSHU

As of January 2013, 1030 products have been approved as FOSHU. As shown in Table 16.4, decisive or functional ingredients of FOSHU products are listed according to the categories of health claims. About 36.6% of FOSHU products claim "maintaining gastrointestinal condition." Oligosaccharides, *Lactobacillus*, *Bifidobacterium*, and dietary fibers are major ingredients for the activities in this category. The other category, representing 23.2% of FOSHU products, makes health claims that they are good for those concerned about high serum

Table 16.3 Health Conditions and Basic Criteria for Clinical Studies Required for FOSHU as a General Rule

Clinical Study	Health Conditions and Basic Criteria
Cholesterol related	To perform a trial for more than 12 weeks with subjects having total blood cholesterol level ranges 200–240 mg/dL and/or LDL cholesterol level ranges 120–160 mg/dL
Triglyceride(TG) related	To perform a trial with subjects having slightly higher TG level (120–200 mg/dL) than normal upper level To perform a trial for more than 12 weeks when the effectiveness of lowering TG level at fast is studied
Blood pressure related	To perform a trial for more than 12 weeks with subjects with high within normal blood pressure levels and/or patients having low to moderate risk of mild hypertension
Blood glucose related	To perform a trial for more than 12 weeks with subjects having blood glucose level in threshold levels at fasting or by glucose tolerance test according to the standards of the Japan Diabetes Society
Body fat related	To perform a trial for more than 12 weeks with subjects having obesity level 1(25 ≤ BMI<30) in accordance with the standards of Japan Society for the Study of Obesity or subjects having comparatively high BMI values within normal range To perform the above trials with subjects stratified by sex and ages

cholesterol/triglycerides and good for those concerned about body fat. Chitosan, soybean protein, phytosterol, dietary fibers, EPA and DHA, etc., are the major active ingredients in this category.

16.8.3 Qualified FOSHU and standardized FOSHU

With the relaxation of the FHC system in February 2005, the qualified and the standardized FOSHU systems were introduced into the framework of the previous FOSHU system, making it easier for applicants to obtain approvals for distributing FOSHU products in the marketplace. Concomitantly, the "disease risk reduction claims" were added to the existing items approved for FOSHU, reflecting the decision of the Codex Alimentarius. The schematic illustration of the new FHC system is exhibited in Table 16.5 and differences between the existing FOSHU and qualified FOSHU systems are shown in Table 16.6. The major differences include:

1. Requirement for elucidating the mechanism of a decisive ingredient
2. Acceptance of P values for statistical analysis

The elucidation of the mechanism in the action of the key or decisive ingredient is not always required for the Qualified FOSHU. For the statistical analysis of the data from clinical studies, the significant difference against the control may

Table 16.4 FOSHU Products (955 Products Approved as of April 1, 2011)

Health Uses	Food Category	Decisive Ingredients (Example)	Model Claim, Statements	Number Approved
GI function	Table sugar	Oligosaccharides	Helps maintain good GI condition Helps improve bowel movement	350 (36.6%)
Cholesterol level	Powdered soft drink	Chitosan	Helps lower cholesterol level	142 (14.9%)
Triacylglycerol Body fat	Refined oil	Medium chain fatty acids	Helps resist body fat gain	70 (7.3%)
	Oolong tea	Polyphenol	For those concerned about body fat	
Blood pressure	Instant powder soup, candy	Peptides	For those with high blood pressure	120 (12.6%)
Bone	Soft drink	Soy isoflavone	Promotes calcium absorption Supports bone health	53 (5.5%)
Teeth	Chewing gum	Mixture of xylitol, calcium monohydrogen phosphate, and Fukuronori extract	Helps maintain strong and healthy teeth	79 (8.3%)
Blood glucose level	Soft drink, instant miso soup	Indigestible dextrin	For those concerned about blood glucose level	141 (14.8%)
Total				**955**
Disease risk reduction	Fish meat sausage	Calcium	(*1)	17 (*2)

*(1) The claim is shown in the text.
*(2) Number approved is as of October 31, 2011.
Data by CAA.

be acceptable with P values<0.1 and the randomized controlled study is not essential for the Qualified FOSHU products. Other data and documents are necessary for the approvals of the Qualified FOSHU products in accordance with the existing FOSHU system, such as a clinical study using higher dosages to confirm adverse effects in the case of overdosing.

Table 16.5 Revision of FHC in February 2005

| | FHC | | |
	FNFC	FOSHU	
Type of system	Standardized	Individual approval FOSHU Standardized FOSHU	Qualified FOSHU
Functional component(s)	Nutrients (12 vitamins and 5 minerals)	Nutrients and other food ingredients	Nutrients and other food ingredients
Claims	Nutrient function claims (structure and function claims)	Specified health claims (structure and function claims) Disease risk reduction claims	Specified health claims (structure and function claims)
Ranking of scientific evidences	A	A, B	C

Note: Ranking of scientific evidences is defined as follows: A, evidence is both medically and nutritionally established from the scientific point of view; B, evidence is confirmed at the level previously required for the approval of existing FOSHU; C, evidence is not established but the efficacy is suggested [2].

Table 16.6 Differences in Criteria between the Existing and Qualified FOSHUs

| Clinical Study | | Randomized Controlled Trial (RCT) | | Non-Randomized Controlled Trial $P < 0.05$ |
		$P < 0.05$	$0.05 \leq P < 0.10$	
Mechanism of action	Clear	Existed FOSHU	Qualified FOSHU	Qualified FOSHU
	Unclear	Qualified FOSHU	Qualified FOSHU	—

Note: Control, placebo; subjects, healthy subjects; compliance with Helsinki Declaration.

The Standardized FOSHU system was also enforced, allowing claims, such as "maintaining good gastrointestinal condition" and "good for those concerned about blood glucose level after meals." The functional ingredients for the former claim include:

1. Indigestible dextrin (3—8 g/day)
2. Polydextrose (7—8 g/day)
3. Xylo-oligosaccharide (1—3 g/day)

4. Fructo-oligosaccharide (3–8 g/day)
5. Soybean oligosaccharide (2–6 g/day)
6. Isomalto-oligosaccharide (10 g/day)
7. Lacto-fructo-oligosaccharide (2–8 g/day)
8. Galacto-oligosaccharide (2–5 g/day)
9. Partially hydrolyzed guar gum (5–12 g/day)

and the ingredient for the later claim is indigestible dextrin (4–6 g/once/day).

The FOSHU (totaling more than 100 products) containing the aforementioned ingredients, having sufficient scientific evidence to support their claims, were approved as Standardized FOSHU products. Thus, clinical studies except for those evaluating the safety are not required for approval. In fact, to classify the present FOSHU claims to the Standardized FOSHU, requires the following criteria:

1. Over 100 kinds of FOSHU products with same decisive ingredients are sold in the market
2. Such FOSHU products need to be in the market at least for 6 years and also more than two companies have to distribute the products that belong to the same claim category.

16.8.4 **FOSHU and the Japanese traditional diets**

Some of the FOSHU products were based on FD materials used in the Japanese traditional diet. For example, fermented soybean (natto), which contains vitamin K_2 (menaquinone-7) that helps the absorption of calcium, was approved as a FOSHU product [5,6]. Soybean is a traditionally natural crop used as a source for fermentation to yield miso, soy sauce, and natto. Soybean isoflavone was shown to be effective in the improvement of mineral absorption [7]. Soy protein as the decisive ingredient proved to control serum cholesterol levels [8,9].

In addition, two soybean isoflavone products in non-conventional form were approved as FOSHU. The issue of maximum upper intake level, concerning safety, however, was raised and discussed at FSC. As a result, an intake of 70–75 mg of aglycone a day was established as the upper limit that was safe on the basis of total intake of isoflavone from daily meals. Since the mean value of soybean isoflavone intake as aglycone from FD was known to be 40 mg for the Japanese people, 30 mg a day was determined as safe for the upper limit intake of aglycone for FOSHU and SCHF products in dosage forms such as tablets and capsules (http://www.fsc.go.jp/hyouka/hy/hy-singi-isoflavone_kihon.pdf).

The other Japanese traditional FD called katsuobushi, or fermented bonito, contains peptides shown to lower blood pressure [10], and made the health claim of "good for those having relatively high blood pressure." Tea polyphenols are also decisive ingredients for "helping maintain strong and healthy teeth" [11]. We expect additional claims will be adopted as FOSHU products based on Japanese traditional diets.

16.8.5 **FOSHU and botanicals**

A few botanical products have been hitherto approved as FOSHU products. These include leaves of guava (*Psidium guajava* L.) as polyphenols for the decisive ingredient claiming good for those concerned about blood glucose level and green tea (*Camellia sinensis* L.) with catechin as the key ingredient claiming it as good for those concerned about body fat. The other botanical species approved for FOSHU is the leaves of Gutta-percha tree (*Eucommia ulmoides* Oliv.) containing geniposide as the key ingredient claiming it as "good for those concerned about relatively high blood pressure." The aforementioned FOSHU products are in the form of tea. We expect FOSHU products based on botanicals to be listed in the non-DR List and confirmed safety and effectiveness will become more available to consumers for maximizing the FOSHU system.

16.8.6 **Disease risk reduction claims**

According to the decision by the Food and Agricultural Organization/World Health Organization Codex Alimentarius Commission, the MHLW introduced the disease risk reduction claims to the existing FOSHU system. The disease risk reduction claims are currently limited to two ingredients, calcium and folic acid, specifying the minimum and maximum limits of daily intakes. The selection of ingredients for the disease risk reduction claims should be entirely based upon concrete scientific evidence, which is comprehensively accepted by scientists. When disease risk reduction claims are expressed on a package label, the disclaimers should be also stated on the same label.

16.8.6.1 Calcium
Daily intake of calcium from the FOSHU products should be in the range of 300−700 mg. Labeling should include the following: the product contains adequate calcium, and intake of a proper amount of calcium contained in healthy meals with appropriate exercise may support the health of the bones of young women and reduce the risk of osteoporosis in the aged.

16.8.6.2 Folic acid
Daily intake of folic acid from the FOSHU products should be between 400 and 1000 μg. Labeling includes the following: the product contains adequate folic acid, and healthy meals containing an appropriate amount of folic acid may help women bear healthy babies and reduce the risk of neural tube defects, such as spondyloschisis.

16.9 **FNFC**

FNFC allows for the nutrient function claims that were enforced by the MHLW, but now it is authorized by the CAA according to the fundamental and decisive scientific evidence. The claims are standardized according to the conclusions derived by the authorities of the CAA. Nutrients acceptable as FNFC must be essential for fundamental activities of human life, which are supported by scientific evidence and extensively acknowledged medically and nutritionally. To make claims about FNFCs, the minimum and maximum daily intakes of individual nutrients are determined as the standard of daily dosage by the CAA. Disclaimers are also required to be placed on the label according to the individual ingredients. This category was enforced since April 1, 2001, under the MHLW.

Twelve vitamins and β-carotene and five minerals are adopted in this category where an individual nutrient function claim is determined. Table 16.7 summarizes these vitamins and minerals. Specified dosage levels of the vitamins and minerals of FNFC and the claims allowed are listed in Tables 16.8−16.11. The labels of FNFC contain statements, such as "Intake of excessive quantities of this product does not heal the illness or improve health. You must follow the dosage as directed." Further statements include: "This product was not investigated individually by the Minister of the CAA unlike FOSHU."

16.10 Safety

Safety is of the utmost importance for HF consumers and manufacturers. The manufacturers, in particular, are expected to take careful action in accordance with the various established regulations and guidelines coupled with upcoming regulations in order to secure the safety of HF products. Consequently, the industry is fully responsible for making every effort and contribution to the health benefits of the consumers.

16.10.1 Risk analysis by the FSC

The occurrence of bovine spongiform encephalopathy in Japan was a major determinant for the establishment of the FSC based on the Food Safety Basic Law

Table 16.7 Vitamins and Minerals Permitted and Not Permitted as FNFC

Permitted (17)	Twelve vitamins	A (and β-carotene), B_1, B_2, B_6, B_{12}, pantothenic acid, biotin, nicotinic acid/nicotinamide, folic acid, C, D, E
	Five minerals	Iron, calcium, copper, zinc, magnesium
Not permitted (8)	No deficiency in Japan	Vitamin K, phosphorus, potassium
	Data not available for the calculation of nutritional parameters based on the national nutritional survey	Iodine, manganese, selenium, chromium, molybdenum

Note: Claims are determined by the CAA for individual nutrient based on the Significant Scientific Agreement Disclaimer that is required for description [2].

Table 16.8 Standards of Daily Dosage of FNFC (Revised in July 2005)

	Vitamins			
	Niacin (mg)	Pantothenic Acid (mg)	Biotin[a] (µg)	Vitamin A[b]
Maximum limit	60	30	500	600 µg (2000 IU)
Minimum limit	3.3	1.65	14	135 µg (450 IU)

	Vitamins			
	Vitamin B_1 (mg)	Vitamin B_2 (mg)	Vitamin B_6 (mg)	Vitamin B_{12} (µg)
Maximum limit	25	12	10	60
Minimum limit	0.30	0.33	0.30	0.60

	Vitamins			
	Vitamin C (mg)	Vitamin D	Vitamin E (mg)	Folic acid (µg)
Maximum limit	1000	5.0 µg (200 IU)	150	200
Minimum limit	24	1.50 µg (60 IU)	2.4	60

[a]Note: Biotin is permitted only for FHC.
[b]β-Carotene as the precursor of vitamin A can be approved as the FNFC of Vitamin A source. In that case the maximum limit is set at 7200 µg the minimum limit is set at 1620 µg [2].

(http://www.fsc.go.jp/english/index.html). To combat food hazards, the FSC is responsible for risk assessment and for adopting systems that are implemented independently from risk management, while the MHLW and MAFF are in charge of risk management. Furthermore, the FSC evaluates all filed documents related to the safety of FOSHU. Also, the FSC performs analysis on hazards induced by FDs in the market.

16.10.2 General foods and safety concern

SCHF is included in the category of the general FD (Figure 16.1). FD in the form or shape of tablets, capsules, and others generally used in the category of supplements in other countries are included in the categories of FNFC and SCHF in Japan. Most of the herbal or botanical products are included in the SCHF and a very few are now found in the FOSHU category. Only statements on the nutritional

Table 16.9 Standards of Daily Dosage of FNFC (Revised in July 2005)

	Minerals	
	Calcium (mg)	Iron (mg)
Maximum limit	600	10
Minimum limit	210	2.25

	Minerals		
	Zinc[a]	Copper[a]	Magnesium[a]
Maximum limit	15 mg (UL[b] minus maximum amount of intake from conventional foods)	6 mg (UL minus maximum amount of intake from conventional foods)	300 mg (calculated by modifying the UL set by the United States comparing the weight difference of the average American and Japanese)
Minimum limit	2.10 mg (30% of NRV) [c]	0.18 mg (30% of NRV)	75 mg (30% of NRV)

[a]Zinc, copper, and magnesium were added in 2004.
[b]UL, tolerable upper intake level.
[c]NRV, nutrient reference value [2].

Table 16.10 Nutrient Function Claims of Vitamins

Vitamin	Nutrient Function Claims
Niacin	Healthy maintenance of skin and mucosa
Pantothenic acid	Healthy maintenance of skin and mucosa
Biotin	Healthy maintenance of skin and mucosa
Vitamin A[a]	Maintenance of eyesight at night Healthy maintenance of skin and mucosa
Vitamin B_1	Produces energy from carbohydrates; healthy maintenance of skin and mucosa
Vitamin B_2	Healthy maintenance of skin and mucosa
Vitamin B_6	Produces energy from protein; healthy maintenance of skin and mucosa
Vitamin B_{12}	Helps in the formation of red cells
Vitamin C	Healthy maintenance of skin and mucosa; helps with the process of anti-oxidation
Vitamin D	Accelerates the absorption of calcium in the bowel; helps with the formation of bones
Vitamin E	Prevents the internal lipid from oxidizing by the process of anti-oxidation; helps with the healthy maintenance of cells
Folic acid	Helps in the formation of red cells Contributes to the healthy growth of the embryo

[a]β-Carotene as the precursor of vitamin A can be approved as the FNFC of vitamin A source. In that case the maximum limit is set at 7200 μg, the minimum limit is set at 1620 μg [2].

Table 16.11 Nutrient Function Claims of Minerals

Mineral	Nutrient Function Claims
Calcium	Necessary nutrient for the formation of the bones and teeth
Iron	Necessary nutrient for the genesis of the red cell
Zinc	Necessary nutrient for keeping the normal condition of degustation Helpful nutrient for the healthy maintenance of skin and mucosa Necessary nutrient for maintaining the normal vital activity participating in the metabolism of proteins and nucleic acids
Copper	Helpful nutrient for the formation of the red cell Helpful nutrient for the normalization of the function of various endogenous enzymes and for the formation of the bones
Magnesium	Necessary nutrient for the formation of the bones and teeth Necessary nutrient for the normalization of the function of various endogenous enzymes; helps energy generation and normalization of blood circulation

From [2].

content are accepted on the labels of products in the general food category and any functional claim is prohibited. As stated before, SCHF is regulated by various and separate laws, which are not included in the framework of the specific laws. Moreover, most of the HF products with dosage forms of tablets or capsules included in the category of SCHF may increase the risk of overdose, resulting in adverse reactions, due to the presence of highly concentrated active ingredients. In fact, some accidental events such as liver disorders have occurred in the past, which were thought to be relevant to the intake of SCHF products. Therefore, efficient countermeasures are required to prevent these accidents from occurring.

On the other hand, for example, cases of potent adverse reactions have been reported in which DR ingredients, etc., were intentionally formulated into FD materials to enhance efficacy. In such cases, products with the confirmed presence of DR ingredients are no longer called and sold as SCHF, but are named, instead, as "non-approved and/or non-permitted DR." Moreover, such product names and DR ingredients are made available to the public. The reported cases often involve privately imported products, including appetite suppressants, tonics adulterated with sildenafil as a DR ingredient, etc.

In 2005, the MHLW announced two guidelines to ensure safety and to guarantee the quality of HF products with dosage forms of tablets, capsules, and related forms:

1. GMP guideline
2. Guideline for the self-investigation of safety of raw materials included in HF products in capsules, tablets, or other forms

Consequently, such guidelines that control requirements for raw materials in GMP and safety assurance may be practiced in other countries.

16.10.3 GMP guideline for HF products

GMP guidelines have been established by the MHLW based on that for DR. The GMP introduced in HF according to the governmental guideline should be performed with their own free will. However, the government strongly recommends the GMP system in the product manufacturing lines for HF. The guideline requires that GMP is not only for finished products, but also for raw materials, including imported raw materials and finished products.

Presently, two organizations, the Japan Health and Nutrition Food Association (JHNFA; http://www.jhnfa.org/) and the Japanese Institute for Health Food Standards (JIHFS; http://www.jihfs.jp/), have established their own GMP regulations and introduced a certifying system for manufacturing facilities. Furthermore, JIHFS and JHNFA confer the products manufactured in the facility certified by their GMP auditing by affixing a GMP certification mark on the product. Since 2012, to ensure the quality of HF products, the MHLW (http://www.mhlw.go.jp/english/) recommends placing the certification mark on the GMP manufactured products (http://www.mhlw.go.jp/topics/bukyoku/iyaku/shoku-anzen/dl/kenkou_shokuhin_gmp.pdf).

16.10.4 Guideline for self-investigation of the safety of raw materials

Raw materials used for manufacturing HF are required to be investigated for safety according to the guideline established by the MHLW. FD products with dosage forms like tablet, capsule, powder, or liquid are subjected to "self-investigation" of the safety of raw materials, in particular, when raw materials are processed by using the methods of extraction, fractionation, purification, and chemical reaction. Raw material manufacturers as well as distributors of finished products are required to evaluate the safety on the basis of the self-investigation system of the raw materials.

The self-investigation of the safety of raw materials is accomplished according to the following steps established by the MHLW as a model system:

Step 1. To identify all raw materials in a finished product.
Step 2. To define raw materials according to the DR or non-DR List.
Step 3. To confirm the identity of an individual raw material by employing reasonable techniques such as profiling and DNA analyses and morphologic characterization or to confirm if cultivation of raw material is carried out under voluntary good agricultural practice system, etc.
Step 4. To confirm if it is equivalent to the existing materials used as ingredients in the conventional FD generally distributed.
Step 5. To study information regarding safety data of the raw materials obtained from Chemical Abstract, PubMed, and RTECS, etc., if reliable data regarding safety, toxicology, and epidemiology are obtainable.

Step 6. To identify undesirable substances (e.g., alkaloids, toxins, hormones, neurotropic, carcinogenic, teratogenic, genotoxic, and other toxic substances and also substances having closely related chemical structures to those of the aforementioned undesirable substances) from the published papers. If sufficient information regarding the previously mentioned substances is not available, then the raw materials should be analyzed to confirm if undesirable substances are present.

Step 7. To perform toxicity studies such as 90 days repeated subchronic toxicity study and *in vitro* genotoxic studies that are required, although this step is applicable only when safety cannot be confirmed by not obtaining data from Step 6.

Step 8. To conclude by certifying the absence of impurities such as heavy metals and microorganisms is required and the manufacturing process is preferably controlled under a proper GMP system.

16.11 Discussion

After the United States, Japan possesses the second largest HF market in the world, which is almost equivalent to the EU market. In fact, the market of HF without FOSHU reached ~$11.9 billion in 2012 (Figure 16.5) and the total market achieved was ~$17.7 billion in 2011. The FOSHU market represents about 32.9% of the total HF market (Figure 16.6). The growth of the HF market has remarkably accelerated since 2001, which was presumably reflected by the deregulations that the Japanese government implemented (Figure 16.5). Therefore, Japan is one of the few countries in the world to attain a large HF market. On the other hand, the systems that the Japanese government developed for HF differ from the systems in Europe, the United States, and various Asian countries. The most remarkable difference is that the positioning of supplements is not legally defined as a system.

The systems of FHC and SCHF are recognized in Japan as frameworks used to classify and determine the labeling of functional claims. Furthermore, both FOSHU and FNFC are allowed for specific functional claims that were at first approved by the Minister of the MHLW and recognized as an exceptional case of the Pharmaceutical Affairs Law. While SCHF is solely positioned as general FD and is distributed in the marketplace, the products are strictly under the control of the Pharmaceutical Affairs Law and are never allowed for any functional claims to be placed on the labels.

HF is classified into FHC and SCHF by the approval of functional claims. Thus, the ingredients, which are occasionally missing from conventional FD or whose physiological and nutritional functions are not expected from general FD, are provided as supplements in dosage forms such as tablets or capsules. By maximizing the use of these supplements, the needed ingredients become available

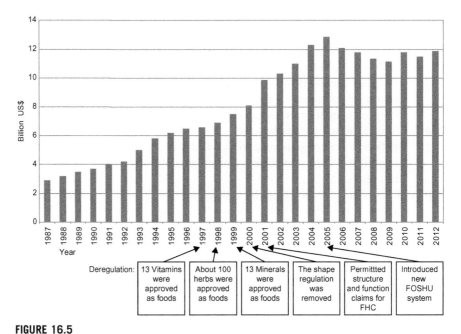

FIGURE 16.5

The changes in retail-sales based market size of HF without FOSHU in Japan.

Source: Personal communication from the Health Industry News, CMP Japan Co., Ltd.

regularly, quantitatively, and continuously for the health benefits of consumers. Unfortunately, the concept or system that distinguishes HF by dosage form is absent in the regulatory system in Japan.

While major foreign countries have attempted or are in the process of establishing a system for supplement, some countries have already completed the system. The establishment of such a system will be crucial for the manufacturing of supplements with respect to evidence-based efficacy, safety, and quality assurance. Although the advanced approval system for FOSHU is present in Japan, the law that specifically regulates SCHF is totally absent. Thus, the regulations are dependent on various laws. Because of the nonspecific regulatory system for the category of SCHF, the products under the category are not allowed to make any efficacy statements on the labels; rather, they are required to handle the issues of safety and quality comprehensively. Furthermore, the HF industry is forced to market SCHF products without mentioning the functions of active ingredients on the labels as well as any advertisement directed to the products, since functional statements are prohibited in this category. In addition, the concerns raised under such circumstances are that, although the majority of products in tablet and capsule forms are distributed as SCHF products, they are infrequently sold like FOSHU products. Since SCHF differs from FOSHU due to the rigorous approval systems, SCHF products that satisfy the legal requirements, such as of the

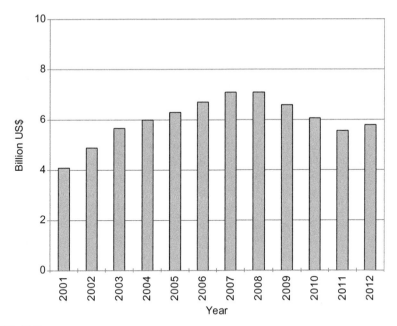

FIGURE 16.6

The changes in retail-sales based market size of FOSHU products in Japan.

Source: Personal communication from the Health Industry News, CMP Japan Co., Ltd.

Pharmaceutical Affairs Law, Food Sanitation Law, and Health Promotion Law and Act against Unjustifiable Premiums and Misleading Representations, based on voluntary judgments, may be freely available without acquiring any approvals from the Japanese government.

With the distribution of HF products in the form of tablets and capsules, specifically in SCHF, overdosing is a major risk that causes crucial problems such as adverse reactions. In addition, exaggerated and false advertising, contamination with toxic substances, and drug interactions are possibly found in the SCHF products, which result in legal and often social problems. Serious HF-related health hazards that have occurred among Japanese consumers have been attributed to SCHF products. This problem situation suggests the need for the establishment of a more stringent law regulating supplements as practiced in many other countries.

The MHLW has shown a strong concern and presented countermeasures. Paying much attention to HF products in the form of tablets or capsules, the MHLW has introduced two guidelines for the management of the manufacturing process with respect to GMP and the self-investigation of the safety of raw materials for the HF industry, requiring industries that are responsible for assuring the safety and quality to aggressively resolve the issues. The primary responsibility for securing the quality and safety of HF is in industries according to the Food Sanitation Law; however, following the guidelines is voluntary for industries.

GMP as a standard for managing the manufacturing process is demanded not only for finished products, but also for raw materials and for imported raw materials and finished products. The two guidelines are considered to be unique and important from the global point of view. Also, in the guideline assuring the safety of raw materials is like a "decision tree," which is a highly sophisticated scheme and highly regarded internationally.

As explained earlier in this chapter, the absence of a regulatory system for supplement implies that the system is not accepted in Japan, even though the significance of supplements has been well recognized globally. On the other hand, the Japanese government has begun to direct attention to the important issue of the products in the form of tablets, capsules, etc. Also, we anticipate the necessity of supplement products to be dealt with and managed by an independent legal system.

Although the Japanese government has presently promoted the systemization of HF from an independent standpoint apart from the global movement of supplement, we ought to plan to establish an exclusive regulatory system for supplements in light of international sharing of information and data for efficacy and safety, global harmonization of the methodology for quality management, and promotion of international trade.

It is important to recognize that Japan is required to show an aggressive position toward the establishment of the regulatory system for supplements to enhance the understandings of the role that supplements play in the midst of the global harmonization, resolving various health concerns such as promoting health and reducing disease risks as well as improving welfare of all mankind. We also think that the roles that Japan plays are significant as proven in developing the concept of the functionalities of FD and presently proposing the new guidelines for safety and quality management by employing an advance method.

The newly introduced system added the qualified and standardized FOSHU. Such "deregulations" are intended to resolve various issues, including the decrease in the number of illegal labeling, advertising, and products with poor quality by bringing HF products under the new FOSHU system. Since regulatory systems for HF in Japan have become rather complex compared to the systems in other countries, the MHLW strongly recommends consumer education concerning roles that the HF products play and fostering supplement advisers, in view of safety and efficacy.

As mentioned above, the Japanese government is concerned about consumer overdosing on HF products on the market and is enforcing HF companies to perform a series of safety tests on their products to ensure the consumer health benefits. However, from the viewpoint of the HF industry, small- and medium-sized companies may suffer a heavy cost burden for conducting safety and relevant tests to ensure the safety of the products, as well as introducing the GMP regulating standards. As a result, we anticipate that some of the companies may have to withdraw from the HF market, while large food or pharmaceutical companies will have opportunities to enter and expand the businesses in the HF market. In fact, most of the FOSHU products approved are from the major companies with good research and development. To surmount such competition is obviously vital. Thus, various organizations including universities/scientific institutes, local commercial organizations, and even the government vigorously collaborate with the small- and medium-sized HF companies, particularly in the area of safety and proper marketing of HF, which will ultimately benefit the consumers and industry and prevent consumers from buying unsafe HF products.

As a basic rule, not all ingredients listed in the non-DR List are safe. To reduce any health hazards to consumers, the effort of manufacturers alone is not sufficient. Of course, consumers need to have accurate information on the FD or HF products that they ingest, although sellers and mass media such as TV, newspapers, magazines, books, etc., ought to provide them with proper information and data, in particular, health-oriented TV programs that play a critical role; for example, the incidence of a popular TV program that took place in January 2007 caused social concern and questioned morals of the mass media by faking scientific evidence for health benefits of FD [12]. In this case, audience ratings were the priority, while the correct information on FD was not conveyed to the audience, including HF users.

To provide reliable data on effects and safety of HF ingredients to consumers and any updated issues including HF related to health hazards on HF, the National Institute of Health and Nutrition (http://www.nih.go.jp/eiken/english/index.html) and Tokyo Metropolitan Health Bureau and Life Culture Bureau (http://www.metro.tokyo.jp/ENGLISH/index.htm) should be included. It is a challenge to make consumers aware of the presence of such websites and resources of information and data on HF.

Appropriate usage and safety of such HF products should be explained by sellers to consumers who are also required to read the label statements carefully and to obtain product information/data. If this cannot be accomplished, a system to support consumer awareness of HF safety and usage should be performed by educating and/or training by health professionals, for example, certified HF specialists, nutritionists, pharmacists, and/or even physicians by employing all communication means available such as radio, TV, drugstores, clinics, etc.

Additional concern is abuse or false information and data from websites on HF, which should be strictly monitored by the government; hence, the government supports the establishment of websites offering accurate information and data to consumers for their health benefits and risks, capitalizing on the aforementioned organizations.

Recently, diversified HF products have been reported by health organizations for the interaction of HF product(s) with over-the-counter and/or prescribed DR that consumers ingest. Since HF products are considered as FD, no clear dosage direction (unlike DR) is stated on labels, as discussed earlier. The overdosing, concentrated ingredients, and simultaneous intake of DR causing health hazards are distinctly controlled by regulations based on the accumulation of scientifically substantiated data on negative HF and DR interaction.

The other measure used to protect health benefits of consumers is post-surveillance of HF, which includes SCHF and FHC. Although the government has initiated the development of methodology for the post-surveillance of HF, an effective method has not yet been under investigation. Also, the MHLW recognizes that procurement and accumulation of information and data on health hazards including FD—DR interaction are urgently required. The present situation is to respond to reports on adverse reactions generated by medical centers and institutes and to handle each case by releasing the reported cases to the public, leading to prohibition of selling these products.

It is our hope that all consumers have health benefits from proper and well-regulated HF products and go on with their daily lives protected from various diseases, which will also help to reduce medical costs incurred to the government.

Acknowledgments

We thank Mr. J. Makino of CMP Japan Co., Ltd. for providing invaluable information. We are also grateful to Mr. S. Ando of Biohealth Research Ltd. for collecting information of regulation on HF.

References

[1] Arai S. Studies on functional foods in Japan. Biosci Biotechnol Biochem 1996;60:9—15.

[2] Ohama H, Ikeda H, Moriyama H. Health foods and foods with health claims. Toxicology 2006;221:95—111.

[3] Sugimura T. Food and cancer. Toxicology 2002;181-182:17—21.

[4] Hosoya N. Health claims in Japan-foods for specified health uses and functional foods. J Nutr Food 1998;1:1—11.

[5] Yamaguchi M, Taguchi H, Gao YH, Igarashi A, Tsukamoto Y. Effect of vitamin K_2 (menaquinone-7) in fermented soybean (natto) on bone loss in ovariectomized rats. Bone Miner Metab 1999;17:23—9.

[6] Tsukamoto Y, Ichise H, Yamaguchi M. Prolonged intake of dietary fermented soybeans (natto) with reinforced vitamin K_2 (menaquinone-7) enhances circulatory γ-carboxylated osteocalcium concentration in normal individuals. J Health Sci 2000;46:317—21 (abstract in English).

[7] Fujikura K, Chiba Y, Yano H, Kobayashi C. Effect of soft drink containing soy isoflavone on urinary bone resorption marker (deoxypyridinoline) in middle aged women. J Nutr Food 2003;6:69—79.

[8] Ishikawa T, Jun CJ, Fukushima Y, Kegai K, Ishida H, Uenishi K, et al. Effect of soy protein drink on serum lipids in subjects with high and normal cholesterol level. J Nutr Food 2002;5:29—40 (abstract in English).

[9] Anderson JW, Johnstone BM, Cook-Newell ME. Meta-analysis of the effects of soy protein intake on serum lipids. N Engl J Med 1995;333:276—82.

[10] Fujita H, Yamagami T, Ohshima K. Effects of an ace-inhibitory agent, katsuobushi oligopeptide in the spontaneous hypertensive rat and in borderline and mild hypertensive subjects. Nutr Res 2001;21:1149−58.

[11] Sakanaka S, Aizawa M, Kim M, Yamamoto T. Preventive effect of green tea polyphenols against dental caries in conventional rats. Biosc Biotech Biochem 1992;56:593−4.

[12] Cyranoski D. Japanese TV show admits faking science. Nature 2007;445:804−5 February 22.

Complementary Medicine Regulation in Australia

17

Dilip Ghosh

Nutriconnect, Sydney, Australia

17.1 Introduction

Throughout the world, the health care environment is changing rapidly and the past decade has seen increasing attention given to Complementary and Alternative Medicine (CAM). The CAM use is undoubtedly increasing among Australians. Over the past 20 years, Australians have adopted complementary medicines (CM) and around half of those questioned in community surveys had taken one or more CM in the previous 12 months [1,2]; however, the evolution of CAM regulation has failed to keep up with this surge in demand. CAM products represent an industry estimated to be worth ~3 billion annually [3]. CAM practitioners in Australia provide up to a half of all health consults and may be the primary care providers for one-third of their users.

17.2 What are CMs?

In Australia, medicinal products containing such ingredients as herbs, vitamins, minerals, nutritional supplements, and homoeopathic and certain aromatherapy preparations are referred to as "complementary medicines" and are regulated as medicines under the Therapeutic Goods Act 1989 (the Act, TGA; http://www.comlaw.gov.au/Series/C2004A03952).

17.2.1 Designated active ingredients

A CM is defined in the Therapeutic Goods Regulations 1990 as "a therapeutic good consisting principally of one or more designated active ingredients mentioned in Schedule 14 of the Regulations, each of which has a clearly established identity and traditional use," which include:

1. An amino acid
2. Charcoal
3. A choline salt

Nutraceutical and Functional Food Regulations in the United States and Around the World.
DOI: http://dx.doi.org/10.1016/B978-0-12-405870-5.00017-7

4. An essential oil
5. Plant or herbal material (or a synthetically produced substitute for material of that kind), including plant fibers, enzymes, algae, fungi, cellulose, and derivatives of cellulose and chlorophyll
6. Homeopathic preparation
7. A microorganism, whole or extracted, except a vaccine
8. A mineral including a mineral salt and a naturally occurring mineral
9. A mucopolysaccharide
10. Non-human animal material (or a synthetically produced substitute for material of that kind) including dried material, bone and cartilage, fats and oils, and other extracts or concentrates
11. A lipid, including an essential fatty acid or phospholipid
12. A substance produced by or obtained from bees, including royal jelly, bee pollen, and propolis
13. A sugar, polysaccharide, or carbohydrate
14. A vitamin or provitamin

17.3 How CMs are regulated in Australia

The TGA regulates therapeutic goods through the following:

- Pre-market assessment
- Post-market monitoring and enforcement of standards
- Licensing of Australian manufacturers and verifying overseas manufacturers' compliance with the same standards as their Australian counterparts

17.3.1 Two-tiered regulatory system

Australia has a two-tiered system for the regulation of medicines, including CMs:

- Higher risk medicines must be registered on the Australian Register of Therapeutic Goods (ARTG), which involves individually evaluating the quality, safety, and effectiveness of the product
- Lower risk medicines containing pre-approved, low-risk ingredients and that make limited claims can be listed on the ARTG

Within the regulatory framework, medicines are classified as either registered ("R") or listed ("L") must be "included" on the ARTG (http://www.tga.gov.au/industry/cm-argcm.htm) before they may be supplied in or exported from Australia, unless exempted.

17.3.1.1 *Registered medicines*

- Registered medicines are assessed by the TGA for quality, safety, and efficacy.

- All prescription medicines are registered.
- Most over-the-counter (OTC) medicines are registered.
- Some CMs are registered.

17.3.1.2 *Listed medicines*
- Listed medicines are assessed by the TGA for quality and safety but not efficacy.
- Some OTC medicines are listed.
- Most CMs are listed.

Some CMs are exempt from the requirement to be included on the ARTG, such as certain preparations of homoeopathic medicines.

The Australian Regulatory Guidelines for Complementary Medicines (ARGCM) provide detail on the regulation of CMs and assist sponsors to meet their legislative obligations.

17.3.2 **Risk management**

The Australian community expects therapeutic goods in the marketplace to be safe, of high quality, and of a standard at least equal to that of comparable countries. Accordingly the TGA approves and regulates products' risks—benefit assessment basis.

All therapeutic goods carry potential risks, some of which are minor, some potentially serious. The TGA applies scientific and clinical expertise to its decision making to ensure that the benefits of a product outweigh any risk.

In assessing the level of risk, factors such as side effects, potential harm through prolonged use, toxicity, and the seriousness of the medical condition for which the product is intended to be used, are all taken into account. The level of TGA regulatory control increases with the level of risk the medicine or device can pose. Risk information is used by the TGA when deciding how to approve a medication for supply. For example, a low-risk product may be safely sold through supermarkets, while higher risk products may only be supplied with a prescription.

The TGA's approach to risk management involves:

- Identifying, assessing, and evaluating the risks posed by therapeutic products
- Applying any measures necessary for treating the risks posed
- Monitoring and reviewing risks over time

The risk—benefit approach assures consumers that the products they take are safe for their intended use, while still providing access to products that are essential to their health needs.

17.3.3 **ARGCM structure and content**

The ARGCM is structured in five parts (http://www.tga.gov.au/industry/cm-argcm.htm).

- Part I provides guidance on the Registration of CMs.
- Part II provides guidance on Listed CMs.
- Part III provides guidance on the evaluation of CM substances for use in Listed medicines.
- Part IV provides general guidance in relation to CM modalities such as homoeopathy, traditional herbal medicine, and aromatherapy. This part also provides information on exempt medicines, combination complementary/pharmaceutical medicines, and the food/medicine interface and contains a glossary of terms.
- Part V provides details of TGA policy guidelines relevant to CMs.

17.3.4 **TGA post-market regulatory activity of CMs**

TGA post-market regulatory activities relate to the monitoring of the continuing safety, quality, and efficacy of listed, registered, and included therapeutic goods once they are on the market. Information on the TGA's approach to managing compliance risk is available at the TGA regulatory framework (http://www.tga.gov.au/about/tga-regulatory-framework.htm). The TGA Office of Manufacturing Quality (http://www.tga.gov.au/about/tga-structure-omq.htm) inspects manufacturers on an ongoing basis for compliance with good manufacturing practice.

The TGA also undertakes Listed CM compliance reviews (http://www.tga.gov.au/industry/cm-basics-regulation-compliance-reviews.htm).

17.3.5 **Adverse events reporting**

Sometimes medicines, including CMs, have unexpected and undesirable effects. The TGA has a strong pharmacovigilance program, which involves the assessment of adverse events that are reported to the TGA by consumers, health professionals, the pharmaceutical industry, international medicines regulators, or by the medical and scientific experts on TGA advisory committees (http://www.tga.gov.au/safety/problem-medicine.htm). If a problem is discovered with a medicine, device, or manufacturer, the TGA is able to take action. Possible regulatory actions vary from continued monitoring to withdrawing the product from the market.

17.4 **Advertising of CMs**

The marketing and advertising of therapeutic goods, including CMs, is to be conducted in a manner that promotes the quality use of the product, reflects social responsibility, and does not mislead or deceive the consumer. The advertising of therapeutic goods in Australia is subject to the advertising requirements of the Therapeutic Goods Act 1989, which adopts the Therapeutic Goods Advertising Code 2007 (TGAC) and the supporting Regulations, the Trade Practices Act 1975, and other relevant laws (http://www.tga.gov.au/industry/legislation-tgac. htm). This is a three-tier system. In addition of the Therapeutic Goods Act 1989 and the Therapeutic Goods Advertising Code 2007, there are co-regulation and self-regulation. TGACC, Complaints Resolution Panel (CRP), and two industry bodies, ASMI (http://www.asmi.com.au/) and CHC (http://www.chc.org.au/), are closely involved with this system. Advertisement in relation to therapeutic goods as defined in the Therapeutic Goods Act 1989 includes "any statement, pictorial representation or design, however made, that is intended, whether directly or indirectly, to promote the use or supply of the goods."

Advertisements for therapeutic goods directed to consumers must comply with the Code whereas the advertisements directed exclusively to healthcare professionals are governed by industry codes of practice and are not subject to this Code. This Code does not apply to bona fide news, public interest, or entertainment programs or information material that complies with the Price Information Code of Practice. In general, an advertisement for therapeutic goods must comply with the statute and common law of the Commonwealth, States, and Territories and also contain correct and balanced statements only and claims that the sponsor has already verified.

Anyone may lodge a complaint about an advertisement for therapeutic goods and all complaints are treated in confidence.

17.5 Attitudes of consumers and healthcare professionals to CMs

Australian consumers have different expectations of CMs compared with conventional medicine. Consumers express satisfaction with CMs despite their expectations generally exceeding the health outcomes achieved. The perception of greater safety is the most guiding factor for this high expectation. However, such a perception may come from a limited knowledge of the potential risks. As most CM use is self-prescribed by consumers, such knowledge gaps result in suboptimal use.

Inadequate communication between consumers and healthcare professionals about CMs is evident in most of the recent studies. The healthcare professionals often underestimate use of CMs and are rarely proactive in enquiring about these products. Limited data suggest that attitudes toward CMs may vary between healthcare professions, e.g., urban versus rural [1,2]. Overall, doctor's attitudes toward complementary and alternative therapies were found to be positive, but still they have strong reservations about the safety and efficacy of CMs. This is reflected in lower rates of referral to practitioners of CM.

Positive attitudes toward CAMs are also reported for nurses [4]. Limited literature indicates pharmacists generally hold positive attitudes toward CMs, considering them useful and safe. However, the data also suggest that many pharmacists question the effectiveness of these products. While pharmacists do recommend CMs, such recommendations appear to be influenced by consumer demand.

17.5.1 Recent controversies

When alternative and CM regulation began in Australia, around 6000 CM products on the market were "grandfathered" onto the ARTG, and those that might have presented a safety hazard were identified and reviewed by an expert group advising the newly created TGA Office for Complementary Medicines. No review was carried out of the evidence supporting therapeutic claims before transferring to new system. Although from 1999, sponsors of products were required to "hold" this evidence as part of an audit or when a complaint was made. Sponsors may list their new products, containing already approved ingredients via the TGA's "Electronic Lodgment Facility." It is evident from recent public outcry that there may be abuse of this facility and it is possible for "rogue" products to slip through [5,6]. The most recent formal commentary on CMs came from the Auditor General's office in August 2011 ([7]; Auditor-General. Audit report No. 3 2011−12).

This report commented that 90% of the CM products in the Australian market are non-compliant, despite the system of self-assessment in place. In this study a small group of 31 CM products was selected at random, for which the labeling, incomplete information, quality, and science substantiation issues were recorded.

In one recent article, Sarris [8] commented on a few serious concerns, one of which is that "some companies may selectively borrow positive evidence from studies of standardized high-quality products to promote their own inferior products (which may not be standardised or have any similar chemical equivalence, and may have little or no active constituents)." Public education on the difference between the relative qualities of products is also highlighted.

17.6 Commentary

The majority of CMs do not meet the same standards of safety, quality, and efficacy as mainstream medicines as they are not as rigorously tested. Information about the level of testing and evidence should be easily accessible by medical practitioners, consumers, and CM practitioners [9].

There is a strong belief that the current Australian regulatory system neither controls CM claims nor supports an evidence-based industry [5]. The onus is now on the Federal Government to overcome all these challenges and position the reforms required. Equally, the CM industry must accept that its future will be based on evidence, not hype. Finally, the challenge for both health professionals and consumers is to learn more about the benefits and risks of CMs and be more open to discussing these matters with each other.

In the absence of sufficient efficacy data (not in all cases), it is essential there be clear and true statements regarding the efficacy and standards of evidence relied on, including accurate labeling. Government agencies such as the TGA and educational bodies such as the National Prescribing Service should ensure that information on the safety, quality, efficacy, and cost-effectiveness of CMs is readily available to consumers and health practitioners.

There should be appropriate regulation of CM practitioners and their activities also. Although Australian governments have undertaken a national regulatory process that encompassed a number of major CAM disciplines (including chiropractic, osteopathy, and Chinese medicine), naturopathy remains largely unregulated [10]. Mechanisms for making complaints about advertising should be robust and penalties enforced. To develop a transparent and sustainable regulatory system, companies must publish all results and only make substantiated claims. In addition, providing intellectual property protection to CAM companies may encourage greater investment in research; however, more accountability of study results is critical [8]. Appropriate and regular education and training is needed for CAM practitioners to further improve clinician understanding and application of evidence-based practice [11].

17.7 Way forward

In December 2003, the Australian and New Zealand governments signed a treaty to establish a single, bi-national agency to regulate therapeutic products, including medical devices and prescription, OTC, and CMs as the Australia New Zealand Therapeutic Products Authority (ANZTPA) [12,13]. After several years in political deadlock, the Prime Ministers of Australia and New Zealand agreed on June 20, 2011, to proceed with a joint scheme for the regulation of therapeutic products. The creation of a joint regulatory scheme across both countries will safeguard public health and safety, while encouraging economic integration and benefiting health professionals, consumers, and industry in both countries. The objective of the new initiative is to establish a trans-Tasman early warning system for advising the public about potential safety concerns associated with medicines and medical devices.

References

[1] D'Onise K, Haren MT, Misan GM, McDermott RA. Who uses complementary and alternative therapies in regional South Australia? Evidence from the whyalla intergenerational study of health. Aust Health Rev 2013;37:104–11.

[2] MacLennan AH, Myers SP, Taylor AW. The continuing use of complementary and alternative medicine in South Australia: costs and beliefs in 2004. Med J Aust 2006;184:27–31.

[3] Access Economics and National Institute of Complementary Medicine. Cost effectiveness of complementary medicines. <http://www.nicm.edu.au/content/view/159/276/>; August 2010 [accessed Mar 2013].

[4] Armstrong AP, Thiébaut SP, Brown LJ, Nepal B. Australian adults use complementary and alternative medicine in the treatment of chronic illness: a national study. Aust NZ J Public Health 2011;35:384—90.

[5] Harvey KJ. A review of proposals to reform the regulation of complementary medicines. Aust Health Rev 2009;33:279—87.

[6] Smith AJ. The efficacy of complementary medicines: where is the evidence? J Pharm Pract Res 2012;42:174—5.

[7] Auditor-General. Audit report No. 3 (2011—12). Performance audit: therapeutic goods 0020 coregulation: complementary medicines. Canberra: Australian National Audit Office; Available from, <www.anao.gov.au/~/media/Uploads/Audit%20Reports/2011%2012/201112%20Audit%20Report%20No%203.pdf>; 2011.

[8] Sarris J. Current challenges in appraising complementary medicine evidence. MJA 2012;196:310—1.

[9] Australian Medical Association Position Statement. Complementary medicine — 2012; 2012.

[10] Wardle J, Steel A, Adams J. A review of tensions and risks in naturopathic education and training in Australia: a need for regulation. J Altern Complement Med 2012;18:362—70.

[11] Leach MJ, Gillham D. Are complementary medicine practitioners implementing evidence based practice? Complement Therap Med 2011;11:128—36.

[12] Ghosh DK, Skinner M, Ferguson L. The role of the therapeutic goods administration and the medicine and medical devices safety authority in evaluating complementary and alternative medicines in Australia and New Zealand. Toxicol 2006;221:88—94.

[13] Ghosh D, Skinner M, Ferguson L. Complementary and alternative medicines in Australia and New Zealand: new regulations. In: Bagchi D, editor. Nutraceuticals and functional foods regulations in the United States and around the world. USA: Elsevier; 2008, Chapter 16, p. 239—48.

Russian Regulations on Nutraceuticals and Functional Foods

18

Victor A. Tutelyan, Boris P. Sukhanov, Alla A. Kochetkova, Svetlana A. Sheveleva and Elena A. Smirnova

Institute of Nutrition of the Russian Academy of Medical Sciences, Moscow, Russia

18.1 Introduction

Lifestyle and diet are the key factors responsible for people's health, working ability, and capability to resist adverse environmental conditions. Dietary nutrients are required to supply the human body with energy and essential substances. They largely determine people's health and physical and creative activity. Changes in living and working conditions of the Russian population caused reduction of the energy consumption and, consequently, of the volume of food intake. At the same time, the intake of essential nutrients was also reduced, while the demand for them remained unchanged [1].

Continuous monitoring of the Russian population's nutrition carried out in Russia since 1983 by the Institute of Nutrition of the Russian Academy of Medical Sciences in collaboration with regional medical research institutions revealed significant disorders of nutritional status in almost all population groups, including insufficient intake of animal protein; animal fat overconsumption; polyunsaturated fatty acids deficit; a high share of carbohydrates in the diet; insufficient intake of the dietary fiber; and deficiency in vitamins C, B_1, B_2, B_6, B_{12}, folic acid, β-carotene, and a number of minerals and microelements (calcium, iron, iodine, fluorine, selenium, and zinc). Nutritional status disorders have a negative impact on health indices and represent a serious risk factor for occurrence and development of many non-communicable diseases. In recent years, the rate of cardiovascular diseases has dramatically increased accounting for 55 per cent of total mortality in Russia. Diseases of the gastrointestinal tract, the syndrome of the intestinal dysbacteriosis, and diabetes are becoming widespread. A number of those who suffer from the reduced immune status and lowered resistivity to natural and technogenic environmental factors have also increased. At the same

Nutraceutical and Functional Food Regulations in the United States and Around the World.
DOI: http://dx.doi.org/10.1016/B978-0-12-405870-5.00018-9

time, dysbiotic dysfunctions of the intestinal flora, which often serve as markers of disadaptation shifts in the human body, have lately become common among adults and children [1−3].

In this respect, the prevention of deficiencies of essential nutrients and microelements and malnutrition-related diseases in all groups of population, particularly among low-income citizens, becomes important.

National food security is of primary importance and an essential issue for the government. The guaranteed access and affordable price of food for everyone in the amount sufficient for an active and healthy life remains a key factor of social stability and a prerequisite to improvement of Russian citizens' life quality.

Russian Federation Food Security Doctrine, approved by Order No.120 of the President of the Russian Federation on 30.01.10, is a base document that secures state guarantees in relation to food. One of the key targets set by the Doctrine is to achieve and maintain availability of safe foods in the amount and range corresponding to adequate dietary allowances associated with an active and healthy lifestyle. In order to promote a healthier public nutrition, the document envisages development of fundamental and applied research in the fields of medical and biological evaluation of safety of new food sources and ingredients; implementation of innovative technologies; organic production technologies; and increase in production of new fortified, dietetic, and functional foods.

On October 25, 2010, the Prime Minister of the Russian Federation signed Decree No. 873-r approving the Russian Federation Policy Framework on Healthy Public Nutrition for the period through to 2020. The main goals outlined by this document include public health maintenance as well as prevention of diseases caused by an inadequate and unbalanced diet. One of the key targets set in order to achieve the outlined goals is to promote production of foods enriched with essential nutrients, specialized baby food products, functional foods, dietetic foods (for medical use and disease prevention), and biologically active food supplements (nutraceuticals).

At present, in view of the creation of the common customs territory and the Customs Union of the Russian Federation, the Republic of Belarus and the Republic of Kazakhstan and Russia's accession to the World Trade Organization, a large-scale revision of the legislation is being carried out in the field of food quality and safety, with the objective to harmonize it with the international legislation and include the new scientific data on food safety. Technical regulations are actively being developed in order to set uniform binding requirements for foods in the Customs Union and to ensure the free transportation of foods released for circulation in the common customs territory [4].

It should be noted that the Russian Federation is continuously refining its national standards that regulate quality of foods used for treatment of various nutritional disorders. Thus, Technical Standardization Committee No. 036 Functional Foods was created in August 2008 for execution of the Federal Law "On Technical Regulation" (No. 184-FZ dt. 27.12.2002), and creation of a system of technical committees for standardization, and management of national standardization of functional foods. This technical committee consisted of three subcommittees (SC): SC1—Functional Foods of Plant Origin, SC2—Functional Foods of Animal Origin, SC3—Evaluation of Functional Foods and Functional Ingredients Efficacy. The main objectives of the Technical Committee 036 are ensuring compliance between the legislative and the regulatory systems of the Russian Federation and their harmonization with the international requirements and legislation of other countries: identifying ingredients with functional potential and their contents in foods; creating a methodology of authentication and prevention of counterfeit; setting the order of informing consumers about functional foods health benefits, and representing interests of the Russian Federation in international organizations.

The work of the national technical committee for functional foods standardization was supported by the Interstate Council for Standardization, Metrology, and Certification which initiated and approved the creation of Interstate Technical Standardization Committee (ITC) No. 526 Functional Foods in 2011. The committee deals with functional foods of plant origin, including fortified foods and probiotic foods; functional foods based on foods of animal origin, including fortified foods and probiotic foods; and functional food ingredients of plant and animal origin, including probiotics, prebiotics, and synbiotics.

Adoption of such important documents as the Food Security Doctrine and the Russian Federation Policy Framework on Healthy Public Nutrition for the period through to 2020 as well as creation of the Technical Standardization Committee on Functional Foods signify significant state support of innovative development in the food industry and in the sphere of production of food supplements, functional foods, and foods for special dietary uses.

18.2 Russian regulations on nutraceuticals

It has been repeatedly demonstrated on large groups of people in Russia that boosting a diet with multivitamin/mineral supplements significantly reduces the incidence of respiratory diseases, especially in spring and autumn, and also of other chronic illnesses. Clinical studies suggested that introducing combinations of multivitamin/mineral supplements, minor compounds, probiotics, and prebiotics in hospital diets speed up recovery while reducing the required dosage of medication. Such diets help in decreasing the length of hospital stay, reducing the frequency of exacerbations in chronic patients, and attenuating their clinical symptoms [5–7].

It is therefore obvious that ways have to be found to compensate nutrients deficiency in diets [8].

One possible solution would be to promote development of the country's food supplement industry. Similar to legislation in the developed countries of Europe,

Asia, and the Americas, the Russian Federal Law No 29-FZ On Quality and Safety of Food Products classifies food supplements as food products. The law defines biologically active food supplements (BAFS) as "natural (identical to natural) biologically active substances designed for consumption with food or for introduction into the composition of food products."

Active development of the Russian market of BAFS started in early 1990s. These products were regarded as one of primary means of reducing essential nutrients deficiencies, including deficiency of minor food components, and as a powerful driver of improving quality of nutrition and life style and reducing the risk of non-communicable diseases. According to Euromonitor International, the Russian food supplement market's turnover stood at $1.4 billion in 2012, which was 12% up from the 2011 levels. Experts forecast that the market will continue to grow at an average annual rate of 5% for the next 5 years, reaching $1.7 billion by 2016. Historically, dietary supplements are manufactured in dosage forms (tablets, capsules, powders, lozenges, etc.), thereby allowing strict qualitative control of active ingredients taking into account consumers' individual requirements.

BAFS circulation in the Russian territory is regulated by Federal Laws On Sanitary and Epidemiologic Well-Being of the Population No. 52-FZ dt. 30.03.1999, On Food Quality and Safety No. 29-FZ dt. 02.01.2000, On Protection of Consumer Rights No. 196-FZ dt. 30.12.2001, and On Advertising No. 38-FZ dt. 13.03.2006 by Russian Federation Government decrees and the national standard GOST R 51074-03 Foods, Consumer Information, General Requirements. Various aspects of circulation, quality, safety, and efficacy of these products are also regulated by a number of other federal documents, such as the Sanitary Regulations and Norms (SanPiN), Methodical Guidelines (MG), and Methodical Recommendations (MR). In addition, a prominent place in the system of BAFS regulation belongs to by orders of the Ministry of Health and of the Federal service on customers' rights protection and human well-being surveillance (Rospotrebnadzor) and decrees of the Chief State Sanitary Doctor of the Russian Federation and other industry-specific documents.

According to documents stated above, manufacturers are responsible for product quality, safety, and efficacy, as well as for the truthfulness and completeness of information about their products. The legal framework for the circulation of food supplements in Russia continues to evolve and improve.

The Russian system for controlling the circulation of food supplements is largely harmonized with those in Europe and the United States, including indications for use of such products. The Russian system is, however, stricter when it comes to the composition of BAFS. In particular, as distinct from other countries, Russia maintains negative lists of substances, compounds, certain tissues, and species of animals and microorganisms that are prohibited for use in food supplements. These include, *inter alia*, medicinal plants with no history of human consumption (the relevant list contains over 190 positions). The Russian legislation also restricts the use of food supplements in children's diets to those considered to be sources of vitamins, individual minerals, probiotics, and prebiotics [9].

Russia's positive list of components approved for use in the manufacture of food supplements comprises 166 positions. It includes amino acids, fatty acids, vitamins, minerals, and biologically active substances and enzymes, and contains information about the recommended and maximum safe intakes. Methods have been developed, approved, and adopted for detecting these compounds and substances in food supplements [9].

According to the requirements of Russia's sanitary legislation BAFS are subject to state registration, which is run by the Rospotrebnadzor. The state registration is designed to assess compliance of the product quality and safety with legal requirements in the Russian Federation and the Customs Union. Information of registered products is stored in the database of state registration certificates (the State Register) [10].

According to the Federal Law On Foods Quality and Safety, BAFS are not drugs and not designated for treatment or diagnosis of human diseases. They fall into the category of foods and serve as an additional source of essential nutrients in human diet. The state surveillance over their production and circulation is under the jurisdiction of Rospotrebnadzor.

Authorization of BAFS manufacturing in each specific manufacturing facility is issued by Rospotrebnadzor agencies and offices on the condition of compliance with sanitary norms and regulations.

State supervision of BAFS production is the responsibility of Rospotrebnadzor in accordance with Russian Federal Laws On Sanitary and Epidemiologic Well-Being of the Population No. 52-FZ dt. 30.03.1999, On Food Quality and Safety No. 29-FZ dt. 02.01.2000, and with the Russian Federation Government Decree Regulation for State Sanitary and Epidemiological Supervision in the Russian Federation No. 569 dt. 21.04.10.

Sanitary rules SR 1.1.21.93-07 Amendments and Addendum 1 to SR 1.1.10.58-01 Organization and Management of Product Compliance with Sanitary Rules and Sanitary Anti-Epidemic (Preventive) Measures in Food Production, together with analytical methods approved by the Rospotrebnadzor, provide guidance on control of the product quality and safety as well as compliance with the legislation in manufacturing dietary supplements. The assessment criteria include those listed in the national sanitary legislation SanPiN 2.3.2.1078-01 Hygienic Safety and Nutritional Value Requirements for Foods with addendum and amendments (SanPiN 2.3.2.2351-08 Addendum and Amendments 7 as well as in MG 2.3.2.721-98 Assessment of Safety and Efficacy of Dietary Supplements.

In addition, according to the Customs Union Agreement on Sanitary Norms, safety of BASF in the customs territory of the Customs Union is also regulated by the Uniform Sanitary, Epidemiological, and Hygienic Requirements for Products Subject to Sanitary and Epidemiological Supervision (Control). The document, now in force, is the SanPiN No. 2.3.2.1290-2003 Hygienic Requirements for Production and Circulation of Biologically Active Food Supplements. It defines mandatory requirements for development and production of BAFS, as well as for their import, storage, transportation, and sale in the territory of the Russian Federation.

Rospotrebnadzor's responsibilities also include post-approval quality and safety control of BAFS currently in circulation.

It would be impossible to secure safe market circulation of BAFS without reliable methods of control of active ingredients or microorganisms content and authentication of components that are claimed to be included into the products. These are major criteria of BAFS quality and efficacy. Procedures developed in Russia allow for quantitative measurement of more than 120 active ingredients (Guidelines No. P 4.1.1672-03 Guidelines for Methods of Quality and Safety Control of Biologically Active Food Supplements).

In recent years, building the system of safety and authenticity control of probiotic microorganisms, used in dietary supplements and foods, has gained high importance. In addition to aforementioned legal and normative documents, the circulation of probiotics in the Russian Federation is regulated by ministerial regulations including the decree by the Chief Sanitary Doctor 149 of 16.09.2003 On Managing Microbiological and Molecular Genetic Expertise of Genetically Modified Microorganisms in Food Production; SanPiN 2.3.2.2340-08 Amendments and Addendum 6 to SanPiN 2.3.2.1078-01 Hygienic Requirements for Safety and Nutrition Value of Foods; MG 2.3.2.1830-04 Microbiological and Molecular Genetic Assessment of Foods Manufactured with the Aid of Genetically Modified Microorganisms; and MG 2.3.2.2789-10 Methodical Guidelines on Sanitary-Epidemiological Assessment of Safety and Functional Effect of Probiotic Microorganisms Used in Food Manufacturing.

The BAFS state registration system, which is currently in force in the Russian Federation, is aligned with best international practices, notably with guidelines of the Codex Alimentarius Commission, and in many aspects is similar to the relevant legislation of Canada, Germany, Great Britain, and the United States (*inter alia*, Federal Food, Drug, and Cosmetic Act of 20.01.99) [10]. The methodology of assessing BAFS properties that define product quality and consumer safety was derived from the existing regulation.

In accordance with the SanPiN No. 2.3.2.1290-2003 and MG No. 2.3.2.721-98 the assessment of BAFS includes the sensory evaluation, the sanitary and epidemiological expertise of consumer information is related to expertise as well as labeling. The choice of evaluation criteria is dependent on individual characteristics of each BAFS: its sensory, physical, and chemical properties and function, role in the modern diet. This approach is shared by many food and nutrition experts around the world.

The consumer information placed on the product label in the column "Indications" is subject to expert approval during the state registration of BAFS. According to SanPiN No. 2.3.2.1290-2003, the column may only contain information about ingredients that are present in the product. It is illegal to mention any therapeutic or curative effects of BAFS on the product label.

SanPiN No. 2.3.2.1290-2003 allows the following statements to be used in description of BAFS efficacy: "for optimization of carbohydrate, lipid, vitamin or other metabolism in various functional conditions"; "for normalization and (or) improvement of functional state of organs and systems" (including products with health-promoting, mild tonic, soothing or other effects observed in different functional states); "for reducing the risk of diseases"; "for normalization of the gastrointestinal tract microflora"; "acting as enterosorbents." The statements according to GOST R 51074-03 Food Products, Consumer Information, General Requirements can be used in product labeling provided the manufacturer possesses documents that prove stated effects. In general, this approach is similar to practices adopted in many countries in labeling of food supplements, although the procedure of approving claimed effects has not been yet fully established.

Manufacturing of BAFS should also comply with other requirements concerning product labeling (SanPiN No. 2.3.2.1290-2003). For example, the trade name of a food supplement should not reflect the product's intended effect. It is not allowed to use the term "environmentally friendly product" either in the product name or in the label information, as well as any other statements that lack legal or scientific evidence.

The full composition of a food supplement with ingredient formulation is listed in the order of decreasing weight of the ingredients, or in percentage points without specifying each ingredient quantity. Any excipients used in the BAFS formulation should also be mentioned after the list of active ingredients (likewise, in the order of decreasing their weight or weight percentage).

Weight and volume of the food supplement per packaging unit should be indicated. Other mandatory information includes: indications for use, dosage, and application; data on contraindications; and indication that the product is not a drug and the phrase "it is recommended to consult a physician prior to use." The list of diseases for which the use of certain foods and food supplements is contraindicated is defined by the Ministry of Health of the Russian Federation.

Important labeling requirements also include the manufacturing date, the expiry date or end of shelf-life, and storage conditions. The responsibility that BAFS developers and manufacturers hold before the consumers of their products implies that they should guarantee the presence of biologically active substances in the product in specified quantities as well as bioavailability of these substances throughout product's shelf-life [9].

Information of the food supplement's state registration should be placed on the product label and accompanied with the registration date and number. In addition, the manufacturer's address and company name, as well as the address and telephone of an organization authorized by the manufacturer (seller) to receive customer claims, should be indicated on the product label. This information should be made available to consumers in any readable form.

The retail sales of food supplements is allowed through pharmacies, pharmacy shops, specialized dietary food stores, and convenience food stores (specialized departments, sections, booths). Food supplements should correspond to the requirements set by normative and technical documentation, and their retail distribution is only allowed in retail packs.

18.3 Russian regulations on functional foods

It is well-known that nutritional status disorders negatively affect life expectancy (at the present time it is 64.0 years for men and 75.6 years for women in the Russian Federation) and mortality rate. It is evident that the prevention of essential macronutrients and micronutrients deficiency and, consequently, of malnutrition-related or hyponutrition-related diseases in different categories of population is a priority. The most reasonable way to compensate nutrient deficiency is to enrich generally available mass consumption foods (bread and other pastries, oils and fatty foods, milk and dairy products, fruit juices and beverages, etc.) with deficient vitamins, minerals, dietary fibers, polyunsaturated fatty acids, and other functional ingredients. In addition, it must be borne in mind that the modern Russian lifestyle characterized by a sharp decrease of physical activity led to a significant reduction of the volume of food intake and, consequently, of micronutrients consumption. As a result, given the daily dietary calories consumption of about 2000−2400 kcal, it became impossible to provide the population of the Russian Federation with essential micronutrients in sufficient amounts. Moreover, as confirmed by the ample evidence collected in the nutrition survey in almost 60 Russian regions, the nutritional density of foods and diets has not changed in the last 50 years characterized by the sharp lifestyle changes. A practical solution to the problem of optimal nutrition is to design functional foods with pronounced physiological effects [11,12].

The Russian market of functional foods and ingredients is currently regulated by general legal documents (Federal Laws On Technical Regulation, On Protection of Consumer Rights, On Sanitary and Epidemiologic Well-Being of the Population, On Food Quality and Safety). Activities of Technical Committee No. 036 Functional Foods resulted in the development of a number of national standards (GOST R) forming a system of technical norms for production and circulation of functional foods in the territory of the Russian Federation and setting requirements for their quality (Table 18.1).

Research and development in the field of functional foods and beverages in Russia is based on the general terminological standard GOST R 52349-2005 Foods- Functional Foods- Terms and Definitions. This document defines *functional food* as a type of food designed for systematic consumption as part of daily diet by all age groups of healthy population for reducing the risk of nutrition-related diseases and for maintaining and promoting the general health.

In 2010 the new Amendment No.1 to GOST R 52349-2005 was issued as part of the national standards development program, which corrected some of the

Table 18.1 List of National Standards Developed by Technical Standardization Committee 036 "Functional Foods"

No.	GOST R Number	Title
1	GOST R 52349-2005	Foods- Functional foods- Terms and definitions
2009–2010		
2	GOST R 53861-2010	Dietetic foods (for medical use and disease prevention)- Dry protein compound supplements- General technical specification
3	Amendment No. 1 to GOST R 52349-2005	Foods- Functional foods- Terms and definitions
4	GOST R 54014-2010	Functional foods- Soluble and Insoluble dietary fiber determination by enzymatic gravimetric method
5	GOST R 54060-2010	Functional foods- Identification- General provisions
6	GOST R 54059-2010	Functional foods- Functional food ingredients- Classification and general requirements
7	GOST R 54058-2010	Functional foods- Carotenoid determination method
2011		
8	GOST R 54637-2011	Functional foods- Vitamin D_3 determination
9	GOST R 54635-2011	Functional foods- Vitamin A determination
10	GOST R 54634-2011	Functional foods- Vitamin E determination
2012		
11	GOST R (draft)	Foods- Functional foods- Information on definitive distinctions and efficacy
12	GOST R (draft)	Foods for special dietary uses- Osmolality measurement
13	Amendment 1 to GOST R 53861-2010	Dietetic foods (for medical use and disease prevention)- Dry protein compound supplements- General technical specification
2013		
14	GOST R (draft)	Functional foods- The biologically active food supplements- Traceability requirements
15	GOST R (draft)	Functional foods- Methods of microbiological analysis
16	GOST R (draft)	Functional foods- Methods of determination of bifidogenic properties
17	GOST R (draft)	Functional foods- Methods of detection and enumeration of probiotic microorganisms

terminology. According to the new revision of GOST R 52349-2005, a functional food is a type of food designed for systematic consumption by all age groups of healthy population as part of the daily diet, which possesses scientifically proven properties, reduces the risk of nutrition-related diseases, prevents or compensates nutrients deficiency in the human body, and maintains and promotes the general health. The same document contains a definition of a *fortified food* as a type of the functional food produced by adding one or several *functional food ingredients* to traditional foods in the amount required to prevent or compensate deficiencies in nutrients and (or) deficit of microflora.

Amendment No.1 to GOST R 52349 also specifies the term *functional food ingredient*, which is a living microorganism; a substance or a complex of substances of animal, plant, microbial, or mineral origin or alternatively, nature-identical substances that are contained in a functional food product in the amount exceeding 15 per cent of daily physiological requirement per one product portion and that are capable of producing scientifically proven and evidenced effect onto one or several physiological functions or onto human body metabolic processes under the condition of the systematic use of the functional food enriched with these ingredients.

Fortified foods with essential nutrients present a significant intervention into the traditional human food patterns, and the need for it is explained by objective changes in lifestyle, range, and nutritional value of the consumed foods; consequently, it may only be done on the basis of accurate, scientifically proven and tested, medical, biological, and technological principles which help to solve the most important problems arising in the process of development, production, and distribution of such foods.

One of the means of preventing macronutrient or micronutrient deficiency-related diseases is fortifying mass consumption foods with vitamins and minerals. The practice of the targeted fortification of foods in Russia started as early as 1939 when the Council of People's Commissars adopted a decree on enrichment of wheat flour produced in the mills with vitamins B_1, B_2, and PP up to the level of their content in the raw materials. This decision was based on the scientific data about the significant loss of vitamins in the processes of grain milling and bran removing. At the present time, the priority is enriching flour and bakery products, dairy products, and oils and fatty foods, as well as beverages, taking into account that these consumer products could be efficiently used for reaching the stated objectives [8].

This way of enriching foods is now widely used in the food production. The medical, biological, and technological aspects of vitamins and minerals use have been thoroughly studied and worked out. The majority of corresponding technical regulation issues have also been settled [13].

Vitamin- and mineral-fortified foods in market circulation should correspond to national legislative requirements in the field of food safety, to the Uniform Sanitary, Epidemiologic, and Hygienic Requirements for Products Subject to Sanitary and Epidemiological Supervision (Control) of the Customs Union as well as to the SanPiN No. 2.3.2.2804-10 Addendum and Amendment No. 22 to the SanPiN No. 2.3.2.1078-01 Hygienic Requirements for Safety and Nutritional Value of Foods. The later document was developed taking into account the results of extensive studies of real-life nutrition, availability of vitamins and minerals to the population, vitamin and mineral content in foods and ready meals, and experience of enriching foods with vitamins and minerals accumulated in Russia and abroad. Moreover, this document has been harmonized with the General Principles for the Addition of Essential Nutrients to Foods (CAC/GL 09-1987) of the Codex Alimentarius Food and Agriculture Organization (FAO)/World Health Organization (WHO) Commission.

According to SanPiN No. 2.3.2.2804-10, fortification of foods with one or several vitamins, macroelements, and/or microelements should correspond to the following requirements:

- Fortification is allowed for mass consumption foods that are commonly and regularly used in the daily diet of adults and children from 3 years old, as well as for foods undergoing refinement and other technological processing leading to significant losses in vitamins and minerals.
- It is recommended to fortify foods with such vitamins and minerals that are objectively underconsumed and/or deficient among the population.
- Foods may be fortified with vitamins and/or minerals regardless of whether the same substances are present in the original product or not.
- Specific enriching micronutrients as well as their dosages and forms are selected according to their safety and efficacy in terms of increasing the nutritional value of the diet.
- While calculating the quantity of vitamins and minerals added to foods with the purpose of their fortification, one should take into account their natural content in the original product or raw materials as well as losses in the course of manufacturing and storage, so that the levels of these vitamins and minerals are not below the reference value throughout the shelf-life of the product.
- Combinations, forms, ways, and stages of fortification should be chosen taking into account the probability of chemical interaction between the enriching ingredients and the original product components, with the view of their optimal preservation during manufacturing and storage.
- Fortification of foods with vitamins and minerals should not impair consumer attributes of the former (for instance, reduce the levels and availability of other nutrient constituents, significantly impair sensory properties of the products, or shorten their shelf-life).
- Fortification of foods with vitamins and minerals should not influence their safety profile.
- Guaranteed content of vitamins and minerals in fortified foods should be mentioned in the labeling of product individual packaging.
- Rationality of inclusion of vitamins and/or minerals into new and specialized foods with the purpose of their fortification should be supported by specific studies confirming food safety and ability to improve the delivery of corresponding vitamins and minerals to the body and to positively affect the health state.

The food groups recommended for fortification with vitamins and/or minerals include:

- Flour and bakery products
- Dairy products
- Non-alcoholic beverages
- Juices and juice-like drinks made of fruits (including berries) and vegetables [juices, fruit and (or) vegetable nectars, fruit and (or) vegetable juice drinks]
- Fat-and-oil products (oils, margarine, butter-like spreads, mayonnaise, sauces)
- Table salt
- Cereals products (breakfast cereals, ready-for-use extruded products, fast-cooked pasta, and other products)
- Food concentrates (kissel, instant drinks and meals, instant cereals)
- Protein products made of cereals, legumes, and other cultures as well as foods designated for certain population groups
- Baby food products
- Dietary foods (for medical use and disease prevention)
- Functional foods
- Foods for special dietary uses, including foods of predetermined composition

It should be mentioned that some food groups should not be enriched with vitamins and minerals, namely unprocessed foods (fruits, vegetables, meat, poultry, and fish), fermented beverages and beverages with alcohol content of more than 1.2 per cent (excluding low alcoholic tonic beverages where vitamins and minerals are added for a different purpose).

Mass consumption products should not be fortified with sodium, choline, inositol, carnitine, taurine, copper, manganese, molybdenum, chrome, and selenium, except for foods for special dietary uses [sport foods, dietary foods (for medical use and disease prevention), foods of predetermined composition], functional foods, baby food products, and biologically active food supplements.

A product can be classified as fortified if its mean daily serving contains 15−50 per cent of human standard physiological reference intake for vitamins and/or minerals. The minimal amount of enriching ingredient added to a food product should reach at least 10% of human standard physiological need. For fortified high-energy foods (with energy value of 350 kcal per 100 g or higher), the content of vitamins and minerals should be 15−50% of human standard physiological reference intake per 100 kcal (one standard product serving).

According to Decree No. 982 of the Government of the Russian Federation dd. December 1, 2009, On Approving the Unified List of Products Subject to Mandatory Certification and the Unified List of Products Requiring a Declaration of Conformity, enriched foods are on the list of products that require a declaration of conformity.

According to Law No. 2300-1 of the Russian Federation dd. 07.02.1992 On Protection of Consumer Rights and its Amendment No. 243 dd.25.10.2007, the manufacturer should supply consumers in due time with sufficient and accurate product information enabling them to make the right choice. Absence of labeling on fortified products may lead to their unintentional or intentional substitution with similar non-fortified products in the course of transportation or retail distribution. The label should bear information about the content of enriching ingredients in absolute values and in relative ones, expressed as a percentage of human standard physiological intake.

Analysis of the Russian market of functional foods shows that in recent years foods fortified with essential nutrients other than vitamins and minerals have been gaining popularity among consumers, while the range of such products is continuously growing due to appearance of new foods and beverages with health claims [14−16].

Development of new products and their formulae involves dealing with the main challenge in this concerning the health claims. For foods enriched with essential nutrients a health benefit will be defined by the presence of functional ingredients in quantities corresponding to human physiological needs as provided for by Methodical Guidelines No. MG 2.3.1.2432−08 Reference values of Energy and Nutrient Physiological Needs for Different Groups of Population of the Russian Federation and as needed for a reliable positive effect.

It is worth mentioning that indicators used in Russia in order to estimate efficacy and reduction of risk of common nutrition-related diseases with the help of functional foods include: body mass index, measurement of energy expenditures, muscle cell glycogen, endurance test, intensity of allergic reactions, blood pressure, high density and low density lipoprotein levels, cholesterol level, glycemic index, fasting plasma glucose, insulin level, condition of the large intestine, composition and properties of the intestinal microflora, bone density, kinetics of calcium excretion, etc. Physiological effect of functional foods is expressed in their beneficial influence onto enhancing physical endurance and immunity, improvement of digestion, and appetite control (in particular, reduction of appetite).

According to the functional food ingredients classification provided in the national standard GOST R 54059-2010 Functional Foods- Functional Food Ingredients- Classification and General Requirements there are certain groups of ingredients that have a scientifically proven effect on certain human body functions and health states, such as the metabolism of essential nutrients, the antioxidant status, the cardiovascular system activity, the gastrointestinal tract activity including composition and biological activity of the intestinal microflora, dental and bone tissue health, and the immune system response.

Another noteworthy aspect is the issue of nutritional and health claims that are placed on the product label and/or used in advertising of foods fortified with essential nutrients. At the present time, manufacturers often accompany food products labeling with information about beneficial effects of foods which is not proved by evidence-based medicine data. This is partly due to the absence of national normative legal documents that would regulate labeling of such products in the Russian Federation (the only exceptions are vitamin- and mineral-fortified foods) [17].

In 2012 a draft of a new national standard GOST R Foods- Functional Foods- Information on Definitive Distinctions and Efficacy was developed. This is the first document establishing rules of evaluation of functional foods nutritional and efficacy claims and their use in labeling and advertising of food products. Moreover, while developing and labeling new functional products Russian manufacturers may follow the requirements of the Guidelines for Use of Nutrition and Health Claims (CAC/GL 23-1997) of the Codex Alimentarius FAO/WHO committee.

18.4 Russian regulations on foods for special dietary uses

Within the framework of the Customs Union, requirements have been developed for foods for special dietary uses, which became part of Technical Regulations of the Customs Union (TR CU) No. 021/2011 On Foods Safety (approved by Customs Union Commission Decision No. 880 dd. 9.12.2011) and No. 027/2012 On Safety of Certain Types of Foods for Special Dietary Uses Including Dietetic Foods for Medical Use and Disease Prevention (approved by Decision No. 34 dd. 15.06.2012 of the Council of the Eurasian Economic Commission). These Technical Regulations were enacted on July 1, 2013.

The TR CU defined foods for special dietary uses as foods that (1) comply with requirements for content and/or ratio between certain substances or all substances and ingredients, and/or (2) undergo modification of content and/or ratio of certain substances as compared to their natural content in such products, and/or (3) contain substances or ingredients (other than food additives and flavors) that are absent in the original product, and/or (4) are accompanied by manufacturer's health promotion and/or maintenance claims, and/or (5) are designated for safe use by certain population categories (see Figure 18.1).

Sport foods are foods with predetermined composition, increased nutritional value, and/or specific efficacy, consisting of a set of products or their certain types and having a specific effect on increasing human adaptive capacities to physical exertion and neuroemotional stress. Such products are used to improve athletes' supply with energy, macroelements and microelements, vitamins, and other necessary biologically active substances with the purpose of enhancing their adaptation abilities and sports performance, rapid post-exercise recovery, and rehydration. Specialized sport nutrition foods may be consumed by athletes prior to, during, or after exercise in accordance with programs developed for specific sports.

Foods for medical use are food products with defined nutritive and energy value and physical and sensory properties designed for use in therapeutic diets.

Dietetic foods are foods designated for carbohydrate, lipid, protein, vitamin, and other metabolism correction with modified content and/or ratio of certain

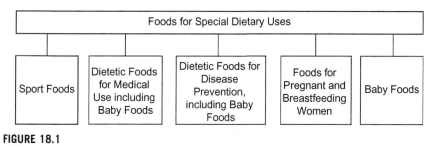

FIGURE 18.1

Types of foods for special dietary uses.

substances compared to their natural content and/or with added substances or ingredients (absent in the original products). In addition, this group includes foods designed for reducing the risk of diseases.

Dietetic foods and foods for medical use include foods for enteral and diabetic nutrition, low-lactose and lactose-free foods, and low amino acid or amino acid-free foods (for example, phenylalanine-free or low phenylalanine products), etc.

Foods for pregnant and breastfeeding women are foods with modified content and/or ratio of certain substances compared to their natural content and/or with added substances or ingredients (absent in the original products) designated for physiological needs of pregnant and breastfeeding women.

Baby foods are foods designated for nutrition of infants and young children (0−3 years old) and preschool (3−6 years old) and school (6 years old and older) age children, that satisfies children's physiological needs and are harmless for a specific age group. Baby foods include adapted infant formulas (breast milk substitutes), early and follow-up milk formulae, supplementary food added to infants diet, formulas for premature and low birth-weight infants, anti-reflux formulas, etc.

Use of foods for special dietary uses is primarily defined by age, body weight, individual food tolerance, sport exercise, acquired disease, or occupational health risk. Objectives and regimes of using foods for special dietary uses are defined on the basis of the specific factors mentioned above in order to achieve maximum efficacy [18].

Foods for special dietary uses circulating in the customs territory of the Customs Union should be harmless throughout their shelf-life provided they are used as intended. Technical Regulations On Safety of Certain Types of Foods for Special Dietary Uses, Including Dietetic Foods for Medical Use and Disease Prevention and On Food Safety list general requirements for safety and nutritive value of foods for special dietary uses; lists of food raw materials allowed for and prohibited from use in manufacturing foods for special dietary uses; and procedures of manufacturing, transportation, distribution, and recycling of foods for special dietary uses. The list of food additives allowed for or prohibited from use in manufacturing foods for special dietary uses is settled in TR CU No. 029/2012 Requirements for Safety of Food

Additives, Flavors and Processing Aids (approved by Decision No. 58 dd. 20.07.2012 of the Council of the Eurasian Economic Commission) and were effective starting July 1, 2013.

Apart from general safety requirements applicable to all food products including foods for special dietary uses, certain types of special dietary foods should comply with additional requirements.

Genetically modified organism (GMO)-containing and GMO-derived raw materials are not allowed for use in manufacturing of baby foods, foods for medical use and disease prevention, and foods for pregnant and breastfeeding women. In addition, in baby foods manufacturing it is not allowed to use preservatives (benzoic acid, sorbic acid and their salts) and plant raw materials grown with pesticides. Only natural food flavors are allowed for aroma and flavor enrichment. It is forbidden to use sweeteners in baby foods, except for certain food ingredients for dietetic foods for medical use and disease prevention. The level of fatty acids trans-isomers in food products for infant foods (breast milk substitutes) should not exceed 4 per cent of the total amount of fatty acids. The list of raw materials prohibited from use in manufacturing food for infants and preschool and school aged children is defined in TR CU No. 021/2011 On Food Safety. Appendix 9 to the TR mentioned above defines forms of vitamins and minerals that may be used for infant food manufacturing.

The dietetic foods for medical use and disease prevention should satisfy the physiological needs of the human body in nutrients and energy intake accounting for the character, the physiological-biochemical nature of nosologies, and the pathogenesis of diseases among population groups, and disease risk factors. They also should comply with existing hygienic requirements in respect to maximum limits defined for contaminants, biologically active substances and compounds, microorganisms, and other biological substances that pose risk to the health of present and future generations [19–21].

Sport foods should not contain any psychotropic, narcotic, poisonous, doping substances and/or their metabolites, as well as other banned substances included in the World Anti-Doping Agency list.

Packaging and labeling of foods for special dietary uses should correspond to the requirements of TR CU No. 027/2012 On Safety of Certain Types of Foods for Special Dietary Uses Including the Dietetic Foods for Medical Use and Disease Prevention, TR CU No. 022/2011 Foods Marking effective from July 1, 2013, and TR CU No. 005/2011 On Packaging Safety effective starting July 1, 2012.

Foods for special dietary uses should be pre-packed and packed in such a way that packaging secures product safety and claimed consumer attributes throughout the whole shelf-life, on the condition that the transportation and storage conditions are maintained. Materials and items used for packaging of foods for special dietary uses should correspond to safety requirements for materials and items in contact with food products, as defined by the TR On Packaging Safety. Perishable dietetic foods for medical use and disease prevention should only be manufactured in small packaging for single-shot use.

Labeling of foods for special dietary uses should additionally contain information about the intended use of the product, target population group, and/or information about modification of the product composition and recommendations for use. Instructions for shelf-life after opening of dietetic foods for medical use and disease prevention should be mentioned in the label. There should be a special warning if the product must not be stored once opened or must not be stored in its packaging once opened.

Labeling of specialized sport foods should contain the following additional information: nutritional and energy value, share of standard physiological need, recommended dosage, modes of preparation (if applicable), and conditions and length of use. The consumer packaging should bear a note: "sport food product".

Compliance of foods for special dietary uses to requirements of the Technical Regulations is assessed in state registration. Foods are released for manufacturing, storage, transportation, and distribution after the state registration. State registration of foods for special dietary uses is issued for indefinite duration by a body authorized by a state member of the Customs Union. A food product for specialized dietary use receives state registration when the corresponding information is entered into the register of foods for special dietary uses.

The register of foods for special dietary uses makes part of the Unified register of registered foods. It consists of national parts of the Unified register of foods for special dietary uses, which are created and maintained by authorized registration bodies of state members of the Customs Union. The Unified register of foods for special dietary uses is kept in an electronic database protected from damage and unauthorized access. Information of the Uniform register of foods for special dietary uses is open to the public and is updated daily on the specific search server in the Internet.

Assessment of conformity of manufacturing, storage, distribution, transportation, and recycling of foods for special dietary uses is effected in the form of state supervision (control) according to requirements established by the relevant Technical Regulations.

Foods for special dietary uses that comply with the safety requirements and that have undergone the conformity assessment procedure should bear a uniform product commercialization mark for the market of the Customs Union state members.

References

[1] Kobelkova I, Baturin A. Analysis of the relationship lifestyle, diet, and anthropometric data to the health of persons working in a particularly harmful production. Vopr Pitaniya 2013;82(1):74−8.

[2] Tutelyan V, Baturin A, Kon I, Safronova A, Keshabyants E, Starovoytov M, et al. Evaluation of nutrition and nutritional status of infants and young children in the Russian Federation. Vopr Pitaniya 2010;79(6):57−63.

[3] Sazonova O, Baturin A. Nutrition and nutritional status of brainworkers with low physical activity. Vopr Pitaniya 2010;79(3):46−50.

[4] Onishchenko G, Slepchenko A, Smolensky V. About the actions taken for the implementation of the Customs Union Agreement on Sanitary Measures and cooperation with the World Trade Organization. Vopr Pitaniya 2013;82(12):70−4.

[5] Spirichev V. Biologically active food supplements are an additional source of vitamins in the diet of healthy and diseased people. Vopr Pitaniya 2006;75(3):50−8.

[6] Spirichev V. Scientific and practical aspects of pathogenetically based application of vitamins in the prevention and treatment purposes. Message 2. Vitamin deficiency - a factor that complicates the disease and reduces the effectiveness of treatment and prevention. Vopr Pitaniya 2011;80(1):4−13.

[7] Tutelyan V, Lashneva N. Biologically active substances of plant origin. Flavones: food sources, bioavailability, the effect on the xenobiotic metabolism enzymes. Vopr Pitaniya 2011;80(5):4−23.

[8] Spirichev V. Scientific and practical aspects of pathogenetically based application of vitamins in the prevention and treatment purposes. Message 1. The lack of vitamins in the diet of modern man: causes, consequences and ways of correction. Vopr Pitaniya 2010;79(5):4−14.

[9] Tutelyan V, Sukhanov B. Biologically active food supplements: modern approaches to quality and safety assurance. Vopr Pitaniya 2008;77(4):4−15.

[10] Sukhanov B, Kerimova M, Elizarova E, Chigireva E. Sanitary and epidemiological examination of foods in the frame of state sanitary and epidemiological surveillance on the customs border and customs territory of the Customs Union. Vopr Pitaniya 2011;80(4):25−31.

[11] Kochetkova A. Functional foods: the general and particular practical problems. Food Ingredients 2012;1:34−7.

[12] Kochetkova A. Actual aspects of technical regulations in the field of healthy foods. Food Ingredients Raw Mater Addit 2013;1:71−4.

[13] Smirnova E, Kochetkova A, Vorobiova I, Vorobiova V. Theoretical and practical aspects of the development of foods fortified with essential nutrients. Food Ind 2012;11:8−12.

[14] Smirnova E, Kochetkova A. Market of functional dairy foods. Dairy Ind 2011;2:63−6.

[15] Mazo V, Kodentsova V, Vrzhesinskaya O, Zilova I. Fortified and functional foods: similarities and differences. Vopr Pitaniya 2012;81(1):63−8.

[16] Samoilov A, Kochetkova A. Functional fat and oil based foods. Spreads with synbiotic. LAP LAMBERT Academic Publishing; 2011.

[17] Bagraintseva O, Mazo V, Kochetkova A, Shatrov G. The use of "functional foods" labeling. Milk Process 2013;2:64−8.

[18] Vorobiova I, Vorobiova V, Kochetkova A, Smirnova E. Foods for special dietary uses: general and specific definitions and characteristics. Food Ind 2012;12:16−8.

[19] Vorobiova V, Shatnyuk L, Vorobiova I, Mikheeva G, Trushina E, Zorina E, et al. Classification and characterization of specialized foods for sportsmen nutrition. Vopr Pitaniya 2010;79(6):64−8.

[20] Tutelyan V, Nikitiuk D, Pozdnyakov A. Optimization of sportsmen nutrition: realias and horizons. Vopr Pitaniya 2010;79(3):78−82.

[21] Zilova I, Nikitiuk D. Analysis of specialized foods intended for sportsmen nutrition (Research carried out in 2007-2010). Vopr Pitaniya 2011;80(2):71−5.

Nutraceutical and Functional Food Regulations in India

19

Raj K. Keservani*, **Anil K. Sharma***, **F. Ahmad**† and **Mirza E. Baig****

**School of Pharmaceutical Sciences, Rajiv Gandhi Proudyogiki Vishwavidyalaya, Bhopal, India*
†*F-34, Okhla, New Delhi, India* ***Pfizer Ltd. Haryana, India*

19.1 Introduction

Nutraceutical a portmanteau of the words "nutrition" and "pharmaceutical," is a food or food product that reportedly provides health and medical benefits, including the prevention and treatment of disease.

A nutraceutical is demonstrated to have a physiological benefit or provide protection against chronic disease. Such products may range from isolated nutrients, dietary supplements, and specific diets to genetically engineered foods, herbal products, and processed foods such as cereals, soups, and beverages. With recent developments in cellular-level nutraceutical agents, researchers and medical practitioners are developing templates for integrating and assessing information from clinical studies on complementary and alternative therapies into responsible medical practice [1].

Functional food is a food where a new ingredient(s) (or more of an existing ingredient) has been added to a food and the new product has an additional function (often one related to health promotion or disease prevention) [1].

The Indians, Egyptians, Chinese and Sumerians are just a few civilizations that have provided evidence suggesting that foods can be effectively used as medicine to treat and prevent disease (see Table 19.1). Ayurveda, the 5000-year-old ancient Indian health science, mentioned benefits of food for therapeutic purpose [2].

Functional food is the generic term for food that has been linked to health benefits. The Institute of Medicine's Food and Nutrition Board (in the United States) has defined functional food as "any food and food ingredients that may provide health benefit beyond the traditional nutrition that it contains". Functional food can be from plant sources or animal sources. The term nutraceutical was coined in the United States and is used to describe foods or food components that have the potential to cure specific disease conditions [3–6].

Historically, in India multiple laws and regulations prescribed varied standards regarding food, food additives, contaminants, food colors, preservatives and labeling.

Table 19.1 Functional Food Components*

Functional Components	Source	Potential Benefits
Carotenoids		
α-Carotene/ β-carotene	Carrots, fruits, vegetables	Neutralize free radicals, which may cause damage to cells
Lutein	Green vegetables	Reduce the risk of macular degeneration
Lycopene	Tomato products (ketchup, sauces)	Reduce the risk of prostate cancer
Dietary fiber		
Insoluble fiber	Wheat bran	Reduce risk of breast or colon cancer
β-Glucan	Oats, barley	Reduce risk of cardiovascular disease; protect against heart disease and some cancers; lower LDL and total cholesterol
Soluble fiber	Psyllium	Reduce risk of cardiovascular disease; protect against heart disease and some cancers; lower LDL and total cholesterol
Fatty acids		
Long chain omega-3 fatty acids-DHA/ EPA	Salmon and other fish oils	Reduce risk of cardiovascular disease Improve mental, visual functions
Conjugated linoleic acid	Cheese, meat products	Improve body composition; decrease risk of certain cancers
Phenolics		
Anthocyanidins	Fruits	Neutralize free radicals; reduce risk of cancer
Catechins	Tea	Neutralize free radicals; reduce risk of cancer
Flavonones	Citrus	Neutralize free radicals; reduce risk of cancer
Flavones	Fruits/vegetables	Neutralize free radicals; reduce risk of cancer
Lignans	Flax, rye, vegetables	Prevention of cancer, renal failure
Tannins (proanthocyanidins)	Cranberries, cranberry products, cocoa, chocolate	Improve urinary tract health; reduce risk of cardiovascular disease
Plant sterols		
Stanol ester	Corn, soy, wheat, wood oils	Lower blood cholesterol levels by inhibiting cholesterol absorption

(Continued)

Table 19.1 (Continued)		
Functional Components	**Source**	**Potential Benefits**
Prebiotics/probiotics		
Fructo-oligosaccharides	Jerusalem artichokes, shallots, onion powder	Improve quality of intestinal microflora, gastrointestinal health
Lactobacillus	Yogurt, other dairy	Improve quality of intestinal microflora, gastrointestinal health
Soy phytoestrogens		
Isoflavones: daidzein genistein	Soybeans and soy-based foods	Menopause symptoms, such as hot flashes, protect against heart disease and some cancers; lower LDL and total cholesterol
Source: International Food Information Council.		

India has recently passed the Food Safety and Standard Act 2006, a modern integrated food law to serve as a single reference point in relation to regulation of food products including nutraceutical, dietary supplements and functional food. The Food Safety and Standard Act still needs to be considerably more substantive with infrastructure and appropriate stewardship for it to match the international standards of the United States and Europe [7]. A significant augmentation is necessary for the act to have a large impact on the Indian functional food and nutraceutical industry like the Dietary Supplements Health Education Act (DSHEA) 1994 has had on the dietary supplement industry in the United States.

The passing of this act in India is a significant first step, but much more has to happen to eliminate the confusing overlap with old laws and regulations. Yet, in India functional foods/nutraceuticals are not categorized separately as in the United States, Europe and Japan. And also, the concept of functional food has somewhat different connotations in different countries. In Japan, for example, functional foods are defined based on their use of natural ingredients. In the United States, however, the functional food concept can include ingredients that are products of biotechnology [8]. In India these functional foods can include herbal extracts, spices, fruits and nutritionally improved foods or food products with added functional ingredients. This chapter deals with the history and present regulatory status of functional foods/nutraceuticals in India.

19.2 Positioning benefits

There are five primary benefit platforms for positioning health and nutrition products [9]. Virtually any functional product can be positioned against any one or a combination of these benefits:

Prevention: Foods that provide health management through disease and symptom prevention fall into this category.

Performance: A product that provides health enhancement through improved physical and mental condition.

Wellness: Wellness benefits are about feeling good and finding balance. This is a holistic approach to health care that includes the body, mind, and spirit.

Nurturing: Foods that can supply a sense of caring for the health and quality of life for others and the associated sense of satisfaction for the caregiver. Marketing a product from this platform would include a focus on growth and development, aging, and healing.

Cosmetics: Cosmetic benefits are about looking good and enhancing self-esteem through improved physical condition and personal appearance.

19.3 Indian market and health

The global nutraceutical market in 2008 was estimated to be $117 billion, of which India's share was only 0.9% [10]. The global market is estimated to reach $177 billion by next year, growing at a healthy Compound Annual Growth Rate (CAGR) of 7%. With increasing penetration of preventative health care products in the Indian market, growing health awareness, higher disposable income and other factors, the Indian nutraceutical industry has shown a promising CAGR of 18% in the last three years. According to one report, the total Indian nutraceutical market in 2015 is expected to be approximately $5 billion. Fast moving consumer goods companies and pharmaceutical companies are major players in the Indian nutraceutical market (Figure 19.1).

19.4 Regulation

The regulatory framework governing foods in major jurisdictions is evolving [11]. Food research, product and process innovation and change in consumer behavior facilitate adaptation of food regulations. Increasingly, the awareness is manifested through consumption of particular foods and dietary supplements believed to contribute to good health and in some cases, to hold therapeutic value in the treatment or prevention of specific diseases. Many of these food products are becoming commonly known as nutraceuticals or functional foods.

Nutraceuticals are products that have the characteristics of both a nutrient and a pharmaceutical. Taken as dietary supplements, they can modulate the symptoms of various disease conditions by providing the additional nutrients our bodies may need to maintain well-being. Food laws in every country are the basis of regulations of all kinds of food including health food, dietary supplements, functional

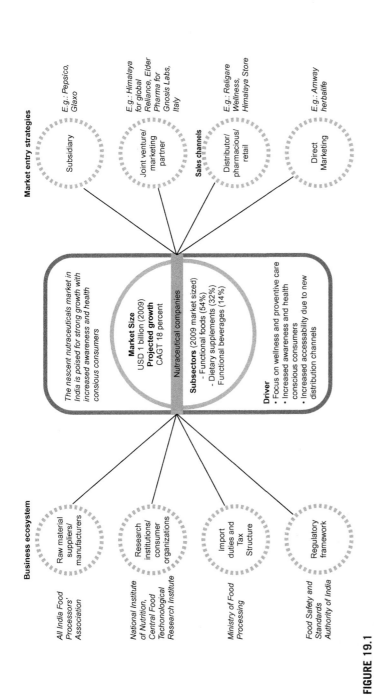

Business ecosystem

All India Food Processors' Association — Raw material suppliers/manufacturers

National Institute of Nutrition, Central Food Techonological Research Institute — Research institutions/consumer organizations

Ministry of Food Processing — Import duties and Tax Structure

Food Safety and Standards Authority of India — Regulatory framework

Market entry strategies

Subsidiary — E.g.: Pepsico, Glaxo

Joint venture/marketing partner — E.g.: Himalaya for global Reliance, Elder Pharma for Gnosis Labs, Italy

Sales channels

Distributor/pharmacious/retail — E.g.: Religare Wellness, Himalaya Store

Direct Marketing — E.g.: Amway herbalife

The nascent nutraceuticals market in India is poised for strong growth with increased awareness and health consious consumers

Nutraceutical companies

Market Size
USD 1 billion (2009)
Projected growth
CAGT 18 percent

Subsectors (2009 market sized)
- Functional foods (54%)
- Dietary supplements (32%)
Functional beverages (14%)

Driver
• Focus on wellness and preventive care
• Increased awareness and health conscious consumers
• Increased accessability due to new distribution channels

FIGURE 19.1

Indian nutraceuticals market snapshot.

food and nutraceuticals as specific guidelines/regulations are framed to regulate health food.

Stephen DeFelice, defined "nutraceutical" as "a food or part of food that provides medical or health benefits including the prevention and treatment of disease."

19.4.1 Overview of regulations

The United States introduced DSHEA in 1994, which allowed considerable flexibility between foods and medicines found in other parts of the world [11]. Under DSHEA a dietary supplement may contain "a herb or other botanical" or "a concentrate, metabolite, constituent, extract or combination of any ingredient from the other categories."

The American food safety regulatory system is far more centralized than the European system. The issue of adulteration became a national issue in 1848 when Congress passed the Drug Importation Act, requiring inspection by the U.S. customs service to prevent the entry of adulterated drugs from abroad. In 1862, in order to address the issue of adulterated food, the chemical division of the United States Department of Agriculture (USDA) was established and then renamed the Bureau of Chemistry (now the Food and Drug Administration, FDA).

The genesis of U.S. food legislation goes back to June 1906 when the Food and Drugs Act was enacted, which prohibited interstate commerce of misbranded and adulterated foods, drinks and drugs. In 1938, the Food and Drugs Act was preempted by the Federal Food, Drug and Cosmetic Act (FDCA). This Act focuses on food misbranding and adulteration and serves as the basic framework for food regulation by the FDA and the USDA. This legislation created food standards, mandated inspections of factories and provided for the issuance of court injunctions in addition to the already existing seizure and prosecution remedies. Since 1938, the FDCA has been amended a number of times and additional supporting laws have been enacted relating to food safety, security threats, and nutrition. Additionally, this federal framework is supplemented by state laws.

The USDA and the FDA are the main authorities in food regulation in the United States. These two federal agencies encompass all phases of the food regulatory system: they evaluate, investigate, regulate, inspect and sanction. However, even in a centralized system like the United States, some argue that the USDA and the FDA should consolidate into one single food agency. The FDA regulates both finished dietary supplement products and dietary ingredients under a different set of regulations than those covering "conventional" foods and drug products.

Under the DSHEA, the dietary supplement or dietary ingredient manufacturer is responsible for ensuring that a dietary supplement or ingredient is safe before it is marketed. The FDA is responsible for taking action against any unsafe dietary supplement product after it reaches the market. All domestic and foreign companies that manufacture, package, label, or hold dietary supplement, including those involved with testing, quality control and dietary supplement distribution in the United States, must comply with the dietary supplement Current Good Manufacturing Practices for quality control.

In addition, the manufacturer, packer, or distributor whose name appears on the label of a dietary supplement marketed in the United States is required to submit to the FDA all serious adverse event reports. Further, the FDA's other responsibilities include product information, such as labeling, claims, package inserts and accompanying literature.

19.4.2 History of food regulations in India

India is the world's second largest producers of fruits and vegetables, but only a small amount of perishable agriculture products are processed at approximately 2% in comparison to 80% in the United States [7]. Barriers to growth in this food sector include poor infrastructure and logistic and tight food regulation. The multiplicity of food regulation policy makers and enforcement agencies prevailing in different sectors of the food industry contributed to considerable confusion among the consumers, producers and retailers and business and is detrimental to the growth of the functional food and nutraceutical industry [12−13]. By the mid-1990s the food processing sector laws were framed in a veritable grid of regulation including a multitude of state laws as well as the following national laws:

- Export (Quality Control and Inspection) Act 1963
- Solvent Extracted Oil Control Order 1967
- The Insecticide Act 1968
- Meat Food Products Order 1973
- Prevention of Food Adulteration Act (PFA) 1954 rules (Ministry of Health and Family Welfare) with last amendments in 1986
- Bureau of Indian Standards Act 1986
- Environmental Protection Act 1986
- Pollution Control Act 1986
- Milk and Milk Products Order 1992
- The Infant Milk Substitutes Feeding Bottles and Infant Food (Regulation of Production, Supply) Act 1992 and Rules 1993
- Food Product Order 1995
- Agriculture Produce Act
- Essential Commodities Act 1995 (Ministry of Food and Consumers Affairs)
- Industrial license
- Vegetable Oil Product Control order 1998

In 1998, the Prime Minister's council on trade and industry appointed a subjective group on food and agriculture industries, which recommended a unified legislation under a single food regulatory authority. Public experts and members of the Standing Committee of Parliament encouraged the convergence of current food laws with single regulatory authorities accountable for public health and food safety in India. Further, integrating all acts and orders relating to food and eliminating multilevel and multidepartmental controls over food, special emphasis was given to nutraceutical and functional food, a poorly defined segment with growing potential and implications on consumer heath.

In 2002, a national nonprofit association had been constituted with main objectives that every food manufacturing company should provide scientific-based support to their products in order to protect the consumers and to promote and defend a regulatory environment conducive to the industry in general as well as consumer protection.

In 2003, a Ministry of Health expert group report indicated the need to create new categories for regulating functional food and dietary supplements in the present food laws. It is recommended that there should be mandatory safety testing for these products. In India voluntary standards are developed by the Bureau of Indian Standards and National Standards body, which comprise representatives from various food sector stakeholder groups. These standards basically deal with product certification, quality, system certification and testing and consumers affairs. Efforts are made to match Indian standards with international ones [14].

In 2005, a number of committees, including the Standing Committee of Parliament on Agriculture, have emphasized the need for a single regulatory body and integrated law.

Finally, the Indian Food Safety Standard Bill 2005, signed into law, promised a major impact on the Indian food processing industry. The Indian Food Safety and Standard Act came into enforcement in 2006 with two main objectives: to introduce a single statute relating to food and to provide for scientific development of the food processing industry.

19.4.2.1 Food Safety and Standard Act 2006

The Food Safety and Standard Act 2006 aims to establish a single reference point for all matters relating to Food Safety and Standards, by moving from multilevel, multidepartmental control to a single line of command. It incorporates the salient provisions of the Prevention of Food Adulteration Act 1954 and is based on international legislations, Instrumentalities and the Codex Alimentarius Commission [12]. The salient features of this Act are as follows:

- The Food Safety and Standards Act 2006 consolidates the eight laws governing the food sector and establishes the Food Safety and Standards Authority (FSSA) to regulate the sector and other allied committees. FSSA will be aided by several scientific panels and a central advisory committee to lay down standards for food safety. These standards will include specifications for ingredients, contaminants, pesticide residue, biological hazards, labels and others.
- Everyone in the food sector is required to get a license or a registration that would be issued by local authorities.

- The law will be enforced through state commissioners of food safety and local level officials.
- The act provides for a graded penalty structure where the punishment depends on the severity of the violation.
- The responsibility of framing and regulating standards for nutraceuticals rests with the Food Safety and Standards Authority of India (FSSAI) as outlined in the Food Safety Act 2006. This authority will be in charge of categories like functional foods, nutraceuticals, dietetic products and other similar products [12–14].
- Food Safety and Standard Act 2006 consists of 12 chapters.

  Chapter IV, Article 22 of the act addresses nutraceutical, functional food, dietary supplements and the need to regulate these products such that anyone can manufacture, sell, or distribute or import these products. These products include novel foods, genetically modified articles of food, irradiated food, organic food and food for special dietary uses, functional food, nutraceuticals and health supplements, whereas Articles 23 and 24 address the packaging and labeling of food and restriction of advertisement regarding foods [12]. According to this act, foods for special dietary uses, functional food, or nutraceutical or dietary supplements are the following:

  1. Foods that are specially processed or formulated in order to satisfy particular dietary requirements that exist because of a particular physiological or physical condition and that are processed as such wherein the composition of these foodstuffs must differ significantly from the ordinary food of comparable nature, if in such ordinary food exists one or more of the following ingredients, namely:
     a. Plants or botanicals in the form of powder, concentrates, or extracts in water, ethyl alcohol, single or in combinations.
     b. Minerals, vitamins, or proteins (amounts not exceeding recommended daily allowance (RDA) for Indians) or enzymes.
     c. Substances from animal origin.
     d. A dietary substance used by human beings to supplement the diet by increasing the total dietary intake.
     e. A product that is labeled as "food for special dietary uses" functional food or nutraceutical dietary supplements, which are not represented for use as conventional food and whereby such products may be formulated in the form of powders, granules, tablets, capsules, liquids, jelly and other dosage forms but not parenterals and are meant for oral administration.
     f. Such product does not include a drug as defined in clause (b) and ayurvedic, siddha, and unani drugs as defined in clauses (a) and (h) of Section 3 of the Drugs and Cosmetics Act 1940 and rules made thereunder.

      g. Does not claim to cure or mitigate any specific disease, disorder, or condition (except for certain health benefits or such promotion claims) as may be permitted by the regulations made under this act.

      h. Does not include a narcotic drug or a psychotropic substance as defined in the schedule of the Narcotic Drugs and Psychotropic Substances Act 1985 and rules made thereunder and substances listed in Schedules E and E1 of the Drugs and Cosmetics Rules 1945.

  2. "Genetically engineered or modified food" means food and food ingredients composed of or containing genetically modified or engineered organisms obtained through modern biotechnology, or food and food ingredients produced from but not containing genetically modified or engineered organisms obtained through modern biotechnology.

  3. "Organic food" means food products that have been produced in accordance with specified organic production standards.

  4. "Proprietary and novel food" means an article of food for which standards have not been specified but is not unsafe: provided that such food does not contain any of the foods and ingredients prohibited under this act and the regulations made thereunder.

19.4.2.2 Benefits of implementation of the Act

- Unification of eight laws, i.e., steps to harmonization
- Alignment of international regulations
- Science-based standards
- Clarity and uniformity on novel food areas
- Help to curb corruptions

19.4.2.3 Problems with implementation of the Act

Regulation broadly defines the intervention of government in industry. Regulation primarily controls the product quality in case of food. Every system of regulation has its own pros and cons. But the benefits from the implementation of the Food Safety and Standards Act overwhelmed the problems that rise due to implementation of this act. Unlike in the United States, where the DSHEA is in place to regulate these products, in India the government is in the process of drafting a law to regulate manufacturing, importing and marketing of health foods, dietary supplements and other nutraceuticals [13−14].

19.5 Emerging opportunities

The United States Pharmacopoeia Convention (USP) decided to have a separate advisory panel on nutraceuticals and will be joining with the Indian scientific community to develop safety standards for the entire range of dietary supplements and nutraceuticals that are currently not under the strict regulatory classification of either drugs or foods [15]. Domestic and multinational companies are vying for position in the $500 million Indian nutraceuticals market, which is growing at 40% annually [16]. The Central Food Technological Research Institute, Mysore, is to be the major agency associated with USP in this regard. The United States and Europe are going to be emerging markets for nutraceutical exports from India, because an existing large market base is already in place and consumers are looking for better and healthier options to prevent lifestyle-related diseases [16,17]. The market potential for the United States and European markets alone for nutraceutical exports from India by 2013 will be to the tune of $90 billion. Companies like Amway India and Herbalife are utilizing direct multilevel marketing to reach new consumers about their products delivered by someone known and trusted. However, most of the large companies have not ventured into nutraceuticals or dietary supplements due to regulatory confusion, lack of adequate awareness and understanding and poor vision of the market [16].

19.6 Regulation of claims pertaining to nutraceuticals

There is currently no official harmonized definition or harmonization of the technical requirements and guidelines for functional food products in Asia. Academics and food companies generally understand functional foods to contain or to be fortified with nutrients or other bioactive compounds that help to maintain and promote health. Functional foods in Asia tend to be regulated under the conventional food category. The following claims are permitted in Asian countries:

> *Nutrient content claims*: Which state the level of certain nutrients on the product label.
> *Nutrient comparative claims*: Which describe the nutrient content or energy value relative to other similar foods.

Disease risk reduction claim established by international standard setting body the Codex Alimentarius (references used by Asian food authorities for the national food legislation) are generally not permitted in Asia. However, these claims are used in northern Asian countries that have established regulations for functional food. However, a health claims regulatory environment is evolving and significant changes in the next 5 years are expected [18–20].

19.7 Licensing and registration requirements

- Every food business operator in the country will have to obtain registration and license in accordance with the procedure laid down in FSSAI (licensing and registration of food business) regulation 2011 [21].

- A manufacturer cannot commence business unless he is registered or has a valid license.
- Petty food manufacturers (annual turnover less than Rs.12 lacs) have to register with the commissioner and manufacturers whose turnover is greater than 12 lacs to obtain a food license from FSSAI office.
- Existing licenses/registration should be converted into FSSAI license/ registration before August 5, 2012 (now extended by a few months).
- An application for the grant of a license shall be made in form B of schedule 2 to the concerned licensing authority; the license shall be issued within 60 days from the date of issue of an application ID number.
- After the issue of application ID number the licensing authority may direct the food safety officer to inspect the premises in the manner prescribed by the FSSAI in accordance with these regulations.
- The licensing authority shall issue a license in format C under schedule 2 of these regulations.
- Registration or license granted under these regulations shall be valid and subsisting, unless otherwise specified, for a period of 1−5 years.

19.7.1 Regulatory requirements for entry in India

As the nutraceutical regulation is evolving in India, with the recent implementation of FSSA, there is a possibility that some of the content is conflicting/confusing, but for Indian industry to take a shape, this has to be streamlined [22−29].

In order to enter the Indian nutraceutical market, some of the very important areas to focus include product evaluation, actual product analysis, procuring licenses and developing India-specific health and label claims.

19.7.1.1 Product evaluation

In Indian conditions, the formulations behave very funnily and get mixed up in classifications. Hence, the due diligence in terms of carving a specific amount for each ingredient and the combination of ingredients becomes very crucial.

In order to perform product assessment as per Indian regulatory definition, it is of utmost importance to examine each active ingredient and additive in the context of permissibility, standards and dosage of vitamins/minerals allowed as per the therapeutic, prophylactic, or RDA for Indians. Also, manufacturers are unclear whether their products will be classified as food or food supplement or drug in the context of the Prevention of Food Adulteration Act 1954 and Rules 1955, Food Safety and Standards Act 2006 and Drugs and Cosmetics Act 1940 and Rules 1945.

The Food Safety and Standards Rules 2011 highlights the regulatory enforcement structure and procedures that central government proposes to make. The structure has a hierarchy from commissioner of food safety to the number of officers and designations such as food safety officer, food analyst, etc., who will be involved in the product analysis process at different points.

Various steps in the product analysis include:

1. Developing extracts of documents and authenticating the same by concerned authority
2. Sample collection (in the presence of witnesses)
3. Sample dispatch to concerned authority (different process for bulk package and single package)
4. Food analysis
5. If analysis is not complete within stipulated period of time, further action plan by designated officer
6. Adjudication proceedings (holding enquiry, appeal procedure, hearing, etc.)

19.7.1.2 Licenses

Although the new FSSA promises to simplify licensing and registration processes for nutraceuticals, the actual process varies depending on the number of parameters. To get the product registered in India, the number of licenses (almost 4–5) will be required depending on the actual product status such as whether:

- The company wants to sell bulk drug or finished formulation
- The company is importing finished product or bulk ingredient
- The product to be imported is with or without an India-specific label and will the claims be developed in India
- The company has packaging license
- It requires a manufacturing license
- It requires a marketing license

The number of documents will have to be furnished by the food importer to the government authority along with registration application dossiers. The interlink through its regulatory product offerings provides regulatory support for the following licensing procedures, which need to be taken care of before launching these products in India:

- Import licensing
- Manufacturing licensing
- Marketing licensing
- Other state and national level clearances/licenses required from the regulatory side

19.7.1.3 Health and label claims

Developing health and label claims specific to Indian regulatory guidelines is the major element to be focused on while entering Indian market. International as well as national clients have a number of questions about Indian labeling and packaging requirements, packing of consignment, need for sample material and declaration for registration, composition of consignment and approach for the same, label content, and structure–function claim and label claim.

Based on the results of a regulatory assessment of the product, India-specific label content and claims are developed. New entrants should also consider some of the health claims used in India and the requirements to be met to make specific product claims.

19.8 Recommendation and conclusion

Although the Food Safety and Standard Act 2006 defines functional food/nutraceuticals legally, there are still further effective regulations; guideline and suitable protocols are required to gain momentum for effective implementation across the nation. Still there is need to clarify and formulate the regulatory framework. If substantiation effectively enforced the Food Safety and Standard Act there is the potential to open up tremendous opportunity for the functional food or nutraceutical industry. There are certain recommendations that could bring about improvements in existing regulations such as:

- To grab a larger pie from this world opportunity, Indian producers of nutraceutical products should unite to form a platform to market India as a brand. There is a need for an increased collaboration on the manufacturing and research and development front among Indian manufacturers. There has to be coordination among all agencies, including policy makers, regulators and manufacturers. The manufacturing, validation, research and development and intellectual property protection needs to be standardized.
- There should be expansion of Indian standards like Indian Pharmacopoeia so that manufacture of functional food/nutraceuticals complies with their safety and quality standards.
- Indian government has still to amend its laws regarding nutrition labeling as the U.S. Nutrition Labeling Education Act 1990 so that consumers become aware of safe and healthy facts regarding functional foods/nutraceuticals. Joint efforts by the government and private agencies in terms of suitable legislation and help from food scientists show there is tremendous potential for processed functional food in India in the future.
- There should be new retailing programs, increased validation and clinical research, heightened awareness due to media and government focus and greater corporate responsibility due to health awareness programs as well as new marketing and communication methods, innovation research and development and product development skills.

To conclude, the passing of the Food Safety and Standard Act 2006 was a significant first step but a lot more has to happen to eliminate the overlap of old laws and regulations. Prior to the FSSA, there were multiple laws and regulations governing food safety and standards. Nutraceuticals were grouped under the PFA. Food was classified as either fortified or proprietary [21]. Later in 2006, all of the existing laws were consolidated to form one single statute in order to ensure systematic and scientific development of the food processing industry. Food was classified under the following heads:

- Novel foods
- Genetically modified food
- Proprietary food
- Standardized food
- Foods for special dietary use
- Functional foods/nutraceuticals/health supplements

The Food Safety and Standards Regulations 2011 notified in the *Gazette of India* came into force on August 5, 2011, to regulate manufacture, distribution and sale of nutraceuticals, functional foods and dietary supplements in India.

References

[1] Nutraceuticals/Functional Foods and Health Claims on Foods: Policy Paper [Health Canada, 1998]. Hc-sc.gc.ca. Retrieved 06.03.11.
[2] Wildman REC, editor. Handbook of nutraceuticals and functional foods. 1st ed. CRC Series in Modern Nutrition. ISBN 0-8493-8734-5; 2001.
[3] International Food Information Council Functional Foods Now. Washington, DC: International Food and Information Council; 1999 [accessed 01.10.08].
[4] Kalra EK. Nutraceutical: definition & introduction. AAPS Pharmsci 2003;5(2): Article 25. doi:10.1208/PS/0500225. [accessed 01.01.09].
[5] Hardy G. Nutraceutical and functional food: introduction and meaning. Nutrition 2000;16:688−9.
[6] Brower V. Nutraceuticals: poised for healthy slice of healthcare market? Nat Biotechnol 1998;16:728−31.
[7] Regulation of functional food in Indian Subcontinent, food and beverages news, <http://www.efenbeonline.com/view_story.asp?type = story&id = 880>.
[8] New Food Words: Functional Foods and Nutraceuticals, Phytochemicals, <http://www.extension.iastate.edu/publications/PMI846.pdf>.
[9] <http://www.agbioforum.org/v3n1/v3n1a05-gilbert.htm>.
[10] <http://www.nutraceuticalsworld.com/issues/2012-11/view_features/tapping-indias-potential/>.
[11] <http://fnbnews.com/article/detnews.asp?articleid = 31115&sectionid = 49>.
[12] <http://www.commonlii.org/in/legis/num_act/fsasa2006234/>.
[13] India together: Legislative Brief at, <http://www.indiatogether.org/2006/feb/laws-foodsafe/htmal/hilite>.
[14] FICCI study on Implementation of Food Safety and Standard Act 2006: An Industry Perspective at, <http://www.indiaenvironmentportal.org.in/Files/food_safety_study.pdf>.
[15] KaK A. Supplementary growth Express Pharma, February 16−29, 2008.
[16] Jacobs K. INDIA: Nutraceutical Market sees 40% growth, Just- Food, 17 June 2008.

[17] Mehta AG. Untapped wealth of nutraceutical exports. The Hindu Business Line; 2008.

[18] <http://www.asiafoodjournal.com/article-5565-definingfiunctionalfoodclaims-Asia.html>.

[19] <http://www.Nutraceuticalsworld.Com/Articles/2008/11/India>.

[20] <http://sitesources.worldbank.org/INTARD/ResourcesHealthEnhancing_Food_ARD_DP_30_final_pdf>.

[21] <http://www.grantthornton.in/html/gt_insight/?p = 1640>.

[22] <http://mofpi.nic.in/ContentPage.aspx?CategoryId = 147>.

[23] The Food Safety and Standards Act, 2006 — Bare act with short comments. Delhi: Professional Book Publishers; 2009.

[24] Nutraceuticals — Critical supplement for building a healthy India, 2009, E & YFICCI report.

[25] Industry Insight — Nutraceuticals, August 2010, Cygnus Report.

[26] Interlink Knowledge Bank.

[27] Expert Interviews.

[28] Global Nutraceutical Industry: Investing in Healthy Living, Frost-FICCI report; 2011.

[29] Chaturvedi S, et al. Role of nutraceuticals in health promotion. Int J PharmTech Res 2011;3(1):442−8.

Regulations on Nutraceuticals, Functional Foods and Dietary Supplements in India

20

Alluri V Krishnaraju, Kiran Bhupathiraju, Krishanu Sengupta and Trimurtulu Golakoti

Laila Nutraceuticals, India

20.1 Introduction

Food is essential for life and is required not only for sustaining growth but also for combating diseases in the human life cycle. All living things are nourished by macro and micro nutrients of food and some foods are known to have health benefits. Foods with health benefits or foods that modulate biological functions of the body and aid prevention and/or relief from pathological states are commonly known as nutraceuticals or functional foods or dietary supplements. India is rated as one of the major food-consuming and -producing nations; a vast variety of foods are produced in India as it falls under a wide climatic range. Developing countries like India achieved remarkable control over communicable diseases but are struggling with rapidly growing lifestyle disorders. At present the Indian community is facing the threat of lifestyle disorders due to the rapid pace of urbanization and lack of awareness of proper food consumption, essential nutrients, and supplements, which are known to rejuvenate the human body. A recent survey suggested that 31% of the population is either overweight or obese, which in turn leads to diabetes and heart disease. India has also recently become the country with the highest percentage of patients with cardiovascular disorders. In addition, increased health care costs cast a huge financial burden on the government and people of India. These factors warrant increased awareness and enhanced use of proper nutritional supplements, which certainly would escalate the pace of nutraceutical trade in India [1]. These lifestyle disorders require long-term interventions like consumption of nutraceutical products and food supplements. In 2010 the Indian nutraceutical market was estimated at $2 billion contributing only 1.5% of the $140.1 billion global market, which includes the dietary supplement (40%) and functional food and beverage market (60%) and is

Nutraceutical and Functional Food Regulations in the United States and Around the World.
DOI: http://dx.doi.org/10.1016/B978-0-12-405870-5.00020-7

expected to reach $5 billion by 2015 [2]. At present, India's nascent market is incorporating traditional herbal ingredients into the nutraceutical segment, such as Chyawanprash ($74.5 million in 2010). Existence of the need for alternative medicine and increasing consumer awareness about conventional nutraceutical ingredients provides a niche for the nutraceutical market. Therefore the growing market for nutraceuticals requires stringent and effective regulations to control import, manufacture, sale, and marketing in India. With an intention to ensure sufficient access to safe and effective food ingredients to the people, the federal government of India established the Food Safety and Standards Authority of India (FSSAI) under the Food Safety and Standards Act 2006. The regulatory framework governing foods, functional foods, and dietary supplements in India is evolving and has taken on a healthy pace after FSSAI was established. The major objective of FSSAI is to set up science-based food standards and regulate manufacture, processing, distribution, sale, and import of foods to ensure the safety of food for human consumption [2,3].

Functional foods or nutraceuticals or health supplements are foods, which are specially processed or formulated to satisfy dietary requirements to address a particular physical or physiological condition or specific diseases and disorders. Minerals, vitamins, proteins, metals, amino acids, ethanol, hydro-alcohol, or water extracts of plants, botanicals, and their compositions are categorized as functional foods [4]. Use of such functional foods is restricted to non-parenteral application only. Application of botanicals, animal products, minerals, and vitamins for specific diseases or disorders is controlled by the Department of Ayurveda, Yoga and Naturopathy, Unani, Siddha, and Homoeopathy (AYUSH) in India. The former Department of Indian Systems of Medicine and Homoeopathy (ISM&H) was transformed into AYUSH by the Indian government in 2003 to provide focused attention for the development of education and research in AYUSH systems of medicine [5].

During the past decade, non-pathogenic microbes that benefit the host by improving its intestinal microbial balance have been included in various foodstuffs including fermented foods. Significant research-based evidence has boosted the demand and consumption of probiotics not only across the globe but also the Indian market, which has resulted in a large influx of probiotic supplements or probiotic fortified food and beverages. Many doctors and other health care professionals are advocating probiotic usage for various health conditions. Probiotics are available in the form of fortified foods (i.e., yogurt, ice creams, and milk beverages) and dietary supplements (i.e., capsules, tablets, and powders). A variety of probiotic products made by companies like Nestle, Mother Dairy, Amul, and Yakult are available in India. Probiotic use is new to India but it is a fast-growing category that necessitates regulatory guidance to monitor and control the production and sale of probiotics as drugs or biologicals or food supplements [6,7]. Likewise, Multi Level Marketing (MLM) or direct selling of supplements and functional food is evolving in India with companies like Amway and Herbalife rapidly expanding their footprint. With the threat of Ponzi schemes and no strong regulation in place to control such activities, the Indian Direct Selling Association (IDSA) and Federation of Indian Chambers of Commerce and Industry (FICCI) have requested the government to establish regulations to enable genuine companies to operate in a healthy environment. Various state governments have started to put in place regulations to control and monitor MLM organizations. With almost 4 million people (70% of them are women) and $1140.9 million in sales, MLM business is reaching a threshold for rapid and sustained growth in India as per The Indian Direct Selling Industry, Annual Survey 2010–11, PHD Chamber of Commerce and Industry and IDSA 2012 [8].

20.2 Food Safety and Standards Act

Indian Parliament passed the Food Safety and Standards Act in 2006 with the intention of integrating all existing food laws into one and to have a single regulatory body to implement these laws. It specifically encompasses eight laws:

- The Prevention of Food Adulteration Act 1954
- The Fruit Products Order 1955
- The Meat Food Products Order 1973
- The Vegetable Oil Products (Control) Order 1947
- The Edible Oils Packaging (Regulation) Order 1998
- The Solvent Extracted Oil, De-oiled Meal, and Edible Flour (Control) Order 1967
- The Milk and Milk Products Order 1992
- Essential Commodities Act 1955 Relating Food

Enactment of a unified law helped to establish a new national regulatory body, FSSAI *vide* notification no. S. O. 2165 (E) dated September 5, 2008, with the mandate to lay down science-based standards for articles of food and to regulate their manufacture, storage, distribution, sale, and import, to ensure availability of safe and wholesome food for human consumption. In addition, the Central Advisory Committee (CAC) was established on October 5, 2009, to ensure close cooperation among the Food Authority, the enforcement agencies, and organizations operating in the field of food safety. The CAC advises the Food Authority on the performance of its duties, in particular the drawing up of a proposal for the Food Authority's work program, the prioritization of work, identifying potential risks, pooling of knowledge, and such other functions specified by regulations [5,9].

The Act provides an imperative resolution to integrate concepts on foods (novel foods, health foods, nutraceuticals, and genetically modified foods). There are provisions in the Act for allocating license for manufacturing safe and quality food with a high degree of consumer confidence and implementing effective, transparent, and accountable regulatory framework within which the industry can work efficiently and comply with the dynamic requirements of the Indian Food Trade Industry and International Trade. Special provisions are also provided in the Act to deal with the intentional violations. It also executes graded penalty depending upon the severity of offense by enforcing legislation by the state governments or the Union Territories through the State Commissioner for Food Safety and local bodies [10].

20.2.1 The Food Safety and Standards Regulations

Followed by notification in the *Gazette of India vide* G.S.R. No:-362-(E) dated May 5, 2011, the Food Safety and Standards Rules 2011 came into force after three months from the date of publication in the official gazette (Table 20.1). These Food Safety and Standards Regulations came into force beginning August 5, 2011 [11].

Table 20.1 Food Standards and Safety 2006, Regulations Notified in the *Gazette of India*

List of Regulations	Exercise of the Powers
*Milk and Milk Products Order (1992), (2009)	Section 99 & Subsection (3) of Section 1 of Food Safety and Standards Act 2006 (34 of 2006)
Licensing and Registration of Food Businesses (2011)	Clause (o) of Subsection (2) of Section 92 & Section 31 of Food Safety and Standards Act 2006 (34 of 2006)
Packaging and Labeling (2011)	Clause (k) of Subsection (2) of Section 92 & Section 23 of Food Safety and Standards Act 2006 (34 of 2006)
Food Product Standards and Food Additives (2011)	Clause (e) of Subsection (2) of Section 92 & Section 16 of Food Safety and Standards Act 2006 (34 of 2006)
Prohibition and Restriction on Sales (2011)	Clause (l) of Subsection (2) of Section 92 & Section 26 of Food Safety and Standards Act 2006 (34 of 2006)
Contaminants, Toxins, and Residues (2011)	Clause (i) of Subsection (2) of Section 92 & Section 20 of Food Safety and Standards Act 2006 (34 of 2006)
Laboratory and Sampling Analysis (2011)	Clause (q) of Subsection (2) of Section 92 & Section 40 and 43 of Food Safety and Standards Act 2006 (34 of 2006)
Food Import Regulations (2013)	Subsection (1) of Section 25 & Section 92 of Food Safety and Standards Act 2006 (34 of 2006)

*Milk and Milk Products Order 1992 was deemed to regulations made under Food Safety and Standards Act 2006 dated on June 29, 2009, notified in the *Gazette of India* Part II-Section (3)-Subsection (ii).
Source: http://www.fssai.gov.in/GazettedNotifications.aspx

20.2.2 **Licensing and registration**

Subsection (1) of Section 31 of the Act deals with Licensing and Registration of Food Business, which states that no person shall commence or carry on any food business except under a license after complying with the Food Safety Management requirements mentioned in Schedule 4. All the Food Business Operators (FBOs) in this country shall be either registered (Petty food business) or licensed (depending up on the scale and volume of business as laid down in Licensing and Registration Regulations 2010). Existing registration/license holders under various orders and Acts shall inform the License/Registration Authority through an application upon expiry of the validity of the existing registration/license. However, FBOs holding registration/license under any Act or order without any specific validity or expiry date should register/license by paying applicable fees [10,12].

20.2.3 **Product approval**

The Act broadly classified functional foods or nutraceuticals or health supplements under the common definition of food as "food for special dietary uses" under Clause 22 of the Act. The approval for food labeled as proprietary foods, whether licensed under previous act/orders or, are intended to be placed on the market and contain novel foods, functional foods, food supplements, irradiated foods, genetically modified foods, foods for special diet or extracts or concentrates of botanicals, herbs, or of animal sources approval shall be made in the format 1(b) under Section 22 of the Act. Product should comply with the specifications given for its approval, which encompass administrative information, technical information, information on dietary exposure, nutritional impact, and potential impact on the customer, efficacy, and details of fee (Figure 20.1). Applications are then subjected to screening and approval procedures, after approval a No Objection Certificate (NOC) is issued by the Product Approval and Screening Committee (PASC) [13].

20.2.4 **Packaging and labeling regulations**

The emerging market for packaged food provides both opportunities and challenges to food processors, importers, food packagers, and labelers to respond to consumers' requirements. Innovation in food product development is a key distinguishing factor for convincing consumers, but packaging and labeling are critical parameters that should comply with regulatory requirement. Labeling requirements were specified in the new Act to provide truthful information of its nutritional values to the consumer.

Section 23 of Food Safety and Standards Act 2006 (Act 34 of 2006) governs Packaging and Labeling Regulations, and mandated the labeling requirements, which encompass name of food, list of ingredients, nutritional information, declaration regarding vegetarian or non-vegetarian origin, food additives, name and complete address of the manufacturer, net quantity, lot/code/batch identification,

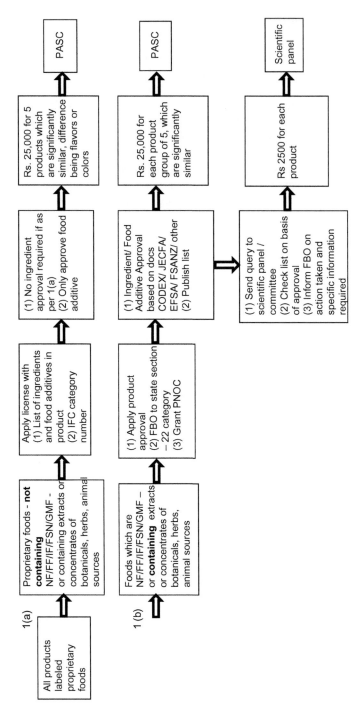

EFSA - European Food Safety Authority, **FBO** - Food Body Organization, **FF** – Functional Foods, **FSANZ** - Food Standards Australia New Zealand, **FSN** – Foods for Special Dietary Uses, **GMF** – Genetically Modified Uses, **IF** – Irradiated Foods, **IFC** – Indian Food Code, **JECFA** - The Joint FAO/WHO Expert Committee on Food Additives, **NF** – Novel Foods, **PASC** - Product Approval and Screening Committee, **PNOC** - Provisional No Objection Certificate.

FIGURE 20.1

Schematic representation of the new product approval.

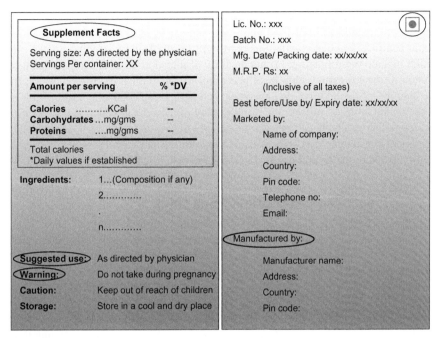

FIGURE 20.2

Labeling requirements.

date of manufacture or packing, best before and use by date, country of origin for imported food, and instructions for use should be imprinted apart from other general label requirements (Figure 20.2). Specific requirements/restrictions were provided in the regulations on the manner of labeling of infant milk substitute and infant food, edible oils and fats, permitted food colors, and irradiated food. Labeling requirements for other products with some exemptions were also specified under the regulations [14].

20.2.5 Food product standards and food additives

The FSSAI justified the ambiguities arising due to a complicated food system (Codex Food Categorization System) by allowing additives as individual products instead of product categories. A comprehensive list of additives that were proven to be safe were provided under Food Safety and Standards (Food Products Standards and Food Additives) Regulations 2011, which include coloring matter, artificial sweeteners, preservatives, anti-oxidants, emulsifying and stabilizing agents, anticaking agents, flavoring agents and related substances, sequestering and buffering agents, and other substances to be used within specified limits [15]. In the new categorization system, products were generalized under broad generic categories and subcategories so that the additives currently allowed to be used in one single product will also be allowed in similar products or proprietary foods falling under same categories. All food additives, subject to the provisions of the Good Manufacturing Practice (GMP) Standard, will be used under conditions of GMP, which include the following: (a) the quantity of the additive added to food shall be limited to the lowest possible level necessary to accomplish its desired effect; (b) the quantity of the additive that becomes a component of food as a result of its use in the manufacturing, processing, or packaging of a food and which is not intended to accomplish any physical, or other technical effect in the food itself, is reduced to the extent reasonably possible; and (c) the additive is of appropriate food grade quality and is prepared and handled in the same way as a food ingredient [16].

20.2.6 Prohibition and restrictions on sales

The Food Safety and Standards (Prohibition and Restrictions on Sales) Regulations 2011, which are based on Section 26 of Food Safety and Standards Act 2006 (34 of 2006), prohibit the sale of certain admixtures and restrict the sale of certain ingredients and products for human consumption [17].

20.2.7 Regulations on contaminants, toxins, and residues

Food Safety and Standards (Contaminants, Toxins, and Residues) Regulations 2011, Section 20 of Food Safety and Standards Act 2006 defines Crop contaminant as "any substance not intentionally added to food, but which gets added to articles of food in the process of their production (including operations carried out in crop husbandry, animal husbandry and veterinary medicine), manufacture, processing, preparation, treatment, packing, packaging transport or holding of articles of such food as a result of environmental contamination." The regulations specify standards for metal contaminants, crop contaminants, and naturally occurring toxic substances, and residues of pesticides, antibiotic, and other pharmacologically active substances [18].

20.2.8 Regulations on laboratory and sampling analysis

Section 43 of Food Safety and Standards Act 2006 notified 68 National Accreditation Board for Testing and Calibration Laboratories (NABL)-accredited food testing laboratories for the purpose of carrying out analysis of samples by the Food Analysts under the Act for a period of one year, with qualified staff and equipment. This accreditation assists the Indian industries to enhance the quality and reliability of Indian goods sold in the domestic and international markets, thereby, improving the growth of Indian economy. The accreditation services are provided for testing and calibration in accordance with International Organization for Standardization (ISO) Standards. The Act endorses an imperative decision to implement mobile laboratories to cater to large public congregations and to inaccessible areas [19–21].

20.2.9 Regulation on authorization of health claims

The FSSAI released the draft Regulation on Labeling (Claims) on December 27, 2012, which provides guidelines to FBOs regarding Nutritional and Health Claims based on science and supported by sound and sufficient scientific evidence. When a function claim is made about the benefit of a nutrient, the food carrying the claim must be at least a source of the said nutrient. All health claims should be accompanied by instructions on "maximum per day serving" of the product. Words such as "complete," "planned," "exhaustive," "total," "absolute," or other synonymous words signifying that the food will provide complete nutrition should not be generally used in foods that are not proposed as a complete diet replacement for weight management or those that are intended to be sold as foods for special medical purposes (Figure 20.3). The draft provides specific guidance for making health claims and provides a list of "nutritional claims including nutrient comparative claims" and "conditions for disease risk reduction claims." The draft says that the final decision on approval will be taken by the Food Authority based on the comments and reviews by the scientific committee and the approved claims will be notified in the gazette [22].

20.2.10 Food recall procedures

Food Authority's Food Recall Procedures Regulations 2009, Section 28 of the Act, guide FBOs on how to carry out a food recall through an efficient, rapid identification as well as removal of unsafe and potentially hazardous food in the market or in the distribution chain, and ensure that unsafe food items are contained and destroyed or rendered safe. A recall plan must be available in written form and shall be made available to the State Food Authority/Food Authority on request. The food business operator is required to comply with the plan developed by the Food Authority when it recalls food [23].

20.2.10.1 Reward scheme for whistle blowers for information on adulterated and unsafe food

Section 95 of the Food Safety and Standard Act 2006 provides an attractive scheme to reward whistle blowers for information on cases of food adulteration and unsafe food regarding the following offenses:

1. Manufacture, storage, distribution, sale, or import of any article of food for human consumption which is
   a. Substandard
   b. Misbranded
   c. Containing extraneous matter or food adulterants
2. Processing and manufacture of food for human consumption in unhygienic or unsanitary conditions
3. Persons aiding or abetting any of the above activities

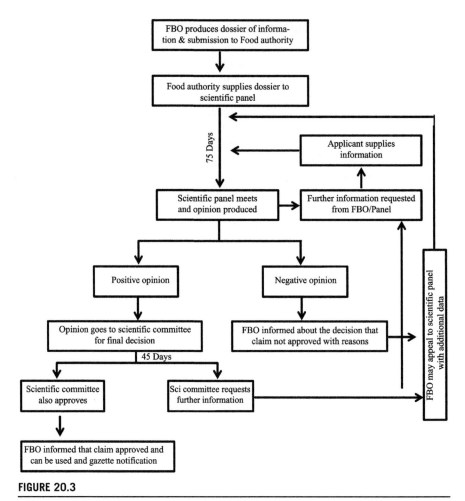

FIGURE 20.3

Flow chart showing process of authorization of claims.

The endeavor of the objective is to encourage inflow of information from all stakeholders regarding adulterated/unsafe foods. The quantum of the rewards will be determined on the basis of the severity of the offense under Food Safety and Standards Act 2006 for each of the previously mentioned categories of the information. The extent of the injury that has occurred to the consumer or may have occurred if the food had been consumed will determine the amount of the reward and the amount may range from Rs 500 to Rs 5,000,000 ($10–100,000). A nodal officer will be nominated in the office of Food Safety Commissioner in each state for receiving information and forwarding the same to the Food Safety Commissioner in the state or CEO in FSSAI headquarters for further necessary action [24].

20.2.11 **Food import and clearance regulations**

Section 25 of the Food Safety and Standards Act 2006 offers regulations for the import of food products:

1. No person shall import any unsafe or misbranded or substandard food or food containing extraneous matter and any article of food in contravention of any other provision of Food Safety and Standards Act or of any rule or regulation made thereunder or any other Act.
2. The Central Government shall, while prohibiting, restricting, or otherwise regulating import of articles of food under the Foreign Trade (Development and Regulation) Act 1992, follow the standards laid down by the Food Authority under the provisions of this Act and the rules and regulations made thereunder.
3. All imports will come under Central Licensing and an existing importer also needs to be registered under the Central Authority and all imported products will be screened for safety parameters by an IT-enabled Food Safety System as a new initiative under the Food Safety and Standards Act. No person shall import unsafe or misbranded or substandard food or food containing extraneous matter.

The Central Licensing Authority has engaged the National Institute of Smart Government (NISG) under the Ministry of Communications and Information Technology for the design and conceptualization of the IT-enabled Imported Food Safety System. As part of this commitment, NISG is assisting the Authority in establishing the operationalization of food import clearance processes through appointment of Authorized Officers at 14 major ports of entry. The draft document provides the provisions under the Act to ensure the safety of imported food, the final notification issued on January 24, 2013, by FSSAI's authorized officer to provide the guidelines/clarifications related to food import clearance process, which include labeling requirements for wholesale packages, testing of proprietary foods, import of dietary supplements, laboratory reports, and import of flavors for wholesale packagers [10,25,26]. The draft notification released on May 27, 2013, provided the food import regulation for FBOs (Figure 20.4).

20.2.12 **Advisory on misbranding/misleading claims/labeling claims**

As per Section 23, Packaging and Labeling of Foods of the Food Safety and Standards Act 2006, the provisions protected under Food Safety and Standards (Packaging and Labeling) Regulation 2011 include definition of health claims, nutritional claims, and risk reduction claims. No person is allowed to manufacture, distribute, sell, or expose for sale or dispatch or deliver to any agent or broker for the purpose of sale, any packaged food products which are not marked and labeled in the manner as may be specified by regulations. The labels should

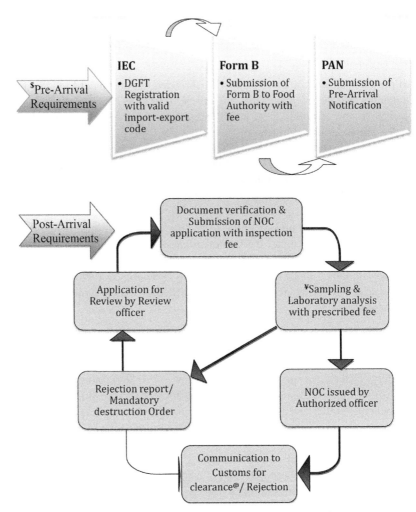

$ **Important requirements:** Standards, Valid documents and Labeling; Packing and storage.
@ **Clearance of food imports for special purpose:** Accredited food imports with valid license; Food export oriented units; and Imports by diplomatic missions.
¥ **Accelerated clearance without sampling and Laboratory analysis:** Import for QA (Quality Assurance), Research & Development; Imports for International trade and Exhibition; and Imports for special sports events.

FIGURE 20.4

Food import process by FBO.

not contain any statement, claim, design, or device which is false or misleading concerning the food products contained in the package or concerning the quantity or the nutritive value implying medicinal or therapeutic claims or in relation to the place of origin of the said food products.

As per Section 24 of Food Safety and Standards Act 2006 regarding restrictions of advertisement and prohibition as to unfair trade practices:

1. No advertisement shall be made for any food, which is misleading or deceiving or contravenes the provisions of this Act, the rules and regulations made thereunder.
2. No persons shall engage himself in any unfair trade practice for the purpose of promoting the sale, supply, use and consumption of articles of food or adopt any unfair or deceptive practice including the practice of making any statement, whether orally or in writing or by visible representation which
    a. Falsely represents that the foods are of a particular standard, quality, quantity, or grade composition.
    b. Making false or misleading representation concerning the need for, or the usefulness.
    c. Give to the public any guarantee of the efficacy that is not based on an adequate or scientific justification thereof.

As per Section 22 of Food Safety and Standards Act 2006, all FBOs as well as any person dealing with food articles including nutraceuticals, health supplements, functional food for which product approval is not taken or operating only on the basis of NOC pending approval of sale cannot make any claim in their advertisement with any health claim, nutraceutical claims, or risk reduction claim and they shall be liable to a penalty for any violations in this regard [27].

20.2.13 Regulation of energy drinks and caffeinated beverages

There are no specific regulations governing standards for energy drinks applicable under the Prevention of Food Adulteration (PFA) Act 1954. The standards of carbonated water under PFA Rules 1955 specify the maximum limits of caffeine to 200 ppm. The second amendment made by Food Safety and Standards Regulations 2011 dated April 18, 2013, provided the limits for caffeine in carbonated beverages with the minimum limit as 145 mg/L and should not exceed 320 mg/L. Based on the executive decisions of an expert panel, energy drinks are now a part of the newly defined category of foods that differ from general purpose and special dietary purpose foods and named as "caffeinated beverage" adopted by the Australian New Zealand Food Authority. The combinations and levels of added substances in energy drinks should be based on evidence of safety rather than efficacy. According to new standards being laid down by the Food Safety and Standards of India (FSSAI), such beverages must also carry a safety warning for consumers stating that such drinks are not recommended for "children, pregnant or lactating women, persons sensitive to caffeine and sportspersons" as well as a warning statement saying "no more than two cans to be consumed per day" [28,29].

20.2.14 Offenses and penalties

Chapter 9 of the Food Safety and Standards Act 2006 provides the provisions relating to offenses and penalties (Table 20.2) and following guidance to adjudicating officer or the tribunal for adjudging the quantum of penalty to be imposed under the Act [30]:

a. The amount of gain or unfair advantage, wherever quantifiable, made as a result of the contravention

Table 20.2 List of Offenses and Penalties under the Food Safety and Standards Act 2006

Description of Offense	Penalty
Selling food not of the nature or substance or quality demanded	Not exceeding Rs 500,000
Penalty for substandard food	Up to Rs 500,000
Penalty for misbranded food	Up to Rs 300,000
Penalty on misleading advertisement	Up to Rs 1,000,000
Food containing extraneous matter	Up to Rs 100,000
Penalty for failure to comply with food safety officer/false and misleading information	Three months of imprisonment/up to Rs 200,000
Penalty for unhygienic processing of food	Up to Rs 100,000
Penalty for possessing adulterant	Rs 200,000–1,000,000 (based on severity)
Punishment for unsafe food	
Does not result in injury	Six months Imprisonment and up to Rs 100,000
Non-grievous injury	One year Imprisonment and up to Rs 300,000
Grievous injury	Six years Imprisonment and up to Rs 500,000
Death	Seven years/life imprisonment and not less than Rs 1,000,000
Unhygienic or unsanitary conditions	Up to Rs 100,000
Carrying out a business without license	Six months imprisonment and up to Rs 500,000
Compensation to consumer	
Death	Not less than Rs 500,000
Grievous injury	Not exceeding Rs 300,000
Injury	Not exceeding Rs 100,000

Source: http://www.fssai.gov.in/portals/0/pdf/food-act.pdf

b. The amount of loss caused or likely to cause to any person as a result of the contravention

c. The repetitive nature of the contravention

d. Whether the contravention is without his knowledge

e. Any other relevant factor

20.3 MLM/direct selling

MLM in dietary supplements and functional foods is fast evolving and companies like Amway and Herbalife have achieved significant success in the last few years. There are over 400 direct selling companies involving close to 4 million direct selling agents with an estimated 2011−2012 revenue of $1140.9 million [8]. The IDSA established in 1996 is a self-regulating industry that mandates its members to operate within the strict provisions of a code of ethics prescribed by the World Federation of Direct Selling Association. There is no clear-cut definition for direct selling; it involves selling usually through explanation and demonstrations by a direct seller. In the Indian constitution, the Seventh Schedule, Entry 33 notifies about trade and commerce and specifically mentions direct selling "the products of any industry where the control of such industry by the Union is declared by Parliament by law to be expedient in the public interest, and imported goods of the same kind as such products." However, subject to this, trade and commerce within the State is in the State List. These regulations do not differentiate illegitimate direct selling corporations involved in pyramid schemes and legitimate direct selling corporations following fair trade. For the first time in India, the Kerala state government issued guidelines on direct selling in 2011 to differentiate legitimate MLM firms from illegal MLM pyramid enterprises. The guidelines define direct selling as "marketing of consumer products/services directly to the consumers generally in their homes or homes of others, at their workplace and other places away from the permanent retail locations, usually through explanation or demonstration of the products by a direct seller or by mail order sales."

Guidelines to be complied by the direct selling entities in Kerala include:

a. The direct selling entity should be legally authorized and file all returns as mandated by law.

b. Should be a valid licensee or a permitted user of trademark which identifies the promoter, goods or services distributed.

c. Should maintain website with complete details of their products/services, terms, conditions, price details of direct sellers and all relevant information about the company and goods. The website shall be updated regularly and shall furnish all necessary information required by District Industries Centre.

d. Should not collect any membership fee from direct seller for enrollment and should provide a valid ID along with description manuals of goods.

e. The compensation to direct sellers shall only be based on the quantum of sale of goods and services.

f. Should ensure that the customer purchasing goods or services from a direct seller shall have at least thirty days from the date of delivery and services to return the same and to receive full fund.

Recently in Kerala, Amway India officials were booked under the Prize Chits and Money Circulation Schemes (Banning) Act 1978 (PCMC Act) based on complaints by distributors in 2011. The FICCI criticized the government's action against Amway and emphasized the requirement of appropriate legislation to harmonize the direct selling segment. Following on Kerala's footsteps, other states are in the process of framing guidelines for direct selling. Until such a time the three Inter Ministerial Groups (IMG) of the Indian government came out with amendments to the PCMC Act or framework for new regulations. Organizations involved in direct selling need to follow the guidelines framed by the individual state to conduct business in India [31,32] apart from existing Acts such as the Sale of Goods Act 1930 (for regulating the sale of goods), the Indian Contract Act 1872 (for the sale of services), the Consumer Protection Act 1986 (to promote and protect the rights of the consumers), and the provisions of the PCMC Act 1978 seeking to ban the promotion or conduct of prize chits and money circulation schemes.

20.4 AYUSH

The Indian government created the Department of ISM&H in March 1995 to upgrade the standards of the Indian Systems of Medicine. ISM&H was renamed the Department of Ayurveda, Yoga, and Naturopathy, Unani, Siddha, and Homoeopathy (AYUSH) in November 2003 to provide more focused attention to the development of education and research in the Indian Systems of Medicine, to upgrade quality control and standardization of AYUSH drugs, and to document traditional wealth made available to the public.

The objectives of AYUSH include upgrading education and research in the ISM&H, to promote cultivation and regeneration of medicinal plants used in these systems, and to evolve Pharmacopoeia Standards for ISM&H drugs [5,33].

Chapter IVA of the Drugs and Cosmetics Act 1940 provides the provisions relating to ayurvedic, siddha, and unani (ASU) drugs, which encompass the herbal formulations traditionally used in the Indian system of medicine. Under this Act all ASU formulations or products were covered under the common term of drug. To ensure and enhance the quality of ASU medicines, the Indian government has notified GMP and Good Agricultural Practice under Schedule "T" of the Drugs and Cosmetics Act 1940, which also ensures that the raw materials used in the manufacture of drugs are authentic, of prescribed quality, and are free from contamination. Drugs and Cosmetics (6th Amendment) Rule 158(B) 2010 specifies that the license shall be attained by the ASU manufacturer by providing adequate evidence for ASU drugs (specified in E1) for safe consumption as nutraceuticals. The Drugs and Cosmetics Act does not specify the ASU products as nutraceuticals [34].

20.5 Probiotic regulatory overview

The Indian probiotic market is relatively nascent with an estimated sale of $12 million supplement sales but growing at over 11% as per a Frost and Sullivan report. Probiotic products are becoming popular among the consumers and health care professionals leading to a strong growth of probiotic supplements and functional foods group. Probiotics are bacteria that help maintain the natural balance of gastrointestinal microbial flora. According to the Food and Agriculture Organization (FAO)/World Health Organization (WHO), in 2002 probiotics are "live microorganisms which when administered in adequate amounts confer a health benefit on the host." Probiotics are sold over the counter as dietary supplements or food products that aid or improve health. Probiotics are categorized distinctly in different countries. Considering the United States, probiotic products are marketed as both foods and biologic drugs and regulated by Dietary Supplement Health and Education Act and Food and Drug administration, respectively. In India probiotics are marketed as "foods for special dietary uses or functional foods" and regulated under the Food Safety and Standards Act 2006. Regulations are not yet in place on evaluation and sale of probiotics in India apart from the initiatives of the Indian Council of Medical Research/Department of Biotechnology in drafting guidelines for evaluation of probiotics in food products that cover the definition of food under the Food Safety and Standards Act 2006 and regulations thereunder. The draft guidelines exclude probiotics, which by definition would come under drugs and beneficial microorganisms not used in foods or genetically modified microorganisms (GMOs). The regulations are yet to be framed for probiotics that are covered under drugs and GMOs to control sale of spurious and ineffective products with false claims [6,7].

20.6 Conclusion

India is proclaimed to be a "relatively nascent market" presently for functional foods and nutraceuticals. Popular among the Indian population, traditional practices such as ayurveda, unani, sidha, and homeopathy document a strong history of medicinal uses of various foods and herbs. Due to increasing awareness and availability of nutraceutical products, consumer demand for products with specific health benefits has been on the rise. A good regulatory framework put in place, such as the Food Safety and Standards Regulations 2011, will enable this market to grow at a rapid pace by encouraging companies to promote safe and effective supplements and functional foods.

The Food Safety and Standards Act 2006, which encompasses the regulatory framework on functional foods and nutraceuticals, has shown its potential by implementing the Food Safety and Standards Regulations 2011 through FSSAI. The Food Safety and Standards Act was initially assumed mostly as a copy of the European regulations, but its framework comes out with stringent and imperative transformations, which define the nutraceuticals as foods. FSSAI recognized several regional food testing laboratories nationwide to test and govern the quality of food and nutraceutical products. The penalties announced in the regulations are strong enough to curtail infringement of food laws and this would allow only safe and quality food and nutraceutical products to the consumers. An interesting part of the Act is the decision made under the regulations to allow the announcement of an award for information providers (whistle blowers) on unsafe foods, which helps to deracinate the adulterated and unsafe foods from the market. Labeling requirements were made transparent for the consumer with more reliable data. The regulations allow the FBOs to obtain licenses to manufacture the nutraceutical products. It provides guidelines for FBOs regarding authorization of Nutritional and Health Claims supported with strong scientific evidence. This can be considered as a major differentiating factor between FSSAI and other established regulatory bodies in Europe, Japan, Australia, and New Zealand, where there is no approval for claims on functional foods and nutraceuticals. It also provides a platform for the FBOs to protect their claims and control over innovation.

Although there are many challenges in implementing food safety acts, FSSAI is moving forward with a resolution to have unique and transparent regulations for foods, food supplements, and nutraceuticals. The regulations ensure safe and effective functional food to the consumer coupled with provisions for the FBO to meet necessary challenges while operating in the Indian functional food segment. This is bound to create significant growth and success for companies in the Indian supplement and functional food space over the coming years.

References

[1] India Development Gateway. Lifestyle Disorders in India causing death, <http://www.indg.in/health/lifestyle-disorders/lifestyle-disorders-in-india-causing-death>.
[2] Frost and Sullivan. Nutraceuticals Critical Supplement for Building a Healthy India, <http://ficci-nutraceuticals.com/files/Nutraceuticals_Final_Report.pdf>; 2012.
[3] Interlink's White Paper. Regulatory Perspective of Nutraceuticals in India, <http://www.interlinkconsultancy.com/pdfs/whitepapers/Regulatory_prespective_Nutraceuticals_whitepaper_Dec2011.pdf>; 2011.
[4] Training Manual for Food Safety Regulators. Vol I — Introduction to Food and Food Processing, <http://www.fssai.gov.in/Portals/0/Training_Manual/Volume%20I-%20Intoduction%20to%20Food%20%20and%20Food%20Processing.pdf>; 2010.

[5] Mukherjee PK, Venkatesh M, Kumar V. An overview on the development in regulation and control of medicinal and aromatic plants in the Indian system of medicine. Bol Latinoam Caribe Plant Med Aromat 2007;6(4):129–36.

[6] Malika A, Sujata S, Ashish B. Comparative insight of regulatory guidelines for probiotics in USA, India and Malaysia: a critical review. Int J Biotechnol Wellness Ind 2013;2:51–64.

[7] ICMR-DBT. Guidelines for Evaluation of Probiotics in Food, <http://icmr.nic.in/guide/PROBIOTICS_GUIDELINES.pdf>; 2011.

[8] The Indian Direct Selling Industry, Annual Survey 2010-11, PHD Chamber of Commerce and Industry and IDSA; 2012.

[9] Food Standards and Safety Authority of India, <http://fssai.gov.in/portals/0/Introduction_to_FSS_Rules_2010.pdf>.

[10] Food Standards and Safety Authority of India. Vol II – Food Safety Regulations & Food Safety Management, <http://www.fssai.gov.in/Portals/0/Training_Manual/Volume%20II-%20Food%20Safety%20Regulators%20and%20Food%20Safety%20Management.pdf>; 2010.

[11] Food Standards and Safety Authority of India. The Food Safety and Standards Regulations, <http://fssai.gov.in/GazettedNotifications.aspx>; 2011.

[12] Food Standards and Safety Authority of India. Licensing and Registration of Food Businesses, <http://fssai.gov.in/Portals/0/Pdf/Food%20safety%20and%20Standards%20(Licensing%20and%20Registration%20of%20Food%20businesses)%20regulation,%202011.pdf>; 2011.

[13] Food Standards and Safety Authority of India. New Product Approval Procedure, <http://fssai.gov.in/Portals/0/Pdf/productApproval(10-12-2012).pdf>; 2012.

[14] Food Standards and Safety Authority of India. Packaging and Labeling, <http://www.fssai.gov.in/Portals/0/Pdf/Food%20Safety%20and%20standards%20(Packaging%20and%20Labelling)%20regulation,%202011.pdf>; 2011.

[15] Food Standards and Safety Authority of India. Food Products Standards and Food Additives, <http://fssai.gov.in/Portals/0/Pdf/Food%20safety%20and%20standards%20(Food%20product%20standards%20and%20Food%20Additives)%20regulation,%202011.pdf>; 2011.

[16] Codex General Standard for Food Additives (Revised 2012), <http://www.codexalimentarius.net/gsfaonline/docs/CXS_192e.pdf>.

[17] Food Standards and Safety Authority of India. Prohibition and Restrictions on Sales, <http://fssai.gov.in/Portals/0/Pdf/Food%20safety%20and%20standards%20(Prohibition%20and%20Restrction%20on%20sales)%20regulation,%202011.pdf>; 2011.

[18] Food Standards and Safety Authority of India. Contaminants, Toxins and Residues, <http://fssai.gov.in/Portals/0/Pdf/Food%20safety%20and%20standards%20(contaminats,%20toxins%20and%20residues)%20regulation,%202011.pdf>; 2011.

[19] Food Standards and Safety Authority of India. List of FSSAI Notified NABL Accredited Food Testing Laboratories, <http://fssai.gov.in/Portals/0/Pdf/list%20of%2068%20labs%20final.pdf>.

[20] Food Standards and Safety Authority of India. <http://fssai.gov.in/Portals/0/Pdf/Eight_Meeting_CAC(06-07-2012).pdf>; 2012.

[21] National Accreditation Board for Testing and Calibration Laboratories, India, <http://www.nabl-india.org/index.php?option = com_content&view = article&id = 139&Itemid = 12>.

[22] Food Standards and Safety Authority of India. Regulation on Labelling (Claims), <http://www.fssai.gov.in/Portals/0/Pdf/covering%20letter%20for%20draft%20regulation.pdf>; 2012.

[23] Food Standards and Safety Authority of India. Regulations on Food Recall Procedure, <http://www.fssai.gov.in/Portals/0/Pdf/Recall_procedure(04-05-2011).pdf>; 2009.

[24] Food Standards and Safety Authority of India. Reward scheme for Whistle Blowers, <http://fssai.gov.in/Portals/0/Pdf/Whistle%20Browers(16-08-2011).pdf>; 2011.

[25] Food Standards and Safety Authority of India. Food Import Regulations, <http://fssai.gov.in/Portals/0/Pdf/Food_Import_Regulations_Draft(07-07-2011).pdf>; 2011.

[26] Food Standards and Safety Authority of India. Guidelines related to Food Import Clearance Process by FSSAI's Authorized Officers, <http://fssai.gov.in/Portals/0/Pdf/Advisories_Final.pdf>; 2013.

[27] Food Standards and Safety Authority of India. Advisory on Misbranding/Misleading claims, <http://fssai.gov.in/Portals/0/Pdf/Adviosry_on_misbranding_&_misleading_claims(04-07-2012).pdf>; 2011.

[28] Food Standards and Safety Authority of India. Regulation of Energy Drinks, and Caffeine (Revised), <http://www.fssai.gov.in/portals/0/standards_of_energy_drinks_pdf>; 2009.

[29] Food Standards and Safety Authority of India, <http://fssai.co.in/india-energy-drinks-to-carry-health-warnings/>.

[30] Food Safety and Standards Act, <http://www.fssai.gov.in/portals/0/pdf/food-act.pdf>; 2006.

[31] Direct selling Guidelines, <http://kerala.gov.in/index.php?option = com_docman&task = doc_download&gid = 3522&Itemid = 2702>; 2011.

[32] FICCI Report on Direct Selling Industry in India. Direct selling in India: Appropriate Regulation Is the Key, <http://www.ficci.com/spdocument/20237/report-mark.pdf>; 2013.

[33] Department of Ayush, New Delhi, India, <http://indianmedicine.nic.in/>.

[34] Ministry of Health and Family Welfare. Drugs and Cosmetics (6th Amendment) Rules, 2010, <http://www.kdpma.com/wp-content/themes/twentyten/pdf/drugs-cosmetics-act/13.pdf>; 2010.

Historical Change of Raw Materials and Claims of Health Food Regulations in China

21

Chun Hu

Nutrilite Health Institute, Buena Park, California

21.1 Introduction

Since the first health food was approved in 1996, about 12061 domestic products and 672 imported health foods have been approved by the end of 2012 in China. The administrative approval of health foods was transferred from the Ministry of Health (MOH) to the State Food and Drug Administration (SFDA) in 2003, and the SFDA introduced a series of policies and provisions in regulating health food registration and re-registration, including a new proposal to change the current scope of health claims and procedures for new claims. Meanwhile, the market size of the health food industry has grown rapidly to RMB 100 billion Yuan in 2010 [1]. A survey conducted by the China Health Care Association and China Academy of Social Science indicated that the market size of health food in China exceeded RMB 260 billion Yuan, and more than half of this market size is attributed to 20% of the brands [2]. In addition to its dramatically expanded market size, China has become one of important raw materials suppliers in the worldwide health food industry. This chapter provides a brief review of historical change in China's health food regulations with a focus on the raw material and claims, and outlines the potential changes in the future.

21.2 Definition of Health Foods in China

The concept of health food (or functional food) was originally introduced in the mid-1990s when the Food Hygiene Law was implemented in China. Article 22 of the Food Hygiene Law described health food as "food with specific health function." The law defined that

Nutraceutical and Functional Food Regulations in the United States and Around the World.
DOI: http://dx.doi.org/10.1016/B978-0-12-405870-5.00021-9

> ...*health food and it instruction must be submitted to health administration authority under the State Council for evaluation and approval. The correspondent hygiene standard, manufacturing, and management system are drafted by health administration authorities....*

Correspondingly, the General Standard of Health (functional) Food (GB16740-97) was implemented in 1997, in which health food was further defined as

> ...*one variety of foods, it generally carries common characteristics of food and modulates body function which is suitable for the consumption by specific population, but is not intended for disease treatment.*

When the SFDA took over the administrative evaluation and approval of health foods, the definition of health food was further described as

> ...*foods that are claimed with specific health function or supplement of vitamin (s) or minerals (s), and suitable for consumption by specific population with specific body function regulation but is not targeted for disease treatment, and shall not cause any acute, sub-acute or chronic hazard to human body... [3]*

The Food Safety Law was introduced in 2009 to supersede the Food Hygiene Law, in which, Article 51 defined health (functional) foods as

> *The State executes strict regulation over food with special health function claim. The relevant regulatory departments shall execute duties and assume responsibilities according to law. Detailed management measures shall be developed by the State Council. Food with special health function claim shall not cause any acute, sub-acute, or chronic hazard to human body. The label and instructions shall not involve in any disease prevention or treatment functions, and the content must be true and clearly indicates the suitable and unsuitable populations, functional components or reference component and its content. Product functions and ingredients must be consistent with label and instructions.*

Therefore, health foods in China typically include both food/supplement with health benefit and nutrient supplements providing vitamin/minerals. Health foods should be used with specific populations to maintain or improve health condition, but should not be used for therapeutic or disease prevention purposes. Health foods must be safe, and consumption of such products will not cause any acute, subacute, or chronic hazard to human body.

Under the current regulation structure, the SFDA is the administrative body responsible for the technical evaluation and approval of health foods (including nutrient supplements), evaluation of advertisement content of health foods, and approval of hygiene permits for the manufacture and supervision of any issues associated with consumption. The MOH is responsible for the approval of novel food ingredients and food additives, drafting national food safety policies and national food safety standards. Other government agencies, such as Administration of Quality Supervision, Inspection and Quarantine is responsible for supervising the quality and safety issues during food processing, monitoring the risk and investigating quality incidence, and conducting import/export inspection of health food products. The Administration for Industry and Commerce is responsible for issuing business licenses and supervising food safety and quality during the distribution and marketing steps.

21.3 Evolution of allowable claims of health foods

When the health food concept was first introduced in the 1990s, only 12 claims were defined [4], including "modulating immunity, anti-aging, improving memory, promoting growth, anti-fatigue, weight loss, hypoxic endurance, anti-irradiation, anti-mutagenesis, inhibiting tumor, modulating blood lipids and improving sexual function." An additional 12 claims were added in 1997, but the number of claims was reduced to 22 in 2000. Three years later, the MOH issued the Technical Protocol of Health Food Detection and Evaluation (2003 edition), in which protocols for 27 functional (health) claims became the official guide for technical evaluation of functional claims for health foods (except for nutrient supplement). After the SFDA took over the administrative authority, it adopted all 27 claims defined in the 2003 protocols and officially added the procedure of nutrient supplement in 2005. Table 21.1 summarizes the change of allowed claims of health foods in China.

The SFDA is responsible for the approval of functional claims, as seen in a typical flow chart [12]. The functional claim is subject to validation by government-appointed laboratories with animal and/or human clinical protocols to confirm whether the applied health claim should be granted. The animal and human clinical test protocols have gone through revision over the past two decades. Since 2003, the 27 claims evaluation has been based on the protocols published by the MOH [7], which was adopted by the SFDA when it took over the administrative responsibility from the MOH in 2003 [8]. Since 2011, the SFDA sought public input on the revision of the 2003 protocols, and nine functional evaluation protocols were finalized and became effective for the products submitted for the function claim approval after May 1, 2012. Evaluation protocols for the nine functions cover the following: " antioxidant," " to assist protecting gastric mucosal injury," "to help lowering blood sugar," "to help relieving visual fatigue," "to help improving iron deficient anemia," "to help lowering blood lipids," "to help lead excretion," "weight loss," and "to clear throat" claims [13]. The revisions of these protocols include the change of animal models selected, duration, biomarkers measured, criteria used to make the assessment, etc. For example, the naturally aging animal model remains the same in the 2012 antioxidant protocol, an ethanol-induced oxidative injury model was added, but

Table 21.1 Allowed Functional Claims 1996−2012

Effective Date	Claims
July 18, 1996	12 claims were modulating immunity, anti-aging, improving memory, promoting growth, anti-fatigue, weight loss, hypoxic endurance, anti-irradiation, anti-mutagenesis, inhibiting tumor, modulating blood lipids, and improving sexual function [4].
July 1, 1997	24 claims were modulating immunity, anti-aging, improving memory, promoting growth, anti-fatigue, weight loss, hypoxic endurance, anti-irradiation, anti-mutagenesis, inhibiting tumor, modulating blood lipids, improving sexual function, modulating blood sugar, improving gastrointestinal function (with specified function), improving sleep, improving nutritious anemia, protecting from chemical induced liver injury, promoting milk secretion, cosmetic (with specified function), improving vision, promoting lead excretion, clearing throat, modulating blood pressure, improving osteoporosis, and nutrient supplement [5].
January 14, 2000	22 claims were modulating immunity, modulating blood lipids, modulating blood sugar, anti-aging, improving memory, promoting lead excretion, clearing throat, modulating blood pressure, improving sleep, promoting milk secretion, anti-mutagenesis, anti-fatigue, hypoxic endurance, anti-irradiation, weight loss, promoting growth, improving osteoporosis, improving anemia, assisting to protect liver from chemical induced injury, cosmetic (acne removal/chloasma removal/improve skin moisture and oil content), improving gastrointestinal function (modulate gut flora/promote digestion/bowel movement/assist to protect gastric mucosa) [6].
May 1, 2003	27 claims are "To improving immunity, to assist lowering blood lipids, to assist lowering blood sugar, antioxidant, to assist to improve memory, to relieve visual fatigue, to promote lead excretion, to clear throat, to assist to lower blood pressure, to improve sleep, to relieve physical fatigue, to improve hypoxia endurance, to assist to protect from irradiation damage, weight loss, to promote growth, to improve bone density, to promote milk secretion, to improve anemia, to assist to protect chemical induced liver injury, to remove acne, to remove chloasma, to improve skin moisture content, to improve skin oil content, to modulate colon flora, to promote digestion, to promote bowel movement, to assist to protect gastric mucosa injury" [7]. SFDA adopted the MOH protocol from 2003 regarding the evaluation and approval of acceptable functional claims of health foods, and kept the 27 claims from the MOH 2003 version [8]. "Blue hat" logo and license number are printed on the product label as "SFDA approved health food" with "G+ license number " indicating domestic product, "J+ license number " for imported product.
July 1, 2005	Required documents for nutrient supplement definition, scope, source materials, and minimal/maximum level of vitamins and minerals for technical evaluation were added [9].
Future (unknown)	SFDA proposed to reduce the number of allowed functional claims to 18 by eliminating 4 ("to promote growth," "to assist protect radiation

(Continued)

Table **21.1** (Continued)	
Effective Date	**Claims**
	damage," "to assist lowering blood pressure," and "to improve skin oil content) and rewording the rest. SFDA actively seeks public and industry input on the proposal in 2011 and 2012 [10,11], although the implementation date remains unknown. The future (if implemented) health claims will be "to help improve immunity, to help lower blood lipids, to help lower blood sugar, to help improve sleep, antioxidant, to help relieve physical fatigue, to help reduce body fat, to help improve bone density, to help improve iron deficient anemia, to help improve memory, to clear throat, to help improve hypoxia endurance, to help protect from alcohol induced liver injury, to help lead excretion, to help milk secretion, to help relieve visual fatigue, to help improve gastrointestinal function (bowel movement, modify colon flora, promote digestion and assist to protect gastric mucosa), and to help improve facial skin health (acne removal, chloasma removal and improve skin moisture content).

bromobenzene- and radiation-induced oxidative models were eliminated. The duration of experiment in the antioxidant model is reduced from 45 to 60 days in the 2003 protocol and 30 to 45 days in the 2012 revision. Biomarkers such as serum 8-isoprostane and carbonyl, the reduced form of glutathione, are introduced in the 2012 protocol. Criteria for animal experiment assessment are much tighter in the 2012 protocol. The new positive criteria require that at least three biomarkers (lipid peroxide product, protein oxidation product, and antioxidant enzyme or non-enzymatic antioxidant substance) are positive, in comparison to the 2003 protocol, which only required that "any antioxidant enzyme and lipid peroxide product results are positive." Similar, criteria for human trial are tighter in the 2012 protocol. This aligns well with the government effort to control the number of approvals and to make sure that only those truly qualified products are approved and licensed in the future.

21.3.1 **New function claim**

Up to 2012, only 27 health (functional) claims were allowed in China, although new function claims have always been discussed in the health food industry and government, and the SFDA originally defined the procedure on the new functional claim application. According to Article 21 of the Interim Administrative Measure of Health Food Registration [3], when an applicant submits health food for a new function claim, the new function protocol is subjected to further validation conducted by government-appointed laboratories, along with the documents typically required for health food; however, the provision did not provide addition details about the procedure of applying for a new claim. No health food has been approved by SFDA for new function claims, and even the government pointed out that there is a high risk associated with new function claims [14].

The SFDA recently published a public consultation document to collect comments on the proposal of the guidance and evaluation procedure of new claims. In the drafted guidance, the new functional claim will meet the following requirements [15]:

a. The new functional claim is intended to regulate body function, improve body health, or reduce the risk of incidence of disease, without any implication of prevention or treatment of any disease;

b. Name and evaluation protocol of new function must be significantly different from current allowed health claims;

c. Scientific evaluation protocol and assessment criteria shall be provided, with sufficient amount of scientific and generally acceptable evidences to support the claim;

d. The new functional claim conforms with current laws and regulations; and

e. Government encourages applicants to develop and submit health food application with new functional claim developed under the guidance of traditional Chinese medicine theory.

The proposed guidance indicates that the new claim must be product specific. If approved, the launched product with a new claim will be under the SFDA's surveillance for 3 years. During the surveillance period, the product license is not permitted to be transferred; meanwhile, the SFDA will not accept any other applicant for the same claim. Once the 3 year surveillance phase expires and if no adverse effect is reported, the SFDA will add such claim into allowed health claims. The proposed application freezing phase provides the original license holder with a technical and marketing advantage by encouraging the business practitioner to conduct R&D activities at their own risk, and allowing them to maximize the benefit of innovation.

21.4 Raw Materials used in health foods

Raw materials in health food are those primary materials associated with the functions of such health food [3]. Raw materials and excipients used in health food production will comply with the national standards (if a national standard does not exist, industry standard or quality specification drafted by the applicant will be submitted). Raw materials and excipients must be safe for human consumption; the amount will not exceed the limit defined in the relevant provisions. Using those banned ingredients and excipients in health foods is prohibited by the SFDA [3]. It was estimated that about one-sixth of health food applications were rejected due to noncompliance of ingredients and excipients or lack of safety and toxicology assessments [16].

In general, common food ingredients are allowed to be used in health foods. Typically raw materials allowed in health foods are those listed in the China Food Composition, National Food Safety Standard—Hygienic Standard of Food Additives (GB 2760-2011), National Food Safety Standard—The Use of Nutritional Fortification Substances in Foods (GB 14880-2012), and food used microbial strains [17] and its sequential update, as well as certain pharmaceutical excipients listed in China Pharmacopoeia [18]. In addition, a series of provisions have been issued by the SFDA to regulate specific active ingredients used in health food, including fungus, probiotic strains, amino acid chelates, wild animal and plants, nucleic acid, ingredient separated by large pore adsorptive resin, melatonin, soy lecithin, ants, unsaturated fatty acids, chitosan, superoxide dismutase, fermented ingredients [9], CoQ$_{10}$[19], soy isoflavone [20], red yeast rice, selenium, chromium, anthraquinone-contained ingredients (such as aloe vera, rhubarb root, fleeceflower root, and cassia seed) [21]. The daily intake, suitable and nonsuitable population, and instruction must be specified when these special ingredients are used in health foods.

China has a long history of using traditional Chinese medicine (TCM). Many herbal medicinal materials are integrated into culinary and medicinal food practice with high consumer acceptance. To differentiate the health benefits of health foods from TCM, the MOH issued three lists on what TCM material is allowed or not permitted in health foods [22] in 2002, which was later adopted by the SFDA [14]. These lists are known as "ingredients as both food and medicine," "TCM ingredient allowed to be used in health food," and "ingredients prohibited in health food" (Table 21.2). The notification also defined that the number of botanical- and animal-originated ingredients in health foods should not exceed 14 [22]. Safety assessment is recommended by following the National Standard for Evaluation Procedure of Food Safety and Toxicology. If an ingredient intended to be used in a health food is newly developed, discovered, or introduced without a consumption history or only limited to a specific region, then such ingredient must be evaluated in accordance with the novel food regulation.

The novel food concept was introduced to China by the MOH when the Administrative Measure of Hygiene of Novel Foods was implemented in 1990, which was superseded by the Administrative Measures of Novel Food at the end of 2007 [23]. Novel foods are those:

i. Animal, plants and microorganisms that are not traditionally consumed in China;
ii. Food materials that are derived from animals, plant, and microorganisms and not traditionally consumed in China;
iii. New varieties of microorganisms that are used during food processing; and
iv. Food material, of which its original composition or structure is modified by the adoption of new technique during manufacturing.

The Administrative Measures of Novel Food focuses more on the food ingredients rather than the finished good compared to the previous version.

Table 21.2 TCM Ingredients Allowed and Prohibited in Food and Health Food [22]

	Items
Ingredients as both food and medicine	*Caryophylli flos, Anisi stellati fructus, Canvaliae semen, Foeniculi fructus, Cirsii herba, Dioscorea rhizoma, Crataegi fructus, Portulacae herba, Zaocys, Mume fructus, Chaenomelis fructus, Cannabis fructus, Aurantii flos, Polygonati odorati rhizoma, Glycyrrhizae radix et rhizoma, Angelicae dahuricae radix, Ginkgo semen, Lablab semen album, Lablab flos album, Longan arillus, Cassiae semen, Lilii bulbus, Myristicae semen, Cinnamomi cortex, Phyllanthi fructus, Citri sarcodactylis fructus, Armeniacae semen amarum, Hippophae fructus, Ostreae concha, Euryales semen, Zanthoxyli pericarpium, Vignae semen, Asini corii colla, Galli gigerii endothelium corneum, Hordei fructus germinatus, Laminariae thallus, Eckloniae thallus, Jujubae fructus, Siraitiae fructus, Pruni semen, Lonicerae japonicae flos, Canarii fructus, Houttuyniae herba, Zingiberis rhizoma recens, Hoventia dulcis semen, Lyii fructus, Gardeniae fructus, Amomi fructus, Sterculiae lychophorae semen, poria, Citri furcus, Moslae herba, Persicae semen, Mori folium, Mori fructus, Citri exocarpium rubrum, Platycodonis radix, Alpiniae oxyphyllae fructus, Nelumbinis folium, Raphani semen, Nelumbinis semen, Alpiniae officinarum rhizoma, Lophatheri herba, Sojae semen preaeparatum, Chrysanthemi flos, Cichorii herba/radix, Brassicae jurhnceae semen, Polygonati rhizoma, Perillae folium, Perillae fructus, Puerariae lobatae radix, Sesami semen nigrum, Piperis fructus, Sophorae flos bud, Sophorae flos, Taraxaci herba, Mel, Torreyae semen, Ziziphi spinosae semen, Imperatae rhizoma, Phragmitis rhizoma, Agkistrodon brevicaduds, Citri exocarpium tangerine, Menthae haplocalycis herba, Coicis semen, Allii macrostemonis bulbus, Rubi fructus, Pogostemonis herba*
TCM ingredient allowed in health food	*Ginseng radix et rhizoma, Ginseng folium, Ginseng fructus, notoginseng radix et rhizoma, Smilacs glabrae rhizoma, Cirsii japonici herba, Ligustri lucidi fructus, Corni fructus, Cyathulae radix, Fritillariae cirrhosae bulbus, Chuanxiong rhizoma, Fetus ceriv, Cervi cornu panthotrichum, Os cervi, Salviae miltiorrhizae radix et rhizoma, Acanthopanacis cortex, Schisandrae Chinesis fructus, Cimicifugae rhizoma, Asparagi radix, Gastrodiae rhizoma, Pseudostellariae radix, Morindae officinalis radix, Aucklandiae radix, Equiseti hiemalis herba, Arctii fructus, Arctii radix, Plantaginis semen, Plantaginis herba, Glehniae radix, Fritillariae ussuriensis bulbus, Scrophulariae radix, Rehmanniae radix, Polygoni multiflori radix, Bletillae rhizoma, Atractylodis Macrocephalae rhizoma, Paeoniae radix alba, Amomi fructus rotundus, Haliotidis concha, Dendrobium nobile Lindl* (with permit), *Lycii cortex,*

(Continued)

Table 21.2 (Continued)

	Items
	Angelicae sinensis radix, Bambusae Caulis in Taenias, Carthami flos, Rhodiolae crenulatae radix et rhizoma, Panacis quinquefolii radix, Euodiae fructus, Achyranthis bidentatae radix, Eucommiae cortex, Eucommiae folium, Astragali complanati semen, Moutan cortex, aloe, Atractylodis rhizoma, Psoraleae fructus, Chebulae fructus, Paeoniae radix rubra, Polygalae radix, Ophiopogonis radix, Testudinis carapax et plastrum, Eupatorii herba, Platycladi cacumen, Rhei radix et rhizoma praeparata, Polygoni multiflori radix praeparata, Acanthopanacis senticois radix et rhizoma seu caulis, Rosa davurica fructus , Lycopi herba, Alismatis rhizoma, Rosae guguose flos, roselle, anemarrhenae rhizoma, Apocyni veneti folium, Ilex folium, Fagopyri dibotryis rhizoma, Rosae laevigatae fructus, Citri reticulatae pericarpium viride, Magnoliae officinalis cortex, Magnoliae officinalis flos, Curcumae longae rhizoma, Aurantii fructus, Aurantii fructus immaturus, Platycladi semen, margarita, Gynostemma herba, Trigonellae semen, rubiae radix et rhizoma, Piperis longi fructus, Allii tuberose semen, Polygoni multiflori caulis, Cyperi rhizoma, Drynariae rhizoma, Codonopsis radix, Mori cortex, Mori ramulus, Fritillariae thunbergii bulbus, Leonuri herba, Centellae herba, Epimedii folium, Cuscutae semen, Chrysanthemi indici flos, Ginkgo folium, Astragali radix, Fritillariae hupehensis bulbus, Sennae folium, Gecko, vaccinum vitis-ideal fructus, Sophorae fructus, Typhae pollen,Tribuli fructus, Propolis, Tamarindus fructus, Ecliptae herba, Rhei radix et Rhizmoa praeparata, Rehmanniae radix praeparata, Trionycis carapax
Ingredients prohibited in health food	*Dysosamatis rhizoma, Rhododendron molle G. Don flos, Eurphorbiae semen, Aristochia debilis radix, Anisodus tanguticus radix, Aconiti radix, Arstolochia fangchi radix, Cariara sinica folium, Strychni semen, Dysosma pleiantha, Hyoscayami semen, Crotonis fructus, mercury, Catharathus rosesus herba, Kansui radix, Arisaematis rhizoma, Pinelliae rhizoma, Typhonii rhizoma, Euphorbiae ebractaolatae radix, mercury chloride, Lycoris radiate bulbus, Aristolochia manshuriensis caulis, Crotalaria sessiliflora aerea, Nerium indicum folium, mercury sulfate, Papaveris pericarpium, Taxus chinensis, Illiccium lanceolatum folium, Strophantus semen and folium, Rhodoendrom molle radix, Iphigenia indica bulbus, Euphorbiae pekinensis radix, Tripterygium hypoglaucum, Fugu fish, Rhodoendri mollis flos, Lytta caraganae Pallas, Derris trifoliata Lour., Digitalis purpurea, Daturae flos, Pharbitadis semen, Aresnolite, Aconti kusnezoffii radix, Periplocae cortex, Peganm harmala herba, Dysosam pleianha radix, Illicium lanceolatum folium, Aconitum szechen radix, Convallaria Keiskei herba, Thevetia*

(Continued)

	Items
Table 21.2 (Continued)	
	peruviana semen, Mylabris, sulfur, arsenic disulfide, Tripterygium wilfordii Hook herba, Belladonnae herba, Veratrum nigrum radix or rhizoma, Bufonis venenum
Nomenclature follows China Pharmacopoeia 2010 and revision [18].	

The concept of *substantial equivalence* was introduced in the 2007 edition. If a given novel food is substantially equivalent to traditional food or food ingredient or approved novel food in terms of species, source, biological characteristics, composition, edible part, amount of usage, application range, and suitable population, and in consistency with both technology and product specification, then it is considered as equally safe and of substantial equivalence. It is the MOH's responsibility to accept novel food application and organize an expert assessment committee for safety evaluation on all applications. The MOH is responsible for administrative review to determine whether novel food license should be granted. Although a novel food ingredient can be used in health food, it is also clear that the MOH does not approve novel food ingredient application only used for health food [24]. Examples of approved novel foods in the past 5 years are listed in Table 21.3. Since some of the recent approved novel foods ingredients have great potential for health food applications, some of these ingredients are already formulated in the approved health foods. For example, *Chlorella pyrenoidesa* and *Dunaliella salina* are found in immune products, plant sterol and plant sterol ester are found in blood lipid-regulating products, docosahexaenoic acid (DHA) algal oil appears in memory products, lutein ester is seen in vision health products and irradiation protection products, inulin is found in the products for "gut flora" or "bowel movement" functions, and conjugated linoleic acid is formulated in weight loss products.

21.5 **Nutrient supplements**

The "nutrient supplement" concept was originally defined in 1997 by the MOH [5]. Under the health food regulation framework, a nutrient supplement was permitted to make claims of "providing single or multiple vitamin(s) and/or mineral(s)." Different from other health claims, a functional evaluation report and a health function claim were not required for nutrient supplement approval.

The SFDA redefined the regulation on nutrient supplement in 2005 with more detailed procedures [9]. Nutrient supplement is defined as a "...product intended to provide vitamin(s) and/or mineral(s) to supplement the shortage from diet, to prevent nutrition deficiency and reduce the risk of certain chronic

Table 21.3 Examples of Novel Foods Approved after 2007

Approved Novel Foods	Date of Approval
Lactobacillus acidophilus DSM13241, xylo-oligosaccharide, sodium hyaluronate, lutein ester, L-arabinose, *Acanthopanax sessiliflorus, Aloe vera* gel	June 6, 20008
Galacto-oligosaccharides, *Lactobacillus paracasei* GM080 and GMNL-33, *Lactobacillus acidophilus* R0052, *Lactobacillus rhamnosus* R0011, Bonepep, isomaltitol, *Lactobacillus plantarum* 299v, *Lactobacillus plantarum* CGMCC No 1258, plant stanol ester, globin peptide	September 9, 2008
Cordyceps militaris	March 16, 2009
Inulin (from *Cichorium intybus* var. *sativum,* Asteraceae), polyfructose (from *Cichorium intybus* var. *sativum,* Asteraceae)	March 25, 2009
γ-Aminobutyric acid, colostrum basic protein, conjugated linoleic acid, conjugated linoleic acid glyceride, *Lactobacillus plantarum* ST-III, *Eucommia ulmoides* oliv. seed oil	September 27, 2009
Tea seed oil (*Camellia sinensis* OK.tez), *Dunaliella salina* and extract, fish oil and extract, diacylglycerol oil, earthworm (*Eisenia foetida* Savigny) protein, milk minerals, milk basic protein	December 22, 2009
DHA algal oil, raffino-oligosaccharide, plant sterol, plant sterol ester, arachidonic acid oil, *Gynura divaricata* (L.) DC, poppy seed oil	March 9, 2010
Camellia chrysantha (Hu) Tuyama leaf, *Inula nervosa* wall. ex DC rhizome, Noni (*Morinda citrifolia* L.) puree, yeast β-glucan (*Saccharomyces cerevisiae*), tissue culture of snow lotus (*Saussurea involucrata*)	May 20, 2010
Sucrose polyesters, corn oligo-peptide powder, phosphatidylserine	October 21, 2010
Haematococcus pluvialis, epigallocatechin gallate (EGCG)	October 29, 2010
Elaeagnus mollis Diels oil, calcium β-hydroxy-β-methyl butyrate	January 18, 2011
Maple seed oil (*Acer truncatum* bunge), peony seed oil (*Paeonia ostii* T. Hong & J.X. Zhang or *P. rockii*)	March 22, 2011
Maca (*Lepidium meyenii Walp*)	May 18, 2011
Hyriopsis cumingii polysaccharide	January 20, 2012
Medium and long chain triacylglyceride, wheat oligo-peptide	August 28, 2012
Cultivated ginseng (*Panax ginseng* C.A. Meyer)	August 29, 2012
Chlorella pyrenoidosa, Linderae aggregate leaf, *Moringa oleifera* leaf	November 12, 2012
Tea (*Camellia sinensis* (L.) O. Kuntze) blossom, *Suaeda salsa* seed oil (*Suaeda salsa* (L). pall), Sacha Inchi oil (*Plukenetia volubilis* L.), Sumac	January 4, 2013

(Continued)

Table 21.3 (Continued)	
Approved Novel Foods	**Date of Approval**
fruit oil (*Rhus chinensis* Mill.), *Cordyceps guangdongensi* fruit body, acai (*Euterpe oleraceae* Mart.), *Phylloporia ribis* (Schumach:Fr) Ryvarden	
Data was collected from the MOH website, www.moh.gov.cn.	

degenerative diseases." The vitamin/mineral must be from the approved ingredients or derived from edible parts of food, and no functional dose of other active substance shall be formulated in the product. This provision provides the detailed amount of vitamin/mineral amount used in the adult nutrient supplement (Table 21.4); the amount used for maternal women and children under 18 years old shall be within 1/3−2/3 of the Recommended Nutrient Intake or Adequate Intake (AI) levels of the correspondent subpopulation groups. Nutrient supplement must be labeled as "nutrient supplement" with only a nutrient supplement claim. Those containing three or more vitamins and minerals are classified as "multivitamins/minerals supplement," and the nutrient content must be labeled as the exact value for its minimal consumption unit. Warning information must be specific that such nutrient supplement shall not be used to substitute medicine, and shall not exceed recommended dose or consume supplement with similar nutrients simultaneously.

Since the approval for nutrient supplements is relatively easy compared to health functional claims, more than one-quarter of recent approved health foods are nutrient supplements (Figure 21.1); this is more obvious in the past five years. Among the approved nutrient supplements, providing calcium and vitamin D are the two most popular claims; more than 800 calcium supplements and 300 vitamin D supplements have been approved (Figure 21.2). For the calcium and vitamin D supplement, typical calcium- and vitamin D-labeled values are shown in Table 21.5. The approved calcium supplements typically provide 400−450 mg calcium daily, while vitamin D supplements provide 3.1−4.6 μg vitamin D daily. This trend is consistent with nutrition survey data showing that dietary calcium is still below the AI level recommended by the Chinese Nutrition Society, even though the dietary intake of calcium has increased in the past two decades. A nine province dietary survey conducted in 2009 indicated that daily calcium intake was only 415.3 mg for men and 367.3 mg for women [25]. In an economically advanced area such as Shanghai, an epidemiological survey on plasma 25-hydroxylvitamin D level indicated that vitamin D deficiency remains relatively high [26]. The nutrient deficiency data suggest that vitamin D and calcium supplement may continue to be a critical growth opportunity for the health food industry in China, particularly for nutrient supplements.

Table 21.4 Source Materials of Vitamin/Mineral and Correspondent Minimum and Maximum Values in Nutrient Supplements [9]

Name	Min	Max	Source Ingredient
Calcium	250 mg/day	1000 mg/day	Calcium acetate, calcium carbonate, calcium caseinate, calcium chloride, calcium citrate, calcium citrate malate, calcium gluconate, calcium lactate, calcium malate, calcium monophosphate, calcium phosphate monobasic, calcium phosphate, calcium sulfate, calcium-L-ascorbate, calcium glycerol phosphate
Magnesium	100 mg/day	300 mg/day	Magnesium carbonate, magnesium chloride, magnesium citrate, magnesium gluconate, magnesium lactate, magnesium phosphate dibasic, magnesium phosphate, magnesium glycerol phosphate
Potassium	600 mg/day	1200 mg/day	Potassium carbonate, potassium phosphate dibasic, potassium chloride, potassium citrate, potassium gluconate, potassium lactate, potassium sulfate, potassium glycerol phosphate
Iron	5 mg/day	20 mg/day	Ferric ammonium citrate, ferric chloride, ferric citrate, ferric carbonate, ferrous citrate, ferrous gluconate, ferrous sulfate, ferrous lactate, heme iron (ferrous porphyrin), hemin (ferriheme), ferrous succinate, ferric pyrophosphate
Zinc	5 mg/day	20 mg/day	Zinc acetate, zinc carbonate, zinc chloride, zinc citrate, zinc gluconate, zinc lactate, zinc sulfate, zinc oxide
Selenium	15 μg/day	100 μg/day	Selenium carrageenan, selenium cysteine, selenium rich yeast, sodium selenate, sodium selenite, selenomethionine
Chromium (Cr^{3+})	15 μg/day	150 μg/day	Chromium trichloride, chromium nicotinate, chromium picolinate, chromium yeast

(Continued)

Table 21.4 (Continued)

Name		Min	Max	Source Ingredient
Copper		0.5 mg/day	1.5 mg/day	Cupric carbonate, cupric citrate, cupric gluconate, cupric sulfate
Manganese		1.0 mg/day	3.0 mg/day	Manganese sulfate, manganese gluconate, manganese chloride, manganese citrate, manganese glycerol phosphate
Molybdenum		20 μg/day	60 μg/day	Ammonium molybdate, sodium molybdate dihydrate
Retinol equivalent (RE) (vitamin A or vitamin A with β-carotene)		200 μg RE/day	800 μg RE/day	All *trans* retinol, vitamin A acetate, retinyl palmitate, all *trans* β-carotene
β-Carotene		1.5 mg/day	5.0 mg/day (synthetic) 7.5 mg/day (natural)	
Vitamin D		1.5 μg/day	10 μg/day	Vitamin D_2 (ergocalciferol), vitamin D_3 (cholecalciferol)
Vitamin E (as α-tocopherol equivalent TE)		5 mg α-TE/day	150 mg α-TE/day	D-α-Tocopherol, DL-α-tocopherol, DL-α-tocopheryl acetate, mixed tocopherols, natural vitamin E (D-α-tocopheryl acetate, D-α-tocopheryl succinate)
Vitamin K		20 μg/day	100 μg/day	Vitamin K_1 (phytonadione), vitamin K_2 (menaquinone)
Vitamin B_1		0.5 mg/day	20 mg/day	Thiamin hydrochloride, thiamin mononitrate
Vitamin B_2		0.5 mg/day	20 mg/day	Riboflavin, riboflavin-5′-phosphate sodium
Vitamin PP	Nicotinic acid	5 mg/day	15 mg/day	Nicotinic acid
	Nicotinamide	5 mg/day	50 mg/day	Nicotinamide
Vitamin B_6		0.5 mg/day	10 mg/day	Pyridoxine hydrochloride, pyridoxine-5′-phosphate
Folic acid		100 μg/day	400 μg/day	Pteroylmonoglutamic acid (folic acid)
Vitamin B_{12}		1 μg/day	10 μg/day	Cyanocobalamin, hydroxocobalamin
Pantothenic acid		2 mg/day	20 mg/day	Pantothenic acid, calcium pantothenate, D-panthenol, D-pantothenate sodium, D-pantothenate calcium

(Continued)

Table 21.4 (Continued)

Name	Min	Max	Source Ingredient
Choline	150 mg/day	1500 mg/day	Choline chloride, choline bitartrate
Biotin	10 μg/day	100 μg/day	D-Biotin
Vitamin C	30 mg/day	500 mg/day	L-Ascorbic acid, ascorbyl palmitate,calcium-L-ascorbate, potassium-L-ascorbate, sodium-L-ascorbate

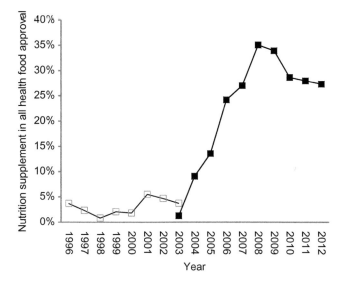

FIGURE 21.1

Percentage of nutrient supplement claims in all approved health foods between 1996 and 2012 (□ = MOH, ■ = SFDA).

21.6 Historical approval

Historically the MOH approved 5016 domestic health foods and 472 imported health foods between 1996 and 2003. The first health food was a bee pollen-based product with blood lipid modulation and immunity function claims. The SFDA had approved 7045 domestic health foods and 200 imported health foods by the end of 2012. There were 12061 domestic and 672 imported health foods approved by authorities in China by the end of 2012 (Table 21.6). It is clear that the majority of approved health foods in China are domestic products. Business

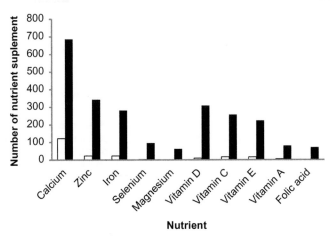

FIGURE 21.2

Number of nutrient supplement providing most common vitamin and mineral (□ = MOH, ■ = SFDA).

Table 21.5 Average of Calcium and Vitamin D Label Value in Approved Nutrient Supplements

	Average Daily Calcium Intake (mg/day)	Average Daily Vitamin D Intake (μg/day)
2004	421.6 ± 144.9	3.4 ± 1.3
2005	425.3 ± 120.2	3.1 ± 1.0
2006	435.4 ± 109.1	3.6 ± 1.3
2007	436.5 ± 121.9	4.1 ± 1.6
2008	427.7 ± 107.5	4.2 ± 1.7
2009	446.2 ± 136.2	4.5 ± 1.8
2010	425.5 ± 130.7	4.4 ± 1.9
2011	445.0 ± 134.0	4.2 ± 1.6
2012	417.7 ± 90.6	3.8 ± 1.1

Value expressed as mean ± SD.

operators in Beijing, Guangdong, and Jiangxi hold the most approved health food licenses; the United States, Hong Kong, and Japan are the top country/regions of origin of imported health products. The SFDA-approved health food license expires in 5 years, and there is a required renewal for another 5 years prior to expiration [3]. The re-registration does not allow the applicant to make changes to the functional claim, although the functional test and correspondent claim should be updated if protocol and category are revised [27]. This procedure will

Table 21.6 Historically Approved Number of Health Foods (Domestic and Imported)

	MOH		SFDA	
	Domestic	*Imported*	*Domestic*	*Imported*
1996	54	2		
1997	981	103		
1998	719	77		
1999	672	88		
2000	763	65		
2001	471	44		
2002	850	50		
2003	606	40	77	4
2004			1505	54
2005			926	27
2006			714	15
2007			377	10
2008			601	19
2009			577	25
2010			770	21
2011			789	13
2012			709	12
Total	5016	472	7045	200

allow the government to dynamically control the number of approved health foods on the market.

Immunity is the most popular claim in all health foods (Table 21.7); approximately 30% of the SFDA-approved health foods (including nutrient supplements) make an "immunity" claim, which might be related to the fact that this function is one of a few claims that does not require human clinical trial (anti-physical fatigue is another popular claim with no human clinical data requirement). This is consistent with the finding showing that "immunity," "anti-fatigue," and "blood lipid regulation" products comprised 63% of all health food (excluding nutrient supplement) between 1996 and 2005 [28]. Clearly, immunity products continue to be the most popular claim in all health foods, followed by nutrient supplement and anti-physical fatigue products. In 2012, 235 immune products, 198 nutrient supplements, 65 anti-physical fatigue products, 43 blood lipid-modulating products, and 36 bone density-improving products comprised 81% of the total approved 709 domestic health foods. On the other hand, there have been very few products approved in other categories. "Promoting lead excretion," "gastric mucosa protection," "acne removal," and "promoting milk secretion" are the least approved categories, with on average less than 3.3 approved per year for the past

Table 21.7 Most Popular Claims between 2003 and 2012 Approved by SFDA

	2003	2004	2005	2006	2007	2008	2009	2010	2011	2012
Immunity	28.6%	29.7%	32.9%	31.6%	29.1%	27.8%	28.5%	29.5%	35.4%	33.1%
Anti-physical fatigue	15.6%	15.0%	16.7%	13.4%	11.9%	9.8%	10.8%	11.6%	8.4%	9.3%
Blood lipids	9.1%	11.7%	5.2%	4.7%	8.5%	5.6%	5.1%	5.3%	5.8%	6.1%
Prevent liver injury	2.6%	4.7%	9.6%	6.7%	9.3%	4.3%	3.0%	6.9%	5.6%	3.4%
Blood sugar	5.2%	4.9%	5.0%	4.2%	4.6%	3.6%	3.5%	3.7%	3.3%	2.5%
Sleep	5.2%	4.3%	4.7%	3.6%	2.8%	3.3%	3.9%	4.5%	3.1%	3.8%
Bone density	5.2%	2.5%	1.8%	4.3%	4.1%	3.1%	2.7%	4.2%	5.9%	5.1%
Bowel movement	6.5%	6.0%	3.6%	3.1%	1.0%	1.5%	1.7%	3.2%	2.7%	2.5%
Memory	1.3%	1.9%	3.1%	3.5%	2.6%	3.3%	4.7%	3.6%	1.0%	1.4%
Chloasma removal	6.5%	3.7%	2.5%	3.4%	2.1%	2.0%	1.4%	1.5%	1.7%	0.7%
Nutrient supplement	3.9%	12.3%	17.4%	23.5%	28.6%	35.4%	35.0%	29.0%	28.5%	27.9%

10 years. Interestingly, neither the MOH or SFDA has ever approved a product with "improving skin oil content" claim, and the SFDA has proposed to eliminate this claim from future framework [11].

For the raw materials used in the health foods, goji berry (*Lycium barbarum*) is the most popular single botanical ingredient used (Figure 21.3). Goji berry is in more than 10% of all approved health food products, followed by American ginseng root (*Panax quinquofolium* L.) and astragalus (*Astragalus membranaceus*). Traditionally, the goji berry is used to nourish the liver and kidney and brighten eyes in TCM [18], and studies have revealed its health benefits on aging, neuroprotection, general well-being, fatigue, endurance, metabolism/energy expenditure, glucose control in diabetics, glaucoma, antioxidant, immunomodulation anti-tumor activity, and cytoprotection [29]. American ginseng root is known to provide *Qi* (energy) and nourish *yin* [18], whereas astragalus is known to provide *Qi* and lift *yang* [18]. *P. ginseng* is known for its adaptogenic effect and benefits for metabolic, endocrine, immune, and cardiovascular health [30]. Among the top 10 most commonly used botanical ingredients, half are from a list of "ingredients as both food and medicine" and half are from the list of "TCM only allowed in health food" (Table 21.2). For the five most widely used botanical ingredients (goji berry, American ginseng, astragalus, ginseng, and poria), only a handful of approved health foods are formulated as single active ingredients; more often these ingredients are formulated in combination with other raw materials (Figure 21.4), especially for those approved by the SFDA, suggesting that more strict regulation and in-depth assessment have been implemented on submitted

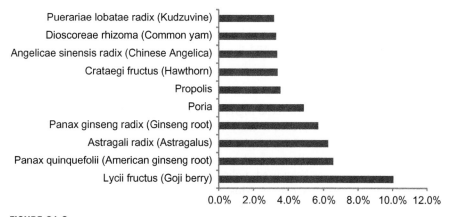

FIGURE 21.3

Top 10 most frequently used botanical ingredients in Chinese health foods 1996–2012 (of which *Lycii fructus*, poria, *Crataegi fructus*, *Dioscorea rhizoma*, and *Pueraria lobata radix* are listed in SFDA/MOH ingredients as both food and medicine. *Panax quinquefolii radix*, *Astragali radix*, *Panax radix et rhizoma*, propolis, and *Angelica sinensis radix* are listed in SFDA/MOH TCM ingredients as only allowed in health foods).

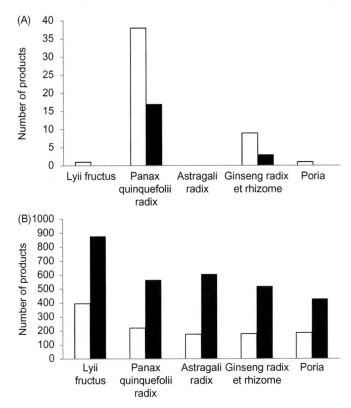

FIGURE 21.4

Top five most widely used botanical ingredients in Chinese health foods 1996–2012 (A) used as a single ingredient and (B) used as a multiple ingredients formula (□ = MOH, ■ = SFDA).

health foods. The majority of these ingredients are associated with a single health claim; less than 30% of products carry two claims (Figure 21.5). On the other hand, among the 201 raw materials listed in Table 21.2, only 42 have appeared more than 100 times in all approved products, 48 have been used less than 10 times, and 11 have never been used, indicating that the selection of the raw materials in health foods is very narrow. Most of the formulas have been developed on the limited choices of "common" materials, suggesting that there is an opportunity to explore those underused materials to create some uniqueness in the functional health food area in the future.

Reference compound (functional composition) is required for the compliance of the product specification of finished goods; it refers to the effective content of functional composition and component. The registration and re-registration of any health food requires the applicant to provide rationale on why the reference

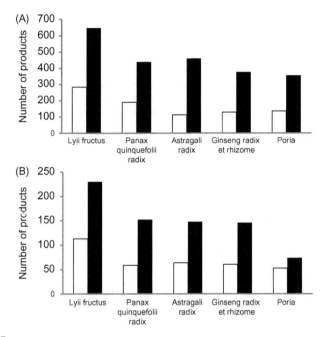

FIGURE 21.5

Number of claims associated with the top five most widely used botanical ingredients in Chinese health foods 1996−2012: (A) one claim and (B) two or more claims (□ = MOH, ■ = SFDA).

compound is selected and how such compound is quantified; meanwhile such reference compound and its content are required to be labeled on the finished product [31]. Saponin, flavonoids, and polysaccharides are the most commonly used functional composition in health foods (Figure 21.6A), which is consistent with the frequency of commonly used botanical ingredients such as ginseng, goji berry, astragalus, etc. Proanthocyanin, for example, is typically related to the claims such as " chloasma removal," "antioxidant," "immunity," and "bone density "etc. (Figure 21.6B). National standards for quantifications of some reference compounds in health foods have been established, including proanthocyanin, CoQ_{10}, soy isoflavone, lycopene, immunoglobulin G, melatonin, α-linolenic acid, eicosapentaenoic acid, DHA, inositol, etc. Otherwise, it is the applicant's responsibility to provide a validated method of quantification of the reference if no national standard has been established.

Immunity claim is the most popular claim; nearly 4000 products have been approved by the end of 2012. In this category, capsule, soft gel, and tablet are the most popular dosage formats (Figure 21.7A), comprising 36.4, 16.0, and 11.7%, respectively, of all immunity products. Crude polysaccharide, total saponin (including ginsenoside), and total flavonoids are the most frequently used reference

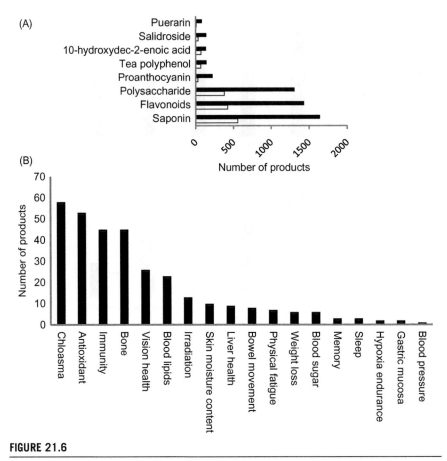

FIGURE 21.6

Commonly used functional reference compounds in Chinese health foods. (A) □ = MOH approval, ■ = SFDA approval. (B) Example of claims products associated with proanthocyanin labeled as a reference compound.

compounds in this category, appearing in 35.1, 28.9, and 19.8% of immunity products, respectively. Goji berry, American ginseng, astragalus, propolis, ginseng root, poria, honey, common yam, jujube, and Chinese angelica are the most commonly used TCM ingredients in this category (Figure 21.7B), which is similar to the commonly used ingredient in all claims (Figure 21.3). In addition, *Ganoderma* (*G. lucidum, G. sinensis*), *Cordyceps* (*C. sinensis*, Paecilomyces hepiali, hirsutella hepiali, *C. militaris*), vitamin C, spirulina, and colostrum are also commonly used in the products with immunity function (Figure 21.7B). In the immunity category, about 900 products carry more than just an immunity claim. Anti-physical fatigue is the most used secondary claim in immune products (Figure 21.7C), followed by blood lipids, liver injury, and irradiation protection claims.

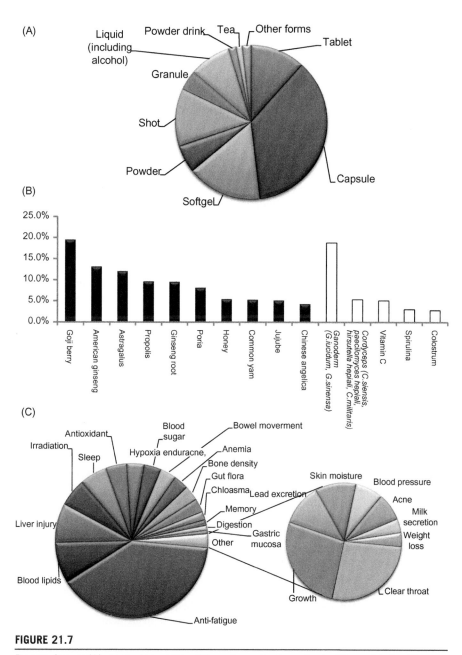

FIGURE 21.7

Dosage format (A), secondary claims (B), and popular ingredients (C) used in health foods with immunity claim.

21.7 Conclusion

In summary, the Chinese health food industry has grown dramatically in the past two decades and become the second largest health foods/dietary supplements market in the world. The Chinese government's 12th 5 year plan outlined the opportunity for the nutrition and health food industry to reach a market size of RMB 1000 billion Yuan by the end of 2015, providing great growth opportunity for this industry. To achieve this, innovation will play an important role, including the further development of novel food resources, exploration of the function—structure relationship of bioactive components, improvement of quality and the technical level of raw materials used in nutrition and health foods, and the development of ingredients with uniqueness by leveraging TCM resources, etc. Solid science and clinical evidence will be the future of health foods, and a more strict evaluation process may be applied to the approval of health food to ensure that only qualified products are approved and marketed; this seems to align with the newly implemented technical assessment protocols. With the implementation of re-registration and fewer claims for health food, it is expected that the number of future health food approvals may be well under control. However, if the new claim procedure is implemented in the future, it will also open the door for business operators to seek opportunities with the "uniqueness" claim. Although significant investment in R&D will be required for this direction, it will elevate the scientific credibility of the industry with good alignment with government policy and guidance.

References

[1] Li J, Li D. Current status of Chinese legal system and standard system for health food. Food Sci 2011;32:318—23.

[2] CHCA. China Health Care Association and Chinese Academy of Social Science published the Blue paper of Development of China Health Food Industry, <http://www.chc.org.cn/news/detail.php?id = 60847>, Beijing, China; 2012.

[3] SFDA. Interim Administrative Measures of Health Food Registration- Decree No.19, 2005, issued April 30, 2005, http://www.sfda.gov.cn/WS01/CL1131/24516.html, Beijing, China; 2005.

[4] MOH. Procedure of Technical evaluation of health foods (MOH oversight division Nofitification 1996 no 38, issued July 18, 1996), Beijing; 1996.

[5] MOH. Notification regarding several issues of health foods administration (MOH oversight division notifiation, 1997, no 38, issue July 1, 1997). Beijing, China; 1997.

[6] MOH. Notification regarding the adjustification of scope of acceptable functional claims and approvals of health foods (MOH oversight division notification No. 20, 2000, issued Jan 14, 2000), Beijing, China; 2000.

[7] MOH. Technical protocols for health foods detection and evaluation (MOH oversight division notification No 42, 2000, issued Feb 24, 2003), Beijing, China; 2003.

[8] SFDA. Implementation of technical protocol of health food detection and evaluation 2003 edition regarding health foods acceptance and evaluation (SFDA Registration Division Notification 13, 2004, issued Jan 17, 2004, <http://www.sda.gov.cn/WS01/CL0847/10232.html>), Beijing, China; 2004.

[9] SFDA. Notification regarding Interim Provision of Nutrient supplement applications and evaluation etc eight provisions (SFDA registration division No 202, 2005, issued May 20, 2005, <http://www.sda.gov.cn/WS01/CL0055/10396.html>), Beijing, China; 2005.

[10] SFDA. Letter to seek public comments on the drafted proposal of adjustment of functional claims scope of health foods (SFDA health food and cosmetic division letter No 322, 2011, issued Aug 1, 2011, <http://www.sda.gov.cn/WS01/CL0780/64433.html>), Beijing, China; 2011.

[11] SFDA. Letter to seek public comments again on the drafted proposal of adjustment of functioanl claims scope of health foods (SFDA health food and cosmetic division letter No 268, 2012, issued Jun 4, 2012, <http://www.sda.gov.cn/WS01/CL0780/72295.html>), Beijing, China; 2012.

[12] Yang Y. Scientific substantiation of functional food health claims in China. J Nutr 2008;138: 1199S−205S.

[13] SFDA. Notification of issuing evaluation protocols related to antioxidant function etc 9 health function evaluations (SFDA Health Food and Cosmetic Division, No 107, 2012; issued April 23, 2012, <http://www.sda.gov.cn/WS01/CL0847/71257.html>), Beijing, China; 2012.

[14] SFDA. Notification regarding matters related to the implementation of interim administrative measures of health food registration (SFDA Registration Divison No 281, 2005; issued May 27, 2005, <http://former.sfda.gov.cn/cmsweb/webportal/W945325/A64003265.html>), Beijing, China; 2005.

[15] SFDA. Letter to seek public comments again on the provisions of new function application and evaluation of health food (consulting version) (SFDA Health Food and Cosmetic Division letter No 490, 2012, issued Nov 5, 2012, <http://www.sda.gov.cn/WS01/CL0780/75896.html>), Beijing, China; 2012.

[16] Liu Q, Bai H. The evolvement and safety supervision of administrative system of China health food registration. Carcinogenesis, tetragenesis, mutagenesis 2012;24:321−4.

[17] MOH. List of micobial strains allowed in foods (MOH General Office No 65, 2010; issued April 22, 2010; <http://www.moh.gov.cn/mohbgt/s10787/201004/47133.shtml>, Beijing, China; 2010.

[18] ChP. In: Commission CP, editor. Pharmacopoeia of People's Republic of China 2010, vol. 1. Beijing: China Medical Science Press; 2010.

[19] SFDA. Notification regarding matters related to the registion submission and evaluation of health food containing Coeznyme Q10 (SFDA Permission Division No 566, 2009, issued Sept 2, 2009; <http://www.sda.gov.cn/WS01/CL0055/41194.html>), Beijing, China; 2009.

[20] SFDA. Notification regarding matters related to the registion submission and evaluation of health food containing soy isoflavone (SFDA Permission Division No 567, 2009, issued Sept 2, 2009; <http://www.sda.gov.cn/WS01/CL0055/41195.html>), Beijing, China; 2009.

[21] SFDA. Notification regarding matters related to registration submission and evaluation of health food containing red yeast and other raw materials (SFDA permission

division No 2, 2010, issued Jan 5, 2010; <http://www.sfda.gov.cn/WS01/CL0847/44918.html>), Beijing, China; 2010.

[22] MOH. Further specification of the raw materials used in the health foods (MOH Oversight Division No 51, 2002, issued Feb 28, 2002; <http://www.moh.gov.cn/mohwsjdj/s9160/200810/38057.shtml>), Beijing, China; 2002.

[23] MOH. Administrative Measures of Novel Food (MOH Decree No 56, 2007, issued Jul 7, 2007; <http://www.moh.gov.cn/mohwsjdj/s3592/200804/16500.shtml>), Beijing, China; 2007.

[24] MOH. Specification on the food additives and novel food administrative permit (MOH Announcement No 14, 2009, issued Oct 13, 2009; <http://www.moh.gov.cn/mohwsjdj/s7891/200910/43139.shtml>) Beijing, China; 2009.

[25] Liu A, Zhang B, Wang H, Du W, Su C, Zhai F. The nutrients intake trend of Chinese population in nine provinces from 1991 to 2009 (VI) Calcium intake trend in Chinese adults aged 18–49 years. Acta Nutrimenta Sinica 2012;34:10–4.

[26] Zhu H, Cheng Q, Gan J, Du Y, Hong W, Zhu X, Li H. Research on the vitamin D status of Shanghai population. Chin J Osteoprosis and Bone Mineral Research 2010;3:157–63.

[27] SFDA. Critical points for the technical evaluation of health food re-registration (SFDA permission division notification No 390, 2010; issued Sept 26, 2010; <http://www.sfda.gov.cn/WS01/CL0055/54296.html>), Beijing, China; 2010.

[28] Shu Y, Liu C, Li L. Study on present situation of certified native functional food products in China. Chinese Journal of Food Hygiene 2006;18:401–5.

[29] Amagase H, Farnsworth NR. A review of botanical characteristics, phytochemistry, clinical relevance in efficacy and safety of Lycium barbarum fruit (Goji). Food Research International 2011;44:1702–17.

[30] Kitts DD, Hu C. Efficacy and safety of ginseng. Public Health Nutrition 2000;3:473–85.

[31] SFDA. Critial points on technical evaluation of health foods (SFDA Permission Division Notfication No 210, 2011; issued May 28, 2011), Beijing, China; 2011.

Regulations on Health/ Functional Foods in Korea

22

Ji Yeon Kim*, **Seong Ju Kim**[†] **and Hyong Joo Lee*****

Department of Food Science and Technology, Seoul National University of Science and Technology, Seoul, South Korea [†]*Ministry of Food and Drug Safety, Chungcheongbuk-do, South Korea* **WCU Biomodulation, Department of Agricultural Biotechnology, College of Agriculture and Life Sciences, Seoul National University, Seoul, South Korea*

22.1 Introduction

With the rapid increase in aging populations across the world, the prevalence of chronic diseases including diabetes, cardiovascular disease, and cancer will continue to rise. These worrying trends have compelled researchers to focus efforts on the identification of potentially useful bioactive components in food, and the relationship between diet and health has been receiving greater media attention. The increasing amount of public information available regarding the health benefits of foods has resulted in consumer interest in health issues becoming a major factor in purchasing decisions. Therefore, the labeling and advertising of products claiming health benefits has become the subject of regulations in order to avoid misunderstanding and exaggeration. The need to protect consumers and ensure their right to accurate information on food functionality has led to strict labeling requirements for functional foods and dietary supplements in many nations since the 1990s, including the United States, Japan, and the European Union. The Health/Functional Food Act (HFFA) introduced in Korea in 2002 [1] initially regulated only the supplement forms of health/functional foods (HFFs). The definition of HFF was then further expanded into various kinds of processed foods in a subsequent revision [1]. This chapter deals with the scope of functional food regulations, the strength of evidence required for their efficacy, safety considerations, and future perspectives in Korea.

22.2 HFFA

The Korean Health and Welfare Committee of the National Assembly proposed the HFFA in November 2000. In August 2002, the HFFA was enacted as a new regulatory framework for the safety, efficacy, and labeling of HFFs, and came into effect starting January 2004. The ultimate goal of the Act was to enhance public health by ensuring the safety of new bioactive ingredients. In 2004, the law defined HFFs as "food supplements such as pills, tablets, capsules and liquids"; in other words, conventional foods were not permitted to make any such claims. However, with increasing pressure from the food industry, after 2008 the definition of HFF was broadly expanded to encompass all types of processed foods.

Now, HFFs are defined as "food manufactured or processed with ingredients or components that possess functionality useful for the human body." However, foods for special dietary usage remain under different regulations according to food hygiene laws.

HFFs are divided into generic and product-specific HFFs. Generic HFFs are defined as products having functional ingredients as listed by the Ministry of Food and Drug Safety (MFDS). These products include vitamins, minerals, and various other functional compounds. All new ingredients that are not generic HFFs are subjected to efficacy evaluation before marketing, on the basis of their ingredients or physiologically active components to ensure the accuracy of the health claims. Authorization is granted on a product-by-product basis by issuing a certificate without regulatory amendments. The HFFA does not define what constitutes "substantiation" for a claim made for an HFF. Instead, it gives the MFDS the exclusive authority to evaluate the safety and efficacy of HFFs prior to their introduction to the market, and it also keeps manufacturers and distributors responsible for providing all evidence regarding the advertised claims of their products, by developing a system for substantiation of claims, or relying on existing information.

Under the HFFA, claims regarding nutritive and other functions, as well as reductions in disease risk that are compatible with those adopted by the Codex Alimentarius Commission in 2004 [2] can be used for labeling and advertising of HFFs. Each claim type is defined as follows:

- *Nutrient-function claims*: These claims describe the physiological role of the nutrient in the growth, development, and normal functions of the body. Such claims apply to the nutrients, which have their own Recommended Daily Allowance (RDA) and must be based on current, university-level nutrition texts as a source of evidence.
- *Other function claims*: These claims concern specific beneficial effects of HFFs in the context of the total diet on normal functions or biological activities of the body. Such claims relate to positive contributions toward health, the specific improvement of a function, or to the modification or preservation of health.
- *Reduction of disease risk claims*: These claims describe the relationship between the consumption of HFFs (in the context of the total diet) and the reduced risk of developing a disease or health-related condition.

Claims regarding the use of HFFs to prevent or cure specific diseases are not permissible. The MFDS intends to develop a more focused definition for such disease-related claims in future.

22.3 **Generic HFFs**

In 2004, 37 generic HFFs including 13 vitamins, 11 minerals, essential amino acids, proteins, dietary fiber, and essential fatty acids were available [3]. The MFDS listed these generic HFFs into an HFF code. The categories were nutritional supplements, health supplements, and ginseng products; however, the HFFA at that time did not require scientific evaluation for functional claims. Therefore, in order to balance with product-specific HFFs and meet the basic outline of the HFFA, a re-evaluation of generic HFFs was needed. Based on these needs, the MFDS re-evaluated 37 generic HFFs for scientific substantiation of their claims from 2003 to 2007. As a result, several generic HFFs including royal jelly, yeast, bee pollen, digestive enzymes, turtles, and eels were eliminated from the HFF category. These products are re-categorized as conventional food which are not able to bear health claim [4].

Several ambiguous expressions in a number of claims were changed into "enhanced function" claims. Dietary fibers were listed separately according to their raw materials and were regarded as separate functional ingredients. In the revised HFF code, the definition of a functional ingredient was also clarified as follows:

"The functional ingredient" is a substance providing health benefits and falls under any of the following subparagraphs: (a) processed raw material originated from animal, plant or microorganism; (b) extract or purified substance of any ingredient described in subparagraph (a); (c) synthetic duplicate of purified substance of any ingredient described in subparagraph (b); or (d) combination of any ingredients described in subparagraph (a), (b), or (c).

In addition, in order to expand the number of generic HFFs covered in the HFF code, new principles for the addition of generic HFFs among product-specific HFFs were introduced. The functional ingredients recognized under the Regulation on Approval of Functional Ingredients for Health Functional Food were permitted to be added to the Code in cases where the functional ingredient was subject to the following:

"In the case(s) where, after being recognized as a functional ingredient, two or more years have passed from the date of the item manufacturing report or the import report or where, after recognition, three or more business persons file the item manufacturing report or the import report; or in cases where the person who has obtained the recognition under the provision of Article 14 (2) of the Health Functional Food Act requests the adding (provided three or more persons obtain the recognition, two-thirds of them shall request the adding together).

However, sometimes after being recognized as a functional ingredient, the applicant requests the protection of data concerning their manufacturing processes, human studies on functionality, and toxicology studies. In these cases, adding of the functional ingredient can be postponed for up to 5 years from the date of the item manufacturing report or import report, dependent upon the validity of the request.

For generic HFFs, nutrient-function claims can only be made for vitamin and mineral supplements, which have their own RDAs. There are well-established health claims describing the physiological role of the nutrient in the growth, development, and normal functions of the body. The quantity of vitamin/mineral supplement per unit must fall between designated upper and lower limits. The lower limit has been set at 30% of the RDA for Koreans, with the upper limit determined based on the risk analysis for each vitamin or mineral. At present, there are 14 vitamins, 11 minerals, proteins, essential fatty acids, and dietary fiber (which is a collective term for non-digestible fiber) available as nutritional supplements. As functional ingredients, 55 ingredients have been listed including ginseng, green tea extracts, and various fibers (Table 22.1).

22.4 Product-specific HFFs

If manufacturers or distributors wish to market HFFs that are not included in the list of generic HFFs, they must lodge an application for a product-specific HFF [5].

If a functional dietary ingredient in the product is not in the list of generic HFFs (and hence is regarded as new), manufacturers or distributors must provide the MFDS with evidence of the standardization, safety, and efficacy of the product. The MFDS then performs a pre-market review within 120 days of receiving the application, focusing on the origin and nature of the ingredients, content of functional components (or index components), processing methods, information on the methods and validation for analysis of functional components, stability data, and purity (in terms of the content of microbial, heavy metals, pesticides, etc.), as well as scientific evidence for the safety and efficacy of the HFF.

The evaluation by the MFDS begins once the applicant has submitted the relevant documents. The MFDS first determines whether the standardization has been performed adequately, and then evaluates the safety, efficacy, and specifications. If the candidate meets the required standards, its efficacy is formally designated as such.

22.4.1 Standardization

The manufacturer submits data on the specific characteristics of the functional ingredient (or, in the case of an unclear functional ingredient, an index ingredient for identifying raw materials), and states the yield and change in content of functional component (or index component) arising from the main manufacturing process. The MFDS then evaluates the adequacy of these data.

Table 22.1 Generic HFFs in HFF Code

Nutrients			
Vitamins		**Minerals**	
Vitamin A	β-Carotene	Calcium	Magnesium
Vitamin D	Vitamin E	Iron	Zinc
Vitamin K	Vitamin B1	Copper	Selenium
Vitamin B2	Niacin	Iodine	Manganese
Pantothenic acid	Vitamin B6	Molybdenum	Potassium
Folic acid	Vitamin B12	Chrome	
Biotin	Vitamin C		
Dietary fiber		Protein	
Essential fatty acids			
Functional Ingredients			
Ginseng	Red ginseng	Plants containing chlorophyll	Chlorella/spirulina
Green tea extracts	Aloe whole leaves	Propolis extracts	Coenzyme Q10
Soybean isoflavone	Omega-3 fatty acids	γ-Linoleic acid	Lecithin
Squalene	Phytosterol/-ester	Alkoxyglycerol	Octacosanol
Japanese apricot extract	CLA	*Garcinia cambogia* extract	Lutein
Haematococcus extract	Saw palmetto extract	Glucosamine	Mucopolysaccharide
N-acetylglucosamine	Guava leaf extract	Banaba leaf extract	Gingko leaf extract
Milk thistle extract	Phosphatidylserine	Evening primrose seed extract	Aloe gel
Ganoderma lucidum fruit body extracts		Chitosan/ oligosaccharide	Probiotics
Fructo-oligosaccharide	Red yeast rice	L-Theanine	Soybean protein
Functional Dietary Fiber			
Guar gum/ hydrolysates	Glucomannan	Indigestible maltodextrin	Oat fiber
Soy fiber	Tree ear	Wheat fiber	Barley fiber
Arabic gum	Corn bran	Inulin	Psyllium husk
Polydextrose	Fenugreek seed		

22.4.2 Safety evaluation

The safety of the active ingredient is assessed by the MFDS, which is responsible for reviewing the submitted data including detailed descriptions on the history of use, manufacturing processes, quantity for consumption, results of toxicity tests, results of human studies, and results of nutritional evaluation and bioavailability. The safety of the active ingredient must be validated scientifically with reference to a decision tree [5] (Figure 22.1). The preparation of safety data can be divided into four categories: (1) cannot be used as a raw ingredient for an HFF; (2) scientific data or historical use describing its manufacturing process, usage, and the amount of intake, database searches of the side effects and toxicity, and evaluating the amount to be consumed; (3) scientific data or historical use, database searches of the side effects and toxicity, evaluating the amount to be consumed, and evaluating the nutritional effects; and (4) database searches of the side effects and toxicity, evaluating the amount to be consumed, evaluating the nutritional effects, toxicity data, and other data relevant for proof of safety.

22.4.3 Efficacy evaluation

The evaluation of HFF efficacy differs from those of medicines, since the target consumers of HFFs are either healthy individuals or individuals in the preliminary stage of a disease or borderline condition in an at-risk group. Therefore, efficacy requirements are less distinct for HFFs than for drugs marketed to patients with a medical condition.

Currently, no formula exists for the extent of studies required to substantiate a claim regarding HFFs, with the MFDS applying a generalized standard of "competent and reliable scientific evidence," which provides manufacturers and distributors with some degree of flexibility in the type of evidence while contributing to the maintenance of consumer confidence in HFFs. To evaluate whether the submitted information constitutes competent and reliable scientific evidence, each study is reviewed independently, followed by a consideration of the strength of the total body of evidence, as follows:

- *Reviewing individual studies*: First, the design type of each study is characterized. Competent and reliable scientific evidence adequate to substantiate a claim generally consists of information derived primarily from human studies. In particular, a randomized, double-blind, parallel-group, placebo-controlled intervention study is considered the gold standard. Other types of scientific evidence such as that from animal studies, *in vitro* studies, anecdotal evidence, meta-analysis, and review articles would generally be considered background information, but alone may not be adequate to substantiate a claim unless there are sufficient animal and *in vitro* studies to explain the biochemical and physiological mechanisms for the beneficial effects on health, or to demonstrate dose−response relationships. Second, the scientific quality of each study is reviewed based on several factors including

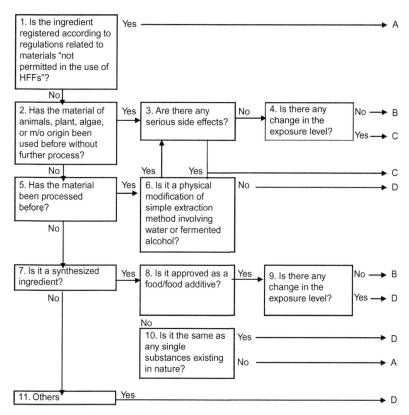

FIGURE 22.1

Decision tree for the preparation of safety data.

the study design and implementation, the study population, data collection, outcome measures, statistical analysis, and confounding variables. A scientific study that adequately addresses most of the above factors would be considered to be of sufficiently high quality.

- *Reviewing the totality of studies*: Although the type and quality of the individual studies are important, each resultant unit of data should be considered in the context of all available information. The strength of the entire body of scientific evidence can be considered based on several criteria, including quantity, consistency, and relevance. The collection of a larger amount of data from independently conducted studies provides more persuasive evidence. It is ideal if the evidence used to substantiate a claim supports the background information, whereas conflicting or inconsistent results question whether a particular claim can be substantiated. Relevant biomarkers may be used as indicators or predictors of the proposed relationship between a component and a health endpoint.

There are concerns that consumer information regarding the health benefits of foods may be too limited if claims that reach the standard required for scientific proof are the only ones allowed for the labeling and advertising of HFFs. The MFDS therefore introduced an evidence-based rating system, where rank is determined according to the type and quality of individual studies, as well as the quantity, consistency, and relevance of the aggregated studies.

Claims related to a reduction of disease require the highest level of evidence, primarily based on well-designed human intervention studies having a valid design for demonstrating a persistent effect of an HFF. It should also be validated by the consensus of significant scientific agreement among experts qualified with the appropriate scientific training and experience to evaluate such claims. However, a wider range of scientific evidence can be used for other-function claims. Although human intervention studies are preferred, animal and *in vitro* studies alone may be sufficient to demonstrate other-function claims of HFFs, especially if they are relevant to or sufficiently close representations of human metabolism. The MFDS defined the following three levels of other-function claims based on these variations in scientific evidence. They are: "convincing," "probable," and "insufficient" (Table 22.2).

The evidence-based ranking system links the ranking of scientific evidence to the wording of relevant claims, where different levels of evidence result in different appropriate statements (see Table 22.2).

22.4.4 Specification

The applicant must also submit samples of the product with documentation regarding the analytical method used for analyzing the functional component to the MFDS, which validates the selectivity, precision, accuracy, linearity, and range of the method. The MFDS then determines the contents of the functional component and, if the decision is made to grant approval for the product as an HFF, confirms the period of conformity and hygiene specifications.

22.4.5 Claims approved for product-specific HFFs

As at October 2012, over 165 functional ingredients have been approved by the MFDS. Although most of the statements for Korean HFFs are for other-function

Table 22.2 Permitted Descriptions for HFFs According to Scientific Evidence (KFDA)

Type of Claim	Level of Scientific Evidence	Permitted Statement
Reduction of disease risk	Significant scientific agreement	"Can help to reduce the risk of... (disease)"
Other function (I)	Convincing	"Can have a beneficial effect on..."
Other function (II)	Probable	"May improve..." "May increase (decrease)..."
Other function (III)	Insufficient	"May improve..., but this requires verification" "May improve..., but the scientific evidence is insufficient"

claims, there exist various types of function claims in product-specific HFFs. The product-specific HFFs are still increasing in number (Figure 22.2). The existing claims are varied, and include body fat reduction and joint health to eye health, postmenopausal woman's health, urinary tract health, digestive health, and memory function (Figure 22.3). Figure 22.4 shows annual market size and data regarding domestic market performance. Although these data do not include imported HFF products, the rapid increase in market size is clear. This is best seen through a comparison of the 2011 and 2010 numbers, showing an overall production increase of 23%. Among the available HFF products in the Korean market, red ginseng accounts for the majority of HFF product sales (Figure 22.5). Among product-specific HFFs, liver health products were by far the most popular (Figure 22.6). These were followed by products marketed as immune health, joint health, and skin health functional ingredients.

22.5 Advisory committees

The MFDS maintains advisory committees for HFFs comprising 30 experts with backgrounds in nutrition, food science, medicines, and consumer unions. These committees comprise six subcommittees that advise the MFDS about regulations, good manufacturing practices, importing and exporting, new active ingredients, standards and specifications, and labeling and advertisements.

22.6 Future perspectives

The ability to determine whether a claim does not mislead consumers is important in establishing a reliable process for product safety and consumer rights. In order

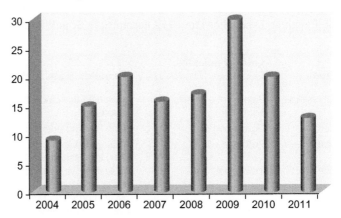

FIGURE 22.2 Annual data for new functional ingredients.

Until 2009, there was an increasing trend in the number of functional ingredients marketed. Since then, the number of new functional ingredients has been decreasing.

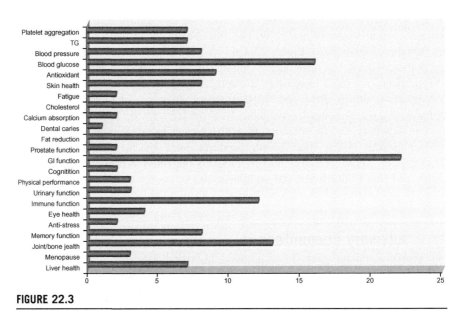

FIGURE 22.3

Number of claims approved as product-specific functional ingredients.

to ensure a successful system of regulation for clear and accurate health claims, it is essential to define the appropriate words and phrases permitted, in order to convey to consumers the different levels of scientific evidence underlying product claims. Our recent survey suggests that the present expressions are inadequate for informing consumers about the scientific evidence underlying other-function

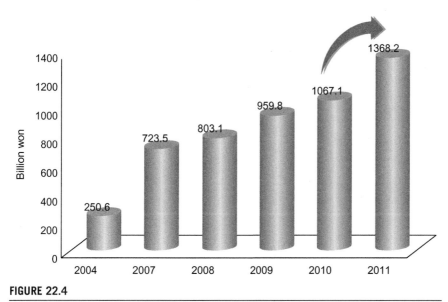

FIGURE 22.4

Annual domestic market performance for HFFs in Korea.

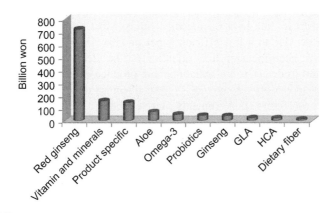

FIGURE 22.5

Popularity of HFF types in 2011 based on domestic performance data in Korea.

claims. Moreover, consumers consider that the present expressions do not describe the efficacy in an easily understandable manner. Some consider that it would be more appropriate to explain the claims rather than provide a conclusive standardized expression. The MFDS therefore needs to develop statements of health claims that consumers can easily understand and provide consumer education to ensure that informed choices can be made about HFFs.

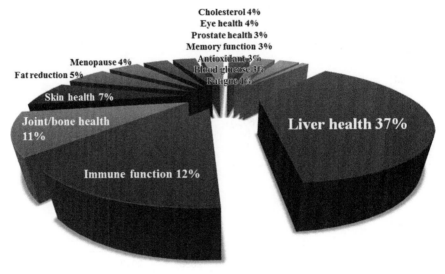

FIGURE 22.6

Best-selling HFF types among product-specific categories in Korea.

From the perspective of research, concerns about biomarkers remain. Food or food components that are ingested and pass through the gut can eventually enter both the bloodstream and liver or intestinal fluids. Most food components exhibit a moderate effect due to a low efficiency per quantity of intake. In recent times, researchers have been focusing their attention on the identification of biomarkers that can be used to detect small variations and produce -omics data. The identification of the most sensitive biomarkers for such -omics data will be of significant value in the near future. Governments and regulatory agencies should also be encouraged to support cutting-edge science in biomarker identification.

References

[1] HFFA. Health/Functional Food Act, 2008. Korea: Ministry of Health and Welfare; 2008.
[2] CCFL. Codex guidelines for use of nutritional and health claims; 2004.
[3] HFFC. Health/functional food code, 2008. No. 2004-64. Korea: Regulation of the Korea Food and Drug Administration; 2008.
[4] FC. Food code, 2010. No. 2010-2. Korea: Regulation of the Korea Food and Drug Administration; 2010.
[5] RHFF. Regulations on the premarket approvals of functional ingredient of product-specific health/functional food and food type health food, 2011. No. 2011-34. Korea: Korea Food and Drug Administration; 2011.

Phytomedicines, Functional Foods, Nutraceuticals, and Their Regulation in Africa

23

Theeshan Bahorun*, **Vidushi S. Neergheen-Bhujun***,†,
Mayuri Dhunnoo* **and Okezie I. Aruoma****

**ANDI Centre of Excellence for Biomedical and Biomaterials Research,
University of Mauritius, Réduit, Republic of Mauritius*
†*Department of Health Sciences, Faculty of Science and ANDI Centre of Excellence for
Biomedical and Biomaterials Research, University of Mauritius, Réduit, Republic of Mauritius*
***American University of Health Sciences, Signal Hill, California*

23.1 Introduction

Plants have always been considered as having medicinal, culinary, and other functional properties. Their prophylactic uses date back to as early as 5000 B.C. Ancient Hebrews, Indians, Chinese, Greeks, and Babylonians were all well-known practising herbalists and more than 850 medicinal plants were even documented in the Ebers papyrus as early as 1550 BC [1]. Despite the technological advances, the World Health Organization (WHO) indicates that 80% of the Indian and African populations still depend heavily on traditional treatments for primary health care [2]. Although the importance of pharmacognosy dropped at one point due to various limitations encountered during the processing of plants, interest was, however, rekindled with developments in extraction, chromatography, electrophoresis, and spectroscopy techniques that have enhanced the importance and value of plant-based foods and drugs [3]. To date, there are around 120 distinct chemical substances derived from plants that are considered as important drugs [3]. The biotechnology and pharmaceutical industries have shown a keen interest and are investing heavily in biopharmaceuticals derived from plant cultures and botanical pharmaceuticals [4]. The factors that still determine the development and supply of botanical drugs and nutraceuticals include among others bioprospection, low cost sources of existing products, standards that govern the introduction of new products, enhancement of acceptance of existing products, pharmacological and toxicological studies complementing

Nutraceutical and Functional Food Regulations in the United States and Around the World.
DOI: http://dx.doi.org/10.1016/B978-0-12-405870-5.00023-2

the regulatory framework governing the sale, manufacture, packaging, labeling, importation, distribution, and storage.

Ever since a relationship was established between diet, health, and well-being, food is no longer merely considered as a means to satisfy hunger and prevent deficiency diseases; it has become a primary vehicle to transport human beings along the road of optimal health and wellness [5]. Alongside growing affluence worldwide, this knowledge has allowed easier access to a safer and more varied diet, all of which ensure longevity. Research has led to the identification and understanding of the potential mechanisms of biologically active components in food, which could improve health and possibly reduce the risk of disease while enhancing our overall well-being. This changing face of food has thus led to the concept of functional foods. Many definitions exist worldwide for functional foods, but there is no official or commonly accepted definition. One view is that any food is indeed functional because it provides nutrients and has a physiological effect. Some foods considered to be functional are actually natural whole foods where new scientific information about their health qualities can be used to proclaim benefits. Many, if not most, fruits, vegetables, grains, fish, and dairy and meat products contain several natural components that deliver benefits beyond basic nutrition. There is also a line of thinking, which considers that only fortified, enriched, or enhanced foods that have a component with a health benefit beyond basic nutrition should be considered as functional. Most definitions also suggest that a functional food should be, or look like, a traditional food and must be part of our normal diet. New promising technologies such as nutrigenomics and imaging techniques, and converging technologies are increasingly being used in nutrition research. Their huge potential will be apparent in the short and medium term; this will further enable the development of foods for targeted population groups with defined risk factors or diseases such as allergy, diabetes, obesity, and cardiovascular disease. Even more innovative is the possibility of merging information about the physiological responses to food with individual genetic information to design personalized food and diets. The ingenuity of functional food technological innovation (e.g., cholesterol-lowering products, sweeteners in chewing gums, fermentation, and genetically modified products) might also contribute to further advances in the development of food products that can support optimum health.

For genetically modified foods, it is, however, notable that their development and use have been sturdily mitigated on the African continent. A large proportion of African countries still brand genetically modified (GM) foods as unsafe and several authorities still refuse to introduce these foods in their food system. In the early 2001, an episode of drought in Southern Africa led to a major food crisis, and Zambia was most affected with 28% of its people on the verge of starvation. Despite this alarming situation, Zambia refused the maize donated by the World Food Program since 70% of the grains were genetically modified [6]. Similarly, Namibia cut off all corn trade with South Africa in 2004 since the latter cultivated GM maize [7].

The reason African governments are so adamant against GM products has been analyzed by Cooke and Downie [6] in the Report of the CSIS Global Food Security Project. They stated that:

> Given the relative paucity of science and research capacities in Africa, it should not be surprising that the arrival of genetic engineering and its application to farming techniques was in many places viewed with suspicion. The absence of a scientific community—outside of South Africa—meant there was no constituency to lead and inform the debate on genetic modification (GM) technology. As a result, those debates quickly became drowned out by non-African voices. On the one hand, some U.S.-based biotech companies were guilty of hyperbolic statements, presenting GM technology as the key to unlocking the problem of food security in Africa. On the other hand, a well-organized group of predominantly Europe-based nongovernmental organizations, or NGOs, backed by elements of the press, made sometimes exaggerated claims about the health and environmental risks of GMOs and their implications for trade and dependence on Western corporate interests.

The reviews by Eicher et al. [8] and Bodulovic [9] present very interesting points of view that still remain valid in the current context. Both papers deduce that:

> The origins of African policy makers' concern over biotechnology are partially a spillover from concerns in Europe about food safety, the environment and generalized public mistrust of multinational seed companies as being manipulative and unscrupulous. It was further observed that the transplanting of European consumer coincided with a regional drought in Southern Africa in the years 2002−2004 necessitating a large amount of food aid. The main supplier of food aid was the United States, which did not have "identity-preserved supply chains" for most of the GM and non-GM maize. Hence African governments became concerned about the potential health, environmental and trade effects of importing food aid. The lack of biosafety regulations and the capacity to evaluate GM and non-GM maize was heightened by the slowness of international organizations to come out and say that GM maize food aid was safe. In the absence of authoritative information, the debate over technical issues turned into sovereignty issues and became a fertile ground for anti-GM activists to fuel the fears of policy makers and general public.

At present South Africa, Burkina Faso, Egypt, Kenya, Mozambique, Nigeria, Uganda, and Zimbabwe have recorded field trials of GM crops in Africa. With the increasing food insecurity, African countries would have to embrace GM food at some point in time. One strong reason supporting Africans opposition to GM crops in their countries is the belief that this would lead to a decline in export sales to Europe, where GM food is not fully accepted. Europe represents a significant source of income for Africa. An estimated €7 billion of farm product is imported by Europe each year [7]. Cultivation of GM goods would not economically hit these countries as badly as predicted. This is because the main foods exported to European countries do not exist anywhere yet in a genetically processed format [7]. The goods that exist in GM format are mostly maize and this is not normally exported to Europe.

The Academy of Science of South Africa in a document on Regulation of Agricultural GM Technology in Africa [10] emphasized that the challenges to the adoption of agricultural biotechnology and GM crops are associated with inadequate investment in scientific research and promotion of innovation, policy makers' indecisiveness, and passive participation of African scientists and the science association. It would then seem clear that exchanging technical information about GM products, training African scientists, creating a public awareness of biotechnology issues, and helping African nations develop their own policies to guide regulatory, legal, and technology transfer issues would represent inward measures for sustainable integration. There is obviously an urgent need to educate Africans on GM food and convince them of its potential benefits.

23.2 African herbal medicine

Africa is blessed with a rich and diverse flora of more than 68,000 plant species including 35,000 endemics. Traditional medicine practice is deep-rooted in the African culture. Information on medicinal plants' use has been obtained from herbalists, herb sellers, and indigenous people in Africa over the years. Under the aegis of the Agence de Cooperation Culturelle et Technique in Paris, ethnobotanical surveys have been conducted in the Central African Republic, Rwanda, Mali, Niger, Federal Islamic Republic of Comoros, Republic of Mauritius, Seychelles, Gabon, Madagascar, Dominica, Tunisia, Togo, Congo, and the Benin Republic. The Organization of African Unity's Scientific Technical and Research Commission (OUA/STRC) has carried out similar surveys in western Nigeria, Uganda, Cameroon, Ghana, and Swaziland. All these surveys have been published and are available in a database developed by OUA/STRC in Lagos, Nigeria. Databases for the flora of the islands of Comoros, Madagascar, Mauritius, Rodrigues, Reunion, and Seychelles have been established. There is also the electronic database of the List of East African Plants and the published phytochemical literature on African plants available in the NAPRECA database.

The southern part of Africa boasts about 3000 botanical drugs with more than 27 million people having access to them while Northern African harbors 10,000 plants, 70% of which are regularly used as medicine [11,12]. The accessibility and affordability of the natural remedies justifies the irregular use. Traditional medicine supporters claim that this practice can cure conditions ranging from cancer to HIV/AIDS. A wide number of plant species from the Alliaceae, Anacardiaceae, Annonaceae, Asteraceae, Balanitaceae, Caricaceae, Caesalpiniaceae, Celastraceae, Compositae Clusiaceae/Guttiferae, Ebenaceae, Euphorbiaceae, Fabaceae, Myrtaceae, Rutaceae, and Zygophyllaceae/Balanitaceae contribute to the management of diseases in the African countries (Table 23.1), and this list is not exhaustive. Some of these treatments have shown their efficacy and they have even been adopted in Western countries. Today, there is an increasing demand for natural therapies in the worldwide market. This is mainly due to a number of side effects observed for a number of pharmaceutical drugs. For instance, certain synthetic drugs are associated with undesirable side effects such as uncoordinated motor skills and drowsiness [13].

Phytomedicine tends to be gentler on the body and in many instances its action lies beyond the treatment of the disease symptom. *Vitex doniana*, for example, not only has an anti-diarrheal activity, but it also helps handle deficiencies in vitamins A and B [14]. The same applies to *Hypoxis hemerocallidea* corm. This plant is known for its hypoglycemic effect and it also reduces systemic arterial blood pressure and heart rate in hypertensives [15]. Herbal medicine is also efficient at treating infectious conditions and they cause fewer side effects than chemical drugs. Nigerian herbalists use *Ageratum conyzoides* as a remedy for HIV/AIDS [16]. Traditional medicine also has anti-fungal activity. A recent investigation conducted among 530 HIV positive patients showed that *A. conyzoides* (goat weed/king grass) leaves, *Allium sativum* (garlic) bulb, *Vernonia amygdalina* (bitter leaves), *Khaya senegalensis* (drywood mahogany) seeds, *Moringa oleifera* (drum stick/horseradish), and *Persea americana* (avocado) seeds resulted in a significant growth inhibition of *Candida* and *Geotrichum* species [17].

23.2.1 African medicinal plants on the market

A large number of African medicinal plants are exploited by pharmaceutical industries for drug production (Table 23.2). The bulk of plants are usually directed to Europe, the United States, and Asia where they are processed [11]. Approximately 26,500 tons of botanical drugs materials were exported to Europe in 1996. Since over 50% of all modern clinical drugs are of natural product origin, considerable attention has been focused on identifying the naturally occurring components in medicinal plants [18]. Characterization, isolation, and the study of action mechanisms of active compounds are of significant interest to the pharmaceutical industry and are essential for ensuring a comprehensive understanding of their overall synergistic action in a complex extract. So far, three classes of plant secondary metabolites have gained significant recognition as therapeutic compounds: alkaloids, terpenoids, and polyphenolics. Quinine is an alkaloid present in the bark of *Cinchona succirubra* and is a powerful anti-malarial agent. A series of synthetic quinines, derived from the natural analog, and much better tolerated than natural quinine [19], have now established applications. Vincristine and vinblastine from *Catharanthus roseus* are among the most potent anti-leukemic drugs in use [20]. Terpenoids have enabled the development of several important pharmaceutical agents: paclitaxel (taxol; anticancer), artemisinin (anti-malarial),

Table 23.1 Medicinal Plants Used in Africa

Plant	Family	Part(s) Used	Therapeutic Use	Country	References
Agathosma betulina	Rutaceae	Leaves	Diuretic, urinary pain reliever, hypertension, and heart diseases	South Africa	[41,42]
Ageratum conyzoides	Asteraceae	Whole plant	HIV/AIDS	Nigeria	[16]
Allium cepa	Alliaceae	Stem, bulb	Induce labor	Uganda	[43]
Aloe barbadensis Linn.	Liliaceae	Leaves	Asthma/bronchitis	Cameroon	[44]
Anacardium occidentale	Anacardiaceae	Leaves	Treatment of diabetes	Nigeria	[16]
Balanites aegyptiaca	Zygophyllaceae/ Balanitaceae	Root bark	Anti-fungal activity	Tanzania	[45]
Carica papaya L.	Caricaceae	Dried seeds	Anti-parasitic activity	Nigeria	[46]
Cassine orientalis	Celastraceae	Leaves	Hypertension and cardiac dysfunctions	Mauritius	[47]
Cassia fistula	Caesalpiniaceae	Pods	Laxative	Mauritius	[47]
Daniellia oliveri	Fabaceae	Leaves	Diarrhea	Northern Nigeria	[48]
Diospyros usambarensis	Ebenaceae	Roots	Anti-fungal	Tanzania	[49]
Garcinia kola	Clusiaceae/ Guttiferae	Bark	Stomach pain, skin infection	Ivory Coast, Liberia, Congo	[50]
Psidium guajava	Myrtaceae	Leaves	Treatment of malaria and fever	South Africa	[51]
Sapium ellipticum	Euphorbiaceae	Leaves	Treatment of cancer	Nigeria	[52]
Sclerocarya birrea	Anacardiaceae	Bark	Treatment of tuberculosis, inflammation, and bacterial-related diseases	South Africa	[53,54]

Species	Family	Part	Use	Country	Ref
Solanum nigrum	Solanaceae	Fruits	Treatment of gastroenteritis and colic pain	Sudan	[56]
Spilanthes filicaulis	Compositae	Flowers	Pneumonia	Cameroon	[44]
Trigonella foenum-graceum L.	Fabaceae	Seeds	Anti-diarrheal, anti-spasmodic, anti-amoeba, dysentery, and anti-diabetics; increase secretion of lactating mothers and facilitate expulsion of placenta	Sudan	[55]
Vernonia amygdalina	Asteraceae	Leaves	Emesis, nausea, diabetes, loss of appetite, dysentery, gastrointestinal tract problems, sexually transmitted diseases, diabetes mellitus, cancer	Nigeria, Cameroon, Zimbabwe	[56]
Xylopia aethiopica	Annonaceae	Fruits	Cough	Cameroon	[44]

Table 23.2 African Plants in the Market for Their Therapeutic Actions

Plant Species	Part Used	Action	Active Component	Country	References
Aloe ferox	Leaves	Laxative	Aloin	South Africa	[57]
Ancistrocladus abbreviatus	Plant	Anti-HIV	Michellamine B	Cameroon, Ghana	[13]
Agave sisalana	Leaves	Corticosteroids and oral contraceptives	Hecogenin	Tanzania	[13]
Cassia spp.	Leaves and pods	Laxative	Senna	Sudan	[58]
Chrysanthemum cinerariifolium	Flowers	Insecticides	Pyrethrins	Ghana, Kenya, Rwanda, Tanzania, South Africa	[13]
Harpagophytum zeyheri; H. procumbens	Roots	Anti-inflammatory, Treatment of arthritis	Glucoiridoids	Namibia, Mozambique, Botswana	[38,59]
Griffonia simplicifolia	Seeds	Treatment of depression, obesity, chronic headache	5-hydroxytryptophan	Ivory Coast, Cameroon, Ghana	[57]
Gloriosa superba	Seeds	Anti-leukemic	Colchicine	Nigeria, Zimbabwe	[57]
Jateoriza palmata	Roots	Laxative	Palmatrin, jateorhizine, columbamine	Tanzania	[57]
Pentadesma butyracea	Fruit	Cosmetic	Fat	Ivory Coast	[60]
Prunus africana	Bark	Prostate gland hypertrophy	Sterols, triterpenes, n-docosanol	Madagascar, Kenya, Cameroon, Zaire, Uganda	[13,38]
Rauvolfia vomitoria	Roots	Tranquilizer and antihypertensive	Reserpine, yohimbine	Nigeria, Zaire, Rwanda, Mozambique	[13,38]
Voacanga africana; V. thouarsii	Seeds	Treatment of heart diseases	Voacamine	Ivory Coast, Cameroon, Ghana Congo	[57]

digitalis sterol glycosides (prescribed for congestive heart diseases), diosgenin (cholesterol lowering), and steroidal saponins (precursors for the synthesis of progesterone-like compounds) used to produce steroid drugs [21,22].

Medicinal plant trade is well documented in some African countries (Table 23.2). It is known, for example, that Cameroon is one of the main exporters of *Prunus africana* bark, with most of the bark going to France and Spain, where it is used by the pharmaceutical industry. In 2005 and 2006 between 1.5 and 2 million kg of *P. africana* bark was harvested annually, valued at approximately $0.5 million to producers, with an export value of $5.5 million. The value of *P. africana* in 1999 was estimated at $0.7 million to Cameroon and $200 million to pharmaceutical companies in consumer countries [23]. Another major plant exported by Cameroon is the seed of *Voacanga africana*, rich in the alkaloid tabersonine, which is used as a central nervous system depressant in geriatric patients. Rooibos tea (*Aspalathus linearis*), "marula" (*Sclerocarya birrea*), Cape aloe (*Aloe ferox*), buchu (*Agathosma betulina*), and Africa potato (*Hypoxis hemerocallidea*) are some of the indigenous species developed as export products by the Republic of South Africa [24]. It is estimated that up to 700,000 tons of plant materials amounting to $150 million are consumed annually in South Africa [25]. Most of the world's supply of devil's claw (*Harpagophytum* genus) comes from Namibia, with lesser amounts from South Africa and Botswana [26]. The roots contain the iridoid glycoside, harpagoside, which is used to treat degenerative rheumatoid arthritis, osteoarthritis, tendonitis, kidney inflammation, and heart disease. In 2002, 1018 tons of dried tubers was exported from southern Africa, causing the harvest of millions of plants. Sales in Germany were estimated to peak at 30 M euros in 2001, corresponding to 74% of the prescriptions for rheumatism. In Madagascar, the export sale of *Catharanthus roseus* and other plants represents major earnings for the country. The roots of *Swartzia madagascariensis* and *Entada africana* are traded from Burkina Faso and Mali to Abidjan in Ivory Coast [27]. Ghana and Ivory Coast remain the major suppliers of *Griffonia simplicifolia* seeds, used for their anti-depressant and healing properties, to Europe.

23.2.2 Difficulties facing herbal medicine

Although herbal medicine presents accrued advantages both in the health care and the economic contexts, this practice is facing several hurdles. The absence of collaboration among traditional medicine practitioners, entrepreneurs, and scientists has been decried and may eventually lead to the loss of valuable ethnomedical knowledge [20,28]. The International Centre for Science and High Technology (Trieste, Italy) has thus initiated the compilation of a compendium on medicinal and aromatic plants for every continent [11]. The very first volume was on Africa and provided information on herbal plants, research and development activities, and the status of their use in traditional practices as well as marketing conditions. This compendium aims to "help policy-makers, the scientific community and entrepreneurs to frame effective policies, formulate projects to augment the R&D activities and create an environment conducive to the growth of plant-based medicine" [11].

It is important to point out that the general belief that plant-derived medications do not generate any side effects is actually not true. There exists a body of evidence that supports this argument [29–32]. Over 4000 phytomedicines have been brought to the attention of the pharmacovigilance unit in Germany during the past 30 years [33]. Many traditional, complementary, and alternative medicines are reported to contain toxic and potentially lethal constituents including aristolochic acids, pyrrolizidine alkaloids, benzophenanthrene alkaloids, lectins, viscotoxins, saponins, diterpenes, cyanogenetic glycosides, and furanocoumarins [34]. The WHO database has over 16,000 suspected case reports with the most commonly reported adverse reactions of hypertension, hepatitis, face edema, angioedema, convulsions, thrombocytopenia, dermatitis, and death [35]. The adverse events related to herbal drugs can be divided into two classes: intrinsic and extrinsic. The intrinsic side effects refer to the negative events caused by the plants. For example, compounds such as pyrrolizidinic alkaloids and cucurbitacins, can have harmful health effects. These adverse effects can be caused by the interaction with other drugs, overdosage, and toxicity [32,36]. The extrinsic side effects are those due to manufacturing problems. These difficulties can range from contamination of the samples to misidentification of the plant species [36]. It is therefore of vital importance that basic research be aligned with translational research to produce safe leads that feed into the drug development pipeline.

Restricted technological advances in several African developing countries hinder extraction, processing, and commercialization of plant bioactive constituents. The review of Nwaka [37] alluded to the set of Millennium Development Goals of the United Nations aimed at alleviating extreme poverty in developing countries; one of the highlights is to halt and begin to reverse the incidence of HIV/AIDS, malaria, and other major diseases by 2015. The expected collaboration with pharmaceutical companies is liable to provide access to affordable essential drugs and can ultimately lead to an increased proportion of populations in developing countries having access to affordable essential drugs on a sustainable basis. Several factors affect access to pharmaceuticals. The drug supply chain has been eminently reviewed, and the following paragraph is seminal [37]:

> *One of the factors of course is availability—the drug must be developed and manufactured. This requires that basic research is properly managed to interface with drug discovery and development. The second major determinant is the accessibility of the new drug in disease-endemic countries.*

This entails effective distribution systems, affordability, sustainable fund support, trained personnel, and informed consumers, among others.

Harvest of botanicals has improved income levels in marginalized communities, but it has also raised questions of sustainability. The low prices paid for plants does not cover replacement and as such, major importers demanding high volumes of plant material are contributing to the decline of medicinal plants species in Africa [38].

23.3 Regulatory status of botanical drugs and functional foods in Africa

Traditional medicinal products, nutraceuticals, and functional foods in Africa have evolved against various ethnological, geographical, environmental, and cultural backgrounds in Africa. They represent utilization of whole, fragmented, or cut plants; algae; lichens; plant foods; and botanical preparations involving various processes, e.g., extraction, distillation, fractionation, purification, concentration, and fermentation. Thus it is of the utmost importance to define and implement rigorous, standardized manufacturing stages/procedures, quality assurance, and quality control techniques. The evaluation of these products and ensuring their safety and efficacy through registration and regulation are challenging issues. The legal situation pertaining to functional foods, nutraceuticals, and phytomedicines differs from country to country. In many African countries, herbal medicines still have not been officially recognized and their regulation and registration have not been properly established or even formulated. In countries with apparent recognition, appropriate budgeting to facilitate the functioning of traditional medicine authorities is inadequate or totally lacking. The traditional medicine community operates outside the framework of national legislation for collection and trade in wild species. Many African countries do not have procedures to register medicinal plant preparations, although these are widely used by the majority of the population. A review discussing the regulatory status of traditional medicines in Africa region [39] indicates that proper classification of herbal or traditional medicinal products within the framework of an efficient regulatory system would comprise "description in a pharmacopoeia monograph, prescription status, claim of a therapeutic effect, scheduled or regulated ingredients or substances, or periods of use." In response to the lack of a proper framework, WHO has developed a number of generic guidelines including generic regulations and law to facilitate the registration, marketing, and distribution of traditional medicines of assured quality in the WHO African region [39]. Table 23.3 provides a list of African countries that have established traditional medicine regulatory frameworks.

The emphasis for the management of diet-related diseases has focused primarily on broad dietary changes for improved health and disease prevention. In this vein, the demand for functional foods and nutraceuticals in industrialized countries has expanded rapidly [40], and a concurrent scenario is occurring in Africa regarding health benefit claims associated with food packaging. In Europe, the United States, Canada, and Asia functional foods and nutraceuticals are regulated as conventional foods, dietary supplements, foods for special dietary use, medical

Table 23.3 Herbal Medicine Regulation in Some African Countries

Country	National Policy on Traditional Medicine/ Complementary and Alternative Medicine	Laws and Regulations on Traditional Medicine/ Complementary and Alternative Medicine	Registration of Herbal Medicine	Pharmacopoeia
South Africa	Issued in 1996 as part of the National Drug Policy	Being developed	None	In development (use of WHO monographs)
Republic of Benin	Issued in 2002	Adopted in 2001	None	In development
Ivory Coast	Issued in 1996	Established in 1999	None	None
Ethiopia	Issued in the Health, Drug, Science, and Technology Policy of 1999	Established in 1999	None	None
Ghana	Issued in 2002	Established in 1992	Started in 1992 through the Food and Drugs Law	Ghana Herbal Pharmacopoeia published in 1992; not legally binding
Guinea	Issued in 1994	Established in 1997	Started in 1994 using legislation similar to that which regulates conventional pharmaceuticals	In development (use of WHO monographs)
Mozambique	Recently approved	Being developed	None	African Pharmacopoeia (1985) is used in place of a national pharmacopoeia and is legally binding

Niger	Being developed	Established in 1997	Introduced in 1997 using legislation similar to the ones regulating conventional pharmaceuticals	In development
Nigeria	Being developed	Established in 1993 and revised in 1999	Introduced in 1993 in Decree No. 15, Reviewed in 1995	In development
Togo	Issued in 1996	Established in 2001	Introduced in 2001 using legislation similar to the ones regulating conventional pharmaceuticals	None
Uganda	Being developed	Being developed	National Drug Authority Statute and Policy of 1993	A contribution of the traditional medicine pharmacopoeia of Uganda (1993)
Tanzania	Issued in 2000	Being developed	None	None
Zambia	National policy on TM/CAM is part of the National Drug Policy, approved in 1997	Being developed	None	None

Source: World Health Organization (2005). National Policy on Traditional Medicine and Regulation of Herbal Medicines - Report of a WHO Global Survey: http://apps.who.int/medicinedocs/en/d/Js7916e/9.1.html. Last accessed: 28th Feb 2013.

foods, or drugs with these distinctions are provided by food manufacturers depending on their marketing strategies or types of claims ascribed to the product. In Africa, the functional foods and nutraceuticals regulatory framework is still in an embryonic state, warranting imperative actions for the setting up of consistent legislations.

23.4 Conclusion

African plants/food plants confer a wide range of health benefits, and many are exploited in the management of health and disease. While it is widely acknowledged that the African population relies heavily on herbal drugs, there is also growing evidence supporting the interest of functional food health benefits. The growth of the pharmaceutical and functional food industry and the expansion of new ranges of natural medicinal/food products will certainly enhance the importance of food and medicinal-based plants in many societies. The concepts of plant-made pharmaceuticals and the latest advances in plant genetic engineering can hold much promise for the production of medicines/food that are inexpensive and yet abundant. A harmonized regulatory framework for such products will definitely provide Africa with an opportunity to develop its botanical drug and functional food industry into one major economic pillar by allowing greater access to regulated markets across the world. Ultimately, how the benefits and risks are addressed by the authorities in African countries remain key elements.

References

[1] Krech S, McNeill JR, Merchant C. Encyclopedia of world environmental history: F-N, vol. *II*. Routledge; 2004, ISBN: 9780415937344.

[2] World Health Organization. Traditional Medicine, <http://www.who.int/mediacentre/factsheets/fs134/en/>; 2008 [accessed 28.02.13.]

[3] Srivastava S, Mishra N. Genetic markers - a cutting-edge technology in herbal drug research. J Chem Pharm Res 2009;1:1−18.

[4] BCC Research Botanical and Plant-Derived Drugs: Global Markets: http://www.bccresearch.com/report/botanical-plan-derived-drugs-bio022e.html>; 2009 [accessed 28.02.13.]

[5] Hasler CM. The changing face of functional foods. J Am Coll Nutr 2000;19:499S−506S.

[6] Cooke JG, Downie R. African perspectives on genetically modified crops: Assessing the debate in Zambia, Kenya, and South Africa. A Report of the CSIS Global Food Security Project. Center for Strategic and International Studies (CSIS), Washington DC, USA: <http://csis.org/files/publication/100701_Cooke_AfricaGMOs_WEB.pdf>; 2010 [accessed 28.02.13].

[7] Paalberg R. Starved for science: how biotechnology is being kept out of africa. USA: Harvard University Press; 2009, ISBN: 978 674 033474.

[8] Eicher CK, Mariddia K, Sithole-Niang I. Crop biotechnology and the African farmer. Food Pol 2006;31:504–27.

[9] Bodulovic G. Is the European attitude to GM products suffocating african development? Functional Plant Biol 2005;32:1069–75.

[10] Academy of Science of South Africa. Regulation of Agricultural GM technology in Africa Mobilising science and Science academics for policymaking, South Africa: <http://www.assaf.co.za/wp-content/uploads/2012/11/K-9610-ASSAF-GMO-Report-Dev-V8-LR.pdf>; 2012. [accessed 1.03.13].

[11] Vasisht K, Kumar V. Compendium of medicinal and aromatic plants Volume I Africa. Earth, Environmental and Marine Sciences and Technologies ICS-UNIDO, Trieste, Italy: <http://institute.unido.org/documents/M8_LearningResources/ICS/80.%20Compendium%20of%20Medicinal%20and%20Aromatic%20Plants-%20Africa%20%28vol.%20I%29.pdf>; 2004. [accessed 28.02.13].

[12] Watson RRR, Preddy V. Botanical Medicine in Clinical Practice. UK: Cromwell Press; 2008, ISBN: 9781845934132.

[13] Okigbo RN, Mmeka EC. An appraisal of phytomedicine in Africa. KMITL Sci. Tech. J 2006;6:83–94.

[14] Ukwuani AN, Salihu S, Anyanwu FC, Yanah YM, Samuel R. Antidiarrhoeal activity of aqeous leaves extract of *Vitex doniana*. IJTPR 2012;4:40–4.

[15] Ojewole JA, Kamadyaapa DR, Musabayane CT. Some in vitro and in vivo cardiovascular effects of Hypoxis hemerocallidea Fisch & CA Mey (Hypoxidaceae) corm (African potato) aqueous extract in experimental animal models. Cardiovasc J S Afr 2007;17:166–71.

[16] Igoli JO, Ogaji OG, Tor-Anyiin TA, Igoli NP. Traditional medicine practice amongst the igede people of Nigeria. Part II. Afr J Trad CAM 2005;2:134–52.

[17] Yongabi KA, Mbacham WF, Nubia KK, Singh RM. Yeast strains isolated from HIV-seropositive patients in Cameroon and their sensitivity to extracts of eight medicinal plants. Afr J Microbiol Res 2009;3:133–6.

[18] Ara I, Bukhari NA, Solaiman D, Bakir MA. Antimicrobial effect of local medicinal plant extracts in the Kingdom of Saudi Arabia and search for their metabolites by gas chromatography-mass spectrometric (GC-MS) analysis. J Med Plants Res 2012;6:5688–94.

[19] Marcus B. Malaria. USA: Inforbase Publishing; 2009, ISBN: 9781438101569.

[20] Elujoba AA, Odeleye OM, Ogunyemi CM. Traditional medical development for medical and dental primary health care delivery system in Africa. Afr J Trad CAM 2005;2:46–61.

[21] Roberts SC. Production and engineering of terpenoids in plant cell culture. Nat Chem Biol 2007;3:387–95.

[22] Son IS, Kim JH, Sohn HY, Son KH, Kim JS, Kwon CS. Antioxidative and hypolipidemic effects of diosgenin, a steroidal saponin of yam (Dioscorea spp.), on high-cholesterol fed rats. Biosci Biotechnol Biochem 2007;71:3063–71.

[23] UNEP. Review of *Prunus africana* from Cameroon. United Nations Environment Programme World Conservation Monitoring Centre: http://www.unepwcmc.org/medialibrary/2011/11/21/0f8687ce/Prunus%20africana%20from%20Cameroon.pdf>; 2008 [accessed 28.02.13].

[24] Street RA, Prinsloo G. Commercially important medicinal plants of south africa: a review. J Chem: <http://dx.doi.org/10.1155/2013/205048>; 2013. [accessed: 28.02.13].

[25] Wiersum KF, Dold AP, Husselman M, Cocks ML. Cultivation of medicinal plants as a tool for biodiversity conservation and poverty alleviation in the Amatola Region, South Africa. In: Bogers RJ, Craker LE, Lange D. editors. Medicinal and Aromatic Plants, 43–57. Springer, Netherlands: <http://library.wur.nl/frontis>; 2006. [Last accessed: 28.02.13].

[26] Stewart KM, Cole D. The commercial harvest of devils claw (*Harpagophytum* spp.) in Southern Africa: the devils in the details. J Ethnopharmacol 2005;100:225–36.

[27] Sofowora A. Medicinal plant research in Africa: Prospects and problems. In: Makhubu LP, Mshana RN, Amusan OO, Adeniji K, Otieno DA, and Msonthi JD. Proceedings of the symposium on African medicinal and indigenous food plants & the role of traditional medicine in health care, 12–28. C.N.P.M.S., Republic of Benin, 1999.

[28] Makhubu L. Traditional medicine in Swaziland. *Afr J Trad CAM Swaziland*: <http://tcdc2.undp.org/GSSDAcademy/SIE/Docs/Vol7/Traditional_Medicine_Swaziland.pdf>; 2006 [accessed 28.02.13].

[29] De Smet PA. Health risks of herbal remedies. Drug Saf 1995;13:81–93.

[30] Brown RG. Toxicity of Chinese herbal remedies. Lancet 1992;340:673.

[31] Rodriguez-Landa JF, Contreras CM. A review of clinical and experimental observations about antidepressant actions and side effects produced by *Hypericum perforatum* extracts. Phytomedicine 2003;10:688–99.

[32] Bateman J, Chapman RD, Simpson D. Possible toxicity of herbal remedies. Scot Med J 1998;43:7–15.

[33] Keller K. Herbal medicinal products in Germany and Europe: experiences with national and European assessment. Drug Inf J 1996;30:933–48.

[34] Adewunmi CO, Ojewole JAO. Safety of traditional medicines, complementary and alternative medicines in Africa. Afr J Trad CAM 2004;1:1–3.

[35] Sahoo N, Manchikanti P, Dey S. Herbal drugs: Standards and regulation. Fitoterapia 2010;81:462–71.

[36] Calixto JB. Efficacy, safety, quality control, marketing and regulatory guidelines for herbal medicines (Phytotherapeutic agents). Braz J Med Biol Res 2000;33:179–89.

[37] Nwaka S. Drug discovery and beyond: the role of public-private partnerships in improving access to new malaria medicines. Trans Royal Soc Trop Med Hyg 2005;99:S20–9.

[38] Cunningham AB. African Medicinal Plants: setting priorities at the interface between conservation and primary health care. Working paper 1. Paris: UNESCO; 1993.

[39] Sharma S, Patel M, MBhunch M, Chatterjee M, Shrivastava S. Regulatory status of traditional medicines in Africa region. IJRAP 2011;2:103–10.

[40] Verbeke W. Consumer acceptance of functional foods: socio-demographic, cognitive and attitudinal determinants. Food Qual Pref 2005;16:45–57.

[41] Duke JA. The green pharmacy herbal handbook: your everyday reference to the best herbs for healing. USA: Rodale; 2000, ISBN: 9781579541842.

[42] Navarra T. The encyclopedia of vitamins, minerals, and supplements. 2nd ed. USA: Infobase Publishing; 2004, ISBN: 9781438121031.

[43] Steenkamp V. Traditional herbal remedies used by South African women for gynaecological complaints. J Ethnopharmacol 2003;86:97–108.

[44] Focho DA, Nkeng EAP, Fonge BA, Fongod AN, Muh CN, Ndam TW, et al. Diversity of plants used to treat respiratory diseases in Tubah, northwest region, Cameroon. Afr J Pharm Pharmacol 2009;3:573–80.

[45] Runyoro DKB, Matee MIN, Ngassapa OD, Joseph CC, Mbwambo ZH. Screening of Tanzanian medicinal plants for anti-Candida activity. BMC Complement Altern Med 2006;6:11.

[46] Okeniyi JA, Ogunlesi TA, Oyelami OA, Adeyemi LA. Effectiveness of dried Carica papaya seeds against human intestinal parasitosis: a pilot study. J Med Food 2007;10:194−6.

[47] Gurib-Fakim A, Guého J, Bissoondoyal MD. *Plantes médicinales de Maurice*, Tome 2. Editions de l'Océan Indien. Mauritius: Rose Hill; 1996, 532.

[48] Ahmadu AA, Zezi AU, Yaro AH. Anti-diarrheal activity of the leaf extracts of daniellia oliveri hutch and dalz (fabaceae) and ficus sycomorus miq (Moraceae). Afr J Tradit Complement and Altern Med 2007;4:524−8.

[49] Hamza OJ, van den Bout-van den Beukel CJ, Matee MI, et al. Antifungal activity of some Tanzanian plants used traditionally for the treatment of fungal infections. J Ethnopharmacol 2006;108:124−32.

[50] Adesuyi AO, Elumm IK, Adaramola FB, Nwokocha AGM. Nutritional and phyto-chemical screening of *Garcinia kola*. Adv J Food Sci Tech 2012;4:9−14.

[51] Olajide OA, Awe SO, Makinde JM. Pharmacological studies on the leaf of Psidium guajava. Fitoterapia 1999;70:25−31.

[52] Sowemimo A, van de Venter M, Baatjies L, Koekemoer T. Cytotoxic activity of selected nigerian plants. Afr J Tradit Complement and Altern Med 2009;6:526−8.

[53] Green E, Samie A, Obi CL, Bessong PO, Ndip RN. Inhibitory properties of selected South African medicinal plants against Mycobacterium tuberculosis. J Ethnopharmacol 2010;130:151−7.

[54] Fotio AL, Olleros ML, Vesin D, Tauzin S, Bisig R, Dimo T, et al. In vitro inhibition of lipopolysaccharide and mycobacterium bovis bacillus Calmette Guérin-induced inflammatory cytokines and in vivo protection from D-galactosamine/LPS -mediated liver injury by the medicinal plant Sclerocarya birrea. Int J Immunopathol Pharmacol 2010;23:61−72.

[55] Khalid H, Abdalla WE, Abdelgadir H, Opatz T, Efferth T. Gems from traditional north-African medicine: medicinal and aromatic plants from Sudan. Nat Prod and Bioprospecting 2012;2:92−103.

[56] Farombi EO, Owoeye O. Antioxidative and chemopreventive properties of *Vernonia amygdalina* and *Garcinia biflavonoid*. Int J Environ Res Public Health 2011; 8:2533−55.

[57] Schmelzer GH, Gurib-Fakim A. Medicinal Plants Volume 11 of Plant resources of tropical Africa. Netherlands: PROTA Foundation; 2008, ISBN: 9789057822049.

[58] Diederichs N. Commercialising Medicinal Plants: A Southern African Guide. Africa: AFRICAN SUN MeDIA; 2006, ISBN: 9781919980836.

[59] European Medicines Agency Assessment report on Harpagophytum Procumbens DC. And/or Harpagophytum zeyhery DECNE, RADIX, London, UK,: <http://www.ema.europa.eu/docs/en_GB/document_library/Herbal_HMPC_assessment_report/2010/01/WC500059019.pdf>; 2009.

[60] van der Vossen HAM, Mkamilo GS. Vegetable oils. Netherlands: PROTA Foundation; 2007, ISBN: 9789057821912.

Regulation of Functional Foods in Selected Asian Countries in the Pacific Rim

24

Jerzy Zawistowski

University of British Columbia, Faculty of Land and Food Systems, Food, Nutrition and Health, Vancouver, British Columbia, Canada

To my wife Ula for her finest gift of immense help and encouragement

24.1 Introduction

Asia is a pioneer in recognizing foods with health benefits, known today as functional foods. In China and other Asian countries, foods have always been considered to be critical for human health and often used in the prevention and even treatment of diseases. Although most of the foods with disease-prevention attributes were lacking scientific support, some foods such as tea have been the subject of intensive research. Tea research performed during the Tang Dynasty (618–907) provided the initial impetus to establish 10 main functions of tea [1]. This historical event could be considered the very first set of health claims for functional foods described by modern standards. The ancient Chinese claim "tea is beneficial to health and relieves fatigue" [1] is a statement very similar to the current Health Canada's Natural Health Products Directorate health claim that states: "tea is accredited for the maintenance of good health and increasing alertness."

The modern concept of functional foods was also born in Asia. In 1980 the Japanese government sponsored a national research program on "systematic analysis and development of functions of food" (1984–1986) and "analysis of functions for adjusting physical conditions of the human body with food" (1988–1990). This research effort led to the identification of the tertiary function of foods. Unlike the conventional (primary, nutrition and secondary, taste/sensory) functions of food, the tertiary function is directly involved in the modulation of human physiological systems such as the immune and digestive [2], which can

Nutraceutical and Functional Food Regulations in the United States and Around the World.
DOI: http://dx.doi.org/10.1016/B978-0-12-405870-5.00024-4

improve or maintain health. It has become clear that food can be designed not only to satisfy primary functions, but also for adjusting conditions of the human body's homeostasis that will regulate health and wellness. This new concept of physiologically functional foods (functional foods) consequently led to the creation of the Japanese regulatory system for this category of foods. In 1993, the Ministry of Health and Welfare established a policy to regulate "Food for Specified Health Use (FOSHU)" or Tokutei Hoken-yo shokuhin, "Tokuho" in its Japanese abbreviation. Under this system, the use of health claims for some selected functional foods is legally permitted [3].

Today, Japan has a well-defined and established regulatory system and vibrant market with over 800 FOSHU products [4]. Following the Japanese example, many global jurisdictions including Asia have succeeded in developing regulations for the production and use of nutraceuticals and functional foods. The regulation of this particular recent food segment is very critical, since it is the fastest growing food market in the world resulting in billions of dollars in global sales.

24.2 Taiwan

24.2.1 Preamble

Taiwan has one of the highest gross national products per capita in Asia, and with a sufficiently large population and a fondness for anything that is novel, this country represents an excellent market for functional foods and dietary supplements [5]. In addition, a well-established regulatory environment for food products including health foods currently exists in Taiwan.

The Executive Yuan, Department of Health (DOH), which is Taiwan's highest authority on health, is responsible for health administration, including the guidance, supervision, and coordination of local health agencies. In January 1999, the DOH issued the Health Food Control Act (HFCA), which was initiated in February and implemented in August 1999. The HFCA has been amended several times including last amendments made in May 2006 [6]. The Act governs all matters relating to health foods, such as health food permits; manufacturing; importing; management of safety and sanitation; labeling and advertising; and inspection of food facilities, manufacturers, and vendors practices, as well as enforcement and sanctions. Although this Act defines only health foods, dietary supplements, and medicine, functional foods match the definition of health foods very well. The HFCA defines "health food" as food products that possess "special nutritious elements" (bioactive components) or "specific health care abilities" to improve health, and/or reduce the risk of disease. This type of food is not intended to be used for mitigation, curing and/or treating human diseases. Any food products that are labeled or advertised as food matching the definition specified in HFCA is governed by the Health Food

Control Law, regardless of whether it is named health foods or functional foods. Subsequently, any foods with health care abilities, including functional foods, must be registered with the DOH.

To qualify for a health food permit, food product has to comply with the following conditions [6]:

- Food should contain clearly identified bioactive components that exert a health benefit within reasonable consumption. The health benefit effect must be supported by scientific proof. If it is not possible to identify specific ingredient(s) that are contributing to the health effect, the beneficial effects should be listed and supporting literature should be provided to the central health authority for evaluation and verification.
- Food must be safe and the health beneficial effect(s) must be harmless to humans with typical consumption as assessed by toxicological techniques.
- The central health authority must approve all methodology that is used to assess efficacy and safety of foods and associated bioactive constituents.

Health foods are permitted to carry seven DOH-approved maintenance claims [6]:

1. Regulating blood lipids
2. Regulating the gastrointestinal tract
3. Regulating the immune system
4. Preventing osteoporosis
5. Maintaining dental health
6. Regulating blood sugar
7. Protecting the liver from chemical damage

The health claim can be made on foods if it is supported by scientific assessments and approved by the DOH. The authorized health claims describe a type of health maintenance that foods may provide in relationship to disease or health-related conditions (e.g., preventing disease). In contrast to the U.S. and some European regulations, the HFCA does not permit a link between a food bioactive ingredient and a disease. Please note that in Taiwanese regulations there are no specific requirements regarding the maximum level of adverse nutrients such as fat, saturated fat, salt, sugar, or cholesterol to qualify food to carry a health claim. This prompted some Taiwanese institutions such as non-governmental organizations (NGOs) to request that the HFCA be amended. The NGOs suggest that a health food product should be evaluated as a whole and the use of excessive amount of adverse nutrients in the food formulation should be restricted [7]. This type of ruling is implemented in the United States under the Nutrition Labeling and Education Act as a prerequisite for food to carry a health claim on a label [8]. This approach makes sense, since not only specific bioactive(s) but also all consumed food constituents play an important role in maintaining the health status of consumers.

24.2.2 Selling health (functional) foods in Taiwan

In order to sell domestic or imported health food products, the manufacturer or importer has to register a food product and then obtain a health food permit from the DOH. The registration process is required to submit an application for review by the Health Authority. A dossier should contain all relevant information including [6,9]:

- A list of ingredients and their specifications.
- Efficacy assessment of the food product as related to the health maintenance claim.
- Assessment of ingredients involved in the efficacy of food. This information should comprise function and its effects relating to the health maintenance claim, ingredient specifications, and methods of analyses. Any research data and literature relevant to efficacy of ingredients should be included.
- Assessment of safety of the food product.
- A summary of the manufacturing process along with supporting documents indicating that food is manufactured in accordance with good manufacturing practices (GMPs). The central competent authorities must certify the GMP standards. Imported health foods must conform to the GMPs of the country of origin.
- A food product label indicating the nutritional content.
- A sample of the food product.
- Application fees, including a permit fee along with review and testing fees are required.

The approval process consists of five steps as described in Table 24.1 [10]. Once all documents are submitted, the DOH conducts an assessment of the dossier and if the review is successful, the manufacturer or importer is granted a permit valid for 5 years to sell the health product. The permit may be extended for another 5 years by applying for renewal within 3 months prior to the expiration of the issued permit. The DOH has the right to revoke the permit if scientific support of product efficacy is in doubt; when the bioactive ingredients, formulation, or method of manufacturing is changed; and when safety of product is in doubt during its validity period.

Under the current Taiwan's regulations, relatively few products have received approval to market as health foods. During the first 4 years since implementation

---

Table 24.1 Approval Process for Health Functional Foods in Taiwan

Step 1: Submission of a dossier to the DOH for examination and registration

Step 2: DOH conducts an initial assessment of the dossier

Step 3: The Health Food Evaluation Committee appointed by the DOH conducts further examination of the dossier to ensure that the efficacy and safety of the product are met

Step 4: If the application is successfully reviewed, the DOH approves the health food product and a permit is granted for 5 years to market the product

of the HFCA, 28 products obtained health food certificates [11]. As of January 2007, only 88 products have been approved. This falls far behind the number of products approved in Japan (over 800) [4] and China (over 3000) [10]. The hypolipidemic foods are the largest group of products, accounting for 36% of the total approved health food products [12]. An example of health (functional) food product that was introduced to the Taiwanese market in 2007 by Uni President is shown in Figures 24.1 and 24.2.

24.2.3 Labeling

In Taiwan, regulations on food labeling are promulgated under a number of laws. Food products, domestic or imported, must conform to the national standards. These standards are similar to international standards for food labeling. This is because Taiwan, as a member of the World Trade Organization, is making a conscious effort to comply with international laws and regulations.

Functional foods that are approved as health foods must be labeled according to the standards promulgated under regulations for food labeling of the Food Administration Act () [13] and HFCA. The regulations apply to conventional foods sold at Taiwanese markets, while HFCA describes specific instruction for health foods. In general, food products must be labeled in Chinese, conspicuously displaying the product name, contents of ingredients and their origin, net weight or volume of the contents, date of production, the expiration date, telephone number, address, and name of the manufacturer. For foods that are imported, a label

FIGURE 24.1

Example of Taiwanese health (functional) food product; sterols containing milk drink for the risk reduction of coronary heart disease. 1. Logo for "health food" with the reference number of the permit. 2. Name of product "Plant sterols milk". 3. Approved health claim: "An animal study shows that consumption of this product may help lower blood total cholesterol."

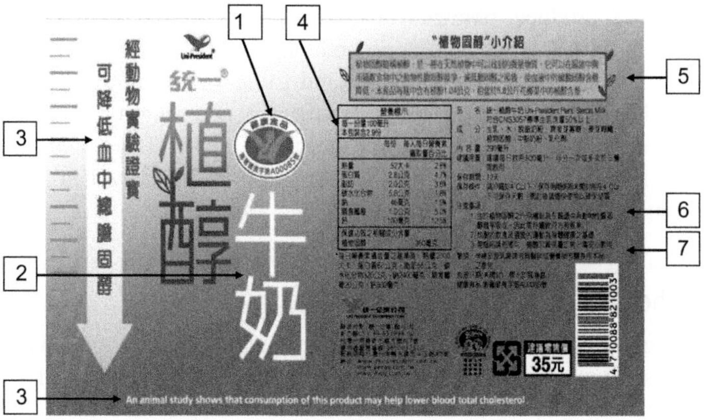

FIGURE 24.2

A label of Taiwanese health (functional) food product; sterols containing milk drink for the risk reduction of coronary heart disease. 1. Logo for "health food" with the reference number of the permit. 2. Name of product "Plant sterols milk." 3. Approved health claim: "An animal study shows that consumption of this product may help lower blood total cholesterol" in Chinese and English. 4. Nutritional information. 5. Information on plant sterols. 6. Instruction of use. 7. Warning "pregnant and breast feeding women should consult their doctor before drinking this product."

should contain the name of country of origin, telephone number, address, and name of the importer. In addition to general labeling standards, health food products must contain the following information on the label:

- The approved health benefit effects (health claim)
- Reference number of the permit and the legend of "health food" and standard logo
- Amount of intake (serving size) and important message for consumption of the health food and other necessary warnings
- Nutritional information
- For health food, which is a blend of two or more ingredients, the ingredients shall be separately labeled

The specifics of food labeling are described in the Regulations for Food Labeling [13].

24.3 Hong Kong
24.3.1 Preamble

In 1997, when Hong Kong became the Special Administrative Region of the People's Republic of China, the basic law guaranteed the independence of the social, economic, political, and judiciary systems for Hong Kong until the year 2047. This enables Hong Kong to work directly with the international community to control trade in strategic commodities, such as food and drugs. Hong Kong has its own food and agricultural import regulations, which are different from those in China mainland [14].

With a population of about seven million and a significant tourism industry, Hong Kong is a substantial market for all kinds of food products [15]. Health foods are commonly sold in the domestic market and include vitamin and mineral supplements, shark liver oil and cartilage, deep-sea fish oil, Chinese medicinal fungi, herbal pills, royal jelly extracts, pollen tablets, and ale extracts [16]. Examples of functional foods are probiotics represented mainly by Yakult drinks and nutritive drinks (tonics), which have been long considered to provide mental and physical benefits, ranging from immune system properties, improving digestive health, and enhancing physical beauty. Bird nest and essence of chicken are two of the most popular products. Dietary supplements in Hong Kong are much more varied with products such as evening primrose oil, glucosamine, and coenzyme Q10 becoming increasingly available in the market [17].

Due to growing consciousness about health food through social dynamics and media, the Hong Kong market is demanding more health foods [18]. Lifestyle underlined by the high-density population and resultant space-constrained living quarters as well as a growing number of women in the workforce, has increased the acceptance of Hong Kongese for a variety of convenient but healthy food habits. "Value for money" is one of most important considerations of Hong Kong consumers. Products with attractive, eye-catching packaging are gaining the acceptance of consumers. The most popular products on the market are beautifying and slimming products for women, and from western health foods, such as fish oil and omega fatty acids. Many dietary supplements and slimming products sold in Hong Kong fall within the general food category. At least two-thirds of the health care products sold in Hong Kong contain traditional Chinese medicine (TCM). TCM is a system of healing that has been practiced by the Chinese for centuries. The Chinese practitioners believe that the body yin and yang imbalance may be corrected by eating the right balance of hot and cold foods. Although TCM is different than functional foods, the underlining principles of the disease prevention rather than treatment are the same. Therefore, the acceptance of TCM by Hong Kongese is the major factor for the popularity of functional foods [19].

In Hong Kong, similar to the other countries, there is no universally accepted definition of health foods or functional foods. Alternative terms such as dietary supplements, nutraceuticals, designed foods, functional foods, and natural health products are used on different occasions to refer to similar products. The growing interest in health/functional foods was reflected by organizing the first International Functional Food Conference in Hong Kong in August 2004. Over 140 companies, universities, and government representatives attended the conference, which offered an excellent forum in which to share ideas about global regulations and standards for functional foods.

24.3.2 Health foods regulations

Hong Kong does not have specific regulations established for health and wellness health care products. Depending on food composition, health foods (functional foods) may be regulated under the following ordinances [20]:

- Pharmacy and Poisons Ordinance (PPO; Cap. 138)
- Chinese Medicine Ordinance (CMO; Cap. 549)
- Public Health and Municipal Service Ordinance (PHMSO; Cap.132)
- Undesirable Medical Advertisements Ordinance (UMAO; Cap. 231)

The law in Hong Kong does not provide a legal definition of health/functional foods products. In the absence of a legal definition, regulation of health food products is based on their ingredient content. Health foods are subject to the same acts and regulations as conventional foods. Retailers are required to provide truthful labeling as regulated by PHMSO (Cap. 132, Sec. 61: False Labeling and an Advertisement of Food or Drugs). Health care food products that cannot be classified as Chinese medicine or western natural health products are regulated as general foods and they are subjected to the regulation of the PHMSO (Cap. 132). The manufacturers and sellers should ensure that their products are fit for human consumption. Health foods should not include medicinal ingredients, or they may be regarded as pharmaceutical products. If they contain medicines or claim to have medicinal effect, they are required to be registered under the Health Department and are regulated by the PPO (Cap. 138). Any product carrying a claim for the diagnosis, treatment, prevention, alleviation, or mitigation of disease or any symptoms of disease is classified as a pharmaceutical product and medicine and is regulated according to the PPO ordinance (Cap. 138, Sec. 2). On the contrary, some Chinese medicines may be regarded as health foods, and they are subject to the CMO regulations (Cap. 549). This Ordinance also controls health products containing Chinese medicines as active ingredients. Although substantiated health claims are allowed, the UMAO (Cap. 231) prohibits advertisements claiming that a product has curative or preventive effects on any of the diseases listed in the schedule (Table 24.2) to the Ordinance [22].

Pharmaceutical products and medicines are defined in PPO (Cap. 138). Chinese medicine is defined in CMO (Cap. 549), and food is defined in PHMCO (Cap. 132).

Foods fortified with vitamins and minerals and some other bioactives may also be considered as functional foods and can be found in the Hong Kong market. The calcium-enriched foods are dominating the calcium supplement market, ranging from dairy products through to biscuits [17].

According to Euromonitor, there were very few probiotic supplements in Hong Kong, and their sales remained negligible in 2003. Nevertheless, Euromonitor expects the future of vitamins and dietary supplements to be dynamic, experiencing growth of 66% in China and 14% in the more mature Hong Kong market between 2003 and 2008. Unfortunately, there is no specific

Table 24.2 Diseases in Respect of which Advertisements are Prohibited or restricted Under the Undesirable Medical Advertisements Ordinance of Hong Kong

1. Any benign or malignant tumor
2. Any viral, bacterial, fungal, or other infectious disease, including tuberculosis, hepatitis, and leprosy
3. Any parasitic disease
4. Any venereal disease, including syphilis, gonorrhea, soft chancre, lymphogranuloma venereum, genital herpes, genital warts, urethritis, vaginitis, urethral or vaginal discharge, AIDS, and any other sexually transmitted disease
5. Any respiratory disease, including asthma, bronchitis, and pneumonia
6. Any disease of the heart or cardiovascular system, including rheumatic heart disease, arteriosclerosis, coronary artery disease, arrhythmias, hypertension, cerebrovascular disease, congenital heart disease, thrombosis, peripheral artery disease, edema, retinal vascular change, and peripheral venous disease
7. Any gastrointestinal disease, including gallstone, cirrhosis, gastrointestinal bleeding, diarrhea, hernia, fistula-in-ano, and hemorrhoids
8. Any disease of the nervous system, including epilepsy, mental disorder, mental retardation, and paralysis
9. Any disease of the genitourinary system, including kidney stone, nephritis, cystitis, any prostatic disease, and phimosis
10. Any disease of the blood or lymphatic system, including anemia, neck glands, bleeding disorders, leukemia and other lymphoproliferative diseases
11. Any disease of the musculoskeletal system, including rheumatism, arthritis, and sciatica
12. Any endocrine disease, including diabetes, thyrotoxicosis, goiter, and any other organic or functional condition related to under or over activity of any part of the system
13. Any organic condition affecting sight, hearing, or balance
14. Any disease of the skin, hair, or scalp

Adapted from [21]. UMAO, Cap. 231.

regulation on the nutrient fortification in food in Hong Kong. However, reference can be taken from the Codex Alimentarius, which issued a general Principles for the Addition of Essential Nutrients to Foods in 1987 and subsequently amended the principles in 1989 and 1991 [22].

24.3.3 Food labeling regulations

Labeling regulations applicable to conventional foods also apply to health foods. The Food and Drugs Regulations require food manufacturers and packers to label their products in a prescribed, uniform, and legible manner. All prepackaged food must be labeled according to regulations described in Composition and Labeling (Cap. 132). The following information in appropriate language (Chinese, English, or both) must be included on the label, except for "exempted items" as provided in the regulations:

- Name of the food, list of ingredients
- Indication of "best before" or "use by" date
- Statement of special conditions for storage or instruction for use
- Name and address of manufacturer or packer
- Count, weight, or volume

The Food Labeling Regulations were amended in 2004. The amendment stated that the food label should declare in the list of ingredients the presence of ingredients, which are known to cause allergy in some individuals. The full list of such substances is included in the amendment. Examples of such substances are cereals containing gluten, eggs and egg products, peanuts and their products, and some food additives as well as others.

Hong Kong does not have any specific regulations regarding the labeling of novel foods products that may encompass such categories as genetically modified foods, functional foods, or other biotech foods. There is no distinction between general food labeling and biotech food labeling requirements. The Working group established under the Center for Food Safety prepared a draft guideline related to the labeling of biotech foods and suggested that

- The labeling of biotech food will comply with the existing food legislation.
- The threshold level applied in the guideline for labeling purpose is 5%, in respect to individual food ingredient.
- Additional declaration on the food label is recommended when significant modifications of the food, e.g., composition, nutrition value, level of antinutritional factors, natural toxicant, presence of allergen, intended use, introduction of an animal gene etc., have taken place.
- Negative labeling is not recommended.

This voluntary guidance applies to prepackaged food and is voluntary and has no legal effect. However, if products are found to have misleading labeling such as negative labeling (e.g., genetically modified organism (GMO) free, free-form GM ingredients), a retailer may be subject to prosecution under Cap. 132, Section 61; False Labeling and Advertisement of Food or Drugs of Public Health and Municipal Services [14]. In 2013, Hong Kong regulators planned to consult the public about launching a mandatory pre-market safety assessment scheme with the specific focus on GMO foods. These regulations may be enacted within 2 years after the consultation period [23].

24.3.4 Nutrition labeling, nutrition claims, and health claims

Until 2008, there were no specific regulations on nutrition labeling for prepackaged food products in Hong Kong. Hong Kong allowed food products to have nutrient claims in any format [14,24]. On March 31, 2008, the government announced an amendment to the Food and Drugs (Composition and Labeling) regulations, PHMSO (Cap. 132), by introducing requirements for nutrition labeling and nutrition claims [25]. These regulations were published in the Gazette on April 3, 2008, with the effective date of July 1, 2010, and covered the nutrition labeling and nutrition claims for pre-packaged food including functional foods.

The nutrition labeling scheme ("one plus seven") requires the listing of energy plus seven core nutrients, including protein, carbohydrates, total fat, saturated fat, trans fat, and sodium, as well as any additional nutrients for which a claim is made. It also requires the nutrient amount to be expressed in absolute amounts either per 100 g (or 100 ml) or per serving size, and energy either in kilocalories or kilojoules [25]. The promulgated regulations are based on the recommendation by the Codex Committee on Food Labeling of 2007 [26].

Under the amended regulation, the following nutrition claims are allowed on pre-packaged foods:

- Nutrient content claims
  - Examples of acceptable claims include: "Low fat," "sugar-free," "high calcium"
- Nutrient function claims
  - Examples of acceptable claims include: "calcium aids in the development of strong bones and teeth, may help to improve bone density," "folic acid contributes to the normal growth of the fetus"
- Nutrient comparative claims
  - Example of acceptable claims include: "reduced fat—25% less than the regular product of the same brand"

A nutrition claim means any representation that states, suggests, or implies that a food has particular nutritional properties relating to energy value and/or food nutrients such as protein, available carbohydrates, total fat, saturated fatty acids, trans fatty acids, sodium, sugars, vitamins, and minerals.

A nutrient content claim describes the energy value or the content level of a nutrient contained in a food, while a nutrient comparative claim means a nutrition claim that compares the energy value or the content level of a nutrient in different versions of the same food or similar foods. The closest claim that links food or food constituents with health is a nutrient function claim that describes the physiological role of a nutrient in growth, development, and normal functions of the body [25]. Nutrition claims are not allowed on foods for children younger than 36 months, and on foods for special dietary uses. In addition, health claims such as reduction of disease risk claims are prohibited [25]. The law also prohibits disease prevention and cure claims made on foods. The advertising regulations [27] state that claims

- Relating to the nutritional and dietary effects of products or services should be handled with care

- Of effects or treatment for conditions of health for which qualified medical attention or advice should reasonably be sought are not acceptable
- Made in food advertisements must be based on scientific evidence or be substantiated

Unfortunately, many health foods can be found in the Hong Kong market, claiming unsubstantiated beneficial effect on health. According to a large survey conducted by the Consumer Council in Hong Kong, seven categories of products including health and functional foods have been implicated in questionable, fake, and ambiguous claims [28]. Health claims such as "anti-cancer" and "anti-aging" benefits for shark liver oil, or "detoxifying the body" claims for dietary supplements are claims that are incapable of substantiation [20].

The need for regulations for health claims was acknowledged during the February 2, 2001, meeting of the Advisory Council on Food and Environmental Hygiene. According to those discussions the Health and Welfare Bureau would study the feasibility of developing a framework to monitor and regulate health claims to protect consumers from misleading information and exaggerated claims [29].

The Director of Health will be empowered to prohibit products from making irresponsible health claims. Initially, the Administration has proposed confining the restriction to food products, and expanding the restriction to cover other products, in light of past experiences. The Administration has proposed that food products claiming health benefits would fall under two categories [20]:

- Those claiming to be able to prevent or cure a specific disease or clinical condition should be first registered with the Director of Health for pre-market approval, with claims properly substantiated by research or trials
- Those claiming to have general beneficial effects on health will be exempted from registration, with the Director of Health retaining the power to determine whether the claim made is a general or specific one

24.4 South Korea

24.4.1 Preamble

In 2004, South Korea introduced the regulatory venue for the "health functional foods." This term encompasses all natural health products and nutritional supplements.

As a result of these changes, Korea, with its population of 48 million, is currently among the largest Asian markets for functional foods. Total market size for health functional foods was estimated at $1.15 billion in 2010 [30]. Major leading products in the Korean market include ginseng, aloe, squalene, chitosan, glucosamine, chlorella, vitamins, and minerals. There is a growing interest in herbal ingredients and plant extracts among local consumers. In addition to domestic products, dietary supplements and natural health products from the United States and Japan account for the largest source of Korean imports. This segment is so significant that the Korean government imposed the regulations on both domestic and imported products. Started on January 1, 2004, all products labeled or advertised as nutritional supplements in the form of capsules, tablets, pills, powder, granule, or liquid fall under the Health Functional Food Act. According to the law, Korean importers for nutritional supplements are required to complete import notification to the Korea Food and Drug Administration (KFDA), which categorizes products into 1 of 32 groups of active ingredients allowable for nutritional supplement [31].

A transition in nutritional practices and products has occurred in Korea over the last 10−15 years [32]. This transition reflected major changes in consumer eating and nutritional patterns. Consumption of food of animal origin increased relative to the consumption of foods of plant origin. Subsequently, intake calories from animal fat were higher over the last decade as compared with traditional Korean food pattern in the past. These current changes led to an increase in people being overweight and obesity and subsequently in the risk of chronic diseases such as diabetes and coronary heart disease [33].

Parallel with these adverse changes in nutrition, the KFDA brought forth positive amendments of food regulations. In 2004, the Korea Health Functional Food Act (HFFA) was introduced to encourage the industry to produce and market health foods that are easy and convenient to use by the fast-paced, average Korean consumer.

24.4.2 KFDA

The first step in the creation of the KFDA was establishing the Korean Food & Drug Safety Headquarters and six regional offices (KFDS) in April 1996. Subsequently, the KFDS was raised to the administration status of the KFDA. The KFDA is the principle government agency with the role of promoting public health by ensuring the safety and efficacy of foods, pharmaceuticals, medical devices, and cosmetics, as well as supporting the development of the food and pharmaceutical industries. The KFDA mission statement in relation to health functional foods is stated as "Overlooking manufacture, importation, and distribution of food and health functional foods and their safety."

The KFDA is responsible for setting and implementing standards and specifications for food in general, functional foods, food additives, food packaging, and equipment [34]. The KFDA headquarters in Seoul consists of four divisions and two of the six departments are dedicated to food-related issues only. KFDA headquarters also oversee six regional KFDA offices. KFDA publishes its food-related regulations, including Food Code, Labeling Standards for Food, Health Functional Food Code (HFFC), and functional foods regulations.

The Food Code, which could be compared to a compendia pharmacopeia, promulgates standards and specifications for manufacturing, processing, usage, cooking, storage of food and equipment, containers, and packaging for food products. The code describes individual standards and specifications for 151 food categories delineated into 20 groups. The food code was updated on June 22, 2005 [34].

The HFFC was enacted in August 2002, and enforced on January 31, 2004. It was amended a number of times including the major revision in 2008. The HFFC contains general standards and specifications governing functional food, individual standards, and specifications for functional food categories [33]. Health functional foods must be in the form of tablet, pill, capsule, granule, powder, or liquid. However, the scope of the revised act in 2008 was extended to include some conventional foods. A food product that meets the criteria for one of the approved categories and listed in the HFFC is permitted to carry a health efficacy claim. Anyone wishing to export a functional food that is not in one of the approved categories specified in the Code can apply to KFDA for the following:

- Recognition of raw materials that have specific health effects (efficacy)
- Recognition of the new category.

Details about recognition procedures, required documents, or any related issues are provided on the KFDA website (www.kfda.go.kr) in the Korean language [34]. There are two divisions within the KFDA related to food: Food Headquarters and Nutrition and Functional Foods Headquarters. Food Headquarters consists of the four following teams:

- Food safety policy
- Food management
- Food import
- Food safety assurance

The focus of those teams is issues related to general policy for food safety management such as the following:

- Supervising the overall food hygiene matters including surveillance for illegal and adulterated food
- Supervising food importation and exportation, and conducting imported food inspections
- Handling all issues relating to genetically modified foods
- Codex Alimentarius activities

The Nutrition and Functional Food Headquarters consists of the following teams:

- Health functional foods team
- Novel food team
- Health functional food standard team
- Nutrition evaluation team
- Food additives team

There are six regional governmental KFDA agencies located in Seoul, Busan (three branches), Gyeongin (three branches), Daegu, Gwangju, and Daejeon [35]. Each agency serves as a field operation base conducting food and drug surveillance and imported food inspections at its local laboratory. In addition there are nine national quarantine stations located in Gunsan, Mokpo, Yeosu, Masan, Tongyeong, Ulsan, Pohang, Donghae, and Jeju.

There are four categories of products that require import notifications:

- Food (meat, dairy, egg, and fish products)
- Food additives
- Apparatus, container packages
- Health functional foods

24.4.3 Food laws

Korea is one of the few Asian countries that recently introduced the functional food regulations. Until 1998, the Korean Ministry of Health and Welfare (MHW) retained all food regulations authorities. With the establishment of the KFDA in 1998, the MHW relinquished most of its authority to the KFDA. It did, however, retain authority to legislate changes to the Food Sanitation Act (FSA) and the HFFA and their implementing the Presidential Decree and Ministerial Ordinance. The FSA defines functional foods as health foods and nutritional supplements [34]. It is noteworthy that most functional foods are sold in tablets, pills, potions, and other medicinal formats that resemble the U.S. dietary supplement defined under the Dietary Supplement Health and Education Act (DSHEA) of 2004. It includes herbal products as well as essential nutrients.

24.4.3.1 Food Sanitation Act and Presidential Decree

The FSA is legislated by the National assembly and is the legal basis for the food safety-related work conducted by the MHW and KFDA. The Act aims for the improvement of national health by ameliorating the quality of nutrition and preventing sanitary hazards during the preparation of food. The Presidential Decree establishes provisions to implement the FSA. It provides more defined guidance on interpretation and implementation of the Act. Ministerial ordinance to the FSA prescribes more detailed guidance on the implementation of the FSA and the Presidential Decree. This ordinance provides relevant instructions for conducting business in Korea, including penalties for the lack of compliance. Standards and regulations for food-relating business in Korea are provided in the form of the Food Code, HFFC, Guidelines, Notices, and others. All of these directives are administered by the KFDA.

24.4.3.2 HFFA and Presidential Decree

The HFFA is legislated by the Korean National Assembly and it represents a legal basis for MHW and its executive branch, KFDA, to oversee the functional foods.

The Act mandates the improvement of national health and consumer protection by ensuring the safety and quality of functional foods and encouraging sound distribution and sales of such products.

The Presidential Decree, issued December 18, 2003, established provisions to implement matters regulated by the HFFA. The Ministerial Ordinance, issued in January 31, 2004, prescribed more detailed guidance on how the HFFA and its Presidential Decree are to be implemented. This ordinance addresses inspection of imported functional food, penalties for violations, and applications for import inspection, among others. The HFFC, Guidelines for Labeling of Functional Food, Guidelines for Advertisement of Functional Food, and relevant Notices provide specific standards and regulations. The administration, amending, and promulgation of standards and regulations are the responsibility of the KFDA [34].

24.4.3.3 HFFA and code

Health functional food is defined as "a processed food with functional raw materials or ingredients that provide useful physiological effect for human body or control the structure/function of the human body" [36]. The functional foods for maintaining human health "shall be processed in forms of tablets, capsule, pill, granule, liquid, powder, paste, syrup, gel or bar, which is easy to intake per serving" [37]. Although the scope of the revised act in 2008 was extended to include some conventional foods, currently only supplements can be labeled with health claims [38].

There is one critical element in this definition that makes it so different from the definition used in the Western regulatory jurisdictions such as Canada, the United States, and the European Union (EU). Korean functional foods are doseable and can be used in the medicinal formats. As previously mentioned, it resembles the form of dietary supplements that are regulated in the United States under DSHEA, rather than functional foods as defined in Canada and the EU.

In 2005, 37 types of health functional foods/nutraceutical ingredients were listed in the Act including vitamins, minerals, essential amino acids, proteins, dietary fiber, essential fatty acids, and other nutrient supplements such as ginseng products, aloe, squalene, and chlorella [31,33]. In the amended HFFC, ingredients are divided into two groups: a group of nutrients including 25 vitamins and minerals, dietary fiber, proteins, and essential fatty acid, and a group of 47 functional ingredients [37].

The group of nutrients are allowed function nutrient claims such as "Vitamin K is necessary for the normal blood coagulation," "calcium is necessary for the normal structure of bones and teeth," or "proteins are necessary for the normal formation of enzymes, hormones and antibodies," It is worthy to know that "harder" health claims for specific types of fibers (e.g., barley fiber), proteins (soybean protein), and essential fatty acids (e.g., omega-3) are approved under the functional group (Table 24.3). Table 24.3 shows all functional ingredients with approved health claims that are listed in the Code [37].

Table 24.3 List of Functional Ingredients with Health Claims

Ingredients	Health Claim
Terpenes	
Ginseng	Support immune function
	Help to relieve from fatigue
Red ginseng	Support immune function
	Help to relieve from fatigue
	Help to maintain healthy blood flow by inhibiting blood coagulation of platelets
Plants containing chlorophyll	Help to maintain healthy skin
	Antioxidant activity
Chlorella	Help to maintain healthy skin
	Antioxidant activity
Spirulina	Help to maintain healthy skin
	Antioxidant activity
	Help to maintain healthy blood cholesterol level
Phenols	
Green tea extracts	Antioxidant activity
Aloe whole leaves	Help to maintain healthy bowel function
Propolis extracts	Antioxidant activity
	Antimicrobial activity in oral cavity
Coenzyme Q10	Help to support antioxidant activity
	Help to maintain health blood pressure
Soybean isoflavone	Help to maintain bone health
Fatty Acid and Lipid	
Edible oil containing omega-3 fatty acid	Help to maintain healthy triglyceride level
	Help to maintain healthy blood flow
Edible oil containing γ-linolenic acid	Help to maintain healthy blood cholesterol level
	Help to maintain healthy blood flow
Lecithin	Help to maintain healthy blood cholesterol level
Squalene	Antioxidant activity
Phytosterol/phytosterol ester	Help to maintain healthy blood cholesterol level
Shark liver oil containing alkoxyglycerol	Support immune function
Edible oil containing octacosanol	Help to improve endurance capacity
Japanese apricot extracts	Help to relieve fatigue
Conjugated linolenic acids	Help to reduce body fat in the overweight adult
Garcinia cambogia extracts	Help to reduce body fat in the overweight adult
Lutein	Help the eye health by maintaining the density of macular pigments
Haematococcus pluvialis extracts	Help to improve eye fatigue
Saw palmetto fruit extracts	Help to maintain prostate health

(Continued)

Table 24.3 (Continued)

Ingredients	Health Claim
Sugar and Carbohydrates	
Glucosamine	Help to maintain healthy joint and cartilage
l-acetylglucosamine	Help to maintain healthy joint and cartilage
	Help to maintain healthy skin moisturizing
Mucopolysaccharide-mucoprotein	Help to maintain healthy joint and cartilage
Dietary Fiber	
Guar gum/guar gum hydrolysate	Help to maintain healthy blood cholesterol level
	Help to maintain healthy postprandial glucose level
	Help to maintain healthy bowel function
	Help to maintain healthy gastrointestinal bacteria population
Glucomannan (konjac mannan)	Help to maintain healthy blood cholesterol level
	Help to maintain healthy bowel function
Oat	Help to maintain healthy blood cholesterol level
	Help to maintain healthy postprandial glucose level
Indigestible maltodextrin	Help to maintain healthy postprandial glucose level
	Help to maintain healthy bowel function
	Help to maintain healthy triglyceride level
Soybean fiber	Help to maintain healthy blood cholesterol level
	Help to maintain healthy postprandial glucose level
	Help to maintain healthy bowel function
Tree ear (*Auricularia auricular*)	Help to maintain healthy bowel function
Wheat fiber	Help to maintain healthy postprandial glucose level
	Help to maintain healthy bowel function
Barley fiber	Help to maintain healthy bowel function
Arabic gum (acacia gum)	Help to maintain healthy bowel function
Corn bran	Help to maintain healthy blood cholesterol level
	Help to maintain healthy postprandial glucose level
Inulin/chicory extracts	Help to maintain healthy blood cholesterol level
	Help to maintain healthy postprandial glucose level
	Help to maintain healthy bowel function
Psyllium husk	Help to maintain healthy blood cholesterol level
	Help to maintain healthy bowel function
Polydextrose	Help to maintain healthy bowel function
Fenugreek seed	Help to maintain healthy postprandial glucose level
Fermentation Microorganisms	
Probiotics	Help to maintain healthy gastrointestinal bacteria population
	Help to maintain healthy bowel function
Red yeast rice	Help to maintain healthy blood cholesterol level

(Continued)

Table 24.3 (Continued)

Ingredients	Health Claim
Amino Acids and Proteins	
Soybean protein	Help to maintain healthy blood cholesterol level
Others	
Aloe gel	Help to maintain healthy skin Help to maintain healthy gastrointestinal function Support immune function
Ganoderma lucidum fruit body extracts	Help to maintain healthy body flow
Chitosan/chito-oligosaccharide	Help to maintain healthy blood cholesterol level
Fructo-oligosaccharide	Help to maintain healthy gastrointestinal bacteria population Help to maintain healthy bowel function Help to absorb calcium
Modified from *[37]*.	

The HFFA refers to two types of functional foods: generic and product-specific health functional foods. If the functional ingredient is on the official HFFC list, food containing this ingredient is classified as generic and the active ingredient does not require extensive characterization and specific proof of its efficacy in functional food products. Many new physiologically active ingredients used in the formulation of functional foods are not listed in the HFFC and require approval by the KFDA.

If the standards or specifications are not included in the HFFC, the manufacturer or importer may prepare their own standards or specifications and submit them for approval to the KFDA [Article 14(2)] [39].

For such ingredients scientific evidence needs to be provided by the Sponsor (food manufacturer) to the KFDA, proving that the new physiologically functional ingredient is safe and effective; this process must occur before marketing of foods including such an ingredient. According to the HFFA [Article 14(1)], the Commissioner of the KFDA has the responsibility to prepare and distribute the HFFC containing standards, specifications, and labeling requirements applicable for manufacturing, usage, and maintenance of health functional foods [33].

24.4.4 Business permit

Anyone who intends to start a manufacturing business for functional foods should first file an application and obtain permission from the KFDA. The same applies to anybody who intends to start importing functional foods. The process requires submitting a business application to the regional branch of the KFDA office. Selling and/or distributing the functional food without manufacturing also requires filing a business application and obtaining approval from the local authority, such as the city office, district office, or county office (in Korean *Si/Gun/Gu*).

24.4.5 Quality and manufacturing process

In order to improve controls over the quality and manufacturing process for health functional foods, the KFDA Commissioner issued a special Notice emphasizing the role of quality controls and GMPs in manufacturing sites involved in the production of health functional foods. According to this notice, effective February 2006, each functional food manufacturing site must assign a Quality Manager responsible for assuring sanitary conditions of the manufacturing facility and provide necessary guidance and supervision to the manufacturing personnel. In addition, following GMPs in the functional food manufacturing sites will be mandatory for venture companies and toll manufacturing facilities. There is no current confirmation available describing the status of implementation of such programs.

24.4.6 Labeling requirements

Since June 1998, the KFDA has been the legal authority for food labeling standards. The Food Safety Division of the KFDA is responsible for establishing labeling standards and KFDA regional offices, and/or provincial government health officials, enforce the labeling standards. Labeling standards for functional foods were established on January 31, 2004. According to the HFFA, the KFDA Commissioner is responsible for preparing and disseminating labeling requirements that are applicable to health functional foods. The food labels should have printed, legible inscriptions including product name, product type, importer's name and address, manufacture date (month and year), shelf-life, contents (weight, volume, or number of pieces), ingredient names and content, nutrients, and other items designated by the detailed labeling standards for food (e.g., radiation-processed product, drained weight for canned products) [33].

It should also be noted, that effective September 2006, the names of all ingredients have to be included on food labels. In addition, food items considered as allergens (eggs, milk, peanuts, crab, tomatoes, etc.) must be indicated on the label in Korean, all imported food products are required to be labeled with the necessary information in the Korean, And stickers may be used instead of manufacturer-printed information in the Korean. The sticker should not cover the original labeling and cannot be easily removable.

For functional foods, however, stickers on packages are not permitted [39]. Manufacturer-printed Korean language labels instead must be used for such products. Nutritional labeling is optional for most food products, with some exceptions. However, foods with functional ingredients must have nutritional labels and these include:

- Health functional foods
- Special nutritional foods
- Health supplementary foods
- Foods with nutrient content claims (e.g., if a product is labeled as "calcium enriched yogurt," the content of calcium must appear on the label)

Korea currently does not allow health efficacy claims on food product labels except for products that meet the criteria for health functional foods. Similarly to the labeling regulations in the United States and Canada, use of such emphatic terms as "non" or "low" are prohibited for food that is naturally low in a particular nutrient (e.g., low-fat apples, low-cholesterol vegetable oil). These terms can be used when a specific nutrient has been reduced through manufacturing processing. When the emphatic term "non" or "low" is to be used for saturated fat, the amount of cholesterol contained in the product should be stated.

24.4.6.1 Labeling standards and information for health functional foods

Labeling standards for health functional foods were established in January 2004 and amended in 2005 and 2007. Labeling standards are applied to both domestic and imported health functional foods [37,40]. It is important to provide consumers with accurate information, to enable them to make appropriate food choices in relation to reducing the risk and dietary management of chronic disease. In accordance with those standards, a manufacturer's printed Korean language label must be on the product. However, Chinese characters or foreign languages in tandem with Korean can be used on the label as long as these characters are not larger than the Korean letters [41]. It should have the following information, in addition to those required for general food products [39,41]:

- Indication by letters or diagram that this is a "Health Functional Food"
- A statement regarding the efficacy claim
- Product name
- Name and address of manufacturer/importer
- Shelf-life and storage instructions
- Net weight and ingredient list
- Description of use and warnings
- Nutritional facts, including specific amounts of nutrients with percentages of the reference values or recommended nutrient amounts
- Functional facts, including specific amounts of functional ingredients or marker ingredients of functional materials
- Directions of use and warnings
- A statement that the product is not a pharmaceutical product that prevents cure or mitigates disease

- Other required information as per the detailed labeling guidelines for functional food

The regulation provides detailed instructions related to methods of labeling, including nutrition and functional information, and nutrient reference values

24.4.7 Regulations pertaining to health functional foods

The following regulations that pertain to health functional foods were established in accordance with the HFFA:

- Enforcement decree of the health functional act, enacted by Presidential decree No. 18164, December 18, 2003, and amended January 12, 2006, and January 18, 2007.
- Enforcement rule of the health functional food act, enacted by the Ordinance of the Ministry of health and Welfare No. 270, January 31, 2004, and amended December 10, 2004; July 3, 2006; and November 20, 2006.
- Regulation on approval of functional ingredients for health functional food, enacted by the Commissioner of the Food and Drug Administration No. 2004-12, January 31, 2004, and amended August 29, 2006, and July 11, 2007.
- Labeling standard for health functional foods [41].
- Regulations on recognition of standards and specifications for health functional foods [42].
- Regulations on recognition of raw materials or ingredients of health functional foods [43].
- Regulations on imported health functional food notification and inspection procedures [44].

24.4.8 Recognition of standards, specifications, and ingredients for health functional foods

The HFFA describes a procedure for the review of health functional foods for which standards and specifications are not yet established (Article 14, Paragraph 4, 43) and/or ingredients are new and require reviewing by the KFDA (Article 15 Paragraph 3, 43). To apply for the recognition of health functional foods standards, specifications, and ingredients, an application accompanied by supporting documents must be submitted to the KFDA Commissioner. If the criteria for health functional food standards and specifications are met, the Commissioner will issue a health functional food recognition certificate to the applicant.

The following criteria must be met:

- Compliance with the HFFA
- Compliance with the regulations on the standards and specifications as well as with regulations pertaining to raw materials and active ingredients used in functional foods

- Benefits for human health
- Scientifically proven safety and functionality of foods
- Intended functionality of the product as linked to bioactive ingredients

It is important to note that the KFDA recognition approval apply for both, domestic as well as imported food products. For food products that are specifically developed for importation, the KFDA will inspect products under the provisions provided in the Guidelines for Inspection of Imported Health Functional Food. An application for the recognition of new ingredients must contain all safety data and animal and human clinical studies that have been conducted under the scientifically sound methodologies and Good Laboratory Practices.

24.4.9 Importation of functional foods to Korea

Imported health functional foods must comply with the same requirements of the HFFA, label and labeling, and other applicable laws as foods that are produced in Korea. The regulations of the importation are promulgated in Article 8 of the HFFA and Article 10 of the Enforcement Regulations Act. The KFDA requires prior notice of imported functional foods and inspection by the designated laboratory either at the regional KFDA or National Quarantine Station [44]. A notification must be submitted to the regional KFDA five or more days before the scheduled date of arrival of food shipments. The inspection is needed to verify whether functional foods are in compliance with the established labeling and advertising standards (Articles17 and 18 of the Act).

If health functional food is determined to be non-compliant, the importer may apply for the conversion of a non-compliant imported functional food for purposes other than human consumption, in accordance with Article 7 paragraph 1 of the regulations [44].

24.5 Malaysia
24.5.1 Preamble

Diet-related chronic diseases have been on the rise in Malaysia as well as other countries of Southeast Asia including Brunei, Indonesia, the Philippines, Singapore, and Thailand. This is especially pronounced among the urban communities. Coronary heart disease was identified as a major cause of death in this region [45]. To combat chronic diseases, the nutritional trends have changed dramatically during the last two decades. There has been greater focus given to the preventive role of nutrition. Being aware of these facts, consumers are now paying greater attention to the nutritive value of their diet. The food industries have also increased efforts to improve nutritional quality of foods, including increasing concentration of desirable and decreasing undesirable nutrients.

New scientific developments in food science offer a range of possibilities to increase the health-promoting properties of specific food products. Examples of such foods are functional foods and nutraceuticals. This category of foods includes foods that are enriched with vitamins and minerals, as well as other food components, such as plant sterol and dietary fiber.

According to the International Life Sciences Institute. Southeast Asian nutritional scientists generally agree that functional foods should follow certain criteria [46]:

- Be in the form of conventional foods and possess sensory characteristics such as appearance, color, texture, consistency, and flavor of food products.
- Contain nutrients and other constituents that confer a physiological benefit above the basic nutritional properties. These components should not be used at the levels for medicinal or therapeutic purposes.
- Possess scientifically proven functional benefits that are delivered upon consumption of regular amounts of these foods.
- Contain "functional" components (bioactives) that may be naturally present or may be added to the food and have been proven to be safe over long-term usage for the intended target population.

24.5.2 Functional foods

Although there are no specific regulation or standards for functional foods and the definition of functional foods is still inconclusive, the Malaysian government is actively proposing international guidance to provide better regulatory control of foods in this category. Working Groups have been established to view recent developments. The "functional claims" criteria have been developed under the general principles of nutrition labeling and claims provision promulgated in the Food Regulations in 1985 and subsequently seen in the Gazette in 2003. These regulations cover requirements for food declaration, nutrient content claims, and nutrition function claims. While there are currently 52 permitted nutrient function claims, disease risk reduction claims are prohibited. However, the market has a wide range of functional foods and products are manufactured locally or abroad in conventional food forms or beverages [47]. Although functional foods may be regulated under the food supplement category, functional foods that are in the food format have to be in line with the current law and regulations regarding food safety and labeling.

Nutrition labeling of foods is one of the strategies assisting consumers in adopting healthy dietary practices. Labeling provides a means for conveying information of nutrient content of the food product, thereby helping consumers make better food choices when planning their daily meals. It is equally important to the food industry, food manufacturers, and retailers to become more aware of the nutritional properties of their products, and to be encouraged to emphasize these properties to consumers [45]. Nutrition labeling is important for traditional foods, but it is even more important for novel foods such as functional foods, where consumers may not be as familiar with those foods or their components. Therefore, frequent questions raised by customers are often centered on safety of functional foods. Dietitians have also raised the safety issue [48]. Their concern is that the existing Malaysian law, which regulates registration of functional foods, does not provide sufficient control of quality and safety of those products, especially in situations where nutrient composition of the functional food product is unknown, and effectiveness is not frequently substantiated by scientific evidence and clinical studies.

It should be noted that functional foods are distributed under the food supplement category and are not covered by the Food and Drug Act [48]. Therefore the less restrictive regulatory environment creates a potential for supplements to be prone to quality control problems. Some dietitians promote the Malaysia Food Pyramid strategy (model), which incorporates a variety of foods comprising all nutrients required for health. This approach proposes a good diet to ensure health rather than single food items or food ingredients to prevent diseases.

In order to control the quality and safety of functional foods, the Malaysian government, under the jurisdiction of Ministry of Health, has assigned three bodies to participate in the implementation of laws concerning functional foods [49]. These include the Department of Food Quality Control, Ministry of Health, the Malaysian National Codex Committee, and the National Pharmaceutical Bureau. It has been proposed that functional foods in the conventional format containing original bioactive components may be registered with the Department of Food Quality Control Ministry of Health. Functional foods in the form of tablets should be registered under the National Pharmaceutical Bureau, as outlined in the Drugs and Cosmetic Act 1984. The Department of Food Quality Control is responsible for reviewing the product label to determine its composition. Acts and regulations used in this confirmation are the Foods Act 1983 and Food Regulations 1985 [48].

24.5.3 Food laws and regulations

The Food Act of 1983 (Act 281 of the Laws of Malaysia) and its regulations, such as Food Regulations of 1985, are the primary legislative documents for foods in Malaysia. Those regulations deal with food hygiene, labeling, imports, exports, advertising, and analytical assessment [50].

24.5.3.1 Specific requirements for nutrition labeling

Under the Food Act (1983) and the Food Regulations (1985), the information that must be listed on the label of food products includes the name of the product and the manufacturer, packer, or importer and the weight or volume of the product as well as the designated food brand and content or list of ingredients provided in accordance to proportion or weight. The content of food additives and storage instructions for the food product must also be provided on the label. Since more than half of Malaysia population is Muslims, labeling requirements for products containing pork and/or alcohol, are very strict. Expiry dates, under the Food Regulations 1985, are only compulsory for certain types of products such as bread, canned foods, and others.

Nutrition labeling means a description intended to inform the consumer of the nutrient content of a food. Although in some countries such as the United States and Canada, nutrition labeling is mandatory for the prepackaged foods, the situation in Malaysia is similar to most Asian countries where nutrition labeling is optional for the majority of food products, with the exception of "special purpose foods." This category includes "functional foods" that are enriched or fortified with specific vitamins or minerals. Many manufacturers, however, voluntarily place nutrition labeling on their food products, although without having a common format for nutrition labeling some errors or inconsistencies can occur.

In August 2000, the Malaysia Ministry of Health announced its intention to amend the current regulations to have mandatory nutrition labeling for a wide variety of foods [51,52]. The proposal was to have mandatory nutrition labeling for a number of core nutrients, namely energy and macronutrients such as carbohydrates including total sugars, protein, and fat for a variety of prepacked foods including functional foods, soft drinks with botanicals, soybean drinks, and others. Along with the four nutrients that must be present on the label, other nutrients, such as minerals and vitamins, are permitted providing that they are listed in the Nutrient Reference Values (Table 24.4). For foods with a nutrition claim, it is mandatory to include a nutrition label and the amount of any other nutrient for which a nutrition claim is made with respect to the food [55]. Examples of such nutrients are dietary fiber, omega-3 fatty acids, and cholesterol. The proposed format for nutrition labeling follows the Codex Alimentarius guidelines [56]. In this respect, the amounts of nutrients are expressed as a percentage of the nutrient reference value (NRV) instead of percentage of recommended daily intake/allowance (RDI and RDA; Table 24.5). In 2003, the Ministry of Health enacted regulations for mandatory nutrition labeling and nutrition claims. Mandatory nutrition information is required for the following food categories [53]:

- Foods that are frequently consumed in significant amounts: prepared cereal, food and bread, milk product, flour confection, canned meat, fish and vegetable, canned fruit and various fruit juices, soft drinks, salad dressing and mayonnaise
- Foods that have been fortified, enriched, vitaminized, supplemented, or strengthened with specific vitamins or minerals
- Foods with nutrition claims
- Special purpose foods including infant formulas and foods for infants and young children

In general the amendments enacted by the Ministry of Health closely follow the guidelines of the Codex Alimentarius, also recommended by Food and Agriculture Organization /World Health Organization with few exceptions. These exceptions are necessary to meet the local needs in Malaysia. The Malaysian

Table 24.4 NRV and the Malaysian RDA Values for Vitamins and Minerals

Name of Nutrient	Units	NRV	RDA
Vitamin A	µg	800	750
Vitamin D	µg	53	N/A
Vitamin E	mg	10	N/A
Vitamin C	mg	60	30
Thiamin	mg	1.4	1.0
Riboflavin	mg	1.6	1.5
Niacin	mg	18	16.7
Vitamin B6	mg	2	N/A
Folic acid	µg	200	200
Vitamin B12	µg	1	N/A
Calcium	mg	800	450
Magnesium	mg	300	N/A
Iron	mg	14	9
Zinc	mg	15	N/A
Iodine	µg	150	N/A
Protein	g	50	N/A

Note: As seen from this table, the NRV values are not very different from the RDA values.
Adapted from [53,54].

Table 24.5 Conditions for Some Nutrient Content Claims (Malaysia)

Component	Claim	Conditions (not less than)
Protein*	Source	10% of NRV per 100 g (solids)
	High	5% of NRV per 100 ml (liquids) or per 100 kcal (At least two times the values for "source of")
Vitamins and minerals	Source	15% of NRV per 100 g (solids)
	High	7.5% of NRV per 100 ml (liquids) or 5 % of NRV per 100 kcal (At least two times the values for "source of")
Dietary fiber	Source	3 g/100 g (solids) or 1.5 g/100 ml (liquids)
	High	6 g/100 g (solids) or 3 g/100 ml (liquids)

Component	Claim	Conditions (not more than)
Fat	Low	3 g/100 g (solids) or 1.5 g/100 ml (liquids)
	Free	0.15 g/100 g (or 100 ml)
Saturated fat	Low	1.5 g/100 g (solids) or 0.75 g/100 ml (liquids) and 10% of total energy of the food
	Free	0.1 g/100 g (or 100 ml)
Cholesterol	Low	0.02 g/100 g (solids) or 0.01 g/100 ml (liquids)
	Free	0.005 g/100 g (or 100 ml)

*NRV for protein is 50 g
Adapted from [53,57].

government has enacted regulations on nutrition labeling and claims [55], thus making it possible to also use them for functional foods. The three groups of nutrition claims for functional foods include:

- Nutrient content claims
- Nutrition comparative claims
- Nutrient function claims
- Claims for enrichment, fortification, or other words of similar meaning as specified in Regulation 26(7) [53]

24.5.3.2 Nutrient content claims

A nutrient content claim describes directly, or indirectly, the level of a nutrient in a food or a group of foods. There are basically two types of nutrient content claims. The first is a negative claim that allows a food to carry "low," "free," or "non-addition" statements in respect to the certain health-adverse food constituents, such as cholesterol, sodium, saturated fat. or trans fat. An example of such a claim is the "cholesterol free" statement listed on foods that contain no more than 5 mg cholesterol per 100 g, for solids or per 100 ml for liquid foods [51].

The second type of claims are considered positive claims that allow label statements such as "good/excellent source," "high," "enriched," and "fortified" in food constituents that are beneficial to health [51]. An example of an expression used in this claim is "a source of" vitamin C. To make this claim, food must contain at least 15% of the NRV value. To claim "high in" vitamin C, food must contain at least 30% of the NRV value.

24.5.3.3 Nutrient comparative claims

Nutrient comparative claims involve similar foods, another brand of the food product, or a different version of the food source. The claims may be expressed on a label in relation to the nutrient content present in the food product, which carries the claim and the food to which it is being compared. The latter food has to be clearly identified. For example, if a new food formulation has lower or reduced sodium content compared with a previous formulation, the label on the new formulation may claim to have "reduced sodium" [51].

24.5.3.4 Nutrient function claims

Nutrient function claim is defined as a nutritional claim that describes the physiological role of the nutrient in the growth, development, and normal function(s) of the body. There are specific prohibitions, however, regarding the use of these claims, which can be summarized below [57]:

- A nutrient function claim shall not imply or include any statement relating to curing, mitigating, or treatment disease.
- No label, which describes any food, shall include any claims relating to the function of a nutrient in the body unless the food for which the nutrient function claim is made shall contain at least the amount of nutrient in the

level to be considered as a source of that nutrient per reference amount as specified in Table 24.5.

- No label on a package containing any food shall bear a nutrient function claim except those permitted in the regulation or with prior written approval of the regulatory authorities.

Permitted nutrient function claims are listed in the *Guide to Nutrition Labeling and Claims*[53]. These include but are not limited to the following:

- Calcium aids in the development of strong bones and teeth
- Protein helps build and repair body tissues
- Iron is a factor in red blood cell formation
- Vitamin D helps the body utilize calcium and phosphorus
- Vitamin B1/thiamine is needed for the release of energy from carbohydrates
- Vitamin B2/riboflavin is needed for the release of energy from proteins
- Niacin is needed for the release of energy from proteins, fats, and carbohydrates
- Folic acid is essential for growth and division of cells
- Vitamin B12/cyanocobalamin is needed for red blood cell production
- Vitamin C enhances absorption of iron from non-meat sources
- Magnesium promotes calcium absorption and retention

These claims show the relationship between nutrients and maintaining the function of the body necessary for good health and normal growth, and they are very similar to the Canadian function nutrient claims (formerly biological role claim nutrients).

Contrary to the Canadian regulations, Malaysian nutrient function claims are allowed on foods with a broader range of bioactives. The examples of ingredients with hypocholesterolemic properties are listed in Table 24.6. The claims can only be made provided the food meets the criteria for "source of" and other conditions [58]. The bioactives such as dietary fiber and plant sterols are constituents of truly functional foods in Europe and the United States. Examples of the model function claims for foods containing discussed bioactives are listed below.

- Inulin helps increase intestinal Bifidobacteria and helps maintain good intestinal environment.
- Oligofructose (fructo-oligosaccharide) helps increase intestinal Bifidobacteria and helps maintain a good intestinal environment.
- Inulin is bifidogenic.
- Oligofructose (fructo-oligosaccharide) is bifidogenic.
- *Bifidobacterium lactis* helps improve a beneficial intestinal microflora.
- Oat or barley soluble fiber (β-glucan) helps lower or reduce cholesterol.
- Plant sterol or plant stanol helps lower or reduce cholesterol.
- Soy protein help to reduce cholesterol.

The other wordings with similar meaning are permitted for the functional claims.

Table 24.6 Conditions for the Malaysian Nutrient Function Claims for Hypocholesterolemic Ingredients

Component	Minimum Amount Required	Other Conditions
β-glucan	0.75 g per serving	i. Source of β-glucan shall be from oat and barley ii. The food to be added with β-glucan shall also contain total dietary fiber for not less than amount required to claim as "source": 3 g per 100 g (solids) 1.5 g per 100 ml (liquids) iii. There shall be written on the label this following statement "Amount recommended for cholesterol lowering effect is 3 g/day"
Plant sterol/plant stanol/plant sterol ester	0.4 g per serving in a "free basis" form	i. Types of plant sterol or plant stanol permitted: "plant sterol/plant stanol, phytosterols/phytostanol, sitosterol, campesterol, stigmasterol or other related plant stanol" ii. Types of plant sterol esters permitted: "campesterol ester, stigmasterol ester and β-sitosterol ester" iii. Maximum amount in daily serving for product added with plant sterol/stanol, plant sterol ester in a "free basis" form is not more than 3 g plant sterol/stanol per day iv. Declaration of the total amount of plant sterol/plant stanols/plant sterol ester contained in the products shall be expressed in metric units per 100 g or per 100 ml or per package if the package contains only a single portion and per serving as quantified on the label v. Only the terms "plant sterols" or "plant stanol" or "plant sterol ester" shall be used in declaring the presence of such components vi. There shall be written on the label of food making such claim a statement a) "Not recommended for pregnant and lactating women, and children under the age of five years"

(Continued)

Table 24.6 (Continued)

Component	Minimum Amount Required	Other Conditions
		b) "Persons on cholesterol-lowering medication shall seek medical advice before consuming this product"
		c) "That the product is consumed as part of a balanced and varied diet and shall include regular consumption of fruits and vegetables to help maintain the carotenoid level"; or
		d) "With added plant sterols"/plant stanols/ plant sterol ester" in not less than 10 point lettering
Soy protein	5 g per serving	Permitted a claim statement:
		"Amount recommended to give the lowering effect on the blood cholesterol is 25 g per day"

Adapted from [53].

24.5.3.5 Nutrient function claims versus health claims

The health claim is constructed to link diet with certain chronic disease conditions. The statement describing the health claim must be specific to the proposed health benefit of the food or the presence of a specific constituent (e.g., bioactive) that, upon ingestion, affects the risk reduction or/and prevention of a particular disease condition. For example, the statement: "Food A is a high source of iron. Iron may help reduce the risk of anemia" is a health claim, because food A, being a high source of iron, can be used in the health claim to link the presence of iron (bioactive) with a disease condition, which is manifested by anemia. The nutrient function claim for iron would state "Iron is a factor in red blood cells formation"; thus this example refers specifically to the function of iron.

Malaysian regulations do not include elaborate provisions for health and nutritional claims. There are, however, several requirements related to health and nutrition claims. For instance, Regulation 18(3) of the Food Act prohibits the description of any food that includes the word "compounded," "medicated," "tonic," or "health" or other descriptions that refer primarily to drugs. According to regulation 26(7) of the Act, food labels cannot claim the food as being "enriched," "fortified," "vitaminized," "supplemented," or "strengthened," or that the food is a source of one or more of vitamins or minerals unless a reference quantity of the food contains no less than the amount of the nutrient in question as specified in Table 24.3 [54]. However, the labels on a food product enriched with essential amino acids, or essential fatty acids, may contain a claim that the food is enriched or supplemented with these nutrients. In practice, claims on "low fat," low or no cholesterol, "high in fiber," "contains fatty acids" (such as omega-3 fatty acids), and various vitamins or mineral enrichment are being made.

24.5.3.6 Forbidden claims

There are some provisions in the current regulations that prohibit the making of certain claims on the label. The amendments include some additional prohibitions [53]:

- Claims stating that any given food will provide an adequate source of all essential nutrients
- Claims implying that a balanced diet, or ordinary intake of foods, cannot supply adequate amounts of all nutrients
- Claims that cannot be substantiated by scientific evidence
- Claims as to the suitability of a food for use in the prevention, alleviation, treatment, or cure of a disease, disorder, or particular physical condition unless they are permitted in this regulation
- Claims that could give rise to doubt the safety of a similar food, or that could arouse or exploit fear in the consumer
- Claims that arouse or exploit fear in consumer
- Disease risk reduction claims

24.5.4 Health and medicinal food products

The National Pharmaceutical Control Bureau (NPCB) regulates all health and medicinal food products. Under the Dangerous Drugs Act 1952 and Control of Drugs and Cosmetics Regulation 1984, the NPCB determines whether the health or medicinal food products should be registered. For imported food products, an import license may be required that can be issued by the Compliance Unit. To register with the NPCB, the exporter or appointed distributor must apply in writing to the NPCB stating the name of the food products, its ingredients, and the content. In addition, claims/usage along with an attached copy of the label/product literature must be made available.

24.5.4.1 Food and drug interface: regulatory process

In Malaysia, all drugs and cosmetic products have to be registered under the Control of Drugs and Cosmetics Regulations 1984 by the NPCB and Drug Control Authority; this is in contrast to foods that fall under the Food Quality Control Division. However, if a food product contains less than 80% of food-based ingredients and more than 20% of the active ingredients, such a product will be regulated by the NPCB [58]. However, some highly potent ingredients, which are below 20% of the total food content, need to be assessed by the committee and may be regulated by the NPCB if necessary.

Malaysian dietary supplements are products formulated to supplement diet in the form of pills, capsules, liquids, or powders and not represented as conventional foods. This practice is contrary to some Asian countries, such as China where bioactives in the medicinal format are defined as functional foods. In Malaysia, dietary supplements may include ingredients such as vitamins, minerals, amino acids, natural substances of plant/animal origin, enzymes, and substances with nutritional/physiological functions. The criteria for manufacturing and registration of dietary supplements are unique. Manufacturing processes must comply with GMP requirements and should conform to established standards of quality; the registration process is through the NPCB (single-stage online registration process). These products can be registered for a maximum period of 5 years. To extend the registration, the manufacturer must provide updated information regarding the safety and efficacy of the product. Products will not be registered if there are public health concerns based on safety considerations [58].

24.6 Indonesia
24.6.1 Preamble

Traditional foods, which may provide health benefits to consumers, have been available in Indonesia for a long time and have the potential to be developed as functional and supplemental foods. Currently, there has been a great deal of attention given to the functionality of foods that are endemic to this country. A good example is the research that has been carried out in Indonesia and Malaysia on crude palm oil. A new study showed that a diet rich in crude palm oil, a trans-free product often procured in Malaysia and Indonesia, may reduce blood triacylglycerol levels [59].

24.6.2 Functional foods

On January 27, 2005, the National Agency of Drug & Food Control (Badan Pengawas Obat dan Makanan; NADFC/BPOM) has issued basic provisions on functional food supervision [60]. This regulation defines a functional food as a processed food, safe for consumption, that contains one or more bioactives with a specific physiological function and proven to have health benefits beyond the basic function of food ingredients and supported by the scientific assessment. Fifteen groups of ingredients have been approved as functional compounds:

I. Vitamin
II. Mineral
III. Sugar alcohol
IV. Unsaturated fatty acid

V. Peptide and protein

VI. Amino acid

VII. Dietary fiber

VIII. Prebiotic

IX. Probiotic

X. Choline, lecithin, and inositol

XI. Carnitine and squalene

XII. Isoflavones (soybean)

XIII. Phytosterols and phytostanols

XIV. Tea polyphenol

XV. Other functional ingredient to be approved latter

Functional foods in Indonesia can be marketed using the same regulatory system that is applied for conventional foods, as long as this category of foods contains food-approved ingredients.

Functional foods may be classified as dietary foods, and/or may be in line with the food fortification program that was launched in 1996 to improve the nutritional status of the community [61]. The most popular constituents of functional foods, promoted by food industries for young children, are the very long essential polyunsaturated fatty acids, such as omega-3 (eicosapentaenoic acid and docosahexaenoic acid) and omega-6 fatty acids as well as calcium.

Furthermore, functional foods enriched with iron, calcium, and traditional herbal components such as ginseng, ginger, and yohimbe are also popular among men and women [62]. It is worthwhile to notice that some of the functional foods approved in the United States have been granted approval for sale in Indonesia. For example, phytosterol-containing foods for heart health such as Nesvita milk were approved as a functional food in Indonesia by the NADFC/BPOM. Functional foods may qualify for some health claims that are allowed in Indonesia. Foods with health claims must have nutrition labeling on the package.

24.6.3 Nutritional labeling

Specific requirements for nutritional labeling are outlined in a compilation of food regulations prepared by the Department of Health Indonesia [62,63]. Nutritional labeling is mandatory only for certain types of foods, such as baby foods, dietary foods, milk products, and other foods, as specified by the Director General. The regulations apply also to foods that make health claims that contain specific nutrients, including energy, protein, and carbohydrate content and levels of vitamins and minerals, as well as fortified and enriched foods with specific nutrients as required by the national legislations.

The regulations apply also to the voluntary labeling of all types of foods. The Indonesian Regulations [63] provide detail conditions for nutrient content and comparative claims for energy, protein, fat and fatty acids content, and enrichment with vitamins and minerals. For instance, no claim for "source" of energy is permitted unless there is at least 300 kcal per reference amount of food. For claims "source of protein," at least 20% by weight of the calories should be derived from protein and there is at least 10 g of protein in the suggested amount of food consumed per day.

24.6.4 Health claims

Ten health claims are permitted under the Indonesian regulations, which are similar to the U.S. claims under the Nutritional Labeling and Education Act regulations. All health claims are disease risk reduction claims. An example provided for calcium and osteoporosis explains acceptable format of such claims: "An active lifestyle and a healthy diet with sufficient calcium intake helps teenagers, and women to maintain healthy bones and reduce the risk of osteoporosis in later life." The permitted health claims include:

1. Calcium and osteoporosis
2. Dietary fat and cancer
3. Dietary saturated fat and cholesterol and coronary heart disease
4. Fiber-containing grain products, fruits and vegetables, and cancer
5. Fruits, vegetables, and grain products that contain fiber, particularly soluble fiber, and risk of coronary heart disease
6. Sodium and hypertension
7. Fruits and vegetables and cancer of the digestive system;
8. Folate and neural tube defect
9. Sugar alcohols do not increase dental caries
10. Soy protein and risk of coronary heart disease

Furthermore, nutrient content and nutrient function claims are allowed on the qualified foods. For example, functional foods with probiotics (Lactobacillus spp. except *L. bulgaricus*, and Bifidobacteria), if qualified, may carry a nutrient content claim or nutrient function claim such as "maintain the health of gastrointestinal tract" [64].

It is worthwhile to note that the regulations prohibit making certain claims, for instance, "good health and longevity can only be maintained by vitamin supplements" or "normal health individuals can look younger and live longer with vitamin supplements."

24.7 Philippines
24.7.1 Preamble

Similar to other countries in Southeast Asia, the Philippines does not have specific regulations for functional foods. However, this country is endowed with numerous natural products that are potential ingredients for functional foods. At least 10,000 plants have been documented and 2000 species are being subjected to research. Without being labeled as a functional food, some of these foods are recognized to have medicinal functions or health benefits beyond the basic nutritional value [65].

24.7.2 Functional foods

The Philippines adapted the Codex Alimentarius definition of functional foods as being a "food that satisfactorily demonstrates that it beneficially affects one or more target functions in the body beyond adequate nutritional effects, in a way which is relevant to either an improved state of health and well-being, or reduction of risk to disease." The term "functional foods" may include foods that are fortified with minerals and vitamins, and which are regulated under the Philippine Food Fortification Act of 2000. The government has implemented this act to make the fortification of certain staple foods mandatory in order to help combat the problem of malnutrition in the country. However, functional foods can also be promoted under this act to be an effective solution to the above problem. Dietary supplements in medicinal format may also be considered as "functional foods." In February 2004, the Bureau of Foods and Drugs of the Department of Health introduced a draft Administrative Order on the "Rules and Regulations Governing Nutrition and Health Claims of Pre-packaged Food Products Distributed in the Philippines" [66].

24.7.3 Nutritional labeling and health claims

According to the Philippine Regulations, nutritional labeling applies to fortified foods [67]. Nutrition information may be presented in a tabulated form on the food package, along with another column that provides the amount of nutrient remaining after cooking, in relation to average or usual serving sizes (expressed in terms of slices, pieces, or specified weight or volume of the food product). The regulation also stipulates the minimum amounts of nutrients that must be present at any one point of the inspection. The methodology used in sampling and analysis of food content must follow the Official Methods of Analysis of the AOAC International. Nutrition regulation will permit nutrition claims such as "high," "rich," "good source," "low," or "free" on foods in relation to the energy, fat, protein, carbohydrates, fiber, sodium, vitamins. and minerals with the nutrient reference value. The criteria for making these claims are based on the Codex guidelines for use of nutrition and health claims [68] in conjunction with the current labeling law. Currently only two health claims are permitted in the Philippines: the first of the claims is associated with calcium to reduce the risk of osteoporosis and the second associates consuming foods low in fat to a reduced risk of cancer. The Philippine regulations also prohibit claims that suggest that a specific food is effective in the prevention, cure, or treatment of any disease or its symptoms.

24.8 Singapore

24.8.1 Preamble

In a developed market such as Singapore consumers are paying more attention to the potential health significance of foods (e.g., functional foods that contain bioactives are believed to promote general well-being or even reduce the risk of chronic diseases). The increased proportion of the aged consumers with the disposable income in this country also contributes to the popularity of functional foods. Older consumers are often more concerned with nutrition and weight maintenance than the younger population.

24.8.2 Functional foods

Although Singapore does not have specific regulations regarding functional foods, the government does feature relevant standards that are applied internationally and have been adopted and/or modified to suit Singapore's conditions. Such controls are in place for public health, security, or safety reasons. Under circumstances where any foodstuff or ingredient falls outside the scope of the Food Regulations, permission is first required from the Food Control Department of Singapore (FCDS). In addition, all new food ingredients that are introduced into the country need to be submitted to the Food Advisory Committee for approval before they can be accepted for use by food processors in Singapore.

For foods that contain novel bioactive components that are not listed as permitted food additives under the Food Regulations Act, all applications related to these products must be submitted to the FCDS for approval. Approval may be granted on a case-by-case basis. Recently, the regulatory authorities have approved phytosterol-containing functional foods such as fat spreads, milk containing no more than 3% total fat (or 1.5 g fat per 100 ml), and yogurt containing no more than 3% total fat per 100 g [69]. These types of products are regulated under the special purpose foods. Specified foods may contain added phytosterols, phytostanols, or their esters, and are allowed to carry the following health claim:

"Plant sterols/stanols have been shown to lower/reduce blood cholesterol. High blood cholesterol is a risk factor in the development of coronary heart disease."

Interestingly, the products must be labeled similar to the EU novel foods regulations, and the content information needs to contain:

- A statement that the product is intended exclusively for people who want to lower their blood cholesterol level
- A statement that patients on cholesterol-lowering medication should only consume the product under medical supervision
- A statement that the product may not be nutritionally appropriate for pregnant and breast-feeding women and children under the age of 5 years

- Advice must be given that the product should be used as part of a balanced and varied diet, including regular consumption of fruit and vegetables to help maintain carotenoid levels
- A statement that the consumption of more than 3 g/day of added phytosterols should be avoided
- A definition of a portion of the food (in grams or milliliters) to be consumed each time (as a serving size) and the number of servings to be consumed per day, with a statement of the phytosterols amount that each portion contains

Figure 24.3 illustrates an approved phytosterol-containing functional food product.

24.8.3 Nutritional labeling and nutrition and health claims

In Singapore, mandatory nutrition information is required for foods that carry a nutrition claim, which suggests that a food have a nutritive property (expressed positively or negatively) in respect to energy, salt, sodium or potassium, vitamins, minerals, fats, fatty acids, cholesterol, amino acids, fiber, starch or sugars, or any other nutrients [70,71]. In 1997 a voluntary program to start nutrition labeling was introduced and the Ministry of Health in Singapore published a *Nutritional Labeling Handbook*, which explains the format of a typical nutrition information panel. This panel should include the serving size of each food as well as the listing or a core group of eight nutrients per serving and as per 100 g or 100 ml [72].

In current Regulations, several nutrient content claims are permitted. Examples of such claims include the words "source" of energy and "protein." The Singapore government has progressed further with nutritional claim regulations as evidenced by actions proposed in 1993 that nutritional claims need to be defined to represent a food with nutritive property, whether general or specific

NESTLE® Omega Plus® ActiCol™ is formulated exclusively for people who want to lower their blood cholesterol level. 2 glasses a day provide the recommended 1.2g of plant sterols that you need.
Consumption of more than 3g/day of added plant sterols should be avoided. Include NESTLE® OMEGA PLUS® with added plant sterols as part of a balanced diet that contains plenty of fruits and vegetables to help maintain carotenoid levels. For persons on cholesterol-lowering medication, seek medical advice. Not recommended for pregnant and lactating women and children under the age of 5 years.

FIGURE 24.3

Nestle Functional milk enriched with calcium, omega-3 and omega-6 fatty acids, and plant sterols.

and whether expressed affirmatively or negatively. This includes reference to energy, salt, sodium/potassium, amino acids, carbohydrates, dietary fiber, cholesterol, fats, protein, and starch or sugars or any other nutrients. Further guidance clarifies the conditions for making claims such as "free," "source," "low," "light," "high," "reduced," etc. In addition, the government introduced the "Healthier Choice" label program in 1997. It is a voluntarily program jointly implemented by the National Heart Association and Ministry of Health [73]. Furthermore, several nutrient function claims are allowed on qualified food products including [74]:

1. Protein helps in tissue building and growth
2. Dietary fiber aids the digestive system
3. Calcium helps build/to support development of strong bones and teeth
4. Vitamin A is essential for the functioning of the eye
5. Docosahexaenoic acid (DHA) and arachidonic acid (ARA) are important building blocks for development of the brain and eyes in infant (only for food for children up to 3 years of age)
6. Probiotics help to maintain a health digestive system

All acceptable nutrient function claims are listed in a guide to food labeling and advertisements [74].

Singapore regulations allow some health claims with a tentative list of 30 acceptable claims (nutrient function and enhanced function claims) prepared by the government and companies that can apply those claims on case-by-case basis. For general food products, permitted claims include those related to probiotics and prebiotics, vitamins and minerals, lactose, protein, and dietary fiber. For infant foods, claims on vitamins, minerals, nucleoproteins, and essential fatty acids can be considered.

In 2011, five disease risk reduction claims related to a healthy diet were approved. The following claims may be made on foods that meet criteria specified in the Fourteenth Schedule [75]:

1. "A healthy diet with adequate calcium and vitamin D, with regular exercise, helps to achieve strong bones and may reduce the risk of osteoporosis. (Here state the name of the food) is a good source of/high in/enriched in/fortified with calcium"
2. "A healthy diet low in sodium may reduce the risk of high blood pressure, a risk factor for stroke and heart disease. (Here state the name of the food) is sodium free/very low in/low in/reduced in sodium"
3. "A healthy diet low in saturated fat and trans fat may reduce the risk of heart disease. (Here state the name of the food) is free of/low in saturated fats, trans fats"
4. "A healthy diet rich in whole grains, fruits and vegetables that contain dietary fiber, may reduce the risk of heart disease. (Here state the name of the food) is low in/free of fat and high in dietary fiber"

5. "A healthy diet rich in fiber containing foods such as whole grains, fruits and vegetables may reduce the risk of some types of cancers. (Here state the name of the food) is free of/low in fat and high in dietary fiber"

The Singapore regulations prohibit making various misleading claims suggesting therapeutic or prophylactic actions of a food in improving human health or curing any disease or conditions affecting human body.

24.9 Thailand
24.9.1 Functional foods

Although functional foods are not a new concept in Thailand, the country has little in the way of specific regulations for this category of foods. In the Thai market, one can find a variety of imported and domestically produced functional foods and dietary supplements. Most of these foods, however, lack scientific evidence to substantiate the health benefits proposed. Furthermore, there is confusion among consumers and regulators concerning what the regulatory aspect of these food products should entail. Based on the Thai Food and Drug Administration definition, functional foods are grouped under the notification for foods for special dietary uses and defined as foods that are similar in appearance to conventional foods, are consumed as part of a normal diet, and exhibit physiological benefits including possibly reducing the risk of chronic diseases. However, sometimes these products cannot be classified as either foods or drugs. Since Thailand does not have a specific regulation for functional foods, most of these foods are now being confused with drugs. A clear-cut identification is necessary in the pre-marketing process in order to classify whether a product is under the Drug Act or the Food Act. To classify a functional food as a drug or a food depends mainly on the type and concentration of active ingredients in the product and the claims made. If a product is classified as a food, three aspects need to be considered together: safety, quality, and efficacy [76].

24.9.2 Nutritional labeling and nutrition and health claims

In Thailand, nutritional labeling is mandatory only for the following categories of foods:

- Foods with nutrition claim, comparative or nutrient function claim
  a. Foods with claims of specific benefits or functions to the body or specific ingredients
- Foods for specific target groups, for instance, school children and the elderly
- Other foods prescribed by the Food and Drug Administration Office

Nutritional labeling may be applied to other foods not mentioned above, on the condition that the stipulated format and regulations are observed. The regulations provide examples of the full format and brief format, which is similar to that of the U.S. Food and Drug Administration [45]. Three types of nutrition claims are identified in the Thai Regulations:

- Nutrient content claims
- Comparative claims
- Nutrient function claims

These claims are similar to those in the Codex Alimentarius guidelines [68]. Examples of nutrient content claim are "source of calcium" and "high in fiber" and "low in fat." Regulations prohibit making a claim of "free" or "low" if the food is naturally free or low in that nutrient.

Examples of nutrient function claims are

- Calcium is an important component of bones and teeth
- Folate is an important component of red cell formation.

In order to make these aforementioned claims, the nutrient must be present in food in certain quantities. Health claims are not permitted under current food regulations; however, the health authorities are examining the draft Codex document on health claims and developing claims that can be used in support of functional foods [77].

References

[1] Ling W. The sprouting and blooming of China's tea culture. Chinese tea culture. Beijing, China: Foreign Languages Press; 2002, Ch. 2, p. 13−5.

[2] Arai S. Studies of functional foods in Japan: state of the art. Biosci Biotechnol Biochem 1996;60:9−15.

[3] The Nutrition Improvement Law Enforcement Regulations, (1996). Ministerial Ordinance 41, July 1991; Amendment to Ministerial Ordinance 33, May 25, 1996.

[4] Tee ES. Nutrition labelling and health claims: Codex guidelines. In: Eighth International Food Data Conference. October 1−3, 2009. Bangkok, Thailand; 2009.

[5] Chung-fang Ch. Does health food guarantee good health? Taiwan Panorama 2007; February 24, 2007.

[6] Health Food Control Act. Department of Health, Executive Yuan. Promulgated on February 3, 1999; amended and promulgated on November 8, 2000; amended and promulgated on January 30, 2002, with amended articles of 2, 3, 14,15, 24 and 28 pursuant to the President's Order Hua Zong Yi Zi No. 09500069821 promulgated on May 17, 2006. (Translated to English by Baker and McKenzi Attorneys-at Law, Taipei, Taiwan); 2006.

[7] Iok-sin L. Group pushes for change to labels on "health food". Taipei Times 2007;2−3 March 13, 2007.

[8] Code of Federal Regulations. Food and Drugs, 21, vol. 2, Chap. I, Part 101, Food Labelling, § 101.14, Health Claims: General Requirements, revised as of April 1, 2002; 2002. p. 62−5.

[9] Chou A. Registration requirements for health food in Taiwan, Report ID:123912, Industry Canada; 2004.

[10] Wong E. Regulation of health foods in overseas places: overall comparison. Hong Kong: Research and Library Services Division Legislative Council Secretariat; 2001, May 15, 2001.

[11] Yen GC. Current development of functional foods in Taiwan. In: Development of functional food and natural health food products for Asian markets. IFT Annual Meeting, Session 81−3, July 2003, Chicago, USA; 2003.

[12] Hwang LS. Recent research and development of functional food in Taiwan. J Med Invest 2007;54:389−91.

[13] Food Administration Act. Council of Agriculture, Executive Yuan, R.O.C. Taiwan; 2007.

[14] GAIN Report − HK6017. Hong Kong, Food and Agricultural Import Regulations and Standards, Country Report 2006, USDA Foreign Agricultural Service, August 1, 2006; 2006. p. 1−18.

[15] Austrade. Food and beverage to Hong Kong, trends and opportunities. Australian Government; 2007.

[16] Austrade. Wellbeing − Hong Kong, market trends. Australian Government; 2007.

[17] Jungbeck K. Hong Kong and China: same country − same expectations? *Euromonitor International*; 2004, October 7, 2004.

[18] Consulate General of India, Hong Kong. Business Opportunities in Hong Kong − Guide to Export Food Products to Hong Kong, January 2002; 2002.

[19] Euromonitor. Hong Kong Functional Food Market In: The World Market for Functional Food and Beverages, January 2004; 2004. p. 100−1.

[20] Wu J. Regulation of health food in Hong Kong. Hong Kong: Research and Library Services Division Legislative Council Secretariat; 2001, May 8, 2001.

[21] CAP 231. Undesirable Medical Advertisements Ordinance of Hong Kong. Diseases and Conditions in respect of which Advertisements are Prohibited or Restricted. Gazette No. E.R. 2 of 2012. Aug. 02, 2012; 2012.

[22] The Government of Hong Kong Special Administrative Region, Food and Environmental Hygiene Department. Health and Functional Food; 2006.

[23] Gain Report − HK1309. Hong Kong. Proposed Regulations of GM Food. USDA Foreign Agricultural Service; 2013, March 8, 2013.

[24] Public Consultation on Proposed Labelling Scheme on Nutrition Information. Hong Kong, Food and Environmental Department Government of Hong Kong Special Administrative Region; 2003.

[25] Legislative Council Brief. Food and Drugs (Composition and Labelling) (Amendment: Requirements for Nutrition Labelling and Nutrition Claim). Regulation 2008. File Ref.: FH CR 1/1886/05. Made by the Director of Food and Environmental Hygiene under section 55 (1) of the Public Health and Municipal Services Ordinance (Cap. 132), Annex A; 2008.

[26] CAC. Joint FAO/WHO Food Standards Programme Codex Committee on Food Labelling. CX/FL 07/35/1. Thirty-fifth Session, Ottawa, Canada, 30 April−4 May 2007; 2007.

[27] Generic Code Of Practice on Television Advertising Standards. Hong Kong SAR, Hong Kong Broadcasting Authority; 2003.

[28] Lai Yeung Wai-ling T. Combating deceptive advertisements and labelling on food products − an exploratory study on the perceptions of teachers. Int J Consum Stud 2004;28:117−26.

[29] Minutes of the 5th Meeting of Advisory Council on Food and Environmental Hygiene. Held on February 2, 2001; 2001.

[30] Export Guide. Functional foods and biosupplements market in South Korea. New Zealand Trade and Enterprise; 2011, May 2011.

[31] Chay Y. Nutritional Supplements, International Natural and Health Products Expo, 2005, Seoul, Korea; 2005.

[32] Sook MS. Food consumption trends and nutrition transition in Korea. Mal J Nutr 2003;9:7−17.

[33] Kim JY, Dai BK, Hyong JL. Regulations on health/functional foods in Korea, Nutrition and Functional Food Headquarters, Korea Food and Drug Administration, Seoul 122−704, South Korea, December 30, 2005; 2005.

[34] GAIN Report - KS5037. Food and Agricultural Imports Regulations and Standards Report (FAIRS). Republic of Korea: USDA Foreign Agriculture Service; 2005. p. 1−46. July 21, 2005.

[35] Korea Food and Drug Administration, Food Import Team. Introduction on Imported Foods System in Korea, Under the Food Sanitation Act, Seoul, Korea; 2004.

[36] Health Functional Food Act. Act No. 6727, Aug. 26, 2002. Amended by Acts No. 7211, March 22, 2004; No. 7428, March 31, 2005; No. 8033, Oct. 4, 2005; No. 8365, Apr. 11, 2007; No. 8852, Feb. 29, 2008; No. 8941, March 21, 2008; No. 9932, Jan. 18, 2010; No. 10128, March 17, 2010; No. 10219, March 31, 2010. Korea Ministry of Food and Drug Safety; 2002.

[37] Health Functional Food Code. Enacted in 2002, Enforced in 2004, Amended and renewed in 2008. English version Sept. 2010. Korea Ministry of Food and Drug Safety, <http://www.kfda.go.kr/files/upload/eng/4.Health_Functioanl_Food_Code_(2010.09).pdf>; 2002 [accessed 25.04.13].

[38] Culliney K. Korea's health claim shake up set for next year. Food Navigator Asia; 2012, May 29, 2012.

[39] Kim J. The Health Functional Food Act −A New Regulatory Framework in Korea. Seoul, Korea: Assistant Director, Health Functional Food Division, Korea Food and Drug Administration; 2004.

[40] Notice #2004-6. Korea Food and Drug Administration, Labelling Standards for Health/Functional Foods, January 31, 2004 (unofficial English translation); 2004.

[41] Labelling Standard of Health Functional Food. Korea Food and Drug Administration Notification No. 2004-6, Jan. 31, 2004. Amended by Notification No. 2005-65, Nov. 11, 2005, and by Notification No. 2007-16, March 22, 2007; 2004.

[42] Notice #2004-11. Korea Food and Drug Administration, Regulations on Recognition of Standards and Specifications for Health/Functional Foods, January 31, 2004 (unofficial English translation); 2004.

[43] Notice #2004-12. Korea Food and Drug Administration, Regulations on Recognition of Raw Materials or Ingredients of Health/Functional Foods, January 31, 2004 (unofficial English translation); 2004.

[44] Notice #2004-8. Korea Food and Drug Administration, Regulations on Imported Health/Functional Food Notification Inspection Procedures, January 31, 2004 (unofficial English translation); 2004.

[45] Tee ES, Tamin S, Ilyas R, Ramos A, Tan W-L, Lai DK-S, et al. Current Status of nutritional labelling and claims in the Southeast Asian region: are we in harmony? Asia Pacific J Clin Nutr 2002;11:S80−6.

[46] Tee ES. Functional food for thought. Briefings from the international conference on functional foods. Malta: The Star Online — health; 2007, May 2007.

[47] Tambi Z. Report Of The Regional Expert Consultation Of The Asia-Pacific Network For Food And Nutrition On Functional Foods And Their Implications In The Daily Diet, (2004). FAO of the Regional Office for Asia and the Pacific, Bangkok, Thailand, 2004. RAP Publication 2004/33; 2004.

[48] Fatimah A. Functional foods from the dietetic perspective in Malaysia (viewpoint). Nutr Diet 2003;1—5.

[49] Fatimah A, Mohd Rizal MR. Regulatory requirements on health claims for nutraceuticals and functional foods in Malaysia. In: Proceedings of Conference on Marketing Nutraceuticals and Functional Foods, January 2000, Singapore; 1999. p. 20—1.

[50] Food Act. Malaysia Food Act 1983 and Food Regulations 1985 (with amendments up to May 1998). Kuala Lumpur: Government of Malaysia; 1983, 1998.

[51] Tee ES. Proposed requirements for nutrition labelling, Malaysia: Part II. In: Proceedings of the national seminar on nutrition labelling: regulations and education, 7—8 August 2000. Kuala Lumpur; 2000. p. 71—80.

[52] Nik Shabnam NMS. Proposed requirements for nutrition labelling, Malaysia: Part I. In: Proceedings of the national seminar on nutrition labelling; regulations and education, 7—8 August 2000. Kuala Lumpur; 2000. p. 73—80.

[53] Guide to Nutrition Labeling and Claims. (2010). Expert Committee on Nutrition, Health Claims and Advertisement. Ministry of Health Malaysia. December 2010.

[54] Tee ES. Proposed new law on nutrition labelling and claims: what should you know. Nutrition Society of Malaysia, Health Claims and Advertisement; 2000. p. 1—9.

[55] Tee ES. Claims and scientific substantiation: efforts in harmonizing in Asia. Conference on Functional Foods, Malta, May 2007; 2007.

[56] FAO/WHO. Food labelling: complete texts (revised 1999). Joint FAO/WHO food standards programme. Rome: FAO/WHO; 1999.

[57] GAIN Report - MY6025. Malaysia, food and agricultural import regulations and standards, Malaysia. USDA Foreign Agricultural Service; 2006. p. 1—97, July 28, 2006.

[58] Tee ES. Labelling guideline. Ministry of Health Malaysia; 2007, February 14, 2007.

[59] Ladeia AM, Costa-Matos E, Barata-Passos R, Guimaraes AC. A palm oil—rich diet may reduce serum lipids in healthy young individuals. Nutrition 2008;24:11—5.

[60] Basic Provisions on Functional Food Supervision. HK.00.05.52.0685. The National Drug & Food Control Republic of Indonesia, DJ 2005-01-27; 2005.

[61] Bogor Agricultural University. National Fortification Commission, Directorate General of Public Health of MOH, Deputy of Food Safety and Hazardous Substances of NADFC, Directorate General of Chemical, Agro-Forestry Based Industry of MOIT, Wheat Flour Producers Association of Indonesia, Cooking Oil Industries Association of Indonesia, Infant Food Producers Association of Indonesia, Iodized Salt Producers Association of Indonesia. Country Investment Plan for Food Fortification in Indonesia, Report, Jakarta; 2003.

[62] Department of Health Indonesia. 3rd ed. Compilation of food regulations, vol. 1. Jakarta: Directorate General of Food and Drug Control, Department of Health; 1994.

[63] Department of Health Indonesia. Government regulation number 69 regarding food labelling and advertisement. Jakarta: Directorate General of Food and Drug Control, Department of Health; 1999.

[64] Lee YK, Shao W, Jin S, Wen Y, Ganguly B, Rahauyu ES, et al. Probiotics regulation in Asian countries. In: Lahtinen S, Ouwehand AC, Salminen S, Von Wright A, editors. Lactic acid bacteria: microbiological and functional aspects. 4th ed. Boca Raton, FL. USA: CRC Press; 2012. p. 716. Ch. 32.

[65] Mallillin AC, Bautista-Batallones C. Review of country status on functional foods: Philippines, Report of the regional expert consultation of the Asia-Pacific network for food and nutrition on functional foods and their implications in the daily diet, FAO Corporate Document Repository, RAP Publication 2004/33; 2004.

[66] Austrade. Wellbeing Philippines, health biotechnology and wellbeing. Australian Government; 2007.

[67] Department of Health Philippines. Administrative Order nr.88-B; rules and regulations governing the labelling of prepackaged food products distributed in the Philippines. Manila: Bureau of Food and Drugs, Ministry of Health, Republic of the Philippines; 1984.

[68] CAC. Guidelines for use of nutrition and health claims. Codex Alimentarius Commission, CAC/GL 23-1997. Adopted in 1997. Revised in 2004. Amended in 2001, 2008, 2009, 2010, 2011 and 2012; 1997.

[69] AVA. Part IV. Special Purpose Foods. Foods containing phytosterols, phytosterol esters, phytostanols or phytostanol esters. 250A. Amended S 195/2011 wef 15/04/2011. In: Food Regulations. G.N.No. S 264/2005 revised edition (30th Nov. 2005). Agri-Food & Veterinary Authority of Singapore. 03/09/2012; 2012.

[70] Government of the Republic of Singapore. The Sale of Food Act, [Fap.283]. Food regulations 1990 and amendments S 398 of 1993. Singapore: Government of the Republic of Singapore; 1993.

[71] AVA. Part III. General Provisions. Nutrition Information Panel. 8A. Amended S 195/2011 wef 15/04/2011. In: Food Regulations. G.N.No. S 264/2005 revised edition (30th Nov. 2005). Agri-Food & Veterinary Authority of Singapore. 03/09/2012; 2012b.

[72] Ministry of Health Singapore. Nutrition labelling: a handbook on the nutrition information panel. Singapore: Department of Health; 1998.

[73] Tan WL. The Singapore experience in nutrition labelling regulations and their education. In: Proceedings of the national seminar on nutrition labelling: regulations and education, 7–8 August 2000. Kuala Lumpur; 2000. p. 31–5.

[74] AVA. A guide to food labeling and advertisements. Singapore: A publication of the Agri-Food & Veterinary Authority; 2011, October 2011.

[75] AVA. Part III. General Provisions. False or misleading statements. 6A and Fourteenth Schedule. Criteria for permitted claims. Regulation 9 (6A). Amended S 195/2011 wef 15/04/2011. In: Food Regulations. G.N.No. S 264/2005 revised edition (30th Nov. 2005). Agri-Food & Veterinary Authority of Singapore. 03/09/2012; 2012.

[76] Charoenpong C, Nitithamyong A. Review of country status on functional foods: Thailand, Report of the regional expert consultation of the Asia-Pacific network for food and nutrition on functional foods and their implications in the daily diet. FAO Corporate Document Repository, RAP Publication 2004/33; 2004.

[77] Kongchuntuk H. Thailand experience in nutrition labelling regulations and education. In: Proceedings of the national seminar on nutrition labelling: regulations and education, 7–8 August 2000: Kuala Lumpur; 2000. p. 36–45.

Overview of Regulations and Development Trends of Functional Foods in Malaysia

25

Teck-Chai Lau

Department of International Business, Universiti Tunku Abdul Rahman (UTAR), Malaysia

25.1 Introduction

There seems to be a significant commercial interest in functional foods, especially in the past couple of years, due to their potential to help reduce levels of diet-related diseases such as cancer, heart disease, and osteoporosis. Functional foods are increasing in popularity and are being marketed for disease prevention and treatment. Consumers are becoming more aware of the benefits in consuming functional foods and have taken an increasingly proactive approach to health, realizing that certain dietary behaviors can reduce their risk for chronic disease. Functional foods, which claim health-boosting effects, can play a significant role in the reduction of health care costs associated with diet-related illnesses [1].

Consumer interest in the relationship between diet and health has increased the demand for information about functional foods. Rapid advances in science and technology, increasing health care costs, changes in food laws on label and product claims, population aging, and rising interest in attaining wellness through diet are among the factors fueling consumer interest in functional foods. Credible scientific research indicates that there are many clinically demonstrated and potential health benefits from food components.

In Malaysia, interest in functional foods is also gaining much impetus with more products being patented and commercialized [2]. There are many functional food products on the Malaysian health food market. Some of these functional foods are traditional or culturally based foods and others are categorized as modified, fortified, or totally new foods. The modern functional food products are easier and more convenient to consume than traditional functional foods. In the Western food market, it is relatively easy to distinguish different food types such as normal food, nutritious food, health food, functional food, and medicine. However, this is not the case in the Asian food market.

Nutraceutical and Functional Food Regulations in the United States and Around the World.
DOI: http://dx.doi.org/10.1016/B978-0-12-405870-5.00025-6

25.2 Western versus eastern perspective on functional foods

There seems to be a clear distinction between the Western perspective and the Eastern perspective on functional foods. Westerners view functional foods as something new and they represent a fast growing segment of the food industry [3]. Based on this understanding, food, drug, chemical, and boutique companies formed basic and developmental research in their effort to bring functional foods to the market as quickly as possible for commercialization.

In contrast, functional foods were part of the Eastern culture for centuries. For example, in traditional Chinese medicine, foods that have medicinal effects have been documented since at least 1000 BC. The Chinese, even in those ancient times, understood that foods have both preventive and therapeutic effects and are an integral part of health. This view is now being increasingly recognized around the world [3].

In Asian societies, functional foods have traditionally been related to health foods [4] and have been heavily influenced by cultural values. Foods with functional properties have been an important part of Asian culture for centuries, even though the term "functional food" is not commonly used [5]. Studies have established that personal values are central to the motivations and determinants of consumer attitudes and consumption behaviors [6,7]. Japan was the first Asian country to introduce both the term functional food and a specific regulatory approval process for such foods [4,8,9]. This was mainly due to the concerns of the Japanese government in 1985 about Japan's aging population and the resultant high health care costs. During that time, the Japanese government actively set out to enhance their citizen's awareness of the link between dieting and health in an effort to save on future health care expenditures [10].

The past two decades have seen the development of the functional food market in the United States and Europe. It was first used in the United States and European markets in the fortification of basic foods such as breakfast cereals and drinks [1]. Today, their use and distribution is more widespread, with an extensive range of functional food ingredients created and used to enhance foodstuffs [11]. Functional food ingredients include dietary fiber, proteins, lactic acid bacteria, vitamins, minerals, fish oils. and plant extracts such as garlic, licorice, and celery [1].

There is no universally accepted definition for functional food [12]. In the context of this chapter, functional food is defined as a category of food that has health-enhancing properties but is not considered a drug, chemical, or vitamin [13,14]. Substances in functional food can have medicinal value, but medicine itself is not functional food [13]. Functional food can be obtained like any other food and does not require any kind of prescription. In Malaysia, there are no specific regulations for health foods or functional foods [5]. Currently, there are only regulations on nutrition labeling and claims. However, the government has appointed the Drug Control Authority (DCA) and the National Pharmaceutical Control Bureau to formulate a separate regulation for dietary supplements [5].

25.3 **Functional foods and the unique Malaysian society**

Malaysia has achieved a large amount of economic and political stability in the last few decades. This has resulted in a rapid change in consumers' food habits and food choice, changing consumers' lifestyles and food consumption patterns [15]. There are more Malaysian consumers than ever who are trying to change their diets and purchase healthy and nutritious food, because health issues have become more prominent since society is becoming more affluent and there is better information on dietary and health matters.

Despite the advancement and openness among Malaysian consumers to try out new health supplements and products, there seems to be a great deal of confusion about the categories of health products available in the market. One of the main problems in Malaysia for producing functional food is the lack of a precise definition for functional food. Due to this, it is difficult for consumers to distinguish between functional food and other types of food such as nutraceuticals, therapeutical food, or pharmaceutical food.

The Malaysian functional foods industry is still at an embryonic stage. More research and development are required to cater to the growing demand of functional foods by increasingly health-knowledgeable consumers. However, the range of functional food is very wide; it can be divided into two main categories [16]:

1. Mass market items: These include traditional food and drink products (e.g., Chinese medicinal products, Essence of Chicken, and soft drinks such as chrysanthemum tea or cooked food) that are sold at hawker centers and in "medicinal restaurants," Malay traditional medicinal products such as Eurycoma longifolia Jack (tongkat ali), Labisia pumila (kacip fatimah), Andrographis paniculata (hempedu bumi), and Orthosiphon stamineus (misai kucing), and Indian traditional medicinal such as ayurvedic medical.
2. Non-indigenous food and drink products (e.g., cultured milk drinks, enriched liquid milk and yogurts, and bottled nutritional drinks): Niche market items that include a range of imported dietary food such as sugar-free products, gluten-free products, fat-free products, high-fiber products, and many more.

Among Malaysian consumers, functional food has been perceived differently and has been associated with traditional medicine of each ethnic group. As a multicultural country consisting of three main ethnic groups—Malays, Chinese, and Indians—each ethnic group has its own beliefs about food and well-being. Furthermore each group has its own popular food with functional properties. For years, Malaysians have supplemented their diets with naturally occurring ingredients. Knowledge of functional foods is often passed down from generation to generation, usually through oral traditions [13,14]. It is important to take note that Malaysia has 8000 species of flowering plants [2] and about 6000 of these have medicinal value [17]. Of these, 1200 are used in traditional medicine.

For the majority ethnic group in Malaysia, the Malays have used herbs and plant roots from the rainforest as traditional dietary supplements for generations [2,18]. Some popular Malay foods are used for enhancing vitality and preventing aging, cancer, diabetes, and hypertension [19,20]. For Malaysians of Chinese ancestry, traditionally the foundation of human health is the therapeutic use of food because food and medicine often come from the same source [21]. In addition, food and medicine are equally important in treating and preventing diseases [21]. Chinese functional foods are also believed to have antioxidant effects and able to prevent cancer [22]. Malaysian Indians share health foods dating back to traditional 2000-year-old Ayurvedic practices [23,24]. These are entrenched in everyday Indian culinary practice. Some Indian spices and herbs are thought to boost energy and provide health benefits because they have phytochemical and antioxidant properties that help suppress multiple myeloma and cancers [25].

Recent years have also seen the convergence of the traditional functional food with modern technology. Currently there are many products under the traditional functional foods category that incorporate the basic or generic ingredients such as botanical extract, aromatic oils, and complex sugars prepared by small- or medium-scale industries. The more advanced or modern functional foods from high-technology food manufacturers have become major players in the market. This category may prove more useful and trendy over the traditional category.

In comparison to other western societies or even some Asian cultures that are monocentric culturally, Malaysia, with its different ethnic compositions, has a unique contribution to the functional food market. Each ethnic group has its own exclusive traditional food and medicinal products. There is a great deal of potential in the underexplored area of medicinal plants or even dietary traditions, which are distinctive to the country and its people. Coupled with the use of modern technique and technology, this means that there is vast potential for a unique functional foods market in Malaysia.

25.4 Functional food research in Malaysia

The bulk of studies on functional food examine the role of functional food in developed Western countries such as the United States, Canada, Finland, Australia, and Sweden in which there is a dominant national culture gearing more toward individualism. The Western philosophy tends to reject traditional values in favor of modern or post-modern consumption values [26–28]. Due to this, the framework developed through these studies may not be suitable for developing countries such as much of Asian societies with many significant minority groups [29].

Although functional food is becoming more popular with future predictions of market growth, there seems to be little research conducted in the Asian setting, especially focusing on functional food in the Malaysian context. However, literature search reveals that there seems to be emerging studies on functional food conducted in Malaysia in the last couple of years. This is encouraging as it added to the vast body of knowledge on functional food research, especially for emerging economy such as Malaysia. In this section we will review some of the research conducted on functional food in Malaysia in the last three years.

In 2011, an exploratory study was conducted to investigate how Malaysian consumers undergoing rapid socioeconomic transition manage their conflicting values in making choices concerning functional foods [30]. Data for the research were collected qualitatively using ethno-consumerist and grounded-theory methodologies. In combination, these two approaches enabled the researcher to conduct research at the "emic" level (within culture). Eventually, the exploratory model was developed to illustrate how the main three ethnic groups in Malaysia manage their values in terms of functional food consumption. The results showed that consumers did not spend much time consciously considering their consumption choices or their values until they were faced with choices or personal values that were inconsistent with cultural, physical, and product characteristics. The finding also revealed that values were managed by prioritization and balancing to suit the participant's health needs and situation [30].

In another study by the same researcher, a consumption model for the multicultural society of Malays, Chinese, and Indians in Malaysia was developed. The study showed that cultural values, personal values, knowledge, convenience, and health motive were significant factors in explaining the consumption of functional food products in multicultural societies. Results also revealed that functional food acceptability was highly dependent on affective factors that were, in turn, influenced by the economic, social, and cultural situation of the consumer market [31].

A study was also conducted to discover the consumption of functional food for Malay Muslim consumers, who are the main ethnic majority group in Malaysia [32]. In the research, data were collected through self-administrated questionnaires. Structural equation modeling was used to develop the consumption of functional food model for the Malaysian Malay Muslims. Based on the results, a conceptual model was developed, which consists of five dimensions. The five dimensions were cultural values, instrumental values (compassion), terminal values (self-accomplishment), knowledge, and health (physical and spiritual health). Functional food consumption was directly influenced by physical and spiritual health factors, which, in turn, were influenced by cultural values, terminal values, and knowledge. Cultural values also influence knowledge and instrumental values. Terminal values (life accomplishment) were influenced directly by instrumental values (compassion) [32].

In 2012 a research was conducted to determine the extent to which selected socioeconomic characteristics and attitudes influence consumers' awareness of functional food [33]. A survey was conducted in the Klang Valley, Malaysia, where 439 respondents were interviewed using a structured questionnaire. The result revealed that most respondents had a positive attitude toward functional food. For the binary logistic estimation, the results indicated that age, income, and other factors such as concern about food safety, subscribing to cooking or health magazines, being a vegetarian, and consumers who have been involved in a food production company, significantly influenced Malaysian consumers' awareness toward functional food [33].

A study by Phuah et al. (2012) was undertaken to determine how willingness to purchase functional food affects Malaysian consumers given the benefits of such types of food to their health and well-being. A survey was conducted in the Klang Valley, Malaysia, and contingent valuation method was used to determine consumers' willingness to purchase. The results indicated that the majority of the respondents had a high level of willingness to pay for functional food. Six latent factors that influence consumers' willingness to purchase functional food were identified using factor analysis. The factors were food safety, consumers' preferences, health consciousness, knowledge toward functional food, product price, and functional food attributes. Using the binary logistic model, the results also showed that gender, income, age, food safety, consumers' preferences, health consciousness, product price, and functional food attributes significantly influenced consumers' willingness to purchase functional food [34].

In the latest study on functional food in Malaysia, the researchers reviewed the definition and regulatory bodies governing functional food across different countries [35]. They also studied functional food growth potential in Malaysia and discussed factors that might influence the purchase of functional food in the Malaysian environment [35].

25.5 Overview of regulatory environment in Malaysia

All food, drinks, and food ingredients imported or manufactured locally are required to comply with the Malaysian food regulation [35,36]. There are several key laws and regulations on food in Malaysia, including:

1. Food Act 1983 and Food Regulation 1985
2. Control of Drugs and Cosmetics Regulation 1984
3. Halal laws and regulations

The main Malaysian government department and agencies that implement and enforce these regulations include:

1. Food Safety and Quality Division (FSQD)
2. DCA, provides the general principles of safety, quality, and efficacy that form the basis for the evaluation and eventual registration of products
3. National Pharmaceutical Control Bureau (NPCB)
4. Committee for the Classification of Food—Drug Interface Products is involved in functional foods and drinks and nutraceuticals

5. Malaysian National Codex Committee ensures Malaysian regulations are developed based on Codex, a legal system in other parts of the world (usually the Developed World)
6. Jabatan Kemajuan Islam Malaysia is the Malaysian national halal certification agency

According to the Malaysian DCA, a product is considered a food product if it contains 80% or more food ingredients either single or in combination, with equal to or less than 20% of biologically active ingredients of natural products with pharmacological and/or therapeutic properties. Due to the complex nature of nutraceutical products and their links to medicines, nutraceutical products can be defined as either food or drugs, depending on their nature, characteristics, and effects on consumers [36]. As a result, functional foods can be considered nutraceutical products in Malaysia due to their ambiguous nature, as there is no official definition of functional foods [15,36,37]. Functional foods are generally understood as foods that contain substances other than nutrients that may have beneficial effects on health beyond their nutritional properties. The great influx of these products into the Malaysian market has led to proposals for international regulation to better control them.

The Committee for the Classification of Food–Drug Interface Products provides official guidelines to determine whether a product is a food or a drug [38]. The guidelines are complex and detailed, and the important ones are extracted to distinguish between a nutraceutical and functional food product [39]:

1. The product is regarded as food products if a product contains 80% or more of food ingredients, either single or in combination, and with equal to or less than 20% of biologically active ingredients of natural products with pharmacological and/or therapeutic properties, and has to be regulated by FSQD. Examples of these are many types of functional foods.
2. If a product contains less than 80% of food-based ingredients and more than 20% of the active ingredients, such product should be regulated by the NPCB. Notwithstanding these general rules for products containing specific ingredients, which possess high potencies, even if they contain less than 20% active ingredients, they shall be reviewed by the Committee and may be regulated by the NPCB as a form of drug [38]. When uncertainty arises about the effectiveness and safety of a product, NPCB would be the preferred authority to regulate. If a product is more than 80% food based, and contains active ingredients that exceed the amount permitted in the Food Regulations 1985, the manufacturer shall be advised to reduce the amount of functional ingredients and to be regulated by FSQD [36,39].

25.5.1 Regulations concerning nutrition labels and health claims

Health claims, nutritional claims, and labeling are permitted on functional foods in Malaysia. The claims follow the guidelines of Codex Alimentarius [40,41].

Malaysia has a well-developed regulatory system for functional foods and natural health products. The alignment of their claims to the guidelines of the Codex Alimentarius is a step that facilitates access to the Malaysian market by investors.

Health claims are made to link a diet to a certain disease condition. The health claims in Malaysia are product specific and must always link the benefit from the product to the particular disease. It could either be a disease risk reduction claim or a disease prevention claim. Regulation in Malaysia does not strictly control these claims. However, certain claims are prohibited such as those that involve words like medicated, compounded, and health. Health claims that cannot be substantiated with scientific evidence are also forbidden [41].

There are three groups of nutrition claims permitted in Malaysia [37,41]:

- Nutrient function claims
- Nutrient content claims
- Nutrition comparative claims

1. Nutrient function claims: Nutrient function claims promote how the nutrient in the product can affect the physiological function of the body in terms of growth and development. Nutrient function claims are permitted on bioactive substances like probiotics and prebiotics in Malaysia. They are not permitted to relate to the curing and treatment of a disease to the product. Claims that cannot be proved scientifically and those that imply that ordinary food cannot provide the needed nutrients of the body are also prohibited [41].

2. Nutrient content claims: Nutrient content claims indicate the level of a nutrient in a food. There are two types in Malaysia: they can either be negative or positive. Negative claims imply that the food is free or less of a particular food component. Positive claims, on the other hand, imply that the food is a source of, or high in, a stated food component or nutrient. A food with a positive claim like "source of" should have at least 15% more than the nutrition reference values of the stated nutrient and 30% more when it has a "high" or "enriched" claim [41,42].

3. Nutrition comparative claims: Nutrition comparative claims are made when another product of similar type is compared to the one carrying the claim. Malaysian regulations require that both the food with the claim, and the one being compared to, be clearly identified. The stated claim should be on the product with the claim. An example of a nutrient comparative claim would be "reduced trans fat" when the product with the claim contains less trans fat compared to a previous type [41,42].

Natural health products are known in Malaysia as dietary supplements. They refer to products that are formulated to supplement diet in the form of pills, capsules, and liquids. Dietary supplements should not be in the form of conventional food. Components of dietary supplements include vitamins, amino acids, and natural substances that can be either animal or plant based [41]. All claims allowed for functional foods are also allowed for dietary supplements. However, all claims are product specific and are subject to a pre-market approval of the NPCB. The manufacturing process should also conform to the Good Manufacturing Practices requirements. Dietary supplements have to undergo a registration process. Each product can be registered for a 5 year period after which new and updated information about the safeness of the product must be recorded before it can be renewed by the NPCB [41].

25.6 Market size, structure, and development trends in Malaysia

The general food and beverage market in Malaysia is estimated at RM30 billion in 2011 [35,36]. Due to the lack of information available for functional foods in Malaysia, it can be assumed that Malaysia has an attractive functional food and beverage niche from its large food and beverage market. Trade sources estimated that functional foods consist of about 40% of total processed and retail packed food and drinks markets [36].

The significant segments include infant and other milk formulas, dairy-based drinks, energy drinks, sport drinks, fruit juices, drinks with Asian herbs, cereals, energy bars, biscuits, baked products, and eggs with omega-3. These segments are mainly locally produced functional products and imported ASEAN-content products, mostly from Thailand with a small percentage from developed countries such as Australia, the United States, or Europe [36]. Imports from developed countries are very small and can only be found in exclusive retailers because of their high prices. As functional products are highly diversified, market segments are very fragmented, which makes it difficult to estimate the overall market size from market observations or trade estimates [36].

Functional foods and drinks surfaced in the Malaysian market during the 1990s when local producers and ASEAN-based multinational food companies competed to introduce new product lines to create new niches, capitalizing on the emerging health trends at that time and market expansion [36]. These companies include Nestlé, Danone, Unilever, Kellogg, and Quaker Oats. Today, healthy living is an increasingly important part of the food market and food marketing [43]. Table 25.1 lists some of commercial products available in the local Malaysian markets [35].

25.6.1 Demand for functional food in Malaysia

According to Stanton et al. (2011), Malaysians tends to consume less processed food products than consumers of developed countries. An increasing trend toward consumption of processed foods is seen among young consumer groups below 40 years of age who seek convenience. These consumers are better educated, informed, and receptive toward new products compared to older consumers. Furthermore, research [36] also revealed that Malaysian consumers generally prefer the fortified version to regular product, if a choice was given, but only on the

Table 25.1 Local Malaysian Functional Food Products

Types	Description	Brand	Producer
Probiotics	Milk drink containing Lactobacillus casei Shirota	Yakult	Yakult Japan
		Nutrigen	Mamee Double-Decker
		Vitagen	Malaysia Milk
Prebiotics	Chocolate malt drink with oligosaccharide-fructo and DHA/EPA	Oligo	Power Root (Malaysia)
Functional drink	Energy drink with kacip fatimah	Per'l kacip fatimah	Power Root (Malaysia)
	Enriched sport drinks with vitamins	Gatorade	Permanis for PepsiCo
Fortified drink	Orange juice contains dietary fiber and vitamins	Sunkist 100% orange juice	F & N
Functional cereal	Low fat whole grain breakfast cereal with minerals and vitamins	Nestlé Fitnesse	Nestlé
Bakery product	Whole meal bread with β-glucan for cholesterol lowering	Gardenia "Breakthru"	Gardenia
	Enriched range of crackers from high fiber, low salt, oat-based	Jacob's	Kraft
Spread	Cholesterols lowering spread made from olive oils, contains vitamins and minerals	Naturel	Lam Soon
Functional egg	Chicken eggs contains omega-3 and vitamins	NutriPlus	NutriPlus

condition that the price difference between the products is not apparent and there is similar quality. It is further discovered that most Malaysian consumers are usually not aware that they are purchasing functional foods, as functional foods are not labeled as such.

Most functional products are not well differentiated and are usually marketed as conventional foods and part of the wider marketing, advertising, and market segmentation of manufacturers. Local functional products or imported ones are commonly distributed via retail channels. The distribution of functional food products uses a similar strategy as conventional products. Table 25.2 lists the main retailers in Malaysia.

25.6.2 Future demand and challenges

The market in Malaysia for functional foods seems bright and is set to grow in the future. There are several key reasons for this [36]:

- Malaysia has a vibrant economy. It is forecasted that it will grow at ~5% per annum from 2013 to 2016, subject to the state of the global economy. These

Table 25.2 Local Malaysian Distribution Channels

Types	Retailers
Hypermarkets	Giant (Dairy Farm International Group)
	Cold Storage (Dairy Farm International Group)
	AEON Big (AEON Group)
	Tesco (UK–Malaysia joint venture)
	Mydin (local hypermarket)
	Econsave (local hypermarket)
Supermarkets	Jusco (AEON Group)
	The Store (local supermarket)

forecasts are very positive for growth in consumption of a wide range of foods and drinks, including that of functional food products.

- Malaysia's current standing is as a middle income nation that is moving toward becoming an advanced developing country. Private consumption is very important to the economy and comprises about 45% of GDP in 2011 [36]. Demand sophistication is also increasing, especially in the area of demands on quality, food health, safety, and nutrition. In most cases, this involves both urban and rural consumers, because rural consumers are more affluent in Malaysia than their counterparts in neighboring countries, such as Thailand and Indonesia.
- Malaysia's population is forecasted to grow to over 30 million persons by 2015.
- There is a sizable middle and upper income group (about 17 million persons; Stanton et al. 2011) with demand for a wide range of processed food and drinks, including health and wellness products of various types.

While there are positive drivers in the Malaysian market, there are several factors that are challenging the development of the market for imported food and drinks [36]:

- Weak value of the Ringgit (local currency) in the global currency markets
- Higher crude oil prices (affecting inward freight costs) and food commodity prices
- Imported retail and wholesale price inflation, especially in areas of the market that cannot be satisfied by local production.

25.7 Conclusion

There is an economically viable market for functional food in Malaysia. However, functional food is still a new concept in Malaysia and consumers are often confused with the definition of functional food. Once society increases its awareness of the specific food components that can affect the various parts of the body functions and the health benefits they provide, this could be a popular product in the future.

The success of functional foods is dependent on the ability of the food industry to develop efficacious products that meet consumer needs. The global potential for functional foods is significant and growing because of increasing health consciousness and the self-care trends associated with aging, knowledgeable, and wealthier consumers. Today's consumer expects foods to be convenient, safe, healthy, and above all tasty. Functional foods in particular are expected to provide a credible health benefit beyond basic attributes to ensure daily and future health. These health messages need to be communicated in a transparent, credible, and understandable manner to the different stakeholders, including expert scientists with a consumer focus, and the various opinion leaders, including health care providers, consumer organizations, and the media.

Governments have a major role to play in the future of functional foods by creating a favorable environment for basic and applied research programs, continuous and truthful consumer education on nutrition science, integration of public health issues, competitive and innovative economic development with ethical and ecological perspectives, the protection of consumers in the short and long term, and regulatory systems for flexible and credible science-based claims on nutrition and health. Furthermore, the government should strengthen the rules and regulations, which are related to food so that they can protect consumers from false claims and high prices made by direct sellers in Malaysia. At the same time, the government or private sector should promote healthy eating and lifestyle campaigns to educate the public and increase their awareness of the benefits of functional food. Finally, functional food science will create many opportunities, but its ultimate success and impact on public health will depend on the consumer's appreciation of products based on objective criteria like taste and convenience and subjective criteria like trust and credibility.

References

[1] O'Regan, E. Give Yourself a Boost with Functional Foods, Irish Independent. Monday, April 5th 1999.

[2] Ahmad S. Research and development on functional foods in Malaysia. Nutr Rev 1996;54(11):S169.

[3] DeBusk, R. (1998). Excerpted from Integrative Medicine: Your Quick Reference Guide.

[4] Kojima K. The Eastern consumer viewpoint: The experience in Japan. Nutr Rev 1996; 54(11):S186.

[5] Tee ES. Functional Foods in Asia: Current Status and Issues, Singapore: International Life Sciences Institute; 2004.

[6] Homer PM, Kahle LR. A structural equation test of the value-attitude-behavior hierarchy. J Pers Soc Psychol 1988;54(4):638−45.

[7] Scott JE, Lamont LM. Relating consumer values to consumer behavior: A model and method for investigation. In: Greer TW, editor. Increasing Marketing Productivity. Chicago: American Marketing Association; 1977. p. 283−8.

[8] Arai S. Global view on functional foods: Asian perspectives. Br J Nutr 2002;88 (2):139−43.

[9] Hasler CM. Functional foods: Their role in disease prevention and health promotion. Food Technol 1998;52(11):63−70.

[10] IFIC. International Food Information Council, <http://ificinfo.health.org>; 1998.

[11] Kuhn, MC. Nutraceuticals in the USA. Foodlink Forum, October 1997.

[12] Weststrate JA, van Poppel G, Verschuren PM. Functional foods, trends and future. Br J Nutr 2002;88(2):233−5.

[13] Hassan, SH. Functional Food Consumption in Multicultural Society, PhD thesis, Australian National University, Canberra; 2008.

[14] Hassan SH, Dann S, Mohd Kamal KA, Nicholls D. Market opportunities from cultural value convergence and functional food: The experiences of the Malaysian marketplace. In: Lindgreen A, Hingley M, editors. "The New Cultures of Food: Marketing Opportunities from Ethnic, Religious and Cultural Diversity". Aldershot: Gower; 2009. p. 223−42.

[15] Arshad F. Functional foods from the dietetic perspective. Jurnal Kesihatan Masyarakat Isu Khas 2002;8−13.

[16] Stanton, Emmsand Sia. Singapore's Markets for Functional Foods, Nutraceuticals and Organic Foods 2008 to 2012; 2008.

[17] Muhamad ZMAM. Traditional Malay Medicinal Plant. Kuala Lumpur: Fajar Bakti Sdn Bhd; 1991.

[18] Rainforest Herbs. The Wisdom of Mother Nature, Rainforest Herbs; 2005.

[19] Nandhasri P, Pawa KK, Kaewtubtim J, Jeamchanya C, Jansom C, Sattaponpun C. Nutraceutical properties of Thai Yor, Morinda citrifolia and Noni Juice extract. Songklanakarin J Sci Technol 2005;2:579−86.

[20] Wang MY, West JB, Jensen CJ, Nowicki D, Su C, Palu KA, Anderson G. Morinda citrifolia (Noni): A literature review and recent advances in noni research. Acta Pharmacol Sin 2002;23(12):1127−41.

[21] Weng W, Chen J. The Eastern perspective on functional foods based on traditional Chinese medicine. Nutr Rev 1996;54(11):S11−6.

[22] Yi D, Yong P, Wenkui L. Chinese Functional Food, Beijing: New World Press; 1999.

[23] Alagiakrishnan K, Chopra, A. Health and health care of Asian Indian elders, Curriculum in Ethnogeriatrics: Core Curriculum and Ethnic Specific Modules. available at:www.stanford.edu/group/ethnoger/; 2001.

[24] (The) Raj Maharishi Ayur-Veda Health Center. Ayurvedic Medicine, New York: Thomson PDR; 2004.

[25] Krishnaswamy K. Indian functional foods: Role in prevention of cancer. Nutr Rev 1996;54(11):S127.

[26] Douglas SP, Craig CS. The changing dynamic of consumer behavior: Implications for cross-cultural research. Int J Res Mark 1997;14(4):379−95.

[27] Finucane ML, Holup JL. Psychosocial and cultural factors affecting the perceived risk of genetically modified food: An overview of the literature. Soc Sci Med 2005; 60(7):1603−12.

[28] Steenkamp JBEM, Burgess SM. Optimum stimulation level and exploratory consumer behavior in an emerging consumer market. Int J Res Mark 2002;19(2): 131−50.

[29] Durvasula S, Andrews JC, Lysonski S, Netemeyer RG. Assessing the cross-national applicability of consumer behavior models: A model of attitude toward advertising in general. J Consum Res 1993;19(4):626−36.

[30] Hassan SH. Managing conflicting values in functional food consumption: The Malaysian experience. Br Food J 2011;113(8):1045−59.

[31] Hassan SH. Functional food consumption models for multicultural society: Malays, Chinese and Indians in Malaysia. Proceedings of the Academy of Marketing Conference, UK; 2011.

[32] Hassan SH. Consumption of functional food model for Malay Muslims in Malaysia. J Islam Mark 2011;2(2):104−24.

[33] Rezai G, Teng PK, Mohamed Z, Shamsudin MN. Functional food knowledge and perceptions among young consumers in Malaysia. World Acad Sci, Eng Technol 2012;63:307−12.

[34] Phuah KT, Rezai G, Mohamed Z, Shamsudin MN. Malaysian consumers' willingness-to-pay for functional food. Proceedings of 2nd International Conference on Management, 11-12 June, Langkawi, Malaysia; 2012.

[35] Lau TC, Chan MW, Tan HP, Kwek CL. Functional food: a growing trend among the health conscious. Asian Soc Sci 2012;9(1):198−208.

[36] Stanton Emms, Sia. Malaysia's Market for Functional Foods, Nutraceuticals and Organic Foods. An Introduction for Canadian Producers and Exporters,. South East Asia: Counsellor and Regional Agri-Food Trade Commissioner; 2011.

[37] Tee ES. Report of ILSI Southeast Asia Region Coordinated Survey of Functional Foods in Asia,. Southeast Asia: International Life Sciences Institute; 2007.

[38] National Pharmaceutical Control Bureau. Retrieved January 31, 2012, from <http://portal.bpfk.gov.my>; 2012.

[39] Ministry of Health. Guide to Classification of Food-Drug Interface Products. Retreived from <http://fsq.moh.gov.my/v3/images/filepicker_users/5ec35272cb78/Penerbitan/risalah/GuidetoFDIproducts.pdf>; 2012.

[40] FAO. The State of World Fisheries and Aquaculture (SOFIA) 2004. Italy: Food and Agriculture Organization; 2004.

[41] Malla S, Hobbs JE, Sogah EK. Functional Foods and Natural Health Products Regulations in Canada and Around the World: Nutrition Labels and Health Claims. Report prepared for the Canadian Agricultural Innovation and Regulation Network (CAIRN); 2013.

[42] Zawistowski, J. Regulation of Functional Food in Selected Asian Countries in the Pacific Rim. pp. 365−401; 2008.

[43] Niva M, Mäkelä J. Finns and functional foods: Socio-demographics, health efforts, notions of technology and the acceptability of health-promoting foods. Int J Consum Stud 2007;31(1):34−45.

World Trade Organization and Food Regulation: Impact on the Food Supply Chain

26

Okezie I. Aruoma

American University of Health Sciences, Signal Hill, California

26.1 Food regulation, supply chain, and the world trade organization

Food regulation is aimed at protecting the consumer's health, increasing economic viability, harmonizing well-being, and engendering fair trade on foods within and between nations. Parallel to this is the adage of the standard food security definition, which observes that the modern retailer-driven food supply chain has generally provided consumers with sustained physical and economic access to sufficient, safe and nutritious food. Many factors that affect the food systems include the climate, available arable land, and technology (from the standpoint of production, preservation, processing, and storage). A modern food control program centers on all stages of the food supply chain, from farm to consumers (Figure 26.1). There is a need to balance the benefits of increased food supplies through technological processes against associated health and economic risks, a need that is of increasing importance given the rising trend in world population, the opening up of boundaries, migration, urbanization, and, with these, changing food habits. Agricultural mechanization and industrial food processing advocated in Figure 26.1 embrace that foods produced and presented to the markets have to meet standards that are universally acceptable. Thus, food legislation ensures a safe supply of commodities and aspires to eliminate fraudulent practices. This is important given the diverse and changing trend of food commodities.

It is interesting that the World Trade Report 2012 examines why and how governments use non-tariff measures (NTMs), including domestic regulation in services. Such measures can serve legitimate public policy goals, such as protecting the health of consumers, but they may also be used for protectionist purposes. The report reveals how the expansion of global production chains and the growing importance of consumer concerns in richer countries affect the use of NTMs.

Nutraceutical and Functional Food Regulations in the United States and Around the World.
DOI: http://dx.doi.org/10.1016/B978-0-12-405870-5.00026-8

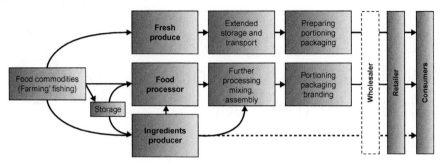

FIGURE 26.1

The food supply chain. This has seen a continued trend in technological innovations from foods leaving the farm and their handling up until reaching the end user, the consumer. Public concern about food safety is placing increasing pressure on government agencies to be more prescriptive and proactive in their regulation of the food industry. There is increased emphasis to optimize the logistics of delivering fresh produce, chilled products, dry groceries, and frozen products to retail outlets and to caterers.

It also reports that such measures represent the main source of concerns for exporters. The focus of the report is on technical barriers to trade (TBT) regarding standards for manufactured goods, sanitary and phytosanitary (SPS) measures concerning food safety and animal/plant health, and domestic regulation in services. While regulatory standards restrict trade in agricultural products, the World Trade Report 2012 finds the existence of standards often has a positive effect on trade in manufactured products, especially in high-technology sectors. Moreover, the harmonization and mutual recognition of standards is likely to increase trade. The report identifies several challenges for international cooperation, and the World Trade Organization (WTO) more specifically. First, the transparency of NTMs needs to be improved. The newly created WTO database Integrated Trade Intelligence Portal will help in improving transparency. Second, more effective criteria are needed to identify why a measure is used. Third, the increase in global production chains calls for deeper integration and regulatory convergence. Last, capacity building is a vital element in improving international cooperation.

Perishable food products, such as meats, fish, and vegetables (processed and packaged and ready-to-eat foods), are finding their way across international boundaries and meeting competition with indigenous manufacturers. Agricultural self-sufficiency ignores the relevance of the whole food chain and how the food chain itself might enhance or weaken food security. Using the standard food security definition, it is clear that the modern retailer-driven food supply chain has generally provided consumers with sustained physical and economic access to sufficient, safe, and nutritious food. Modern supply chains have vulnerabilities, but are not necessarily more risky than alternative, or historic, supply chain systems (the intricacy of this can be read by the various contributors in the segments in Figure 26.1). There may, however, be a trade-off between resilience and efficiency in the issue of redundant physical capacity. The competitiveness of markets, domestically and internationally, is of great importance. Many consumers worldwide worry that food derived from genetically modified organisms (GMOs) may be unhealthy and hence regulation on GMO authorizations and labeling have become more stringent. Today, there is a higher demand for non-GM products, and these products could be differentiated from GM products using the identity preservation system (IP) that could apply throughout the grain processing system. IP is the creation of a transparent communication system that encompasses hazard analysis at critical control points (HACCP), traceability, and related systems in the supply chain. This process guarantees that certain characteristics of the lots of food (non-GM origin) are maintained "from farm to fork." In pursuance of the risk assessment governing GMOs, manufacturers should be able to supply evidence to satisfy the competent authorities that they have taken appropriate steps to avoid the presence of adventitious or technically unavoidable materials. There may be a case to tolerate the presence of organisms that is inherent because of the nature and geographical origin of the crop, the current structure of the supply chain, and current industrial practices and premises, as far as the regulatory thresholds are met. The appropriate steps taken to avoid the adventitious or technically unavoidable presence of GM material will depend on, *inter alia*, the origin, nature, and composition of the food or food ingredient. Risk analysis has become important to assess conditions and make decisions on control procedures on their consideration of prerequisites in the evaluation of GM food [1].

The WTO, formed in 1995, is the only global international organization dealing with the rules of trade between nations. At its heart are the WTO agreements, negotiated and signed by the bulk of the world's trading nations and ratified in their parliaments. The goal is to help producers of goods and services, exporters, and importers conduct their business (Figure 26.2). These agreements are the legal ground rules for international commerce. Essentially, the agreements are contracts that guarantee Member Countries important trade rights. The instruments also bind governments to keep their trade policies within agreed limits to everybody's benefit. The agreements were negotiated and signed by governments, but their purpose is to help producers of goods and services, exporters, and importers conduct their business. Although the government's regulatory mechanisms oversee the analyses of public health problems and their association to the food supply, food safety is still mainly the responsibility of the consumer, and they need to be well informed.

The activities of the WTO can be summarized as follows:

- Administering trade agreements
- Acting as a forum for trade negotiations
- Settling trade disputes
- Reviewing national trade policies
- Assisting developing countries in trade policy issues through technical assistance and training
- Cooperating with other international organizations.

The World Trade Organization (WTO) is the only international organization dealing with the rules of trade between nations. WTO agreements are negotiated and signed by the bulk of the world's trading nations and ratified in their parliaments

The agreements are legal ground rules for international commerce that aims to guarantee member countries trading rights. In the same vein, trade policies are kept within an agreed limits to mutual benefits

The WTO's overriding objective is to help trade flow smoothly, freely, fairly and predictably. By lowering trade barriers, the WTO's system breaks down other barriers between peoples and nations

Agreements are intended to help producers of goods, exporters and importers conduct their business

FIGURE 26.2

The role of the WTO and international trade. The liberalization that has been achieved through GATT/WTO negotiations is especially noteworthy in light of the fact that negotiations occur through time between the governments of various countries. This feature raises the possibility that the market access implied by existing tariff commitments may be altered by tariff commitments made at some point in the future.

According to the 2012 WTO report:

The SPS and TBT agreements are "post-discriminatory" agreements. Although they include non-discrimination obligations, they also contain provisions that go beyond a "shallow integration" approach. They promote harmonization through the use of international standards and include obligations that are additional to the non-discrimination obligation. This includes, for instance, the need to ensure that requirements are not unnecessarily trade restrictive. Some question the appropriateness of these post-discriminatory obligations, arguing that the assessment of a measure's consistency with such requirements is difficult without WTO adjudicators "second-guessing" a member's domestic regulatory choices.

Further,

The challenges to international co-operation on NTMs. First, the transparency of NTMs must be improved and the WTO has a central role to play with its multiple transparency mechanisms. Secondly, more effective criteria are needed to identify why a measure is used. Better integration of economic and legal analysis may help achieve this goal. Thirdly, the increase in global production sharing poses additional challenges for the multilateral trading system, calling for deeper integration.

While it is clear that developed countries monitor their food systems with the expected outcome in providing a consistent and constant supply of safe and wholesome food, many developing countries battle with an unsafe and inadequate food supply. This leads to heavy economic losses (particularly scarce foreign exchange) and health hazards ranging from malnutrition to food-borne illnesses. Interestingly, while economic globalization can easily equate with the expansion of international trade, it is clear that some developing countries continue to struggle to become fully integrated into the world trading system. It is critical therefore to improve awareness, simplify rules, improve skills and infrastructure, adapt food safety monitoring to local conditions, and help formulate risk management systems for niche products from developing countries [2–7]. As argued by Henson et al. [5]: There are concerns about power relations through the supply chain; supermarket demand for high-value fresh produce can provide opportunities for the enhancement of small-scale producer livelihoods and, at the same time, attention has focused on the processes through which small-scale producers are integrated into, or excluded from, supermarket supply chains and the associated impact of stricter food safety and quality standards and logistics requirements.

Resilience itself is increasingly a commercial issue. Many of the risks involved are in firms' interests to guard against since this directly affects their business or reputation. Business continuity planning has grown in recent years, but there is potential for further improvement. Contingency planning by the government, and the need to work closely with the food industry, remains important to overcome any infrastructure, information, and coordination failures.

The confusion between quality and safety, over regulation, selective enforcement, lack of integration of food laws and regulations in the overall legislative system, the multiplicity of responsible agencies, and the mismatch between the required standards are among the major issues that the World Bank has come to realize. So poor economics, poor infrastructure, and lagging skills have negative impact on trade for the developing countries. Improving food safety along the standards of the developed economies, however, may carry considerable costs and price food out of reach of the poor. Indeed, improving awareness, simplifying rules, improving skills and infrastructure, adapting food safety monitoring to local conditions, and help in formulating management systems, e.g., HACCPs, for niche products for developing countries are within the context of seeking harmonization in world trade.

Among the identified factors affecting food safety compliance within small and medium-sized enterprises (SMEs) and the implications for regulatory and enforcement strategies in the UK include lack of money (where SMEs focus on immediate survival rather than potential benefits derived over the long term) and lack of time. Time has tended to prevent the identification and interpretation of regulations, thereby preventing further action taken by SMEs. More interesting, however, is the realization that food sector SMEs do not see these steps as part of their business operation [8]. This was viewed as the duty of the external agencies such as the Environmental Health Practitioners who inspect food businesses according to criteria contained within Code of Practice issued under Section 40 of the UK's Food Standards Agency Act. Yapp and Fairman [8] argued that it is this reactive attitude rather than a lack of time that prevents identification and interpretation of regulations by SMEs. Lack of experience, lack of access to information (seen as a problem with overprovision of information resulting in confusion about relevance), lack of support (where SMEs perceive that support is biased toward larger companies), lack of interest (where SMEs focus upon business survival rather than compliance with regulations), and lack of knowledge (where SMEs have poor awareness of the relevance of legislation) were noted. Yapp and Fairman [8] brought attention to the view that the principal objective of a compliance law enforcement system is to secure conformity with the law by means of ensuring compliance or by taking action to prevent potential law violations without the necessity to detect, process, and penalize violators. The principal objective of deterrence law enforcement systems is to secure conformity with law by detecting violations of law, determining who is responsible for their violation, and penalizing violators to deter violations in the future. It is clear that these considerations are not dissimilar from the experience of other SMEs in the world market.

The intent of the Global Agreement on Tariffs and Trade (GATT) has been one of streamlining international trade. In the early phases of GATT, most attention was given to trade and trade conflicts between large trading blocks and markets such as the United States, European Community, Japan, etc., and too little to the interests of smaller developing countries. So the developing countries (or the less developed countries) have been left to adopt the rules that were specifically created for these large markets and given little time to adjust their institutions to ensure that export products are in compliance with, to them exotic, food safety rules. Thus, the introduction and application of food safety tools and rules need to be affordable and build on local food management customs rather than simply imposing standards that are expensive to monitor [3,5,9–11]. Such an application has its benefit as Baker [12] observed: an HACCP-based food safety system that was integrated with restaurant policies, operations, documentation, and communication strategies could have avoided the costly exercise of removing a highly profitable menu item, the associated loss of market momentum, heightened regulatory scrutiny, and the potential to raise questions of consumer confidence. The WTO SPS agreement relates to three main issues: food safety, animal health, and plant health. In each case, the agreement identifies an international body as providing the basic standards against which disputes over national regulations would be judged.

Under the WTO, SPS agreements and the codes of practices issued by the Codex Alimentarius Commission constitute the benchmark for international harmonization, which guarantees the trade of safe food [7,13]. These need to be enacted and enforced in each country's food legislation.

The market access agenda in industrial countries extends to trade-impeding regulations such as environmental and health standards and restrictive rules of origin, as well as restrictions and regulations that limit the ability of developing countries to sell services abroad, especially through the temporary movement of workers [14]. It is clear that trade liberalization needs to be complemented by a number of other policies such as sound macroeconomic management, effective regulation (e.g., financial services), and improved customs and tax administration. This needs to embrace the policies and institutions to support social objectives and safeguard the interests of the poorest in society. Further, global trade rules need to be defined from a developmental perspective if they are to serve as tools for poverty reduction. In this context, capacity-building measures should include bolstering the ability of stakeholders in developing countries to participate in the development and implementation of trade-related policies and global trading rules.

Compliance with the decision of the WTO remains an area of concern given the tendency of developed and powerful countries continuing to threaten the survival of the WTO system through lengthy and costly legal interpretations and appeals of dispute settlements without serious efforts by all parties to find win−win solutions. The most flagrant trade violations have been eliminated as a result of improved transparency in the process. Consumers are, in general, better off since they have greater diversity and increased safety in what they can buy. The guidelines of the World Bank in its 2002 handbook on the magnitude of border barriers and trade liberalization, upon which the foregoing comments were based, can be summarized as seeking to foster:

- Effective market access that has a wider dimension than border restrictions on goods
- Reciprocal liberalization for developing countries that achieves improved access to markets abroad and greater openness at home, but there are major political economy constraints to be overcome
- Trade liberalization to form only a small part of the comprehensive domestic reforms that are needed to deliver poverty-reducing growth
- Better analysis of the costs and benefits of global trade rules for developing countries
- Integrating developing countries more effectively into the global economy.

26.2 Sanitary and phytosanitary agreements

Global harmonization of food safety regulations will undoubtedly help to ensure fair competition among countries in terms of trade and, at the same time, it will enable all populations to enjoy the same degree of food safety. This was one of the ideas behind the Uruguay Round of Multilateral Trade Negotiations, which resulted in the creation of the WTO in 1995, including a number of agreements, e.g., the agreement on the application of SPS and the agreement on Technical Barriers to Trade.

The Agreement on the Application of Sanitary and Phytosanitary Measures (the SPS agreement) came into force with the formation of the WTO 1995. The SPS agreement was aimed at controlling issues affecting food safety measures. Although some constraints had been applied by the original GATT in 1947, it was believed that various countries were using food safety concerns to justify maintaining or erecting food regulations against imported foods, which were a barrier to trade. One major qualification is contained in Article 3(3) of the SPS agreement which states that

> *Members may introduce or maintain sanitary or phytosanitary measures which result in a higher level of sanitary or phytosanitary protection than would be achieved by measures based on the relevant international standards, guidelines or recommendations if there is a scientific justification, or as a consequence of the level of sanitary or phytosanitary protection a Member determines to be appropriate in accordance with the relevant provisions...*

To help with interpretation, the term "scientific justification" was further defined as follows: there is a scientific justification if, on the basis of an examination and evaluation of available scientific information in conformity with the relevant provisions of this agreement, a Member determines that the relevant international standards, guidelines, or recommendations are not sufficient to achieve its appropriate level of sanitary or phytosanitary protection. The term "sanitary or phytosanitary measure" is defined as

> *any measure applied to protect human, animal, or plant life or health from certain risks, including risks arising from: (i) the spread of pests, diseases, disease-carrying organisms or disease-causing organisms, (ii) the presence of additives, contaminants, toxins or disease-causing organisms in foods, beverages or feeds and (iii) diseases carried by animals, plants or products thereof.*

The SPS agreement, therefore, includes a broad scope of activities related to food safety as well as the protection of animal and plant health. As such, it applies to all sanitary and phytosanitary measures that may affect international trade. Exporters must meet the quality and safety demanded by import market consumers. For this reason, international trading rules are in place to ensure that public standards are applied fairly and equally to domestic and imported products. WTO members supported the following SPS principles:

- Transparency
- Equivalence
- Harmonization

- Science-based measures
- Regionalization
- National sovereignty
- Dispute resolution.

Hence, nations are required to publish their regulations and provide a mechanism for answering questions from trading partners. Member nations must accept that SPS measures of another country are equivalent if they result in the same level of public health protection, even if the measures themselves differ. The same level of health protection should apply to both domestic and imported products. Regulations should be such that they cannot impose requirements that do not have a scientific basis for reducing risk. The concept of pest- or disease-free areas within an exporting country is recognized. Exports can be allowed from such areas, even if other areas of an exporting country still have the disease or pest. Countries may choose a risk standard that differs from the international standard. This recognizes that individual nations are unwilling to subscribe to uniform international standards for all hazards. There is a clearly defined mechanism for resolving disputes between countries in a timely manner. The dispute settlement panel is expected only to state whether the SPS measures under question have a scientific basis and are consistently applied. Further, Member Nations recognize the desirability of common SPS measures.

Three international organizations are recognized as sources of internationally agreed-upon standards:

1. The Codex Alimentarius Commission
2. The International Office of Epizootics
3. The International Plant Protection Convention

The Codex Alimentarius develops food safety standards, which serve as a reference for international food trade. This was set up in the 1960s as a joint instrument of the United Nations Food and Agriculture Organization and the World Health Organization with the primary mission to protect the health of consumers and ensure fair practices in international food trade. Thus, the Codex Alimentarius Commission adopted standards for commodities, codes of practice, and maximum limits for additives, contaminants, pesticides residues, and veterinary drugs. Following the conclusion of the Uruguay Round in 1994, the role of Codex Alimentarius Standards was strengthened. The WTO agreement on SPS measures considers that WTO members applying the Codex Alimentarius standards meet their obligations under this agreement. Additional information and updates can be obtained from http://www.wto.org.

Traditional trade protection has been reduced by the 1994 GATT agreement, which means that SPS measures assume greater importance in determining market access [4,7,15−17]. The challenges and issues of food safety standards for export across trade barriers impinges on the following context:

1. Importance of fresh food product trade by region and the kinds of issues that arise from those products

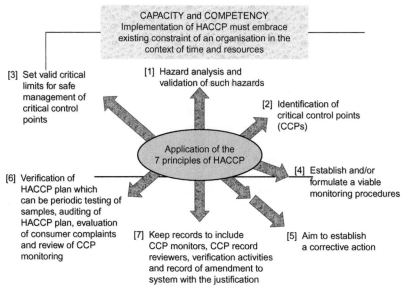

FIGURE 26.3

Application of the principles of HACCP. Food supply chain management systems in ensuring safety and quality are interfaced by a prerequisite requirement (e.g., GMP, GHP), food safety assurance plan (e.g., HACCP plan), a quality system, and a cultural/managerial approach (e.g., ISO 9000, TQM, etc.).

2. Role of farm-to-table approaches and HACCP (Figure 26.3) in ensuring safety
3. Role of the public sector in WTO Member Countries and the less developed countries in facilitating trade
4. Potential role of the SPS agreement in resolving disputes and determining equivalency of standards between high- and low-income countries

26.3 HACCP

The health status of a population can be evaluated by use of the microbiological risk assessment (MRA) for a product or product group to which a pathogen is associated. On a population basis, a calculation of risk can predict the expected number of specific illnesses or deaths per 100,000 population per year attributable to the pathogen/food in question, or risk can be defined as the probability of a specific adverse outcome per exposure to the food. An MRA can give an absolute or a relative indication of the health status, i.e., provide an absolute numerical expression of the risk at population level, respectively, a relative or benchmarked expression (e.g., a ranking). This applies whether the food product

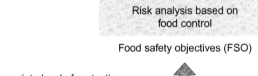

Food safety controls and food safety management.
Government's regulatory mechanisms oversees the
analyses of public health problems and their
association to the food supply chain

FIGURE 26.4

Risk assessment criteria. ALOP is the level of protection deemed appropriate by the
Member (country) establishing a sanitary or phytosanitary measure to protect human,
animal, or plant life or health within its territory. FSO refers to the maximum frequency
and/or concentration of a hazard in a food at the time of consumption that provides or
contributes to the ALOP. FSO is just one of the options to give guidance to food safety
management the expected management of risks. Performance objective (PO) is the
maximum frequency and/or concentration of a hazard in a food at a specified step in the
food chain before the time of consumption that provides or contributes to an FSO or
ALOP, as applicable. Performance criterion (PC) is the effect in frequency and/or
concentration of a hazard in a food that must be achieved by the application of one or
more control measures to provide or contribute to a PO or an FSO. Control measure (CM)
is any action and activity that can be used to prevent or eliminate a food safety hazard or
to reduce it to an acceptable level (it can be microbiological specifications, guidelines on
pathogen control, hygiene codes, microbiological criteria, specific information, e.g.,
labeling, training, education, and others).

originates from one country or is imported into it. Applicable definitions in risk
analysis-based food control are presented in Figure 26.4.

The food industry is not only responsible for producing safe food but also for
demonstrating in a transparent manner how food safety has been planned, for
example, hazards that have been considered in production and the measures that
have been put in place to ensure the safety of products. This is done through the
development of HACCP studies and HACCP plans as part of the food safety
assurance system. HACCP is a straightforward and logical system of control

based on the prevention of problems—a common sense approach to food safety management.

HACCP merits in food safety management can be realized only if the people charged with its implementation have the knowledge and expertise to apply it effectively. This could be a combination of horizontal and vertical partnerships, as in the poultry industry. The resulting network involved all kinds of parties in the industry from breeding farms down to the packing stations, along with feed producers, veterinarians, and a quality service organization. The need is to guarantee the quality of poultry products to the consumer and to provide information for associated transparency. Seminal discussions on the workings of HACCP and application to several food sectors are widely reviewed in the literature (e.g., [18–25]; http://www.codexalimentarius.net, http://www.wto.org, and http://www.fao.org).

"Hazard" is "a biological, chemical or physical agent in, or condition of, food with the potential to cause an adverse health effect" [19]. Hazard analysis, therefore, requires that both the likelihood of occurrence and severity of that hazard are considered, in effect an assessment of risk. Validation of the hazard analysis is an important element and, probably, the key principle in the whole HACCP system and the one which many find difficult to apply. Critical control points (CCPs) can be designated a prerequisite so as not to undermine the whole process. For example, failure to wash hands if not designated as a prerequisite hygiene program can create problems if regarded as a CCP and it is not adhered to. It is no accident that HACCP evolved at the food-processing step of the farm-to-table supply chain (see Figure 26.1). It is at this step that effective controls, such as cooking, drying, acidification, or refining, are available to eliminate significant hazards. Two typical examples are pasteurized dairy products and canned foods. Here food safety is ensured by process control, not by finished product testing. HACCP is of critical importance to the food service sector as it helps to ensure that the whole production line of the food chain is acceptable, which is necessary to improve public health. For HACCP to be effective when targeting the specific needs of the retail food establishment, it must be compatible with the products sold, the clients served, and the facilities and equipment used during food production. As argued by Sperber [23]:

> HACCP cannot provide greater transparency in the food supply chain in the context of this type of opaque regulatory environment. Rather, greater transparency, and improved public health protection, must be realized through the development of voluntary science based systems, especially involving the food processing industry, where the very idea of HACCP was conceived and implemented.

Regulatory action is an appropriate response when food markets fail to deliver the level of safety required to satisfy social public health goals.

> In the field of food safety economics, social welfare analysis of policies focuses on the regulation of markets to increase social welfare (i.e. improvements in public health) in situations where markets fail, while the political economy

(private) approach focuses on the position of interest groups in the process of regulation [26].

Food safety management as in good manufacturing practices (GMP), good hygiene practices (GHP), and HACCP provisions are specific to the available facility, the processing line, and the exact product composition and processing. So, for a specific food product, microbiological risk assessment considers all foods consumed in a country, whether produced in that country or imported; it involves all different production facilities, a multitude of production lines and product compositions, and processing. MRA takes a generic, population-level view on the overall production and marketing of a food product. Risk assessment is a science-based investigation consisting of four steps: hazard identification, exposure assessment, hazard characterization, and risk characterization as outlined in Figure 26.5, which is based on the framework adopted by the Codex Alimentarius Commission [27].

Hazard identification identifies the issues of concern and provides the focus of the risk assessment. The exposure assessment generates estimates of the likelihood and magnitude of exposure to the hazard, setting the stage for the next two steps of the assessment, hazard characterization and risk characterization, in which the exposure outputs are translated into a measure of risk [27].

Policy makers may argue that the primary responsibility for food safety lies with the private sector, whereas the definition of basic standards, monitoring, and policing are the responsibility of the public sector. It is clear that the heightened concern for food safety has meant that both public and private regulations and activities are jointly instrumental in the delivery of a safer food supply [21−33]. Governmental risk managers may choose to implement specific risk management measures (standards, microbiological criteria, hygiene code, labeling, education, etc.) in addition to a food safety objective (FSO). Such measures may be relevant to all or the majority of the supply chain, so they should be included in all cases. Alternatively, such measures may be essential additions to the target without which the appropriate level of protection (ALOP) may not be met. FSO is just one of the options to guide food safety management through the expected management of risks. As there are often many links in a food supply chain (see Figure 26.1), it is clear that establishing and/or defining several operational targets along the chain will help ensure that the chain as a whole operates to meet the FSO at consumption. HACCP is a major contributor to overall quality assurance system in international food trade [30−33]. The role of governments and of the WTO regarding private standards also needs clarification. The 2012 WTO report [32] is worth perusing. The term non-tariff measures may include any policy measures other than tariffs that can impact trade rows. NTMs may be imposed on imports and exports at the border. For imports, this may extend to import quotas, import prohibitions, import licensing, and customs procedures and administration fees. The category of NTMs that covers exports may extend to

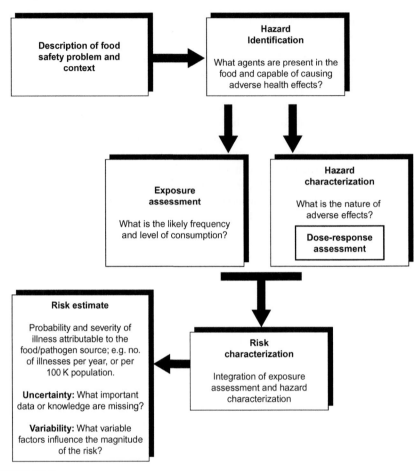

FIGURE 26.5

Risk assessment is a science-based investigation consisting of four steps: hazard identification, exposure assessment, hazard characterization, and risk characterization in accordance with the framework adopted by the Codex Alimentarius Commission [33].

export taxes, export subsidies, export quotas, export prohibitions, and voluntary export restraints. The behind-the-border measures include domestic legislation covering health/technical/product/labor/environmental standards, internal taxes or charges, and domestic subsidies, so this example is imposed internally in the domestic economy. Negotiations on domestic regulation in services have turned out to be very difficult to conclude, mainly because of concerns with regulatory autonomy. Last, capacity building could make a more significant contribution to improving international cooperation on public policies.

Acknowledgment

Adapted from a review article Aruoma O.I. (2006). The impact of food regulation on the food supply chain. *Toxicology* 221, 119–127. Updated from the review article Aruoma O.I. (2008). World Trade Organization and food regulation: impact on the food supply chain. In *Nutraceutical and Functional Food Regulations*, First Edition, Elsevier Inc.

References

[1] Zepeda C, Salman M, Ruppanner R. International trade, animal health and veterinary epidemiology: challenges and opportunities. Prev Vet Med 2001;48:261–71.

[2] Anyanwu RC, Jukes J. Food systems and food control in Nigeria. Food Pol 1991;16:112–26.

[3] Brown CG, Longworth JW, Waldron S. Food safety and development of the beef industry in China. Food Pol 2002;27:269–84.

[4] Henson S, Loader R. Barriers to agricultural exports from developing countries: the role of sanitary and phytosanitary requirements. World Dev 2001;29:85–102.

[5] Henson S, Masakure O, Boselie D. Private food safety and quality standards for fresh produce exporters: The case of Hortico Agrisystems, Zimbabwe. Food Pol 2005; 30:371–84.

[6] Juke D. Developing a food control system. The Tanzanian experience. Food Pol 1988;13:298–304.

[7] Jukes D. The role of science in international food standards. Food Control 2000; 11:181–94.

[8] Yapp C, Fairman R. Factors affecting food safety compliance within small and medium-sized enterprises: implications for regulatory and enforcement strategies. Food Control 2006;17:42–51.

[9] Chen J. Challenges to developing countries after joining WTO: risk assessment of chemicals in food. Toxicology 2004;198:3–7.

[10] Key N, Runsten D. Contract farming, smallholders and rural development in Latin America: the organization of agro-processing firms and the scale of out-grower production. World Dev 1999;27:381–401.

[11] Nguz K. Assessing food safety system in sub-Saharan countries: an overview of key issues. Food Control 2007;18:131–4.

[12] Baker DA. Use of food safety objectives to satisfy the intent of food safety law. Food Control 2002;13:371–6.

[13] Boutrif E. The new role of Codex Alimentarius in the context of WTO/SPS agreement. Food Control 2003;14:81–8.

[14] World Bank. Development, trade and the WTO. A Handbook. Washington, DC: The World Bank; 2002.

[15] Kastner JJ, Pawsey RK. Harmonising sanitary measures and resolving trade disputes through the WTO-SPS framework. Part I: a case study of the US-EU hormone-treated beef dispute. Food Control 2002;13:49–55.

[16] Kastner JJ, Pawsey RK. Harmonising sanitary measures and resolving trade disputes through the WTO-SPS framework. Part II: a case study of the US-Australia determination of equivalence in meat inspection. Food Control 2002;13:57–60.

[17] Unnevehr LJ. Food safety issues and fresh food product exports from LDCs. Agric Econ 2000;23:231−40.

[18] Azanza MPV. HACCP certification of food services in Philippine. inter-island passenger vessel. Food Control 2006;17:93−101.

[19] Codex Alimentarius Commission. Hazard analysis and critical control point (HACCP) system and guideline for its application. Annex to CAC/RCP 1997, 1−1969. Rev 3.

[20] Jeng H-YJ, Fang TJ. Food safety control system in Taiwan. The example of food service sector. Food Control 2003;14:317−22.

[21] Mortimore S. How to make HACCP really work in practice. Food Control 2001;12:209−15.

[22] Sperber WH. Auditing and verification of food safety and HACCP. Food Control 1998;9:157−62.

[23] Sperber WH. HACCP and transparency. Food Control 2005;16:505−9.

[24] Sperber WH. HACCP does not work from farm to table. Food Control 2005; 16:511−4.

[25] Sun Y-M, Ockerman HW. A review of the needs and current applications of hazard analysis and critical control point (HACCP) system in foodservice areas. Food Control 2005;16:325−32.

[26] Martinez MG, Fearne A, Caswell JA, Henson S. Co-regulation as a possible model for food safety governance: opportunities for public−private partnerships. Food Pol 2007;32:299−314.

[27] Lammerding AM, Fazil A. Hazard identification and exposure assessment for microbial food safety risk assessment. Int J Food Microb 2000;58:147−57.

[28] Jukes D. Regulation and enforcement of food safety in the UK. Food Pol 1993; 18:131−42.

[29] Varzakas TH, Chryssochoidis G, Argyropoulos D. Approaches in the risk assessment of genetically modified foods by the Hellenic Food Safety Authority. Food Chem Toxicol 2007;45:530−42.

[30] WTO. Agreement on the Application of Sanitary and Phytosanitary Measures. World Trade Organization; 1995, <www.wto.org>.

[31] WTO. Review of operation and implementation of the Agreement on the Application of sanitary and Phytosanitary Issues. World Trade Organization; 1999, <www.wto.org>.

[32] WTO. World Trade Report 2012. Trade and public policies: A closer look at non-tariff measures in the 21st century. <www.wto.org>; 2012.

[33] Lammerding AM, Fazil A. Hazard identification and exposure assessment for microbial food safety risk assessment. Intl J Food Microbiol 2002;58:147−57.

Regulations on Pet Food

Functional Ingredients in the Pet Food Industry: Regulatory Considerations

Nikita McGee, Jennifer Radosevich and Nancy E. Rawson

AFB International, St. Charles, Missouri

27.1 Introduction

The incorporation of nutraceutical and functional ingredients into pet foods, treats, and supplements has accelerated in recent years. As veterinary care and diet quality has improved, pet owners are faced with the care of aging pets, and the desire to maximize their quality of life [1]. While many functional ingredients have begun to appear in pet foods based on their purported benefits in human foods, a number of products have entered the market containing functional ingredients such as omega-6 fatty acids and antioxidants, which are supported by clinical studies to promote cognitive or other clinical benefits in companion animals [2−5]. These trends reflect those occurring in the human food industry, but the regulations and considerations governing ingredient use in the pet food industry differ from those governing the human food industry. Accordingly, the manufacturer must consider both the relevance of supporting research to the species being targeted as well as the appropriate regulatory requirements.

The pet food industry is regulated worldwide with many countries mandating specific rules and requirements for pet food ingredients. The rules and regulations for most countries are published and accessible to the general public. However, successful interpretation of the regulations is greatly impacted by knowledge and prior experiences. The information provided here is generally applicable for all classes of ingredients.

This overview is intended to provide general information as it pertains to the regulation of pet food ingredients in the United States, Canada, European Union (EU), China, and Latin America. Resource information is also provided to direct the reader to detailed information.

Nutraceutical and Functional Food Regulations in the United States and Around the World.
DOI: http://dx.doi.org/10.1016/B978-0-12-405870-5.00027-X

Table 27.1 Regulatory Body (Agency) Per Country

United States	FDA: Food and Drug Administration
	USDA: United States Department of Agriculture
Canada	CFIA: Canadian Food Inspection Authority
EU	EC: European Commission
China	MOA: Ministry of Agriculture
Latin America	SAGARPA—SENASICA: National Agro-Alimentary Health, Safety, and Quality Service

27.2 Regulatory bodies

Canada, EU, China, and Latin America have one main regulatory body governing pet food ingredients (see Table 27.1). The United States serves as an exception as two regulatory bodies govern the pet food and pet food ingredient industries.

27.3 Regulatory approval of the manufacturing facility

The majority of countries require that pet food and pet food ingredient companies obtain a license or permit to operate within their country. At a minimum, countries require foreign manufacturing facilities to register with the regulatory body. In addition to registration, some countries require foreign facilities to obtain facility approval and maintain the certification via a facility renewal process (see Table 27.2).

Table 27.2 Regulatory Facility Approval Per Country

United States	Domestic and foreign facilities: Registration of each facility via the Food Facility Registration (FFRM) program.
	State level: Some states require additional license and/or inspection.
Canada	Domestic facilities: Registration of the Canadian facility
	Foreign facilities: No formal registration process.
	For rendering plants, a CFIA Facility Questionnaire must be completed for each Canadian customer; the Questionnaire must be signed by the foreign facility's regulatory body to allow import into Canada.
EU	Domestic facilities: License required to operate
	Foreign facilities: Annual inspection by the foreign facility's regulatory body to ensure compliance with EU regulations; inspection conducted by USDA.
China	Domestic facilities: License required to operate
	Foreign facilities: Annual inspection by the foreign facility's regulatory body to ensure compliance with EU regulations
Latin America	Domestic facilities: License required to operate
	Foreign facilities: Registration rules vary depending upon the characteristics of the product produced at the facility

27.4 **Regulatory approval of the pet food ingredient**

The United States, Canada, EU, and China have identified ingredients that are approved for use in their country as it pertains to the pet food industry. The approved ingredients vary per country as well as the frequency in which the lists are updated (see Table 27.3).

In addition to the use of approved ingredients, China and Latin America require the pre-market authorization of pet food and pet food ingredients for the further manufacture of pet food.

Table 27.3 Regulatory Ingredient Approval List(s) Per Country

United States	Pet food ingredients must consist of approved ingredients.
	States require compliance with AAFCO.
	Approved-Ingredient Lists
	AAFCO: Association of American Feed Control Officials
	Note: Not publically available, fee required for access
	www.aafco.org
	GRAS: Generally recognized as safe
	21 CFR 582: Substances Generally Recognized as Safe (GRAS)
	http://www.ecfr.gov
	21 CFR 570; 21 CFR 571; 21 CFR 573: Food Additives
	http://www.ecfr.gov
Canada	Pet food ingredients must consist of approved ingredients.
	Approved-Ingredient List
	CFIA Feeds Regulations (Schedule IV and V)
	http://laws-lois.justice.gc.ca/eng/regulations/SOR-83-593/index.html
EU	Pet food ingredients must consist of approved ingredients.
	Approved-Ingredient List
	EU Register of Feed Additives: EC No. 1831/2003
	List:
	http://ec.europa.eu/food/food/animalnutrition/feedadditives/comm_register_feed_additives_1831-03.pdf
	Regulation:
	http://eur-lex.europa.eu/LexUriServ/LexUriServ.do?uri = OJ:L:2003:268:0029:0043:EN:PDF
China	Pet food ingredients must consist of approved ingredients.
	Product requires pre-market authorization submission to regulatory body.
	Approved-Ingredient List
	http://www.moa.gov.cn
Latin America	Pet food ingredients must consist of approved ingredients.
	Product requires pre-market authorization submission to regulatory body.
	Approved Ingredient List
	Access restricted to business operators

The new pet food ingredient approval process is difficult to summarize on a broad scale as the differences per country and ingredient category are vastly different. Links are provided as a resource to obtain detailed information for each country.

27.5 Globally accepted pet food ingredients

Although the approved-ingredients lists vary by country, many ingredient categories are common among countries such as vitamins, preservatives/acidity regulators, amino acids, and flavorings. Examples of globally accepted pet food ingredients in the United States, Canada, EU, China, and Latin America in each of these categories are provided as follows.

- Vitamins: Vitamin C (ascorbic acid), Vitamin E (tocopherols)
- Preservatives/acidity regulators: Benzoic acid, potassium sorbate, citric acid
- Amino acids: Methionine, thiamine
- Flavorings: Natural flavors such as rosemary

27.6 Regulation trends in pet food ingredients

Knowledge of the characteristics of each pet food ingredient is becoming more important to the industry as consumers take interest in product trends. Regulations are quickly evolving to include legislation regarding the use of genetically modified organism (GMO) ingredients, determination of gluten status, and the definition of natural.

27.6.1 GMO

Restrictions regarding use of GMO ingredients are controlled on a state level in the United States. The addition of such legislation is becoming more popular and continually being debated in the United States. The majority of states differ in their definition of GMO and the restrictions on usage in the pet food industry. For this reason, compliance with GMO legislation in the United States may become quite challenging in the near future.

Pet food ingredients for use in the EU currently require non-GMO declaration. Canada, China, and Latin America have not entered GMO restrictions into legislation.

27.6.2 Other trends in regulations

There is much activity in the United States regarding the definition and appropriate uses for the terms gluten free and natural. The Food and Drug Administration

(FDA) recently published the definition for the term gluten free. However, regulatory bodies in the United States have not defined the term natural.

27.6.3 **Health claims regulations**

Within the United States, the Federal Food, Drug, and Cosmetic Act (FFDCA) requires that all animal foods, like human foods, be safe to eat, produced under sanitary conditions, contain no harmful substances, and be truthfully labeled. The Center for Veterinary Medicine (CVM)–FDA reviews specific claims on pet food, such as "maintains urinary tract health," "low magnesium," "tartar control," "hairball control," and "improved digestibility." Guidance for collecting data to make a urinary tract health claim is available in Guideline for Industry 55 on the CVM portion of the FDA internet site. In addition to these regulations, the American Veterinary Medical Association (AVMA) encourages the pet food industry to act responsibly by only making health or therapeutic claims that are supported by quality scientific evidence. Veterinarians should assess relevant product information through principles of evidence-based medicine prior to using or recommending wellness or therapeutic pet foods. In the interest of pet safety, the AVMA recommends that the FDA requires all pet food products with implied or explicit health or drug claims to include a prominent statement on the label indicating that these claims have not been evaluated by the FDA.

Links

International Feed Industry Association
www.ifif.org
Association of American Feed Control Officials
www.aafco.org
American Feed Industry Association
www.afia.org
Food and Drug Administration [United States]
www.fda.gov
FFDCA [United States]
http://www.fda.gov/AnimalVeterinary/Products/AnimalFoodFeeds/PetFood/default.htm
United States Department of Agriculture [United States]
http://www.aphis.usda.gov/
Canadian Food Inspection Authority [Canada]
www.inspection.gc.ca
European Commission [EU]
http://ec.europa.eu/index_en.htm
Ministry of Agriculture [China]
www.moa.gov.cn

SAGARPA—SENASICA: National Agro-Alimentary Health, Safety, and Quality Service [Latin America]
http://www.senasica.gob.mx/
Center for Veterinary Medicine, FDA
http://www.fda.gov/AnimalVeterinary/
CVM Guideline for Industry 55:
http://www.fda.gov/AnimalVeterinary/GuidanceComplianceEnforcement/GuidanceforIndustry/ucm053415.htm

References

[1] Taylor J. Antioxidant Update: Adding more than shelf life to petfood. Petfood Industry. June 04, 2012. <http://www.petfoodindustry.com/PrintPage.aspx?id = 46274>.

[2] Siwak CT, Tapp PD, Head E, Zicker SC, Murphey HL, Muggenburg BA, et al. Chronic antioxidant and mitochondrial cofactor administration improves discrimination learning in aged but not young dogs. Prog Neuro-Psychopharmacol Biol Psychiatry 2005;29(3):461—9.

[3] Milgram NW, Head E, Zicker SC, Ikeda-Douglas C, Murphey H, Muggenberg BA, et al. Long-term treatment with antioxidants and a program of behavioral enrichment reduces age-dependent impairment in discrimination and reversal learning in beagle dogs. Exp Gerontol 2004;39(5):753—65.

[4] Cotman CW, Head E, Muggenburg BA, Zicker S, Milgram NW. Brain aging in the canine: a diet enriched in antioxidants reduces cognitive dysfunction. Neurobiol Aging 2002;23(5):809—18.

[5] Zicker SC, Wedekind KJ, Jewell DE. Antioxidants in veterinary nutrition. Vet Clin Small Anim 2006;36:1183—98.

Validation Approach

Validation Approach in Nutraceutical Industry

28

Chandra S. Yeevani and Deb Kumar Nath

Global QA, Apotex Inc, Toronto

28.1 Background

The process of establishing documented evidence through testing a specific process, equipment, facility, or utility system consistently produces an output meeting its predetermined specifications and quality attributes.

Validation is part of good manufacturing practices (GMPs) within the regulatory framework of the Food and Drug Administration (FDA), European Union (EU), International Conference on Harmonization (ICH), and World Health Organization (WHO) current GMP (cGMP) guidelines. It is mandatory to comply with regulations for a pharmaceutical drug product or drug substance manufacturer. For a nutraceutical manufacturer it is not mandatory to follow similar guidelines as that for a drug substance manufacturer. However, due to widespread uptake of dietary supplements and natural products by consumers, and its inherent health risks because of the introduction of new ingredients, the combination of ingredients, formulations, route of administration, and lack of adequate patient awareness while consuming along with other pharmaceutical products, it is desired to have better control over their processes. The nutraceutical industry has been steadily growing over the past 10 years, subjecting it more to FDA initiatives for ensuring public health and safety.

In between drug and food products, with direct effect on illness treatment, the nutraceutical industry is getting hard pressed by the FDA to implement GMPs because of the similarities between the pharmaceutical and nutraceutical operations.

This chapter provides guidance on the minimum validation requirements based on industry practice and available regulations on this topic.

28.2 Validation approach: V-model

Figure 28.1 depicts the V-model used for critical systems requiring qualification [1]. Flow of the qualification studies and inter-relationships with each other are also illustrated.

Nutraceutical and Functional Food Regulations in the United States and Around the World.
DOI: http://dx.doi.org/10.1016/B978-0-12-405870-5.00028-1

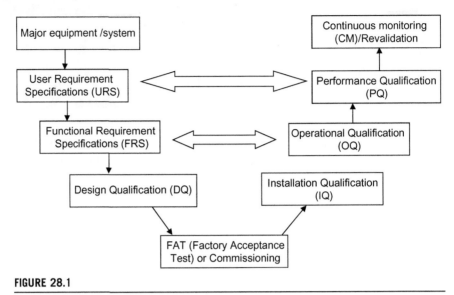

FIGURE 28.1

This concept is taken from the ISPE—Good Automated Manufacturing Practice life cycle approach. However, it was modified to suit the pharmaceutical and food industries.

The qualification and validation process should establish and provide documented evidence that the premises, the supporting utilities, the equipment, and the processes have been designed in accordance with the requirements of GMP. This normally constitutes design qualification (DQ). The premises, supporting utilities, and equipment have been built and installed in compliance with their design specifications. This constitutes installation qualification (IQ). The premises, supporting utilities, and the equipment operate in accordance with their design specifications. This constitutes operational qualification (OQ). A specific process will consistently produce a product meeting its predetermined specifications and quality attributes. This constitutes performance qualification (PQ).

28.2.1 Validation master plan

A validation master plan (VMP) is a document that summarizes the firm's overall philosophy, intentions, and approach to be used for establishing validation policy and presents an overview of the entire validation operation, its organizational structure, its content, and its planning. The core of the VMP is the list/inventory of the items to be validated and the validation schedule.

The qualification and validation process should establish and provide documented evidence that the premises, the supporting utilities, the equipment, and the processes have been designed in accordance with the requirements of GMP. This normally constitutes DQ. The premises, supporting utilities, and the equipment have been built and installed in compliance with their design specifications.

This constitutes IQ. The premises, supporting utilities, and the equipment operate in accordance with their design specifications. This constitutes OQ. A specific process will consistently produce a product meeting its predetermined specifications and quality attributes. This constitutes process validation (PV).

The purpose of the VMP is shown here, but not limited to the following:

1. *Management*: Know what the validation program involves with respect to time, people, and money, and understand necessity of the program
2. *Validation team*: Know their tasks and responsibilities
3. *Regulatory inspections*: Understand the firm's approach to validation and the set up and organization of all validation activities

All validation activities should be summarized and compiled in a matrix format. The matrix should provide an overview of the content of the activities.

The VMP format and content should include [2]:

- Introduction: validation policy, scope, location, and schedule
- Organizational structure: personnel responsibilities
- Plant/process/product description: rationale for inclusions or exclusions and extent of validation
- Specific process considerations that are critical and those requiring extra attention
- List of products/processes/systems to be validated, summarized in a matrix format using the validation approach
- Re-validation activities, actual status, and future planning
- Key acceptance criteria
- Documentation format
- Reference to the required standard operating procedures (SOPs)
- Time plans of each validation project and subproject.

28.2.2 User requirements specification

A user requirement specification (URS) is a requirement specification that describes what the equipment or system is supposed to do, thus containing at least a set of criteria or conditions that have to be met. This document can be in the form of a template or protocol based on the complexity of the equipment or system design and approved by the system owner as well as by the Quality Assurance (QA) group.

URS should contain following items, but not be limited to the following:

- Introduction
- Operational requirements
- Process controls
- Facility requirements
- Constrains

- Life cycle: development, testing [Facility Acceptance Test (FAT) requirements], delivery, support, and warranty
- Glossary
- References
- Drawings/layouts

28.2.3 **FAT**

Facility acceptance testing may be performed on select equipment and systems based on criticality of the manufacturing process. Protocol detailing all the aspects of FAT should be established and pre-approved by the customer as well the manufacturer. Upon completion of the tests as per pre-approved protocol the results should be reviewed and approved prior to shipment. Factory acceptance testing results and reports can leverage subsequent validation activities, for example, IQ/OQ.

Site acceptance testing may be performed on select systems based on criticality to the manufacturing process and extent of customization. Also, site acceptance testing can be performed to verify that the equipment is installed and operates according to the intended use.

28.2.4 **Validation protocols**

The validation protocol should be numbered, signed, and dated, and should contain at a minimum the following information, but not limited to [2,3]

- Objectives, scope of coverage of the validation study
- Validation team membership, their qualifications, and responsibilities
- Type of validation: prospective, concurrent, retrospective, re-validation
- Number and selection of batches to be included in validation study
- A list of all equipment to be used; their normal and worst case operating parameters
- Outcome of IQ, OQ for critical equipment
- Requirements for calibration of all measuring devices
- Critical process parameters and their respective tolerances
- Description of the processing steps: copy of the master documents for the product
- Sampling points, stages of sampling, methods of sampling, sampling plans
- Statistical tools to be used in the analysis of data
- Training requirements for the processing operators
- Validated test methods to be used in in-process testing and for the finished product
- Specifications for raw and packaging materials and test methods
- Forms and charts to be used for documenting results
- Format for presentation of results, documenting conclusions and for approval of study results.

28.2.5 IQ

During IQ, a complete analysis of the system prior to start up must be performed. The protocol should provide a systematic method to check the critical system attributes prior to or following commissioning activities with a predetermined acceptance criteria.

A detailed discussion of the system will be written. This discussion should include a description of the purpose of the system and all important major/minor components of the system. A list of applicable SOPs will be verified. This includes system operation, maintenance, cleaning, and/or sanitization. The system design must be reviewed during or after commissioning to verify that the system installed is the same as the one specified in URS or DQ. System engineering drawings (P&ID), manuals, data sheets, and specifications will be used to docu ment proper component installation and placement.

The system must be evaluated for proper connection and installation of critical supporting services (e.g., water, steam, electric, etc.) and components (e.g., filters, coils, piping, valves, gauges, controls, etc.). Calibration or accuracy checks of control, monitoring, and recording instrumentation (e.g., pressure gauges, temperature sensors, timers, etc.) that could impact product quality will be documented. Calibration requirements include confirmation of calibration of calibrating equipment with reference to the appropriate national standards.

IQ may be conducted simultaneously with commissioning activities, but cannot be concluded prior to the successful completion of commissioning. Preventative maintenance requirements should be documented for installed equipment. At this stage new equipment and the preventative maintenance requirements should be added to the preventative maintenance schedule. Any deviations encountered during the IQ that could impact process integrity or product reproducibility should be identified, investigated, and documented (including justification, correction, and any necessary re-qualification studies).

Installation qualification protocol should include following elements, but not be limited to the following:

Introduction of the system/equipment/facility

* A process description
* Overview of the system/equipment (functional description)
* Critical instruments and devices (list of major equipment and instruments)
* Risk assessment

Test Plans should inlude following verification items, but not be limited to the following:

* Pre-execution approval
* SOPs
* Major component description
* Major instruments calibration verification
* Major devices testing verification

- Product contact parts/surfaces test
- List of spare parts
- Preventive maintenance plan and shedule
- Control/computer system and Programmable Logic Controllers (PLCs)
- Inspection and installation sheets of utilities such as power, water, air, etc.
- Pneumatic inspection
- Mechanical inspection
- Installation checks
- Deviations
- Sumary and final coclusions
- Post approval from the owner, engineering, and QA

28.2.6 OQs

Upon satisfactory completion execution of the system IQ protocol, the OQ should be performed. The OQ is used to test predetermined dynamic attributes of a system. During OQs all testing equipment should be identified and calibrated before use. Test methods should be authorized and implemented and the resulting data collected and evaluated. Draft SOPs for the equipment and services operation, cleaning activities, maintenance requirements, and calibration schedules should be available. The OQ protocol should describe the operational tests, measurements, and control tolerances of key parameters that are critical for the proper operation of the system. Test objectives, procedures, and acceptance criteria will be defined for all tests. Execution of the OQ will involve the testing and measuring of key system operational parameters.

This activity will include monitoring and evaluating operational data obtained from instruments, indicators, gauges, sensors, and physical attributes of any material being tested. All possible scenarios with operation ranges will be challenged to test the system capability.

OQ test procedures and acceptance criteria must be based on System Functional Specifications (FRS), URS, or engineering studies, or commissioning results. Any deviations encountered during the OQ should be identified, investigated, and documented (including justification, correction, and any necessary re-qualification studies). Upon satisfactory completion of IQ and OQ of the equipment/plant should be "released" to the next stage in validation.

Operational Qualification protocol should include following elements, but not be limited to the following:

- Pre-execution approval
- Introduction
- Participants log
- Test plans
- Calibration of test instruments

- Alarms and fault confirmation test: the appropriate alarms or responses are generated
- Power failure and restoration test
- Computer system: disaster recovery review
- Security access testing
- Interference test
- Process operational tests
- List of deviations/investigations
- Final summary and conclusions
- Post excution approval from owner, engineering, and QA

28.2.7 **PQ**

Upon successful completion of system IQ and OQ protocols, the PQ will be performed for those critical systems or processes requiring it. It is used to test a system or process across applicable ranges to demonstrate reproducible performance when integrated with other systems and/or materials being tested. The PQ protocol should test those systems where performance or process parameters are known and would affect the product in any manner critical to its quality. The PQ protocol should integrate the procedures, personnel, materials, equipment, and process. Test objectives, procedures, and acceptance criteria will be defined prior to execution of the PQ.

A sufficient number of replicate studies will be performed to demonstrate the ability of the system to achieve reproducible results. Testing may include analysis for chemical, physical, and microbiological constituents. The ability of the system or process to perform the intended function within the defined upper and lower process variable limits will be measured.

PQ test procedures and acceptance criteria must be based on functional specifications, URS, and product specifications of regulatory requirements. Any deviations encountered during the PQ should be identified, investigated, and documented (including justification, correction, and any necessary re-qualification studies). PQ protocol should include following elements, but not be limited to the following.

As an example of manufacturing area/utility qualification

- Pre-approval
- Introduction of the system (description based on the URS/DQ)
- Predetermined acceptance criteria
- Regulatory requirements: classification of the area (for example, ISO grade 9)
- Test plan
- Temperature mapping studies
- Non-viable particle count
- Viable particle count
- Smoke studies (air flow patterns)

- Risk assessment: in terms of containment based on the materials that will be used
- Differential pressure between process rooms and non processing rooms
- Three-phase monitoring system for water
- Sampling plans and selection of sampling location for previously mentioned tests
- Deviations and investigation reports
- Summary of the conclusion
- Post approval from related departments and QA

28.2.8 PV

Although the new PV guidelines for USFDA, dated 2011, or EU GMP-related PV are not applicable to nutraceutical manufacturing, it is advised to follow the principles and policies specified in these guidelines in order to produce a product consistently of same quality and achieve better customer satisfaction [4,5].

Based on FDA guidelines, PV is defined as the collection and evaluation of data, from the process design stage through commercial production, which establishes scientific evidence that a process is capable of consistently delivering quality product. PV involves a series of activities taking place over the life cycle of the product and process.

FDA guidance describes PV activities in three stages:

Stage 1—Process design: The commercial manufacturing process is defined during this stage based on knowledge gained through R&D development and scale-up activities.
Stage 2—PQ: During this stage, the process design is evaluated to determine whether the process is capable of reproducible commercial manufacturing.
Stage 3—Continued PV: Ongoing assurance is gained during routine production that the process remains in a state of control

PV is a product-specific validation conducted after all necessary preceding qualification activities (for example, IQ and OQ) have been performed on all the pertinent systems and processes involved in the overall product manufacturing process. It will be conducted using commercial manufacturing and packaging documents. Acceptance criteria for PV studies are based on product and process parameters as established during the R&D stage and as per regulatory requirements.

Based on HPFBI (Health Canada) Validation Guidelines for Pharmaceutical Dosage Forms: GUI-0029, the following types of PV approaches are described:

Prospective validation: The validation protocol is executed before the process is put into commercial use. It is generally considered acceptable that three consecutive batches/runs within the finally agreed parameters, giving a product of the desired quality, would constitute a proper validation of the process.

Concurrent validation: This validation activity takes place along with commercial batches.

Retrospective validation: In many establishments, processes that are stable and in routine use do not have to undergo a formally documented validation process. Historical data may be utilized to provide necessary documented evidence that the processes are validated.

28.2.9 Cleaning validation

Cleaning validation (CV) is an equipment-specific validation activity conducted on processes used to clean qualified product contact equipment, which is used in the manufacture of drug products [6]. Cleaning procedures should be established for each type of equipment used in the manufacturing and packaging process.

A formal risk assessment should be written and approved by the QA department. The risk associated with handling of nutraceutical products can be considered as high risk because of the following factors:

- Complexity of the materials
- Material origin
- Lack of toxicity data
- Health hazard assessment
- Multiple ingredients in a single batch
- Lack of adequate sensitive product-specific analytical test methods
- Lack of forced degradation data
- Mostly unknown therapeutic doses
- Lack of regulations

Based on these factors it is essential to validate cleaning activities that are used in a multiproduct nutraceutical manufacturing facility. Cleaning procedures should be validated when there is a changeover to a different class of product. The residual limits must be established following cleaning activities based on the maximum allowable carry over (MACO) related to following risk factors:

a. Therapeutic dosage of the active nutraceutical material
b. Toxicity (LD_{50} values)
c. Solubility of the material in cleaning agent

Acceptable risks (limits for CV) based on International Society for Pharmaceutical Engineering (ISPE) risk MaPP guide [7] can also be used. As per ISPE Risk MaPP, health-based limits should be developed based on the toxicological and pharmacological data or data from clinical studies can be used to establish the limits of Acceptable Daily Exposure (ADE). Please refer to the ISPE Risk MaPP guide for more information on this topic.

Validated analytical methods are a critical prerequisite for substantiating the validity of all CV studies.

28.2.10 **Revalidation and change control**

Modifications, relocation, and major changes to qualified (validated) equipment should follow an approved change control process before commencement of any changes to the equipment. This formal review should include consideration of re-qualification of the equipment. Minor changes or changes having no direct impact on final or in-process product quality should be handled through the documentation system of preventative maintenance program.

28.3 **Definitions**

Change control: A written procedure that describes action to be taken if a change is proposed (a) to facilities, materials, equipment, and/or processes used in the fabrication, packaging, and testing of drugs, or (b) that may affect the operation of the quality or support system.

CV: The documented act of demonstrating that cleaning procedures conducted on equipment used in fabricating/packaging will reduce to an acceptable level all residues (products/cleaning agents). It will also demonstrate that routine cleaning and storage of equipment does not allow microbial proliferation.

Concurrent validation: A process where current production batches are used to monitor processing parameters. It gives assurance of the present batch being studied, and offers limited assurance regarding consistency of quality from batch to batch.

Critical process parameter: A parameter, which if not controlled, will contribute to the variability of the end product.

Equipment qualification: Studies that establish with confidence that the process equipment and ancillary systems are capable of consistently operating within established limits and tolerances. The studies must include equipment specifications, IQ, and OQ of all major equipment to be used in the manufacture of commercial scale batches. EQ should simulate actual production conditions, including "worst case"/stressed conditions.

IQ: The documented act of demonstrating that process equipment and ancillary systems are appropriately selected and correctly installed.

Major equipment: A piece of equipment that performs significant processing steps in the sequence of operations required for fabrication/packaging of drug products. Some examples of major equipment include tablet compression machines, mills, blenders, fluid bed dryers, heaters, drying ovens, tablet coaters, encapsulators, fermentors, centrifuges, etc.

OQ: The documented action of demonstrating that process equipment and ancillary systems work correctly and operate consistently in accordance with established specifications.

PV: Establishing documented evidence with a high degree of assurance that a specific process will consistently produce a product meeting its predetermined specifications and quality characteristics. PV may take the form of prospective, concurrent, or retrospective.

VP: A written plan of action stating how PV will be conducted. It will specify who will conduct the various tasks and define testing parameters, sampling plans, test methods, and specifications and will specify product characteristics and equipment to be used. It must specify the minimum number of batches to be used for validation studies, and it must specify the acceptance criteria and who will sign/approve/disapprove the conclusions derived from such a scientific study.

References

[1] ISPE-GAMP guide.
[2] Validation Guidelines for Pharmaceutical Dosage Forms -GUI-0029 — Health Canada.
[3] EU GMP Qualification and validation Annex 15 to the EU Guide to Good Manufacturing Practice.
[4] Validation Master Plan Installation and Operational Qualification Non-sterile process validation Cleaning validation (Pics).
[5] USFDA Guidance for Industry - Process Validation: General Principles and Practices; 2011.
[6] Health Products and Food Branch Inspectorate Guidance Document Cleaning Validation Guidelines -GUIDE-0028.
[7] Risk Based Manufacturing of Pharmaceutical Products- ISPE publication; Sept 2010.

Adverse Event Reporting

VII

VII

Global Adverse Event Reporting Regulations for Nutraceuticals, Functional Foods, and Dietary/Food/ Health Supplements

29

Andrew Shao

Herbalife International of America, Inc., California

29.1 Introduction

The policy that underlies the regulation for a consumer product should be one that appropriately balances benefit and consumer access and choice with safety (risk), regardless of the category under which the product is regulated. Furthermore, irrespective of the pre-market requirements for new products (notification, registration, or pre-market approval), the single most effective means of assessing product safety and acceptability in the marketplace is via post-market surveillance. Monitoring product performance in the marketplace through collection and investigations of consumer inquiries, complaints, and adverse reactions is truly the most effective means of assuring quality and safety.

For example, in the United States, which has a comprehensive pre-market approval process for prescription drugs, more than 75 approved drugs were removed from the market between 1969 and 2002 due to safety issues identified via post-market surveillance [1].

It is, however, important to recognize how post-market surveillance and adverse event (AE) monitoring can and should be used. In the few countries that require AE reporting for foods and/or supplements, the respective regulatory agencies utilize the information for signal detection, not causality analysis. The value of collecting AE reports for public health lies with the ability of regulators to use these reports broadly to identify quality, safety, or manufacturing issues with products in the marketplace, not to establish causality between a given product and a reported adverse reaction [2]. Regulators do not individually "rate" or score individual reported AEs. Rather, they consider all incidents in totality in context to identify signals. Agricultural or manufacturing errors, product

Nutraceutical and Functional Food Regulations in the United States and Around the World.
DOI: http://dx.doi.org/10.1016/B978-0-12-405870-5.00029-3

contamination, and tampering are examples of issues that can be identified through the collection of AEs. Equally important is the distinction between post-market surveillance and adverse event reporting. The latter is a regulatory requirement in some countries, typically involving the collection, documentation, and reporting of AEs to regulators. Post-market surveillance is a broader field, incorporating the collection and analysis of consumer inquiries and complaints, in addition to AEs, and using this information to resolve issues and practice continuous improvement. Companies in the food and supplement industries that practice post-market surveillance do so not only to comply with applicable AE reporting requirements, but also to best serve the interests of their consumers, while also managing liability.

In this chapter, the various AE reporting regulations are summarized for those countries in which reporting to regulators is mandatory. Further recommendations are provided for robust post-market surveillance as well as available voluntary industry guidelines.

29.2 Global AE monitoring and reporting regulations

AE monitoring and reporting requirements for foods and supplements vary greatly around the world. Most countries have not established regulatory requirements, particularly for conventional foods. In some countries reporting is practiced voluntarily by some companies and also provided through health care professionals. Some countries have established specific requirements for AE reporting (e.g., United States and Canada), while others require companies to follow requirements established for drugs (e.g., Peru and Ecuador). Table 29.1 summarizes the basic aspects of the requirements in those countries that have established them.

29.2.1 United States

In late 2006 the U.S. Congress passed the Dietary Supplement and Non-prescription Drug Consumer Protection Act, requiring manufacturers of dietary supplements (and over-the-counter drugs; OTC) in the United States to report to the U.S. Food and Drug Administration (FDA) all serious AEs they receive within 15 business days and to maintain records of all AEs they receive up to 6 years [3]. Prior to the law's enactment in 2007, reporting of AEs to the FDA involving supplements or OTC drugs was solely voluntary, with many reports coming directly to the FDA from consumers or health care professionals without manufacturers being aware. Each year since going into effect, the number of mandatory (serious) AE reports the FDA has received from the industry has increased dramatically, a reflection of improved industry compliance. Between 2008 and 2011, the FDA received over 4300 total AE reports from the industry, and the agency has indicated this has helped in the identification and protection of the public from unsafe products [2].

Table 29.1 Global Adverse Event Monitoring and Reporting Requirements

Country	Legislation/Regulation	Basic Aspects
United States	Dietary Supplement and Non-prescription Drug Consumer Protection Act	Requires manufacturers of dietary supplements to report all serious adverse events received to the U.S. FDA and to keep records of all AEs received
Canada	Natural Health Product Regulations	Requires NHP licensees to report to HC all serious AEs received within or outside Canada related to a given product
Australia	Australian Guidelines for Complementary Medicines	Requires manufacturers of complementary medicines to report all serious and unexpected AEs to the TGA
Peru	Pharmacovigilance and Technovigilance for Pharmaceutical Products and Medical Devices	Requires manufacturers of health products to report all AEs to the National Authority for Pharmaceuticals, Medical Devices, and Health Products
South Korea	Functional Health Foods Act	Requires manufacturers of health supplements to report to the Korean FDA all AEs and maintain reports
France*	French Act on Regional Health Governance	A proposal by the French Agency ANSES to require the reporting of AEs involving fortified foods and food supplements to the agency

Proposed.

29.2.2 **Canada**

In Canada the requirements are similar to those in the United States. Licensed natural health product (NHP) manufacturers are required to report to Health Canada (HC) all serious AEs received within 15 business days [4]. However, where Canada diverges from the United States is the jurisdiction of the requirement. HC requires that all serious AEs arising from any country, but related to a given product, be reported to the agency. HC has provided comprehensive guidance documents to assist the industry in compliance with both domestic and foreign AE reporting [5].

29.2.3 **Australia**

The Therapeutic Goods Administration (TGA), which regulates supplements under the category of Complementary Medicines [6], relies on pharmacovigilance requirements [7] for AE reporting in the absence of specific requirements for complementary medicines. The AE reporting requirements are similar to those required in the United States and Canada, but they also include a requirement to conduct an investigation and collect follow-up information.

29.2.4 South Korea

According to the Functional Health Foods Act, all manufacturers of health supplements must report to the Korean Food and Drug Administration harmful effects regarding health supplements and maintain records for up to 2 years [8].

29.2.5 Peru

In Peru, where supplements are regulated under the same law as drugs and devices, all suspected AEs must be reported to the National Authority for Pharmaceuticals, Medical Devices, and Health Products [9]. Companies must also file annual safety reports, which include all investigational products with the Authority.

29.2.6 France

Although harmonized regulations on AE monitoring and reporting have not yet materialized in the European Union, some individual Member States have taken the initiative to propose new regulations. In France, the French Agency for Food, Environmental, and Occupational Health & Safety (ANSES) proposed in 2009 new regulations to require the reporting of AEs involving fortified foods and food supplements to the agency [10]. Termed nutritional vigilance or "nutrivigilance" by ANSES, the proposal is still being evaluated, but appears to be gaining momentum based on some compelling pilot data collected by the agency [11].

29.3 Post-market surveillance versus AE reporting

Compliance with AE reporting requirements ensures companies meet the "letter" of the law—receiving, documenting, tableting, and submitting AEs. Post-market surveillance goes beyond this to assure the overall quality and consistency of products in addition to managing company liability by monitoring the performance and safety experience for a given product in the marketplace.

In contrast to drugs, for dietary supplements the majority of adverse reactions are "spontaneously" reported by consumers directly to the manufacturer (vs professionals) rather than to regulators. Many foods and supplements contain multiple ingredients that do not adhere to any national compendial standards. Unlike drugs, which have proprietary safety and efficacy data and for which some amount of toxicity is expected, for foods and supplements toxicity is unexpected and not tolerated in most markets. In sum, companies are often in a better position to provide safety and quality assessments based on post-market surveillance than regulators.

A robust post-market surveillance system involves comprehensive investigation of quality and adverse reaction incidents. This includes collection, documentation, and categorizing of incidents, followed by causality analysis and corrective action or risk mitigation efforts where applicable. This process, by and large, falls outside the scope of most mandatory AE reporting requirements. For the handling and mitigating of consumer complaints related to product quality, some regulatory agencies have incorporated requirements into Good Manufacturing Practice (GMP) regulations. For example, in the United States, the U.S. FDA has promulgated requirements for complaint handling in the Current GMP regulation published in 2007 [12].

The consumer complaint and adverse reaction information collected and analyzed post-market for foods and supplements not only helps protect the company from liability but can also inform valuable product improvement and development efforts. In effect, this serves a similar purpose to much of the pre-market safety testing required for drugs.

29.4 **Guidance for industry and regulators**

Regulatory requirements for AE reporting worldwide are relatively new, and for broader post-market surveillance have yet to be imposed. It thus follows that many companies in the industry, as well as regulators lack the context and understanding of how to collect, analyze, and report AEs. To help address this void in understanding the International Alliance of Dietary/Food Supplement Associations (IADSA) has developed and published an AE complaint handling guideline. IADSA's 48-page voluntary *Global Guide to the Handling of Adverse Event Complaints Guidelines for Supplement Companies*

> *is a handbook designed...as a training manual offering guidance to companies on designing and putting in place procedures to ensure that any complaints of adverse events received, are dealt with in a logical, functional and comprehensive manner...[13].*

The Guideline does not impose mandatory requirements or regulations, but is intended to be used as a voluntary tool to help companies develop and implement their own internal AE collection procedures. It can be utilized in both markets—those that have established specific reporting requirements or those that have not.

It is becoming more evident that for foods and supplements, robust post-market surveillance is critical to assure product safety and quality in the marketplace. This approach helps build consumer confidence, and helps justify less resource-intensive market entry requirements (i.e., product notification rather than product registration). This in turn helps to promote better consumer access to safe, well-made products. Regulatory agencies in a few select countries have established AE reporting requirements for foods and supplements. To properly protect consumers and maintain open access, more countries need to establish such regulations. Furthermore, additional guidance is needed to assist both the industry and regulators with AE reporting and post-market surveillance.

References

[1] Wysowski DK, Swartz L. Adverse drug event surveillance and drug withdrawals in the United States, 1969-2002: the importance of reporting suspected reactions. Arch Intern Med 2005;165:1363.

[2] Frankos V, Street D, O'Neill R. FDA regulation of dietary supplements and requirements regarding adverse event reporting. Clin Pharmacol Therap 2009;87:239—44.

[3] 109th Congress of the United States. Public Law 109—462: Dietary Supplement and Nonprescription Drug Consumer Protection Act. Retrieved January 30, 2013, from <http://www.fda.gov/RegulatoryInformation/Legislation/FederalFoodDrugandCosmetic ActFDCAct/SignificantAmendmentstotheFDCAct/ucm148035.htm>; 2006.

[4] P.C. 2003-847: Natural Health Products Regulations. Retrieved January 30, 2013, from <http://gazette.gc.ca/archives/p2/2003/2003-06-18/html/sor-dors196-eng.html>; 2003.

[5] Health Canada. Guidance Document for Industry - Reporting Adverse Reactions to Marketed Health Products: Adverse Reaction Reporting by Market Authorization Holders. Retrieved January 30, 2013, from <http://hc-sc.gc.ca/dhp-mps/pubs/medeff/ _guide/2011-guidance-directrice_reporting-notification/index-eng.php#a13>; 2011.

[6] Therapeutic Goods Administration. Australian Guidelines for Complementary Medicines (ARGCM) Part II: listed complementary medicines. Canberra, Australia: Australian Government Department of Health and Aging; 2011.

[7] Therapeutic Goods Administration. Australian Adverse Drug Reactions Bulletin. Retrieved January 30, 2013, from <http://www.tga.gov.au/hp/aadrb.htm>; 2011.

[8] South Korean Ministry of Health and Welfare. Ministry of Health and Welfare Decree No. 142, 2012.8.1: Enforcement Regulations of the Functional Health Foods Act. Seoul, Korea: Ministry of Legislation, National Law Information Center; 2012.

[9] The Peruvian National Authority for Pharmaceuticals, Medical Devices and Health Products. Title V of the pharmacovigilance and technovigilance system. Lima, Peru: Peruvian Government; 2011.

[10] ANSES. L'Agence nationale de sécurité sanitaire de l'alimentation, de l'environnement et du travail. Retrieved January 30, 2013, from <http://www.anses.fr/index. htm>; 2012.

[11] ANSES. Dispositif national de vigilance sur les compléments alimentaires: Bilan de la phase pilote et propositions pour la mise en place du dispositif national de nutrivigilance. France: ANSES, Maisons-Alfort Cedex; 2010.

[12] FDA. Title 21 — Food and drugs, Subchapter B — Food for human consumption, Part 111 — Current good manufacturing practice in manufacturing, packaging, labeling, or holding operations for dietary supplements, Subpart O — Product complaints. Retrieved January 30, 2013, from <http://www.accessdata.fda.gov/scripts/cdrh/cfdocs/cfcfr/ CFRSearch.cfm?CFRPart=111&showFR=1&subpartNode=21:2.0.1.1. 11.15>; 2012.

[13] IADSA. Global guide to the handling of adverse event complaints: guidelines for supplement companies. Brussels, Belgium: International Alliance of Dietary/Food Supplement Associations; 2012.

Intellectual Property, Branding, Trademark, and Regulatory Approvals in Nutraceuticals and Functional Foods

Intellectual Property, Branding, Trademark, and Regulatory Approvals in Nutraceuticals and Functional Foods

30

Leighton K. Chong, Lawrence J. Udell and Bernard W. Downs

Udell Associates, Castro Valley, California

30.1 Introduction

Intellectual property (IP) can be a relatively mundane topic. However, the information presented in this chapter opens up a variety of creative issues that not only demonstrate how to acquire and position IP (and IP relationships), but how to utilize that IP properly. This information will be practical, innovative, and useful in bridging the gap between the creation and development of IP and successfully marketing it using powerful and protective branding strategies that better insulate the long-term return on investment (ROI). Having been involved in the big wide world of IP creation, management, and marketing for more than 70 years, the authors offer a unique perspective based on a wealth of successful experiences. How do you transfer to paper diversified experiences in working with inventors and entrepreneurs in a multitude of industries?

Let us first explore and understand the values of IP and its relationship to the industry of nutraceuticals. The knowledge of this very special business started many years ago with the inception of "alternative" and complementary medicine, even before it was discussed openly and generally accepted.

It is important to understand that IP in this industry does not only encompass what is normally recognized as patents and trademarks. The business of harnessing and commercializing the power of nature relies on the ability to transform nature's creations into proprietary compounds marketed in the form of liquids, powders, capsules, tablets, and topicals. The resulting IP is comprised of formulas, methods, and procedures for combining those derivatives in a unique manner into products that have the *novel* ability and characteristics to positively affect the user. Novel product technologies are derived from the soil and sea and every

Nutraceutical and Functional Food Regulations in the United States and Around the World.
DOI: http://dx.doi.org/10.1016/B978-0-12-405870-5.00030-X

imaginable plant, insect, reptile, mammal, marine creature, etc. For example, plants are the source of a plethora of phytonutrients; various proteins, with therapeutic and analytical benefits, are being isolated from insects [1]; reptiles are a source for a number of novel substances (i.e., antivenom and immune complexes, etc.); mammals contribute a host of therapeutic agents from proteins to undenatured collagen [2]; algae, seaweed, and mussels and fish and krill oils, etc., offer a host of therapeutic options; and glandular extracts provide therapeutic naturopathic interventions and everything in between. The scientists who are creating these potentially remarkable products are a special group imbued with a lust for finding the secrets that will reduce needless pain and suffering and increase the opportunities to improve health, possibly thrusting themselves into fame and fortune in the process.

However, their motivation for discovery is not totally aligned with recognition, but truly enhanced by reaching out to touch the future and the lives of potentially millions of people they will never know. All of the results and motivation are enhanced by the knowledge that their creations will (hopefully) be protected by the existing IP laws that exist in the majority of countries throughout the world along with powerful branding and marketing strategies.

This is the primary focus of this chapter. After spending untold millions of dollars in R&D, the benefits to the researchers and their companies are weighed in the potential value of their inventions and their subsequent transformation into commercial products.

30.2 Nutraceuticals, patent rights, and bioprospecting

As defined in Wikipedia, "nutraceutical" is a word phrase combination of "nutrition" and "pharmaceutical" and refers to foods, and more generally supplements, intended to have a therapeutically beneficial effect on human health. It can also refer to individual chemicals that are present in common foods (and therefore may be delivered in a non-drug form). Many such nutraceuticals are phytonutrients, which are plant compounds possessing both health-protecting and health-promoting properties. Nutraceuticals are often used in nutrient premixes or nutrient systems in the food and pharmaceutical industries. Nutraceuticals that are in or from foods are also sometimes called "functional foods," but are probably more accurately identified as components of functional foods. For example, red wine (containing resveratrol) is an antioxidant [3] and an anticholesteremic [4], broccoli (containing sulforaphane) is a cancer preventive [5], and soy and clover (containing isoflavonoids) are nutraceuticals used to improve arterial health in women [6], inhibit undesirable blood vessel formation in tumorous tissues (antiangiogenic) [7], and also possess antitumor effects [8,9].

Except for newly discovered foods, which must be an extremely rare occurrence given the course of human history and foods, all nutraceuticals by definition involve food or food nutrients that are already known to be used by the world's populations and cultures. Any "discovery" in nutraceuticals then would consist of identifying, isolating, and making a composition containing the active ingredient or nutrient in an already known food substance found to have a *previously unknown* beneficial effect on human health or formulated with other ingredients in an effective but previously unknown manner and/or for a previously unknown benefit or purpose.

Commercial entities commonly seek to obtain IP rights in discoveries and utilize strong branding strategies in order to reap the benefits of their investment of time and resources in research and to incentivize discoverers. Under the IP regimes existing in the world today, the primary form of IP protection for discoveries is to obtain a patent, i.e., to obtain a government patent grant for the exclusive right to a discovery. The patent has a limited term, typically 20 years from the first application. But, in seeking to obtain a patent for identifying, isolating, and making a composition of an active ingredient, from an already known food substance, the inventor is only entitled to patent claim coverage of only that contribution found by the inventor to be both "new" and "non-obvious" over all prior public knowledge about the already known food substance.

Thus, attempts by food and pharmaceutical companies to secure patents on nutraceuticals will invariably raise tensions between a company's claim for exclusive patent rights in making a new and non-obvious discovery and the common heritage of the public to continue to access what was already known about foods. The more well known an existing food substance and its benefits are, the less likely is a company's claim to exclusive rights to a nutraceutical made from the food's active ingredient to be broadly defined, granted, and/or upheld. Conversely, if a food substance and its benefits are not considered to be widely known, or perhaps are hitherto undocumented, the more likely it will be that a company can obtain broad patent rights in them. Knowledge about a food substance and its benefits are often undocumented when it is known only to a primitive or indigenous people and transmitted only through folklore or sacred practices.

The practice of companies conducting research in remote areas for naturally occurring biological substances that may be found to have benefits undocumented or unknown to the world at large is called "bioprospecting." The ethics, fairness, and potential harm from current practices in biotechnology research are being increasingly questioned as Third World governments, indigenous peoples, and ethnic populations worldwide have developed a growing awareness and interest in controlling the use of and sharing the benefits of bioprospecting research.

An increasingly complex national and international policy framework arising out of the 1992 World Convention on Biodiversity (CBD) seeks to have companies conduct bioprospecting research under formal agreements entered into with the sovereign governments, indigenous peoples, and local communities who control or occupy the lands on which such research is to be conducted [10]. Article 15 of the CBD addresses the terms and conditions for access to genetic resources and traditional knowledge, informed consent, and benefit sharing. It recognizes the sovereignty of states over their natural resources and provides that access to these resources shall be subject to the prior informed consent of the contracting party (signatory country) providing such resources. It also provides that access will be based on mutually agreed terms in order to ensure the sharing of benefits arising from the commercial or other utilization of these genetic resources with the contracting party providing such resources.

In 1999, work was begun to operationalize the mandates of the CBD, which resulted in the issuance of the "Bonn Guidelines on Access to Genetic Resources and Benefit Sharing" in April 2002 by the CBD Secretariat. These were adopted unanimously by CBD treaty members [11]. The Bonn Guidelines were adopted to assist contracting parties, governments, providers, users, and other stakeholders in developing overall access and benefit-sharing strategies and in identifying the steps involved in the process of obtaining access to genetic resources and benefit sharing. More specifically, the Guidelines are intended to help them when establishing legislative, administrative, or policy measures on access and benefit sharing and/or when negotiating contractual arrangements for access and benefit sharing.

While a signatory to the CBD, the United States has not yet ratified the treaty. The U.S. State Department has formed a coordinating agency as the U.S. National Focal Point under the CBD Clearing-House Mechanism [12]. For the time being, the U.S. National Focal Point has deferred responsibility to the various national agencies having jurisdiction over regions or lands under federal control in which bioprospecting may be permitted, such as national park lands, federal conservation trust lands, Indian reservations, etc. If the United States ratifies the CBD Treaty, the U.S. National Focal Point and designated national authorities would have a duty to assist participating parties in bioprospecting negotiations. Furthermore, these authorities would need to advise and assist state agencies (as regional authorities) in the development of policies regulating bioprospecting in compliance with the CBD framework in an effort to ensure the equitable sharing of benefits with all involved parties from the commercialization of these resources.

The Bonn Guidelines place great emphasis on the obligation for research users to seek the prior informed consent of bioaccess providers. They also identify the basic requirements for mutually agreeable terms and define the main roles and responsibilities of, and stress the importance of, the substantive involvement of all stakeholders. They also cover other elements such as incentives, accountability, means for verification, and dispute settlement. Finally, they enumerate suggested elements for inclusion in material transfer agreements and provide a suggested checklist for monetary and non-monetary benefit sharing.

The focal point of the ongoing arbitration of rights in bioprospecting has centered on intellectual property rights (IPR) and, specifically, patent rights in discoveries derived from bioprospecting. The Conference of Parties organized for ongoing negotiations of Contracting Parties under CBD requested the World

Intellectual Property Organization (WIPO), an international agency dealing with world IP issues under the auspices of the United Nations, to conduct studies on the interface of IPR protocols with bioprospecting mandates. WIPO formed the Intergovernmental Committee (IGC) on Genetic Resources, Traditional Knowledge, and Folklore in 2000 to conduct, among other subjects, studies of the interface of bioprospecting mandates and IPR rights [13]. At least 17 countries are active participants on the WIPO/IGC Committee and many non-governmental organizations and activist groups have observer status to the Committee's deliberations.

Patents are government-sanctioned grants of exclusive rights to inventors (and their companies) to commercially exploit their new discoveries, mainly through deriving profits from the sale of products based on those discoveries. When discoveries are based on genetic resources or traditional knowledge obtained from bioprospecting, issues arise as to whether the researchers had proper permission to access biological materials in the field, including obtaining the informed consent of the providing party affected, whether they derived their inventions from already known or traditionally known substances or healing practices, whether their use of the biological materials and information was proper, and whether there was an equitable sharing of benefits with the local authorities who permitted access for such research.

The WIPO/IGC studies have raised the issue whether the patent systems of participating countries should be changed to implement CBD mandates by requiring disclosure of bioprospecting agreements in patent applications [14]. A majority of IGC members favor imposing an affirmative duty on research companies to state that their research was conducted pursuant to and in legal compliance with a valid bioprospecting agreement, subject to sanctions including patent invalidity if the requirement is not met. More conservative IGC members (such as Japan and the United States) have expressed concern that such a substantive bioprospecting disclosure requirement could have unintended adverse consequences on patent systems and the objectives of rewarding innovation and investment in research.

A WIPO draft statement on a bioprospecting disclosure requirement was recently transmitted to the CBD Conference of the Parties in May 2006 [14].

The biotechnology industry in the United States has generally recognized the implications of the developing international framework on bioprospecting under the CBD, and the significant milestone reached by unanimous approval of the Bonn Guidelines for member countries, users, and providers. Although the Bonn Guidelines are not yet legally binding, biotechnology companies have had to weigh the possible losses that might be incurred through research foregone and benefits relinquished and the considerable costs of compliance against the risks of incurring local protests, possible legal sanctions, loss of public goodwill, and invalidation of patents. Many are concluding that compliance with international mandates on bioprospecting is inevitable, or at least would be good business practice to achieve some degree of compliance with mandates that are likely to become widely adopted.

As an example, the Biotechnology Industry Organization (BIO), representing more than 1100 biotechnology companies, academic institutions, state biotechnology centers, and related organizations in 50 U.S. states and 31 other nations, recently adopted its recommended Guidelines on bioprospecting to its members [15]. The BIO Guidelines identify "best practices" that should be followed by companies engaging in bioprospecting. BIO advises its members that the Guidelines provide only a "road map," but that they "have extensive discretion to shape their conduct to meet whatever requirements countries impose with respect to bioprospecting activities." A member company is not required to follow the Guidelines and those Guidelines are not enforceable against a member company. However, it is conceivable that companies that do not engage in conduct consistent with the Guidelines might be subject to criticism for not following "best practices." The biotechnology industry thus acknowledges that the proper conduct of bioprospecting activities requires the use of formal bioprospecting agreements with the relevant national and local authorities.

U.S. Patent Laws require that an applicant disclose, in any patent application, information known to the applicant that is material to the patentability of the patent claims submitted. The relevant Patent Office Rules, 37 Code of Federal Regulations, state in pertinent part, as follows:

§ 1.56 Duty to disclose information material to patentability

(a) ... Each individual associated with the filing and prosecution of a patent application has a duty of candor and good faith in dealing with the Office, which includes a duty to disclose to the Office all information known to that individual to be material to patentability as defined in this section.

The duty to disclose information exists with respect to each pending claim until the claim is cancelled or withdrawn from consideration, or the application becomes abandoned.

The duty to disclose all information known to be material to patentability is deemed to be satisfied if all information known to be material to patentability of any claim issued in a patent was cited by the Office or submitted to the Office in the manner prescribed by §§ 1.97(b)−(d) and 1.98. However, no patent will be granted on an application in connection with which fraud on the Office was practiced or attempted or the duty of disclosure was violated through bad faith or intentional misconduct.

The Office encourages applicants to carefully examine:

(b) The closest information over which individuals associated with the filing or prosecution of a patent application believe any pending claim patentably defines, to make sure that any material information contained therein is disclosed to the Office.

The applicant's duty of disclosure under existing U.S. Patent Office rules would require the disclosure of relevant information as to prior uses, traditional knowledge, and/or source materials where relevant to a particular patent application. Likewise, the written description and enablement requirements of the U.S. Patent

Laws, Section 112, would require U.S. applicants to identify relevant source materials on which claims to invention are made. However, it is important to keep in mind that biotech discoveries are often made following many steps or levels of extraction, modification, refinement, or synthesis away from a naturally occurring material or traditional remedy or practice product, such that the importance of the initial material investigated may no longer be relevant to the patentability of the invention. Nevertheless, a patent applicant may deem it safer to disclose source materials initially investigated in the patent application or during patent prosecution rather than take the risk of having to litigate the issue later.

30.3 Branding: a hypothetical case scenario

Now That We Have Patent Protection, We Can Market Our Natural Creations With The Confidence That They Are Completely Protected... Right? (Not Quite ... There's More)

Great disappointment and potentially huge legal defense expenses await the inventors of patent-protected nutraceutical products and/or technologies if they attempt to take their creations to market in a big way solely on the protective strength of the patents, without establishing powerful identity, branding, and trademark strategies. The key factor here is "in a big way." Just ask the original IP-based suppliers of products like Cat's claw, Ginkgo, St. John's wort, *Garcinia cambogia*, *Hoodia*, *Citrus aurantium*, grape seed extracts, chromium nicotinate, noni, mangosteen, and countless others. The evidence-based technology and patent protection are only half the story when it comes to propelling and insulating the market success of a patented nutraceutical product. To complete the equation required for long-term market success, strong branding strategies are a must.

   The principle seems to be that the greater the market impact and success an invention/creation has, the greater the lure for copy-cat knock-off suppliers to bring competing products to market. "But I have a patent and am protected," you might say. The patent provides a basis to defend the invention, not to prevent it from being infringed, although, the hope is that patent protection should discourage infringement. But, to savvy knock-off artists, a commercially successful patented product can offer an enormous opportunity for copy-cat windfall profits, especially if the patent holder is a financially fragile start up venture...an easy mark.

   How so? The following is an (oversimplified) example. Suppose you discovered a botanical species (call it "Onlyess importantium") that was a powerful antioxidant and had a profound effect on turning off chronic inflammation. For the purpose of this illustration, *Onlyess importantium* has already been sold as a botanical detoxifier since before 1994 (no New Dietary Ingredient approvals by the Food and Drug Administration (FDA) are required). So the species has been commercially available, but for other applications. Further investigation finds that one previously inconspicuous component in *Onlyess importantium* made the most significant contribution to achieving the greatest therapeutic effect on inflammation; call it an "inflammanoid." So, you developed a solvent-free extraction method to achieve a 50% concentration of the inflammanoid (and inflammanoid glycosides), which naturally occurs at about 6% of the total composition. When tested, this product had an oxygen radical absorbance capacity (ORAC) value off the charts and turned off the inflammation process like a switch, promoting profound and rapid healing (which required conservatively a $250,000 investment in a clinical study). It is sensational. You go through the process of applying for and receiving patents on the extraction method, the novel composition of the 50% inflammanoid, and for its application as an antioxidant and downregulator of (chronic) inflammation (requiring a significant investment as well). Keep in mind that you are still incurring business *and* research expenses during the period of research, presentations, and publication (hopefully) of the study, prior to its intended market launch.

Now, you have to tell the world about your invention. So, you create a new raw material product and call it "AntioxInflam." In one go-to-market strategy, you contract with an infomercial company to promote AntioxInflam™, a proprietary source of inflammanoid. The infomercial costs about $200,000 to produce and about $1.5 million per month to run. Plus, you have to invest in manufacturing a few thousand bottles to start and arrange for fulfillment, unless the infomercial company takes care of this. All of this costs money.

The infomercial is a success and the phones at the call center are ringing off the hook. Your inventory requirements at the fulfillment center skyrocket. Assuming that you are in strict compliance with FDA and Federal Trade Commission (FTC) guidelines and are making only supportable structure function claims, you are selling 100,000 bottles a month at $34.95 each. You are making a boatload of money and helping an enormous number of people. Life is good! However, you have now jumped high onto *everybody's* radar, especially opportunistic competitors.

Keep in mind that once you make AntioxInflam and *Onlyess importantium* household names, competitors can enter the market, but do not need to make any claims. They also do not need to duplicate your extraction methods. They do not need to infringe on your patents to sell products. They only need to manufacture a reasonable concentrate, extracted the old (non-infringing) way and supply a very cheap generic/commodity version of *Onlyess importantium* to as many branded finished product marketers as possible. Recall that these copy-cat merchants have *none of the costs* associated with R&D nor creating and building the market. So, eliminate the R&D and market building costs (that you incurred), factor in the significantly cheaper raw material costs, and even with a product costing a fraction of your original product, these copy-cat merchants can make a significantly higher profit percentage than you on a much lower suggested manufacturer's retail price.

Again, they only need to supply other finished product companies with their version of the raw material for which *no claims* need to be made. You already popularized the *Onlyess importantium* and have an exponentially higher breakeven point than them. They do not have to do anything except supply the market. And, depending on the magnitude of your success, the market will be flooded with counterfeit versions all appearing to be enough like the original (yours) that consumers will buy those much cheaper products in hopes of receiving similar benefits. Knock-off artists are experts at creating the illusion of quality without actually fulfilling that value promise. This results in ever increasing numbers of consumers switching from your product to counterfeit versions, cratering your sales and potential for ROI. They also undermine and destroy the credibility of the product (because "it's not working"). The knock-off counterfeiter just *loves* your infomercial. And, they will enter the market in increasing numbers as long as the opportunity to sell *Onlyess importantium* exists.

But, the market pie can only be sliced in so many pieces before the ROI potential is destroyed for everyone, including the knock-off artists. They will even put each other out of business in vicious price wars. By the way, purchasing agents of major companies are often paid a bonus based on the amount of money saved in lowering product costs, and most purchasing agents (and executives and investors in profit sensitive public companies) are not scientists with an understanding of analytical chemistry or the differences between different sources of the same raw materials. So, if it looks the same and costs less, chances are the lower priced product will replace higher priced versions, and lost consumer confidence is the irrevocable casualty. It has been proposed that this phenomenon played a major role in the nutraceutical market crash of the mid to late 1990s.

Now for the dilemma. How long and to what magnitude do you continue to invest in your market-building promotions once competing low-cost counterfeit market predators have a plethora of alternative products blitzing consumers? In this scenario, the answer is not very long. And, now the activities and costs of defending your patents become an excruciating requirement against diminishing sales and profits. In addition, your ability to monitor the influx and potential of patent infringers diminishes against the increasing magnitude of counterfeit product marketers and time. For all the companies that do not make any claims, there will be some that do. But, as the crowd of competitors gets larger, the ability to identify these violators gets more difficult. And, to aggravate further the situation, your "benefit-sharing" partners of the source technologies are becoming increasingly cynical over the sudden reduction in fortunes and do not understand what has happened, because the product category (*all Onlyess importantium* products) *appears* to be gaining in popularity and experiencing explosive growth. Potential suspicion, friction, blame, and retribution by them could inflame things further. So, can this process be averted? Absolutely yes . . . but how?

30.3.1 Protecting your patents with coded identities and branding . . . the right way

Not to overstate the obvious, but the traditional role of branding is to establish a value promise that is linked to a *unique* "branded" trade name (and supposedly a uniquely superior product or service). This certainly sounds simple enough. However, most people view branding in a very one-dimensional way, limited to a consumer product or service label (i.e., Band-Aid®). In the nutraceutical, cosmeceutical, functional food, and beverage categories, branding can incorporate the value promise/reputation of the company (brand), the finished product (brand), the ingredient (brand), and/or any branded valuations/validations (i.e., Good Housekeeping, ANSI, USP, GRAS, Kosher, Organic, GMP, etc.). This strategy validates the value promise with "layers of branding" representing layers of validation.

In this aspect, the most important factor is differentiating the unique identity and therefore the superiority of the product. In regard to where competition is directed (at you), this means beginning with your raw material. So, even the species of *Onlyess importantium* must now somehow be differentiated from all the other sources. Since you have developed a different product, even though you have trade named the finished raw material (AntioxInflam), you still need to identify that *your* unique raw material has more different or special properties than any of its generic counterparts or else there will be severe market consequences.

When research is published (in scientific journals and industry and lay publications), editors and publishers are most often very reluctant to publish branded trade names for fear of appearing to endorse products and losing the credibility of being unbiased. For this reason, they (at least the respected publications) almost unanimously publish only the generic name of an ingredient. Therefore, consider that when your "independent" research is published, if the benefits of AntioxInflam are attributed only to the *Onlyess importantium* species, then your potential competitors will be given the most powerful market weapon to use against you ... scientific support for *generic Onlyess importantium*. And, to any industry or lay publication that reviles using branded trade names, attributing the research benefits to just *Onlyess importantium* bolsters the opportunities for market exploitation with counterfeit products by all of your knock-off competitors.

Therefore, you must give your raw material a unique identity in addition to the generic name. One tactic is to assign a product identity or a research code to the material (Ex. XYZ123). Now AntioxInflam is officially a *unique* XYZ123 form of *Onlyess importantium*. By the way, this code must appear in the research publications, at least in the "Materials and Methods" section. If publishers and editors refuse to identify the brand name, they will be very reluctant to delete the product identity research code in association with the generic name, which only you have.

This tactic is also valuable for creating other branding options when and/or if needed. As an example, if one very large company wants to have an exclusive for a period of time in a specific channel, you have the means to create a separate brand name exclusive to that customer, as long as you link the brand to the XYZ123 research code. This enables that different brand to base the product claims on original validated research. And you have successfully created your own competition, which is a far superior option to having a vicious counterfeit competitor. The more you establish your other patent-protected brands of XYZ123 *Onlyess importantium* in the market, the more of a disincentive you can create for knock-off products, as the market can appear to be filling up with competitive products, reducing the knock-off profit potential and ROI. Now your science, IP, significant investments, and the product's ultimate value promise are all better insulated from market piracy.

This strategy also applies to multiple ingredient formulas in finished products. Research and patents on finished products increases their market potential exponentially. Up to this time, the vast majority of research and patent protections in the natural products industry has been done by ingredient manufacturers on single ingredients. Research and patents on finished products amount to only a very small fraction of that done on nutraceutical ingredients. Yet, it offers the greatest revenue potential and return on investment.

There are other effective strategies that improve the potential for long-term success in the nutraceutical industry. However, discussions regarding those methods are more the subject of an advanced marketing course on branding and outside the scope of this chapter. Suffice it to say, this information provides a basic knowledge and guide on how to create a successful strategic agenda for developing, validating, and marketing nutraceutical IP from source to consumer.

30.4 **Conclusion**

Prior to the Dietary Supplement Health and Education Action of 1994 (DSHEA), the nutritional products industry was characterized more by flash-in-the-pan product popularity that usually lasted at most for about 2 years. These popular products sparked a fad-like following that usually dissipated with the next new product fad. During those years preceding 1994, there was not a very significant incentive to do the kind of research that would reveal a nutritional product's powerful health correcting and/or promoting properties. The regulations prohibited making just about any kind of important and/or informative claims. So the market opportunities were relatively small and the arsenal of IP was comparatively sparse. Since the passage of DSHEA, the opportunities to do exciting research, develop new product technologies, and obtain important IP are almost endless. We believe that this growing industry will find, through extensive R&D, some remarkable and beneficial new materials that will continue to benefit humankind in ways that will amaze the medical and scientific communities. Furthermore, we expect that in the future, nutraceutical technologies and IP will be increasingly administered with and/or in place of conventional pharmaceutical medicine. The next phase of market growth is dependent on valid scientific research for new product technologies, patents, stronger more effective branding and trademark strategies in product marketing, and international regulatory compliance.

References

[1] Quincozes-Santos A, Andreazza AC, Nardin P, Funchal C, Goncalves CA, Gottfried C. Resveratrol attenuates oxidative-induced DNA damage in C6 glioma cells. Neurotoxicology 2007;28:886−91.

[2] Pal S, Naissides M, Mamo J. Polyphenolics and fat absorption. Int J Obes Relat Metab Disord 2004;28:324−6.

[3] Dinkova-Kostova AT, Fahey JW, Wade KL, et al. Induction of the phase 2 response in mouse and human skin by sulforaphane-containing broccoli sprout extracts. Cancer Epidemiol Biomarkers Prev 2007;16:847−51.

[4] Clarkson TB. Soy, soy phytoestrogens and cardiovascular disease. J Nutr 2002;132:566S−9S.

[5] Fotsis T, Pepper M, Adlercreutz H, et al. Genistein, a dietary-derived inhibitor of in vitro angiogenesis. Proc Natl Acad Sci USA 1993;90:2690−4.

[6] Barnes S, Sfakianos J, Coward L, Kirk M. Soy isoflavonoids and cancer prevention. Underlying biochemical and pharmacological issues. Adv Exp Med Biol 1996;401:87−100.

[7] Wietrzyk J. The influence of isoflavonoids on the antitumor activity of vitamin D3. Postepy Hig Med Dosw (Online) 2007;61:253−60.

[8] Convention on Biodiversity, available online at <http://www.biodiv.org/convention/default.shtml▷.

[9] Bonn Guidelines on Access to Genetic Resources and Benefit Sharing, Secretariat of Convention on Biological Diversity, April 2002, available at <http://www.biodiv.org/doc/publications/cbd-bonn-gdls-en.pdf▷.

[10] CBD, National Focal Points, see listing at <http://www.biodiv.org/convention/default.shtml▷.

[11] WIPO, IGC Committee, Genetic Resources and Intellectual Property: An Overview, <http://www.wipo.int/tk/en/genetic/background/index.html▷.

[12] WIPO Technical Study on Disclosure Requirements Concerning Genetic Resources et al., documents available at <http://www.wipo.int/tk/en/genetic/proposals/index.html#resources▷.

[13] Guidelines for BIO Members Engaging in Bioprospecting, July 2005, available at <http://www.bio.org/ip/international/200507guide.asp▷.

[14] Beall EL, Lewis PW, Bell M, Rocha M, Jones DL, Botchan MR. Discovery of tMAC: a Drosophila testis-specific meiotic arrest complex paralogous to Myb-Muv B. Genes Dev 2007;15:904−19.

[15] Bagchi D, Misner B, Bagchi M, Kothari SC, Downs BW, Preuss HG. Effects of orally administered undenatured type II collagen against arthritic inflammatory diseases: a mechanistic exploration. Int J Clin Pharmacol Res 2002;22:101−10.

Index

Note: Page numbers followed by "*f*" and "*t*" refers to figures and tables respectively.

Food Science and Technology International Series

Amerine, M.A., Pangborn, R.M., and Roessler, E.B., 1965. Principles of Sensory Evaluation of Food.

Glicksman, M., 1970. Gum Technology in the Food Industry.

Joslyn, M.A., 1970. Methods in Food Analysis, Second Ed.

Stumbo, C.R., 1973. Thermobacteriology in Food Processing, Second Ed.

Altschul, A.M. (Ed.), New Protein Foods: Volume 1, Technology, Part A—1974. Volume 2, Technology, Part B—1976. Volume 3, Animal Protein Supplies, Part A—1978. Volume 4, Animal Protein Sup-plies, Part B—1981. Volume 5, Seed Storage Proteins—1985.

Goldblith, S.A., Rey, L., and Rothmayr, W.W., 1975. Freeze Drying and Advanced Food Technology.

Bender, A.E., 1975. Food Processing and Nutrition.

Troller, J.A., and Christian, J.H.B., 1978. Water Activity and Food.

Osborne, D.R., and Voogt, P., 1978. The Analysis of Nutrients in Foods.

Loncin, M., and Merson, R.L., 1979. Food Engineering: Principles and Selected Applications.

Vaughan, J.G. (Ed.), 1979. Food Microscopy.

Pollock, J.R.A. (Ed.), Brewing Science, Volume 1—1979. Volume 2—1980. Volume 3—1987.

Christopher Bauernfeind, J. (Ed.), 1981. Carotenoids as Colorants and Vitamin A Precursors: Technological and Nutritional Applications.

Markakis, P. (Ed.), 1982. Anthocyanins as Food Colors.

Stewart, G.G., and Amerine, M.A. (Eds.), 1982. Introduction to Food Science and Technology, Second Ed.

Iglesias, H.A., and Chirife, J., 1982. Handbook of Food Isotherms: Water Sorption Parameters for Food and Food Components.

Dennis, C. (Ed.), 1983. Post-Harvest Pathology of Fruits and Vegetables.

Barnes, P.J. (Ed.), 1983. Lipids in Cereal Technology.

Pimentel, D., and Hall, C.W. (Eds.), 1984. Food and Energy Resources.

Regenstein, J.M., and Regenstein, C.E., 1984. Food Protein Chemistry: An Introduction for Food Scientists.

Gacula Jr. M.C., and Singh, J., 1984. Statistical Methods in Food and Consumer Research.

Clydesdale, F.M., and Wiemer, K.L. (Eds.), 1985. Iron Fortification of Foods.

Decareau, R.V., 1985. Microwaves in the Food Processing Industry.

Herschdoerfer, S.M. (Ed.), Quality Control in the Food Industry, second edition. Volume 1—1985. Volume 2—1985. Volume 3—1986. Volume 4—1987.

Urbain, W.M., 1986. Food Irradiation.

Bechtel, P.J., 1986. Muscle as Food.

Chan, H.W.-S., 1986. Autoxidation of Unsaturated Lipids.

Cunningham, F.E., and Cox, N.A. (Eds.), 1987. Microbiology of Poultry Meat Products.

McCorkle Jr. C.O., 1987. Economics of Food Processing in the United States.

Japtiani, J., Chan Jr., H.T., and Sakai, W.S., 1987. Tropical Fruit Processing.

Solms, J., Booth, D.A., Dangborn, R.M., and Raunhardt, O., 1987. Food Acceptance and Nutrition.

Macrae, R., 1988. HPLC in Food Analysis, Second Ed.

Pearson, A.M., and Young, R.B., 1989. Muscle and Meat Biochemistry.

Penfield, M.P., and Campbell, A.M., 1990. Experimental Food Science, Third Ed.

Blankenship, L.C., 1991. Colonization Control of Human Bacterial Enteropathogens in Poultry.

Pomeranz, Y., 1991. Functional Properties of Food Components, Second Ed.

Walter, R.H., 1991. The Chemistry and Technology of Pectin.

Stone, H., and Sidel, J.L., 1993. Sensory Evaluation Practices, Second Ed.

Shewfelt, R.L., and Prussia, S.E., 1993. Postharvest Handling: A Systems Approach.

Nagodawithana, T., and Reed, G., 1993. Enzymes in Food Processing, Third Ed.

Hoover, D.G., and Steenson, L.R., 1993. Bacteriocins.

Shibamoto, T., and Bjeldanes, L., 1993. Introduction to Food Toxicology.

Troller, J.A., 1993. Sanitation in Food Processing, Second Ed.

Hafs, D., and Zimbelman, R.G., 1994. Low-fat Meats.

Phillips, L.G., Whitehead, D.M., and Kinsella, J., 1994. Structure-Function Properties of Food Proteins.

Jensen, R.G., 1995. Handbook of Milk Composition.

Roos, Y.H., 1995. Phase Transitions in Foods.

Walter, R.H., 1997. Polysaccharide Dispersions.

Barbosa-Canovas, G.V., Marcela Góngora-Nieto, M., Pothakamury, U.R., and Swanson, B.G., 1999. Preservation of Foods with Pulsed Electric Fields.

Jackson, R.S., 2002. Wine Tasting: A Professional Handbook.

Bourne, M.C., 2002. Food Texture and Viscosity: Concept and Measurement, second ed.

Caballero, B., and Popkin, B.M. (Eds.), 2002. The Nutrition Transition: Diet and Disease in the Developing World.

Cliver, D.O., and Riemann, H.P. (Eds.), 2002. Foodborne Diseases, Second Ed.

Kohlmeier, M., 2003. Nutrient Metabolism.

Stone, H., and Sidel, J.L., 2004. Sensory Evaluation Practices, Third Ed.

Han, J.H., 2005. Innovations in Food Packaging.

Sun, D.-W. (Ed.), 2005. Emerging Technologies for Food Processing.

Riemann, H.P., and Cliver, D.O. (Eds.), 2006. Foodborne Infections and Intoxications, Third Ed.

Arvanitoyannis, I.S., 2008. Waste Management for the Food Industries.

Jackson, R.S., 2008. Wine Science: Principles and Applications, Third Ed.

Sun, D.-W. (Ed.), 2008. Computer Vision Technology for Food Quality Evaluation.

David, K., and Thompson, P., (Eds.), 2008. What Can Nanotechnology Learn From Biotechnology?

Arendt, E.K., and Bello, F.D. (Eds.), 2008. Gluten-Free Cereal Products and Beverages.

Bagchi, D. (Ed.), 2008. Nutraceutical and Functional Food Regulations in the United States and Around the World.

Singh, R.P., and Heldman, D.R., 2008. Introduction to Food Engineering, Fourth Ed.

Berk, Z., 2009. Food Process Engineering and Technology.

Thompson, A., Boland, M., and Singh, H. (Eds.), 2009. Milk Proteins: From Expression to Food.

Florkowski, W.J., Prussia, S.E., Shewfelt, R.L. and Brueckner, B. (Eds.), 2009. Postharvest Handling, Second Ed.

Gacula Jr., M., Singh, J., Bi, J., and Altan, S., 2009. Statistical Methods in Food and Consumer Research, Second Ed.

Shibamoto, T., and Bjeldanes, L., 2009. Introduction to Food Toxicology, Second Ed.

BeMiller, J. and Whistler, R. (Eds.), 2009. Starch: Chemistry and Technology, Third Ed.

Jackson, R.S., 2009. Wine Tasting: A Professional Handbook, Second Ed.

Sapers, G.M., Solomon, E.B., and Matthews, K.R. (Eds.), 2009. The Produce Contamination Problem: Causes and Solutions.

Heldman, D.R., 2011. Food Preservation Process Design.

Tiwari, B.K., Gowen, A. and McKenna, B. (Eds.), 2011. Pulse Foods: Processing, Quality and Nutraceutical Applications.

Cullen, PJ., Tiwari, B.K., and Valdramidis, V.P. (Eds.), 2012. Novel Thermal and Non-Thermal Technologies for Fluid Foods.

Stone, H., Bleibaum, R., and Thomas, H., 2012. Sensory Evaluation Practices, Fourth Ed.

Kosseva, M.R. and Webb, C. (Eds.), 2013. Food Industry Wastes: Assessment and Recuperation of Commodities.

Morris, J.G. and Potter, M.E. (Eds.), 2013. Foodborne Infections and Intoxications, Fourth Ed.

Berk, Z., 2013. Food Processing Engineering and Technology, Second Ed.

Singh, R.P., and Heldman, D.R., 2014. Introduction to Food Engineering, Fifth Ed.

Han, J.H. (Ed.), 2014. Innovations in Food Packaging, Second Ed.

Madsen, C., Crevel, R., Mills, C., and Taylor, S. (Eds.), 2014. Risk Management for Food Allergy

Jackson, R.S., 2014. Wine Science: Principles and Applications, Fourth Ed.

Printed and bound by CPI Group (UK) Ltd, Croydon, CR0 4YY

08/05/2025

01864858-0001